Lecture Notes in Computer Science 15679

Founding Editors

Gerhard Goos
Juris Hartmanis

AF167651

The series Lecture Notes in Computer Science (LNCS), including its subseries Lecture Notes in Artificial Intelligence (LNAI) and Lecture Notes in Bioinformatics (LNBI), has established itself as a medium for the publication of new developments in computer science and information technology research, teaching, and education.

LNCS enjoys close cooperation with the computer science R & D community, the series counts many renowned academics among its volume editors and paper authors, and collaborates with prestigious societies. Its mission is to serve this international community by providing an invaluable service, mainly focused on the publication of conference and workshop proceedings and postproceedings. LNCS commenced publication in 1973.

Irene Finocchi · Loukas Georgiadis

Editors

Algorithms and Complexity

14th International Conference, CIAC 2025
Rome, Italy, June 10–12, 2025
Proceedings, Part I

 Springer

Editors
Irene Finocchi
Luiss Guido Carli
Rome, Italy

Loukas Georgiadis
University of Ioannina
Ioannina, Greece

ISSN 0302-9743   ISSN 1611-3349 (electronic)
Lecture Notes in Computer Science
ISBN 978-3-031-92931-1   ISBN 978-3-031-92932-8 (eBook)
https://doi.org/10.1007/978-3-031-92932-8

This Springer imprint is published by the registered company Springer Nature Switzerland AG
The registered company address is: Gewerbestrasse 11, 6330 Cham, Switzerland

If disposing of this product, please recycle the paper.

Preface

The 14th International Conference on Algorithms and Complexity (CIAC 2025) was held at Luiss University in Rome from June 10–12, 2025. CIAC serves as an important venue for researchers working on computational complexity and the design, analysis, experimentation, and application of efficient algorithms and data structures. These proceedings include all contributed papers presented at the conference.

In response to the call for papers, the Program Committee received 110 submissions, from which 44 papers were selected for inclusion in the scientific program, distributed across two proceedings volumes. Two papers were honored with the Best Paper Award, sponsored by Springer:

- Atoms versus avoiding simplicial vertices, by Karl Boddy, Konrad K. Dabrowski, and Daniel Paulusma.
- Realizing graphs with cut constraints, by Lucas de Oliveira Silva, Vítor Gomes Chagas, Samuel Plaça de Paula, Greis Yvet Oropeza Quesquén, and Uéverton dos Santos Souza.

In addition to the contributed papers, the conference featured plenary lectures by:

- Susanna F. de Rezende (Lund University, Sweden)
- Paolo Ferragina (Sant'Anna School of Advanced Studies, Italy)
- Sophie Huiberts (Clermont Auvergne University, France)

The Program Committee conducted all work electronically, implementing a lightweight double-blind reviewing process with an average of three reviews per submission. Paper selection was based on originality, quality, and relevance to theoretical computer science.

We extend our sincere gratitude to all authors who submitted papers, the Program Committee for their dedication, and the external reviewers for their valuable contributions to the evaluation process. A special thanks to the Organizing Committee for their efforts and to Luiss Guido Carli University for hosting and supporting the conference.

We hope that CIAC 2025 fostered stimulating discussions and collaborations, advancing research in algorithms and complexity.

June 2025

Irene Finocchi
Loukas Georgiadis

Organization

Program Committee

Davide Bilò	University of L'Aquila, Italy
Keerti Choudhary	IIT Delhi, India
Paloma T. de Lima	IT University of Copenhagen, Denmark
Adrian Dumitrescu	AlgOrEsEArch L.L.C., USA.
Thomas Erlebach	Durham University, UK
Bruno Escoffier	Sorbonne Université, France
Venkata Gandikota	Syracuse University, USA
Loukas Georgiadis	University of Ioannina, Greece
Archontia Giannopoulou	National and Kapodistrian University of Athens, Greece
Petr Golovach	Bergen University, Norway
Kristoffer Arnsfelt Hansen	Aarhus University, Denmark
Shunsuke Inenaga	Kyushu University, Japan
Evangelos Kipouridis	Saarland University, Max Planck Institute for Informatics, Germany
Jakub Łącki	Google Research, USA
Alexandra Lassota	Eindhoven University of Technology, The Netherlands
Wolfgang Mulzer	Freie Universität Berlin, Germany
Yuto Nakashima	Kyushu University, Japan
Charis Papadopoulos	University of Ioannina, Greece
Anastasios Sidiropoulos	Ohio State University, USA
Kostas Tsichlas	University of Patras, Greece
Nithin Varma	Chennai Mathematical Institute, India
Anthony Wirth	University of Sydney, Australia
Alexander Wolff	Universität Würzburg, Germany

Additional Reviewers

Bellitto, Thomas	Calinescu, Gruia
Belovs, Aleksandrs	Carmel, Amir
Bentert, Matthias	Chandramouleeswaran, Harish
Berndt, Sebastian	Christopoulos, Konstantinos
Bhattacharya, Anup	Cook, James

Cristi, Andrés
Das, Shantanu
de Castro Mendes Gomes, Guilherme
Deng, Chengyuan
Dey, Dipan
Di Fonso, Alessia
Di Stefano, Gabriele
Dudeja, Aditi
Dumas, Jean-Guillaume
Dutta, Pranjal
Dutta, Sagnik
Duyster, Anouk
Döring, Simon
Dürr, Christoph
Ebbens, Matthijs
Eden, Talya
Eiben, Eduard
Fairbairn, David
Fink, Simon D.
Fleszar, Krzysztof
Ghosal, Pratik
Ghosh, Anirban
Gibney, Daniel
Gledel, Valentin
Greger, Matthias
Gualà, Luciano
Hamm, Thekla
Hanaka, Tesshu
Hegemann, Tim
Hermo, Montserrat
Herold, Martin
Huang, Chien-Chung
Inamdar, Tanmay
Jabal Ameli, Afrouz
Jaffke, Lars
Jeż, Łukasz
Jin, Ce
Kalavas, Andreas
Kawahara, Jun
Kimura, Kei
Klemz, Boris
Knop, Dušan
Kobayashi, Yasuaki
Konstantopoulos, Christos
Kontogiannis, Spyros

Korhonen, Tuukka
Kosinas, Evangelos
Koutecký, Martin
Kowaluk, Miroslaw
Krithika, R.
Kryven, Myroslav
Kurita, Kazuhiro
Köhler, Noleen
Lahiri, Abhiruk
Leucci, Stefano
Ligthart, Koen
Lin, Wenqing
Lucke, Felicia
Makris, Christos
Mandal, Soumen
Martin, Barnaby
Mavropoulos, Filippos
Mertz, Ian
Mizuki, Takaaki
Mizutani, Yosuke
Mnich, Matthias
Morawietz, Nils
Mpanti, Anna
O'Rourke, Joseph
Okada, Yuto
Otachi, Yota
Oz, Mert Sinan
Palit, Diptaksho
Park, Eunku
Perez, Anthony
Pothitos, Nikolaos
Radoszewski, Jakub
Ramamoorthi, Vijayaragunathan
Rattan, Gaurav
Ruangwises, Suthee
Saggi, Lakshay
Sahlot, Vibha
Sahu, Abhishek
Satti, Srinivasa Rao
Saurabh, Saket
Saurav, Kumar
Savicky, Petr
Schaeffer, Luke
Schlöter, Jens
Seara, Carlos

Sharma, Roohani
Sieper, Marie Diana
Sioutas, Spyros
Skambath, Malte
Straziota, Alessandro
Tengse, Anamay
Verma, Shaily
Volkovich, Ilya
Walzer, Stefan
Wang, Jiaheng

Wang, Yanheng
Wasa, Kunihiro
Wolf, Samuel
Wu, Zihang
Yang, Haodong
Yoshinaka, Ryo
Zhang, Zhuo
Zink, Johannes
Zisopoulos, Charilaos
Zoros, Dimitris

Contents – Part I

Exact and Approximate High-Multiplicity Scheduling on Identical Machines

Klaus Jansen$^{(\boxtimes)}$(ID), Kai Kahler$^{(\boxtimes)}$(ID), and Esther Zwanger

Department of Computer Science, Kiel University, Kiel, Germany
kj@informatik.uni-kiel.de, kai.kahler@web.de, stu222121@mail.uni-kiel.de

Abstract. Goemans and Rothvoss (SODA'14) gave a framework for solving problems which can be described as finding a point in int.cone($P \cap \mathbb{Z}^N$) $\cap\, Q$, where $P, Q \subset \mathbb{R}^N$ are (bounded) polyhedra. The running time for solving such a problem is $\langle P \rangle^{2^{O(N)}} \langle Q \rangle^{O(1)}$. This framework can be used to solve various scheduling problems, but the encoding length $\langle P \rangle$ usually involves parameters like the makespan or deadlines (which can be very large compared to the processing times). We describe three tools to improve the framework by Goemans and Rothvoss:

- Problem-specific preprocessing can be used to greatly reduce $\langle P \rangle$.
- By solving a certain LP relaxation, one can obtain bounds for the points in P. Combined with the classical result by Frank and Tardos (J. Comb. '87), these yield a more compact encoding of P in general.
- A result by Jansen and Klein (SODA'17) changes the running time of the algorithm by Goemans and Rothvoss to $|V|^{2^{O(N)}} \langle P \rangle^{O(1)} \langle Q \rangle^{O(1)}$, where V is the set of vertices of the convex hull of $P \cap \mathbb{Z}^N$. We provide a new bound for $|V|$ that is similar to the one by Berndt et al. (SOSA'21) but better for our setting; this gives an alternative way to improve the framework.

For example, applied to the scheduling problems $P||\{C_{\max}, C_{\min}, C_{\mathrm{envy}}\}$, these tools improve the running time from $(\log(C_{\max}))^{2^{O(d)}} \langle I \rangle^{O(1)}$ to the possibly much better $(\log(p_{\max}))^{2^{O(d)}} \langle I \rangle^{O(1)}$. Here, $p_{\max}$ is the largest processing time, d is the number of different processing times, $C_{\max}$ is the makespan and $\langle I \rangle$ is the encoding length of the instance.

On the complexity side, we use reductions from the literature to provide new parameterized lower bounds for $P||C_{\max}$. Finally, we show that the big open question asked by Mnich and van Bevern (Comput. Oper. Res. '18) whether $P||C_{\max}$ is FPT w.r.t. the number of job types d has the same answer as the question whether $Q||C_{\max}$ is FPT w.r.t. the number of job and machine types $d + \tau$ (all in high-multiplicity encoding). The same holds for objective $C_{\min}$.

Keywords: scheduling · parameterized complexity · approximation algorithms · high-multiplicity · bin packing

Supported by the German Research Foundation (DFG) project "Fine-grained complexity and algorithms for scheduling and packing" (JA 612/25-1).

I. Finocchi and L. Georgiadis (Eds.): CIAC 2025, LNCS 15679, pp. 1–17, 2025.
https://doi.org/10.1007/978-3-031-92932-8_1

1   Introduction

Distributing jobs or tasks among workers or machines is one of the most natural optimization problems arising in practice. Such *scheduling problems* have been systematically studied for over half a century (see e.g. [25] for a great introduction to scheduling). Different machine models, job characteristics and objectives make for a plethora of different problems; some of them can be solved efficiently by scheduling jobs in a certain order, but many are NP-hard. Hence, a large part of research on scheduling is concerned with approximation algorithms; on the other hand, there are also many pseudo-polynomial algorithms. Another – regarding the whole history of scheduling – more recent research direction are *parameterized algorithms*. Such algorithms encapsulate the complexity in certain parameters of the problem, and should these parameters be quite small, the algorithms can be very efficient [7].

In this work, we primarily focus on the parameter d, the number of different *job types*. The exact definition may vary as the problems become more complex (with weights, classes or deadlines), but in a simple setting where a job j only has a processing time p_j, the parameter d describes the number of different processing times. This parameter choice seems natural in an industrialized world, where workers and machines have a large degree of specialization and hence do not work on too many different types of jobs or tasks. We use the standard three-field notation introduced by Graham et al. [14]; for a detailed description of our problems and parameters, see Sect. 2.

With the result by Goemans and Rothvoss [12] (or the one by Jansen and Klein [18]), one can solve problems that can be modelled by two polyhedra P and Q (with P bounded) such that solutions of the problem directly correspond to a point $y \in$ int.cone$(P \cap \mathbb{Z}^N) \cap Q$. Many scheduling problems fall into this category, e.g. makespan minimization on identical machines ($P||C_{\max}$). However, when described by such a *PQ-representation* (in short PQ-R), the makespan $C_{\max}$ appears in the system $Ax \leq b$ that represents the polytope P and thus greatly influences the running time.

Our Contributions. We describe three tools that can be used to improve the framework by Goemans and Rothvoss [12] such that the running time does not depend on $\|b\|_\infty$, which in the case of scheduling problems eliminates parameters like the makespan or the largest due date from the running time. Due to space constraints, some of our proofs and applications of our results are only contained in the full version [17].

The first tool (covered in Sect. 3) is a balancing result by Govzmann et al. [13] that allows pre-scheduling many of the jobs, leaving us with a makespan value that is bounded by $2dp_{\max}$, where $p_{\max}$ is the largest processing time. Applying this preprocessing before using the algorithm by Goemans and Rothvoss yields the following running times for solving $P||\{C_{\max}, C_{\min}, C_{\text{envy}}\}$:

Theorem 1. *The optimization problems* $P||\{C_{max}, C_{min}\}$ *can be solved in time* $(d \log(p_{max}) + \log(2dp_{max}^2))^{2^{O(d)}} \langle I \rangle^{O(1)}$ *and the optimization problem* $P||C_{envy}$ *in time* $p_{max}(d \log(p_{max}) + \log(2dp_{max}^2))^{2^{O(d)}} \langle I \rangle^{O(1)}$.

We also show lower bounds for $P||C_{max}$ that somewhat resemble the running time of our algorithm, though there is still a gap (for a proof, see [17]):

Theorem 2. *Let* $\varepsilon > 0$. *Unless the ETH fails,* $P||C_{max}$ *cannot be solved in time* $\langle I \rangle^{O(d^{1-\varepsilon})} p_{max}^{O(1)}$, $\langle I \rangle^{O(1)} d^{O(d^{1-\varepsilon})} p_{max}^{O(1)}$, $\langle I \rangle^{o\left(\frac{d}{\log(d)}\right)} p_{max}^{O(1)}$ *or* $p_{max}^{O(d^{1-\varepsilon})}$.

Note that the encoding length $\langle I \rangle$ includes $\log(p_{max})$ and d. Moreover, we show that the balancing result can be used to speed up an additive approximation scheme by Buchem et al. [3] in the case where the number of jobs $\|n\|_1$ is larger than the number of machines m times p_{max} (see [17]):

Theorem 3. *There is an additive approximation scheme with error at most* εp_{max} *and running time* $(mp_{max})^{O(\frac{1}{\varepsilon})}$ *for the problems* $P||\{C_{max}, C_{min}, C_{envy}\}$.

In the full version [17], we also give a matching lower bound for objective C_{max} via ETH:

Theorem 4. *Let* $\delta > 0$. *Then* $P||C_{max}$ *cannot be approximated with additive error at most* εp_{max} *in time* $(mp_{max})^{O((\frac{1}{\varepsilon})^{1-\delta})}$, *unless the ETH fails.*

The second tool (covered in Sect. 4) involves solving a special relaxation of an integer linear program (ILP) associated with a given PQ-R. A proximity result by Cslovjecsek et al. [5] then gives an upper bound for the points in P, which allows us to reduce the coefficients in the system $Ax \leq b$ describing P with the famous result by Frank and Tardos [11]. Together, this then yields an algorithm for solving PQ-Rs in a running time independent of $\|b\|_\infty$:[1]

Theorem 5. *If a problem has a PQ-R* (P, Q, m) *given by a polytope* $P = \{x \in \mathbb{R}_{\geq 0}^N \mid A^{(P)}x = b^{(P)}\}$ *and* $Q = \{x \in \mathbb{R}_{\geq 0}^N \mid A^{(Q)}x = b^{(Q)}\}$ *with* $M^{(P)}$ *and* $M^{(Q)}$ *constraints, respectively, it can be solved in time*

$$\left(\left(M^{(P)}M^{(Q)} \log\left(\max\left\{\left\|A^{(P)}\right\|_\infty, \left\|A^{(Q)}\right\|_\infty\right\}\right)\right)^{2^{O(N)}} + 2^{O((M^{(Q)})^2)}\right)$$

$$(\langle P \rangle \langle Q \rangle \log(m))^{O(1)}.$$

The third tool (covered in Sect. 5) is an upper bound for the number of vertices of the integer hull of a polytope that is similar to the one by Berndt et al. [2] but better for our specific purpose:

Theorem 6. *The integer hull of a polytope* $P = \{x \in \mathbb{R}_{\geq 0}^N \mid Ax = b\}$ *has at most* $N^M O(M \log(M\Delta))^N$ *vertices, where* M *is the number of constraints and* $\Delta = \|A\|_\infty$.

---

[1] Of course, the entries in b still appear in the encoding length of P, but only logarithmically.

Note that this does not depend on $\|b\|_\infty$. Previously known bounds either depend on $\|b\|_\infty$ or are exponential in $\log(\|A\|_\infty)$. Combined with the algorithm by Jansen and Klein [18], this gives an alternative way of solving PQ-Rs:

Theorem 7. *Let* $P = \left\{ x \in \mathbb{R}_{\geq 0}^N \,\middle|\, A^{(P)}x = b^{(P)} \right\}$ *with* $M^{(P)}$ *constraints and* $Q = \left\{ x \in \mathbb{R}_{\geq 0}^N \,\middle|\, A^{(Q)}x = b^{(Q)} \right\}$ *with* $M^{(Q)}$ *constraints. Then the PQ-R* (P, Q, m) *can be solved in time*

$$\left(N^{M^{(P)}} M^{(P)} \log \left(\left\| A^{(P)} \right\|_\infty \right) \right)^{2^{O(N)}} (\langle P \rangle \langle Q \rangle \log(m))^{O(1)}.$$

Both Theorem 5 and Theorem 7 can be used to obtain algorithms for various scheduling problems that are more efficient than a straightforward application of the results by Goemans and Rothvoss [12] and Jansen and Klein [18] with previously known bounds for the vertices of the integer hull. The full version [17] also includes PQ-Rs for various scheduling problems, MINSUM-WEIGHTED-BINPACKING (MSWBP) and uniform n-fold ILPs.

The theorems also yield $(\log(p_{\max}))^{2^{O(d)}} \langle I \rangle^{O(1)}$-time (respectively for objective C_{envy} with an additional $p_{\max}$-factor) algorithms for $P||\{C_{\max}, C_{\min}, C_{\text{envy}}\}$, but with worse constants than in Theorem 1. One might ask why such running times are interesting, as the parameters $p_{\max}$ and d are still entangled. But with an inequality that can be found in an exercise from [7] (Hint 3.18) and has been used by Koutecký and Zink [23], one can bound $(\log(\Delta))^{2^{O(d)}}$ by $2^{2^{O(d)}} \Delta^{o(1)}$ (see [17]). So Theorem 1 almost answers the open question by Mnich and van Bevern [27] for an algorithm solving $P||C_{\max}$ that is fixed-parameter tractable (FPT) w.r.t. d. This question had been partially answered by Koutecký and Zink [23], who gave an $f(d)n^{o(1)}\langle I \rangle^{O(1)}$-time algorithm, where f is some computable function. Actually, one can analyze the function $(\log(\Delta))^{2^{O(d)}}$ even further. Define the tower function for any $\alpha \in \mathbb{N}$, $\beta \in \mathbb{R}$ recursively as follows: $\text{Tower}(\alpha, \beta) = \beta$ if $\alpha = 0$ and $\text{Tower}(\alpha, \beta) = 2^{\text{Tower}(\alpha-1,\beta)}$ otherwise. For example, $\text{Tower}(1, \beta) = 2^\beta$ and $\text{Tower}(2, \beta) = 2^{2^\beta}$. We generalize the inequality used by Koutecký and Zink [23] to the following:

Lemma 1. *For* $\alpha, \beta, \gamma \in \mathbb{N}_{>0}$, *we have* $\beta^\gamma \leq \max \left\{ \text{Tower}(\alpha, \gamma)^\gamma, \beta^{\log^{(\alpha)}(\beta)} \right\}$.

Proof. Case 1: $\gamma \geq \log^{(\alpha)}(\beta)$. Then $\text{Tower}(\alpha, \gamma) \geq \beta$ and $\beta^\gamma \leq \text{Tower}(\alpha, \gamma)^\gamma$.
Case 2: $\gamma < \log^{(\alpha)}(\beta)$. Then $\beta^\gamma < \beta^{\log^{(\alpha)}(\beta)}$.

So inserting $\beta = \log(p_{\max})$, $\gamma = 2^{O(d)}$, we get for any $\alpha \in \mathbb{N}_{>0}$:

$$\log(p_{\max})^{2^{O(d)}} \leq \max \left\{ \text{Tower}(\alpha, 2^{O(d)})^{2^{O(d)}}, \log(p_{\max})^{\log^{(\alpha+1)}(p_{\max})} \right\}$$

Interestingly enough, $\log^{(\alpha+1)}(p_{\max}) \leq 1$ for $\alpha = 4$, as long as $p_{\max} \leq 2^{65536}$. So one could argue that for any reasonable value of $p_{\max}$ $(\leq 2^{65536})$, the running time $\log(p_{\max})^{2^{O(d)}} \langle I \rangle^{O(1)}$ is FPT w.r.t. d. Of course, it is a bit hypocritical to argue about "reasonable" values (which would be encountered in practice) if at the same time the running time then includes a tower of height 5. Also, the assumption $p_{\max} \leq 2^{65536}$ directly yields an FPT running time. It still seems to us that this is an interesting observation, as this might be as close to an FPT running time as one can get.

Finally, the full version [17] contains a reduction from $Q||\{C_{\max}, C_{\min}\}$ to $P||\{C_{\max}, C_{\min}\}$, producing instances with $d+\tau$ job types, where τ is the number of machine types. This shows that the big open question whether $P||C_{\max}$ is FPT w.r.t. parameter d (see [27]) and the question whether $Q||\{C_{\min}$ is FPT w.r.t. parameters d and τ have the same answer:

Theorem 8. *The following statements are equivalent:*

1. $P||\{C_{\max}, C_{\min}\}$ *is FPT w.r.t. parameter d.*
2. $Q||\{C_{\max}, C_{\min}\}$ *is FPT w.r.t. parameters d and τ.*

Related Work. In their paper [12], Goemans and Rothvoss showed the following result:

Theorem 9 (Goemans & Rothvoss [12]**).** *Given rational polyhedra $P, Q \subset \mathbb{R}^N$ where P is bounded, one can find a vector $y \in int.cone(P \cap \mathbb{Z}^N) \cap Q$ and a vector $x \in \mathbb{N}^{P \cap \mathbb{Z}^N}$ such that $y = \sum_{c \in P \cap \mathbb{Z}^N} cx_c$ in time $\langle P \rangle^{2^{O(N)}} \langle Q \rangle^{O(1)}$ or decide that no such y exists. Here, $\langle P \rangle$ and $\langle Q \rangle$ are the encoding lengths of P and Q, respectively.*

They then briefly showed how this can be used directly to solve various scheduling problems. For example, they get an algorithm for $P||C_{\max}$ that runs in time $(\log(C_{\max}))^{2^{O(d)}} \langle I \rangle^{O(1)}$. The structural result has been generalized by Jansen and Klein:

Theorem 10 (Jansen & Klein [12]**).** *Given rational polyhedra $P, Q \subset \mathbb{R}^N$ where P is bounded, one can find a vector $y \in int.cone(P \cap \mathbb{Z}^N) \cap Q$ and a vector $x \in \mathbb{N}^{P \cap \mathbb{Z}^N}$ such that $y = \sum_{c \in P \cap \mathbb{Z}^N} cx_c$ in time $|V|^{2^{O(N)}} \langle P \rangle^{O(1)} \langle Q \rangle^{O(1)}$ or decide that no such y exists. Here, $\langle P \rangle$ and $\langle Q \rangle$ are the encoding lengths of P and Q, respectively and V is the set of vertices of the integer hull of P.*

Kowalik et al. [24] showed that the doubly-exponential dependency on N is necessary, unless the ETH fails. Knop et al. [22] gave a framework for solving configuration ILPs and depending on the objective, the algorithms have different running times. As part of their framework, they also generalize the algorithm by Goemans and Rothvoss [12] to multiple types of P-polytopes, which allows modelling uniform and unrelated machines quite efficiently. The right-hand-sides of the systems describing the polytopes (and hence such terms as the makespan or due dates) still appear in the running time in the same way as they do in the algorithm by Goemans and Rothvoss [12]. Knop et al. [20] then show how this framework can be used to obtain parameterized algorithms for various scheduling problems. For identical machines however, their running time that is not parameterized by $p_{\max}$ does not provide an improvement over the running time by Goemans and Rothvoss [12]. It should be noted that during the publication process of this paper, our framework has been generalized to the multi-type setting by Jansen and Kahler [16], improving the framework by Knop et al. [22] in the same way as our framework improves the one by Goemans and Rothvoss [12] (eliminating the right-hand-side from the running time).

When one considers p_{max} a parameter, there are plenty of other algorithms, mostly using ILPs as a subroutine. Mnich and Wiese [28] use a balancing result in combination with solving an ILP in small dimension. Knop and Koutecký [21] use the normal assignment ILP and solve it with an n-fold algorithm. The framework by Knop et al. [20] uses huge n-fold ILPs and proximity; a similar approach was described by Koutecký and Zink [23]. Govzmann et al. [13] bound the coefficients in the configurations and use an algorithm for ILPs with few constraints.

As mentioned above, Koutecký and Zink [23] gave an algorithm for $P||C_{max}$ with running time $2^{2^{O(d)}} \|n\|_1^{O(1)}$, brilliantly combining the algorithm by Goemans and Rothvoss [12] with the result by Frank and Tardos [11] to bound the coefficients in terms of $\|n\|_1$ (the total number of jobs). Moreover, they stated in a footnote (without proof) that the result by Goemans and Rothvoss gives a $(\log(p_{max}))^{f(d)}(\log(\|n\|_1))^{O(1)}$-time algorithm for $P||C_{max}$. The running time stated by Goemans and Rothvoss [12] is $(\log(\max\{C_{max}, \|n\|_1\}))^{f(d)}$ and it is clear that $\|n\|_1$ does not have to be part of the basis, as $\|n\|_1$ only appears in the encoding of Q, but it is not clear at all how the dependency on C_{max} can be avoided if one uses the algorithm by Goemans and Rothvoss. Using either of our three tools, however, yields a running time like the one claimed by Koutecký and Zink [23]. As mentioned above, Mnich and van Bevern [27] stated the parameterized complexity of $P||C_{max}$ in high-multiplicity encoding with parameter d as an open problem.

2   Preliminaries

In this section, we introduce the studied problems, notation and terminology. We denote by $[k] := \{1, \ldots, k\}$ the numbers up to k and by $v_{max} := \|v\|_\infty = \max_{i=1,\ldots,N}\{|v_i|\}$ the largest (absolute) entry in a vector $v \in \mathbb{R}^N$. As it is custom in parameterized complexity, we sometimes write $f(k)$ to represent *any* computable function that only depends on the parameter k. Hence, it might not be explicitly specified and it might change, e.g. by consuming additional factors that depend only on k. A problem is fixed-parameter tractable (FPT) w.r.t. parameter k if it can be solved in time $f(k)\langle I \rangle^{O(1)}$. For any problem instance, we always assume that all input numbers are integral. This includes the matrix and right-hand-side entries in polyhedra.

Applications. To show how Theorem 5 and Theorem 7 can be useful, we apply them to various problems, mainly from the area of scheduling, to obtain faster algorithms. Following the notation introduced by Graham et al. [14], we denote scheduling problems by a triple $\alpha|\beta|\gamma$, where α describes the machine setting, β describes a list of additional job constraints and γ describes the objective. In this short version, we only consider the problems $P||\{C_{max}, C_{min}, C_{envy}\}$.

For $P||\{C_{max}, C_{min}, C_{envy}\}$, we are given a job vector $n \in \mathbb{N}_{>0}^d$, a corresponding processing time vector $p \in \mathbb{N}_{>0}^d$ and a number m of identical machines. Formally, the task is to define an assignment $\pi : [d] \mapsto ([m], [\|n\|_1])$ that maps job types to machines (and multiplicities) such that the maximum load $C_{max} = \max_{i\in[m]}\{L_{\pi,i}\}$ over the machines is minimized (for objective C_{max}), the minimum load $C_{min} = \min_{i\in[m]}\{L_{\pi,i}\}$ is maximized (for objective C_{min}) or the difference between C_{max} and C_{min} is minimized (for objective C_{envy}).[2] The load $L_{\pi,i}$ of a machine $i \in [m]$ w.r.t. schedule π is defined as the sum of all processing times of the jobs assigned to i, i.e., $L_{\pi,i} := \sum_{\substack{j\in[d] \\ \pi(j)=(i,k)}} kp_j$. Of course, we may not schedule more jobs of a type than we are given. An important concept in scheduling are *configurations*. A configuration is a selection of jobs (represented by a d-dimensional vector) that can be scheduled together on a machine. In the simple case of $P||C_{max}$, the set of configurations is defined as $\mathcal{C} := \{c \in \mathbb{N}^d \,|\, p^T c \leq u\}$ when all jobs have to be completed by time u.

As the reader might have noticed, the input to our scheduling problems is given by vectors of dimension d and multiplicities. Classically, in the low-multiplicity encoding, the $\|n\|_1$ jobs would all be listed together with their processing times and other characteristics, even if all jobs were identical. In this paper, all inputs are high-multiplicity encoded. The encoding length of an instance (or similarly for a polytope) I is denoted by $\langle I \rangle$. e.g. for $P||C_{max}$, we have $\langle I \rangle \leq O(d\log(p_{max} + n_{max}) + \log(m))$ as opposed to the classical encoding, which might have size $O(\|n\|_1 \log(p_{max}))$. Note that one can assume that $m \leq \|n\|_1$, because more than $\|n\|_1$ machines are not necessary, so in the classical encoding, the part that encodes the number of machines vanishes. Depending on the given instance, $\|n\|_1$ might be exponential in $\langle I \rangle$ and m might be as well. This is the reason why the $f(d)\|n\|_1^{O(1)}$-time algorithm by Koutecký and Zink [23] is not FPT in the high-multiplicity setting.

Note that the scheduling problems described above are optimization problems. However, the framework can only really model decision problems. Fortunately, the optimization problems can be easily solved by solving several decision problems as part of a binary search over some interval $[0, v]$. The key ingredient is that the optimum is always bounded by some value v that depends on the numbers in the input and hence $\log(v) \leq \langle I \rangle^{O(1)}$.[3]

Throughout this paper, we only consider the decision problems, and as our algorithms all have an $\langle I \rangle^{O(1)}$-term in the running time anyway, this directly also yields algorithms with the same running time for the optimization problem, except in the C_{envy}-case. There, we get an additional factor, which is p_{max}. Hence, we will always assume that we are given thresholds ℓ and/or u for the load values when solving these problems. Clearly, (the logarithms of) these are then also part

---

[2] Note that for the problems considered in this paper, the order or starting times do not matter; only the load values (and the schedulability of early jobs) are important.

[3] Note e.g. in the case of $P||C_{max}$ that $\log(v) = \log(\sum_{j=1}^{d} p_j n_j) \leq \log(dp_{max}n_{max}) \leq \langle I \rangle^{O(1)}$.

of the encoding. A detailed justification for only considering decision problems is given in the full version [17].

Reducing Coefficients. If the coefficients of an inequality (or equality) are much larger compared to the number of variables and the variables are also bounded by a rather small value, the coefficients can be reduced with the following classical result by Frank and Tardos:

Theorem 11 (Frank & Tardos [11]). *For every $w \in \mathbb{R}^N$ and $\Delta \in \mathbb{N}$, there exists a $\bar{w} \in \mathbb{Z}^N$ such that $\|\bar{w}\|_\infty \leq (N\Delta)^{O(N^3)}$ and $sign(w^T x) = sign(\bar{w}^T x)$ for every $x \in \mathbb{Z}^N$ with $\|x\|_1 \leq \Delta - 1$. Moreover, $\bar{w}$ can be computed in time $N^{O(1)}$.*

Corollary 1 (Stated in a similar form by Etscheidt et al. [10]). *For every $w \in \mathbb{R}^N$, $b \in \mathbb{R}$, $\Delta \in \mathbb{N}$, one can compute $\bar{w} \in \mathbb{Z}^N$, $\bar{b} \in \mathbb{Z}$ with $\|\bar{w}\|_\infty, |\bar{b}| \leq (N\Delta)^{O(N^3)}$ in time $N^{O(1)}$ such that for every $x \in [-\Delta, \Delta]^N$, $w^T x \leq b \iff \bar{w}^T x \leq \bar{b}$.*

A proof is given in the full version [17]. Note that due to the equivalence, zeros are mapped to zeros, negative entries are mapped to negative entries and positive entries are mapped to positive entries. This can be seen by setting x to be a unit vector in Theorem 11.

Another way to greatly reduce a given problem instance (and the set of configurations) is provided by the following result by Govzman et al. [13]. It is the key ingredient used in the full version [17] to improve the running time of the additive approximation scheme from [3] in a high-multiplicity setting and can also be used to reduce the coefficients in a PQ-R. For completeness, we include a proof in the full version [17], as the result has not been published yet. A similar but exponential bound had already been shown by Mnich and Wiese [28].

Lemma 2 (Govzmann et al. [13]). *For $P||\{C_{max}, C_{min}, C_{envy}\}$, there exists a kernel where the number of jobs of a specific type on a specific machine is bounded by $2p_{max}$. So the load of every machine is bounded by $2p_{max}^2 d$. The kernelization runs in $O(d)$ time.*

PQ-Rs. "PQ-Rs" are formally defined as follows:

Definition 1 (PQ-Representation). *Given a problem X, a PQ-R of X is a triple (P, Q, m) of polytopes $P, Q \subset \mathbb{R}^N$ and a number m such that for every instance I of X, I is positive if and only if there exists a $y = y^{(1)} + \ldots + y^{(m)} \in int.cone(P \cap \mathbb{Z}^N) \cap Q$ such that for all $i \in [m]$, $y^{(i)} \in P \cap \mathbb{Z}^N$.*

Problems that have a PQ-R can be solved with the algorithm by Goemans and Rothvoss [12] or the algorithm by Jansen and Klein [18]:

Proposition 1 (See [18] and [12]). *A problem with a PQ-R (P, Q, m), where $P, Q \subset \mathbb{R}^N$, can be solved in $|V|^{2^{O(N)}} \langle P \rangle^{O(1)} \langle Q \rangle^{O(1)} (\log(m))^{O(1)}$ or in $\langle P \rangle^{2^{O(N)}} \langle Q \rangle^{O(1)} (\log(m))^{O(1)}$, where V is the set of vertices of the integer hull of P.*

Bounds on the Vertices of the Integer Hull. Clearly, the number of vertices of the integer hull plays a big role in the running time of Proposition 1. Formally, the integer hull of a polytope $P \subset \mathbb{R}^N$ is defined as the convex hull of $P \cap \mathbb{Z}^N$. There are several upper and lower bounds for the number of vertices of the integer hull, e.g. the one by Berndt et al. [2], which improved the one by Aliev et al. [1]:

Theorem 12 (Berndt et al. [2]). *The number of vertices of the integer hull of a polytope $P = \left\{ x \in \mathbb{R}^N_{\geq 0} \mid Ax = b \right\}$ is at most $(NM \log(M\Delta))^{O(M \log(\sqrt{M}\Delta))}$, where $\Delta = \|A\|_\infty$ and $\bar{M}$ is the number of constraints.*

Note that this does not depend on the right-hand-side but has a stronger dependency on the largest coefficient in the matrix compared to the older result by Hartmann [15] and Cook et al. [4], which appears to be nearly optimal in general [29]:

Theorem 13 (Hartmann [15] and Cook et al. [4]). *Consider a rational polyhedron $P = \{x \in \mathbb{R}^N \mid Ax \leq b\}$ with $A \in \mathbb{Z}^{M \times N}$ and let $\Delta = \|A\|_\infty$. Then the integer hull of P has at most $(MN\phi)^{O(N)} = (MN \log(\max\{\Delta, \|b\|_\infty\}))^{O(N)}$ vertices, where $\phi = O(N \log(\max\{\Delta, \|b\|_\infty\}))$ is the maximum encoding length of the inequalities.*

3 Tool 1: The Balancing Result by Govzmann et al.

How Proposition 1 can be useful for solving scheduling and packing problems can be seen at the example of $P||\{C_{\max}, C_{\min}, C_{\text{envy}}\}$. Remember that we are given load thresholds ℓ and/or u, depending on the objective:

Lemma 3. *$P||\{C_{\max}, C_{\min}, C_{\text{envy}}\}$ has a PQ-R with m set to the number of machines,* $Q_{C_{\max}} = \left\{ \begin{pmatrix} c \\ t \end{pmatrix} \in \mathbb{R}^{d+1}_{\geq 0} \;\middle|\; c = n \right\},$

$Q_{C_{\min}} = Q_{C_{\text{envy}}} = \left\{ \begin{pmatrix} c \\ t_1 \\ t_2 \end{pmatrix} \in \mathbb{R}^{d+2}_{\geq 0} \;\middle|\; c = n \right\}$ *and respectively*

$$P_{C_{\max}} = \left\{ \begin{pmatrix} c \\ t \end{pmatrix} \in \mathbb{R}^{d+1}_{\geq 0} \;\middle|\; p^T c + t = u \right\},$$

$$P_{C_{\min}} = \left\{ \begin{pmatrix} c \\ t_1 \\ t_2 \end{pmatrix} \in \mathbb{R}^{d+2}_{\geq 0} \;\middle|\; \begin{pmatrix} p^T & 1 & 0 \\ -p^T & 0 & 1 \end{pmatrix} \begin{pmatrix} c \\ t_1 \\ t_2 \end{pmatrix} = \begin{pmatrix} \ell + 2p_{\max} \\ -\ell \end{pmatrix} \right\} \quad and$$

$$P_{C_{\text{envy}}} = \left\{ \begin{pmatrix} c \\ t_1 \\ t_2 \end{pmatrix} \in \mathbb{R}^{d+2}_{\geq 0} \;\middle|\; \begin{pmatrix} p^T & 1 & 0 \\ -p^T & 0 & 1 \end{pmatrix} \begin{pmatrix} c \\ t_1 \\ t_2 \end{pmatrix} = \begin{pmatrix} u \\ -\ell \end{pmatrix} \right\}.$$

Proof. We only show the claim for $P||C_{\max}$, so let $P := P_{C_{\max}}$ and $Q := Q_{C_{\max}}$. The other proofs work analogously, with the only difference being that we can restrict ourselves to configurations that have load at most $\ell + 2p_{\max}$ in the case of $C_{\min}$, as we show in the full version [17]. First note that the t-entry fulfills the role of a slack variable, making it so we have an equality constraint. Clearly, P is bounded. Consider an instance I of $P||C_{\max}$ and suppose that it is positive, i.e., there is a schedule of all the jobs with makespan at most u. For each machine $k \in [m]$, create a $(d+1)$-dimensional vector $y^{(k)}$ such that $y_i^{(k)}$ is the number of jobs with processing time p_i scheduled on machine k for all $i \in [d]$ and $y_{d+1}^{(k)} = u - \sum_{i=1}^d p_i c_i$. Clearly, summing up the all the vectors $y^{(k)}$ projected to the first d variables yields $\begin{pmatrix} n_1 \dots n_d \end{pmatrix}^T$, since all jobs are scheduled. Moreover, we constructed m such vectors. Since the $y^{(k)}$ are integral, we have $\sum_{k=1}^m y^{(k)} \in \text{int.cone}(P \cap \mathbb{Z}^d) \cap Q$.

Conversely, suppose there is a vector $y \in \text{int.cone}(P \cap \mathbb{Z}^d) \cap Q$. Since $y \in \text{int.cone}(P \cap \mathbb{Z}^d)$, y can be written as a conic combination of integral vectors in P. Let S be the multi-set of these integral vectors, i.e., $\sum_{y^{(k)} \in S} y^{(k)} = y$. Each $y^{(k)} \in S$ corresponds to an integral $d+1$-dimensional vector. For each $y^{(k)} \in S$, pick an (idle) machine $k \in [m]$ and for every $i \in [d]$, schedule $y_i^{(k)}$ jobs of type i on k. Since y is a sum of m vectors, we know that $|S| = m$ and since $y \in Q$, $\sum_{y^{(k)} \in S} y_i^{(k)} = n_i$ holds for every $i \in [d]$. So indeed we schedule all the jobs and have enough machines to do so. Since each $y^{(k)} \in S$ also lies in P, the jobs scheduled on the same machine have total processing time at most u. So the constructed schedule is feasible and the given instance I is positive.

So a straightforward application of Proposition 1 to the above formulations would yield algorithms running in time $|V|^{2^{O(d)}} \langle P \rangle^{O(1)} \langle Q \rangle^{O(1)} (\log(m))^{O(1)}$, where $|V|$ is the number of vertices of P's integer hull or in the other case running time $\langle P \rangle^{2^{O(d)}} \langle Q \rangle^{O(1)} (\log(m))^{O(1)}$. Note however that without further consideration, bounds for $|V|$ either contain a term $(\log(p_{\max}))^{O(\log(p_{\max}))}$ or $\log(\ell) \le \log(u)$, depending on whether we use Theorem 12 or Theorem 13. Moreover, $\langle P \rangle$ also depends on $\log(\ell) \le \log(u)$. So these do not yet yield a running time $(\log(p_{\max}))^{2^{O(d)}} \langle I \rangle^{O(1)}$. This is where the balancing result by Govzmann et al. (Lemma 2) can help: By paying $O(d)$ time (which vanishes in the other terms), we can assume that $\ell \le u \le 2dp_{\max}^2$, as there are at most $2p_{\max}$ jobs of each type on any machine. Then the running time $\langle P \rangle^{2^{O(d)}} \langle Q \rangle^{O(1)} (\log(m))^{O(1)}$ becomes

$$(d \log(p_{\max}) + \log(u))^{2^{O(d)}} \langle I \rangle^{O(1)} = (d \log(p_{\max}) + \log(2dp_{\max}^2))^{2^{O(d)}} \langle I \rangle^{O(1)}$$

and with the considerations about optimization (see [17]) in mind, we get the following result:

Theorem 1. *The optimization problems $P||\{C_{\max}, C_{\min}\}$ can be solved in time* $(d \log(p_{\max}) + \log(2dp_{\max}^2))^{2^{O(d)}} \langle I \rangle^{O(1)}$ *and the optimization problem $P||C_{\text{envy}}$ in time* $p_{\max}(d \log(p_{\max}) + \log(2dp_{\max}^2))^{2^{O(d)}} \langle I \rangle^{O(1)}$.

4   Tool 2: Proximity

In some cases, we might not have such a fine preprocessing algorithm like the one by Govzmann [13] for our problem but we still want to reduce the encoding length of P. In this section, we show how a similar preprocessing can be done for general PQ-Rs. Each PQ-R also has an equivalent ILP and solving a relaxation of this ILP allows us to remove the dependency on the right-hand-side (e.g. u in the case of $P||C_{\max}$):

Lemma 4. *Let (P, Q, m) be a given PQ-R with $P = \left\{ x \in \mathbb{R}^N_{\geq 0} \mid A^{(P)}x = b^{(P)} \right\}$ and $Q = \left\{ x \in \mathbb{R}^N_{\geq 0} \mid A^{(Q)}x = b^{(Q)} \right\}$. Then the points $y \in \mathrm{int.cone}(P \cap \mathbb{Z}^N) \cap Q$ have a one-to-one correspondence to the solutions of the following ILP, projected to block $m+1$ and the other blocks correspond to vectors $y^{(1)}, \ldots, y^{(m)} \in P \cap \mathbb{Z}^N$ such that $y = y^{(1)} + \ldots + y^{(m)}$:*

$$
\begin{pmatrix}
\boxed{\begin{matrix} 0 & 0 & \cdots & 0 & A^{(Q)} \\ I & I & \cdots & I & -I \end{matrix}} \\
\boxed{A^{(P)}} \\
& \boxed{A^{(P)}} \\
& & \ddots \\
& & & \boxed{A^{(P)}} \\
& & & & \boxed{0}
\end{pmatrix}
x =
\begin{pmatrix}
b^{(Q)} \\
\mathbf{0} \\
b^{(P)} \\
b^{(P)} \\
\vdots \\
b^{(P)} \\
\mathbf{0}
\end{pmatrix}
$$

$$x \in \mathbb{N}^{(m+1)N}$$

Proof. Suppose $y = y^{(1)} + \ldots + y^{(m)} \in \mathrm{int.cone}(P \cap \mathbb{Z}^N) \cap Q$ such that for all $i \in [m]$, $y^{(i)} \in P \cap \mathbb{Z}^N$. Then $x := (y^{(1)} \ldots y^{(m)} y)^T$ is a solution of the above ILP. Now suppose we have a solution $x = (y^{(1)} \ldots y^{(m)} y)^T$. Then by the local, non-negativity and integrality constraints, we have $y^{(i)} \in P \cap \mathbb{Z}^N$ for all $i \in [m]$. By the first set of constraints, we have $y \in Q$ and by the second set of constraints, we have $y = y^{(1)} + \ldots + y^{(m)}$. Hence, $y \in \mathrm{int.cone}(P \cap \mathbb{Z}^N) \cap Q$.

The ILP in Lemma 4 has a so-called n-fold structure, as indicated by the framing (see e.g. [8]). We use a proximity result by Clsovjecsek et al. [6] to bound the vectors in P, which allows us to reduce the coefficients in the right-hand-side.[4] The key idea is to solve a relaxation of the ILP corresponding to the

---

[4] We cite the first version of their paper here, because it is much easier to see how the approach for solving the relaxation can be adapted to the case where many blocks are identical.

PQ-R. In general, this *convexified relaxation* of an n-fold ILP looks like this:

$$\max \sum_{i=1}^{n} (c^{(i)})^T x^{(i)}$$

$$\sum_{i=1}^{n} C^{(i)} x^{(i)} = b^{(0)}$$

$$x^{(i)} \in Q^{(i)} := \text{int.hull}\left(\{x \in \mathbb{R}^t \,|\, B^{(i)} x = b^{(i)}, \, 0 \leq x \leq u^{(i)}\}\right) \quad \forall i \in [n]$$

where the $C^{(i)} \in \mathbb{Z}^{r \times t}$ are the blocks at the top, $b^{(0)} \in \mathbb{Z}^r$ is the corresponding right-hand-side, the $B^{(i)} \in \mathbb{Z}^{s \times t}$ are the blocks on the diagonal, the $b^{(i)} \in \mathbb{Z}^s$ are the corresponding right-hand-sides, the $c^{(i)} \in \mathbb{Z}^t$ are the parts of the objective function and the $u \in \mathbb{Z}^t$ are upper bounds for the variables. Cslovjecsek et al. [6] show that the Lagrangean dual (w.r.t. the first r constraints) of this problem can be solved with at most $r^{O(1)} \log^2(ntK)$ calls to a separation oracle, where K is the largest number appearing in the objective function, matrix, right-hand-side and upper bounds. The separation oracle has to solve

$$L(\lambda) = \max_{\substack{x^{(i)} \in Q^{(i)} \\ i=1,\ldots,n}} \left\{ \sum_{i=1}^{n} (c^{(i)})^T x^{(i)} + \lambda^T \left(\sum_{i=1}^{n} C^{(i)} x^{(i)} - b^{(0)} \right) \right\}$$

for a given $\lambda \in \mathbb{R}^r$. In general, this can be done by solving ILPs corresponding to the n blocks $Q^{(i)}$ separately, which may take time $nt^{O(t)}(st \log(K))^{O(1)}$ using the algorithm by Kannan [19] n times.[5] However, note that that if for two blocks i and j, we have $c^{(i)} = c^{(j)}$, as well as $B^{(i)} = B^{(j)}$, $C^{(i)} = C^{(j)}$, $u^{(i)} = u^{(j)}$ and $b^{(i)} = b^{(j)}$, the ILP has the same optimal solution(s).

When we have computed a solution of the Lagrangean dual, we still have to compute a solution of the primal, i.e., the convexified relaxation. This part is described in more detail in the second version of the paper by Cslovjecsek et al. [5]: While solving the dual with the cutting plane algorithm, each call to the separation oracle yields an optimizer $z^{(j)} \in \mathbb{Z}^{nt}$ of the maximum in iteration j. One then only has to find a vector $x^* \in \mathbb{R}^{nt}$ in the convex hull of the $z^{(j)}$ that also fulfills the linking constraints $\sum_{i=1}^{n} C^{(i)} x^{(i)} = b^{(0)}$ and maximizes the objective $\max \sum_{i=1}^{n} (c^{(i)})^T x^{(i)}$. This LP has many variables (equal to the number of calls to the separation oracle) but only $r + 1$ constraints. So it can be solved e.g. with the algorithm by Megiddo [26] (applied to the dual of this LP; it also provides a primal solution) in a time that is polynomial in the number of oracle calls and $2^{O(r^2)}$. Altogether, for τ different blocks, this yields running time

$$\underbrace{r^{O(1)} \log^2(ntK)}_{\text{number of oracle calls}} \underbrace{\tau t^{O(t)} (st \log(K))^{O(1)}}_{\text{time for one oracle call}} + \underbrace{(r^{O(1)} \log^2(ntK))^{O(1)} 2^{O(r^2)}}_{\text{time for solving the LP}}$$

$$\leq \tau t^{O(t)} 2^{O(r^2)} (sr \log(Kn))^{O(1)}$$

---

[5] Cslovjecsek et al. [6] use the algorithm by Eisenbrand and Weismantel [9] instead, but here it makes more sense to use the algorithm by Kannan [19], as our running time will be exponential in t anyway.

and we get the following result:

Proposition 2. *Let τ be the number of different $\left(c^{(i)}, B^{(i)}, C^{(i)}, b^{(i)}, u^{(i)}\right)$-pairs, n the number of blocks, r, s, t the block dimensions and K the largest number appearing in the objective function, matrix, right-hand-side and the upper bounds. Then the convexified relaxation can be solved in time*

$$\tau t^{O(t)} 2^{O(r^2)} (sr \log(Kn))^{O(1)}.$$

Cslovjecsek et al. [5] show that the convexified relaxation has nice proximity properties, namely they show the following:

Theorem 14 (Cslovjecsek et al. [5]). *For every solution x^* of the convexified relaxation, there exists a solution z^* of the corresponding integer program such that $\|x^* - z^*\|_1 \leq (r\Delta(s\Delta)^{O(s)})^{O(r)}$, where Δ is the largest absolute value in the matrix.*

By first solving the convexified relaxation using Proposition 2 and then applying the proximity bound from Theorem 14, we can bound the solutions of the ILP. For a full proof, see [17].

Theorem 15. *Let $P = \left\{x \in \mathbb{R}^N_{\geq 0} \mid A^{(P)}x = b^{(P)}\right\}$ with $M^{(P)}$ constraints and $Q = \left\{x \in \mathbb{R}^N_{\geq 0} \mid A^{(Q)}x = b^{(Q)}\right\}$ with $M^{(Q)}$ constraints, and consider the PQ-R (P, Q, m). Then by solving the convexified relaxation of the corresponding ILP in time*

$$2^{O((M^{(Q)}+N)^2)} \left(M^{(P)} \log \left(\max\left\{\left\|A^{(P)}\right\|_\infty, \left\|A^{(Q)}\right\|_\infty,\right.\right.\right.$$
$$\left.\left.\left.\left\|b^{(P)}\right\|_\infty, \left\|b^{(Q)}\right\|_\infty\right\} m\right)\right)^{O(1)}$$

and modifying P, we can assume that the points in P have ℓ_∞-norm at most

$$\left((M^{(Q)} + N)M^{(P)} \max\left\{\left\|A^{(P)}\right\|_\infty, \left\|A^{(Q)}\right\|_\infty\right\}\right)^{O((M^{(Q)}+N)M^{(P)})}.$$

We now have all the tools to prove Theorem 5. The key idea is to combine Theorem 15 with the result by Frank and Tardos (Corollary 1). This reduces the coefficients in the inequalities describing P. In particular, it makes their size independent of the original right-hand-side. Again, for a full proof, see [17].

Theorem 5. *If a problem has a PQ-R (P, Q, m) given by a polytope $P = \left\{x \in \mathbb{R}^N_{\geq 0} \mid A^{(P)}x = b^{(P)}\right\}$ and $Q = \left\{x \in \mathbb{R}^N_{\geq 0} \mid A^{(Q)}x = b^{(Q)}\right\}$ with $M^{(P)}$ and $M^{(Q)}$ constraints, respectively, it can be solved in time*

$$\left(\left(M^{(P)}M^{(Q)} \log \left(\max\left\{\left\|A^{(P)}\right\|_\infty, \left\|A^{(Q)}\right\|_\infty\right\}\right)\right)^{2^{O(N)}} + 2^{O((M^{(Q)})^2)}\right)$$
$$(\langle P \rangle \langle Q \rangle \log(m))^{O(1)}.$$

Note how this result also yields an algorithm for $P||\{C_{\max}, C_{\min}, C_{\text{envy}}\}$ with running time $(\log(p_{\max}))^{2^{O(d)}} \langle I \rangle^{O(1)}$, but with worse constants.

5 Tool 3: More Fitting Bounds for the Vertices of the Integer Hull

As mentioned above, given a polytope $P = \{x \in \mathbb{R}^N \mid Ax \leq b\}$, there are upper bounds for the number of vertices of P's integer hull. But they either depend on $\log(\|b\|_\infty)$ (like Theorem 13) or are exponential in $\log(\|A\|_\infty)$ (like Theorem 12). For using the algorithm by Jansen and Klein [18] – which takes time $|V|^{2^{O(d)}} \langle P \rangle^{O(1)} \langle Q \rangle^{O(1)} \log(m)^{O(1)}$ to solve a PQ-R– it would be nice to have a bound that does not depend on $\log(\|b\|_\infty)$ but is also not exponential in $\log(\|A\|_\infty)$. Fortunately, the proof by Berndt et al. [2] can be adapted to get the following result:

Theorem 6. *The integer hull of a polytope* $P = \{x \in \mathbb{R}^N_{\geq 0} \mid Ax = b\}$ *has at most* $N^M O(M \log(M\Delta))^N$ *vertices, where* M *is the number of constraints and* $\Delta = \|A\|_\infty$.

Proof (Sketch). Denote by P_I the integer hull of P. It is well-known that P has at most $\binom{N}{M}$ vertices (the basic feasible solutions). Eisenbrand and Weismantel [9] have shown that for each vertex z^* of P_I, there exists a vertex x^* of P such that $\|x^* - z^*\|_1 \leq M(2M\Delta + 1)^M$.[6] The idea of the proof by Berndt et al. [2] is to use this proximity result to remove the dependency on $\|b\|_\infty$. Each vertex of P_I is associated with at least one vertex of P; if we were to draw balls around each vertex of P with the radius set to the proximity bound, each vertex of P_I would be in at least one of these circles. Instead of rasterizing the whole P_I, Berndt et al. only rasterize the regions around the vertices of P. This yields at most $O(\log(M(2M\Delta + 1)^M))^N = O(M \log(M\Delta))^N$ boxes per vertex of P, each of which contains at most one vertex of P_I. So P_I has at most

$$\binom{N}{M} O(M \log(M\Delta))^N \leq N^M O(M \log(M\Delta))^N$$

vertices. Berndt et al. then proceed to reduce the exponent so that it does not depend on N, which might in some cases be rather large compared to Δ and M.

We can use this result to solve PQ-Rs with the algorithm by Jansen and Klein [18]:

Theorem 7. *Let* $P = \{x \in \mathbb{R}^N_{\geq 0} \mid A^{(P)}x = b^{(P)}\}$ *with* $M^{(P)}$ *constraints and* $Q = \{x \in \mathbb{R}^N_{\geq 0} \mid A^{(Q)}x = b^{(Q)}\}$ *with* $M^{(Q)}$ *constraints. Then the PQ-R* (P, Q, m) *can be solved in time*

$$\left(N^{M^{(P)}} M^{(P)} \log\left(\left\|A^{(P)}\right\|_\infty\right)\right)^{2^{O(N)}} (\langle P \rangle \langle Q \rangle \log(m))^{O(1)}.$$

---

[6] To be precise, they showed that for each optimal solution of the associated LP, there exists an optimal solution of the corresponding ILP with this distance. But by perturbing the linear objective function, every vertex of P_I can be made a unique optimum.

Proof. We use the $|V|^{2^{O(N)}} (\langle P \rangle \langle Q \rangle \log(m))^{O(1)}$-time algorithm by Jansen and Klein [18] (see Proposition 1), so we need to bound $|V|$, the number of vertices of P's integer hull. For this, we use Theorem 6, which yields

$$|V| \le N^{M^{(P)}} O\left(M^{(P)} \log\left(M^{(P)} \left\| A^{(P)} \right\|_\infty \right) \right)^N.$$

In total, we get the following running time:

$$|V|^{2^{O(N)}} (\langle P \rangle \langle Q \rangle \log(m))^{O(1)}$$

$$= \left(N^{M^{(P)}} O\left(M^{(P)} \log\left(M^{(P)} \left\| A^{(P)} \right\|_\infty \right) \right)^N \right)^{2^{O(N)}} (\langle P \rangle \langle Q \rangle \log(m))^{O(1)}$$

$$= \left(N^{M^{(P)}} M^{(P)} \log\left(\left\| A^{(P)} \right\|_\infty \right) \right)^{2^{O(N)}} (\langle P \rangle \langle Q \rangle \log(m))^{O(1)}$$

Observe that this also yields an algorithm for $P||\{C_{\max}, C_{\min} C_{\text{envy}}\}$ with running time $(\log(p_{\max}))^{2^{O(d)}} \langle I \rangle^{O(1)}$ (again, with worse constants than in Theorem 1).

6 Conclusion and Future Work

In this work, we investigated three tools that improve the running time of the algorithm by Goemans and Rothvoss [12]. The first tool could be considered a problem-specific preprocessing, the second tool uses a special relaxation of an n-fold ILP and proximity and the third tool utilizes the framework by Jansen and Klein [18] together with new customized bounds for the vertices of the integer hull. Overall, each tool provides a $(\log(p_{\max}))^{2^{O(d)}} \langle I \rangle^{O(1)}$-time algorithm for $P||\{C_{\max}, C_{\min}, C_{\text{envy}}\}$, which is FPT w.r.t. d if the processing times are given in unary. The second and third tool also yield similar algorithms for other problems. The second tool provides a bound for the variables in P. In some cases, there are natural bounds for the variables in P already, e.g. $\|n\|_1$ for $P||\{C_{\max}, C_{\min}, C_{\text{envy}}\}$. Of course, one can also use such bounds instead of proximity, if they produce a better result. On another note, it would be interesting to apply the improved framework to other scheduling and packing problems like the ones considered by Goemans and Rothvoss [12] and to see how eliminating the right-hand-side influences the running times.

References

1. Aliev, I., De Loera, J.A., Eisenbrand, F., Oertel, T., Weismantel, R.: The support of integer optimal solutions. SIAM J. Optim. **28**(3), 2152–2157 (2018). https://doi.org/10.1137/17M1162792
2. Berndt, S., Jansen, K., Klein, K.: New bounds for the vertices of the integer hull. In: SOSA, pp. 25–36. SIAM (2021). https://doi.org/10.1137/1.9781611976496.3

3. Buchem, M., Rohwedder, L., Vredeveld, T., Wiese, A.: Additive approximation schemes for load balancing problems. In: ICALP. LIPIcs, vol. 198, pp. 42:1–42:17. Schloss Dagstuhl - Leibniz-Zentrum für Informatik (2021). https://doi.org/10.4230/LIPIcs.ICALP.2021.42
4. Cook, W., Hartmann, M., Kannan, R., McDiarmid, C.: On integer points in polyhedra. Comb. **12**(1), 27–37 (1992). https://doi.org/10.1007/BF01191202
5. Cslovjecsek, J., Eisenbrand, F., Hunkenschröder, C., Rohwedder, L., Weismantel, R.: Block-structured integer and linear programming in strongly polynomial and near linear time. In: SODA, pp. 1666–1681. SIAM (2021). https://doi.org/10.1137/1.9781611976465.101
6. Cslovjecsek, J., Eisenbrand, F., Weismantel, R.: N-fold integer programming via LP rounding. CoRR abs/2002.07745 (2020). https://doi.org/10.48550/arXiv.2002.07745
7. Cygan, M., et al.: Parameterized Algorithms. Springer, Cham (2015). https://doi.org/10.1007/978-3-319-21275-3
8. Eisenbrand, F., Hunkenschröder, C., Klein, K.: Faster algorithms for integer programs with block structure. In: ICALP. LIPIcs, vol. 107, pp. 49:1–49:13. Schloss Dagstuhl - Leibniz-Zentrum für Informatik (2018). https://doi.org/10.4230/LIPIcs.ICALP.2018.49
9. Eisenbrand, F., Weismantel, R.: Proximity results and faster algorithms for integer programming using the Steinitz lemma. ACM Trans. Algorithms **16**(1) (2019). https://doi.org/10.1145/3340322
10. Etscheid, M., Kratsch, S., Mnich, M., Röglin, H.: Polynomial kernels for weighted problems. J. Comput. Syst. Sci. **84**, 1–10 (2017). https://doi.org/10.1016/J.JCSS.2016.06.004
11. Frank, A., Tardos, E.: An application of simultaneous Diophantine approximation in combinatorial optimization. Comb. **7**(1), 49–65 (1987). https://doi.org/10.1007/BF02579200
12. Goemans, M.X., Rothvoss, T.: Polynomiality for bin packing with a constant number of item types. J. ACM **67**(6) (2020). https://doi.org/10.1145/3421750
13. Govzmann, A., Mnich, M., Omlor, S.: Faster algorithms for parallel and related machine scheduling (2023, manuscript)
14. Graham, R., Lawler, E., Lenstra, J., Kan, A.: Optimization and approximation in deterministic sequencing and scheduling: a survey. In: Hammer, P., Johnson, E., Korte, B. (eds.) Discrete Optimization II, Annals of Discrete Mathematics, vol. 5, pp. 287–326. Elsevier (1979). https://doi.org/10.1016/S0167-5060(08)70356-X
15. Hartmann, M.E.: Cutting planes and the complexity of the integer hull. Ph.D. thesis, Cornell University (1989)
16. Jansen, K., Kahler, K.: Faster algorithms for multitype cone and polytope intersection. Manuscript (Personal Communication)
17. Jansen, K., Kahler, K., Zwanger, E.: Exact and approximate high-multiplicity scheduling on identical machines. CoRR abs/2404.17274 (2024). https://doi.org/10.48550/arXiv.2404.17274
18. Jansen, K., Klein, K.M.: About the structure of the integer cone and its application to bin packing. Math. Oper. Res. **45**(4), 1498–1511 (2020). https://doi.org/10.1287/moor.2019.1040
19. Kannan, R.: Minkowski's convex body theorem and integer programming. Math. Oper. Res. **12**(3), 415–440 (1987). https://doi.org/10.1287/MOOR.12.3.415
20. Knop, D., Koutecký, M., Levin, A., Mnich, M., Onn, S.: High-multiplicity N-fold IP via configuration LP. Math. Program. **200**(1), 199–227 (2023). https://doi.org/10.1007/s10107-022-01882-9

21. Knop, D., Koutecký, M.: Scheduling meets N-fold integer programming. J. Sched. **21**(5), 493–503 (2018). https://doi.org/10.1007/s10951-017-0550-0
22. Knop, D., Koutecký, M., Levin, A., Mnich, M., Onn, S.: Parameterized complexity of configuration integer programs. Oper. Res. Lett. **49**(6), 908–913 (2021). https://doi.org/10.1016/j.orl.2021.11.005
23. Koutecký, M., Zink, J.: Complexity of scheduling few types of jobs on related and unrelated machines. In: ISAAC. LIPIcs, vol. 181, pp. 18:1–18:17. Schloss Dagstuhl - Leibniz-Zentrum für Informatik (2020). https://doi.org/10.4230/LIPIcs.ISAAC.2020.18
24. Kowalik, L., Lassota, A., Majewski, K., Pilipczuk, M., Sokolowski, M.: Detecting points in integer cones of polytopes is double-exponentially hard. In: SOSA, pp. 279–285. SIAM (2024). https://doi.org/10.1137/1.9781611977936.25
25. Lenstra, J.K., Shmoys, D.B.: Elements of scheduling. CoRR abs/2001.06005 (2020). https://doi.org/10.48550/arXiv.2001.06005
26. Megiddo, N.: Linear programming in linear time when the dimension is fixed. J. ACM **31**(1), 114–127 (1984). https://doi.org/10.1145/2422.322418
27. Mnich, M., van Bevern, R.: Parameterized complexity of machine scheduling: 15 open problems. Comput. Oper. Res. **100**, 254–261 (2018). https://doi.org/10.1016/j.cor.2018.07.020
28. Mnich, M., Wiese, A.: Scheduling and fixed-parameter tractability. Math. Program. **154**(1–2), 533–562 (2015). https://doi.org/10.1007/s10107-014-0830-9
29. Zolotykh, N.Y.: On the number of vertices in integer linear programming problems. CoRR math/0611356 (2006). https://doi.org/10.48550/arXiv.math/0611356

Unit Refutations in Horn Constraint Systems

Piotr Wojciechowski and K. Subramani[✉]

LDCSEE, West Virginia University, Morgantown, WV, USA
{pwojciec,k.subramani}@mail.wvu.edu

Abstract. This paper is concerned with algorithmic procedures for checking if a Horn Constraint System (HCS) has unit refutations of linear feasibility. A Horn constraint is a linear constraint of the form $\mathbf{a} \cdot \mathbf{x} \geq b$, where each $a_i \in \{0, 1, -1\}$ and at most one of the a_is is positive. Horn constraints generalize Horn clauses and find applications in a number of practical domains such as program verification and econometrics. Our main contribution is showing that the problem of checking if an infeasible HCS has a unit refutation is in **P**. To this end, we provide a certifying algorithm which runs in $O(m \cdot n^2 \cdot (m+n))$ time, where m and n are the number of constraints and number of variables in the input HCS respectively. Our algorithm is based on a combination of new and existing insights regarding Horn constraints.

1 Introduction

This paper is concerned with the design and analysis of algorithms for extracting *unit refutations* from infeasible Horn Constraint Systems (HCS). A Horn constraint is a linear constraint of the form $\mathbf{a} \cdot \mathbf{x} \geq b$, where each $a_i \in \{0, 1, -1\}$ and at most one a_i is positive. A conjunction of Horn constraints constitutes a Horn Constraint System and can be represented in matrix form as: $\mathbf{A} \cdot \mathbf{x} \geq \mathbf{b}$. HCSs find applications in a wide variety of domains including scheduling [7], operations research [2,5], and program verification (abstract interpretation) [1,14].

A refutation of a (Horn) constraint system is a proof of unsatisfiability, i.e., it certifies the infeasibility of the given instance. Certifying algorithms enhance confidence in the output provided by algorithmic implementations. This feature is particularly useful in mission-critical software such as the ones used in flight-control systems. In this paper, we focus on a specific type of refutation called *unit refutation*. Unit refutations are distinct from general refutations in that every inferential step must include a one variable constraint (also called an absolute constraint). Unit refutations in linear constraint systems are analogous to unit resolution in Boolean formulas in conjunctive normal form (CNF) [10]. While unit resolution is **complete** for Horn formulas, unit refutation is **incomplete**, even for HCSs. Although incomplete, unit refutations are interesting since they provide domain specific refutations. In other words, a unit refutation proves that an HCS is infeasible with its current set of absolute constraints. However, a

© The Author(s), under exclusive license to Springer Nature Switzerland AG 2025
I. Finocchi and L. Georgiadis (Eds.): CIAC 2025, LNCS 15679, pp. 18–33, 2025.
https://doi.org/10.1007/978-3-031-92932-8_2

unit refutation does not prove the infeasibility of the underlying system of non-absolute constraints. Note that an unrestricted refutation does not necessarily use absolute constraints. Thus, an unrestricted refutation does not necessarily depend on the current variable domains. In other words, unit refutations are a useful tool for proving that the current variable domains are responsible for the infeasibility of an HCS.

Our approach for checking if an infeasible HCS has a unit refutation is certifying. This means that both positive and negative certificates are produced as part of the output. Thus, if the input HCS has a unit refutation, then the algorithm will return proof that the unit refutation exists. Additionally, if the input HCS does not have a unit refutation, then the algorithm will return proof that a unit refutation does not exist.

2   Statement of Problems

In this section, we formally describe the problems under consideration in this paper. Our principal focus is on a specialized linear program called a Horn program or Horn Constraint System (HCS).

Definition 1. *A linear constraint* $\mathbf{a} \cdot \mathbf{x} \geq b$ *is said to be a* **Horn Constraint**, *if: 1. The entries in* $\mathbf{a}$ *belong to the set* $\{0, 1, -1\}$. *2.* $\mathbf{a}$ *contains at most one positive entry. 3.* b *is an integer.*

Definition 2. *A linear constraint system* $\mathbf{A} \cdot \mathbf{x} \geq \mathbf{b}$ *is said to be a* **Horn Constraint System (HCS)**, *if it is a conjunction of Horn constraints.*

If a Horn constraint has only one non-zero coefficient, then it is called an **absolute constraint**. If this coefficient is positive, then this constraint is called a **positive absolute constraint**. If a Horn constraint has no variables with positive coefficient, then it is called an **all-negative** constraint. We will use n to refer to the number of variables in an HCS, and m to refer to the number of constraints.

The **Linear Refutability Problem (LRP)** is concerned with providing proof that a given HCS does not have any linear (rational) solutions. Such a proof is known as a refutation. These refutations consist of a sequence of inferences that results in a contradiction of the form $0 \geq b$ where $b > 0$. In case of LRP, we use a single inference rule (Rule 1).

$$\text{ADD} : \frac{\sum_{i=1}^{n} a_i \cdot x_i \geq b_1 \qquad \sum_{i=1}^{n} a_i' \cdot x_i \geq b_2}{\sum_{i=1}^{n} (a_i + a_i') \cdot x_i \geq b_1 + b_2} \qquad (1)$$

We refer to Rule (1) as the **ADD rule**. It is easy to see that Rule (1) is sound since any assignment that satisfies the hypotheses also satisfies the consequent. Additionally, the ADD rule is complete [4].

In this paper, we study a restricted version of the ADD rule, known as the unit-ADD (UADD) rule. In the UADD rule, at least one of the constraints must be an absolute constraint. Rule (2) represents the UADD rule.

$$\text{UADD}: \frac{a_i \cdot x_i \geq b_1 \quad - a_i \cdot x_i + \sum_{j \neq i} a_j \cdot x_j \geq b_2}{\sum_{j \neq i} a_j \cdot x_j \geq b_1 + b_2} \tag{2}$$

A linear refutation using only the UADD rule is called a **unit linear refutation**. The problem of finding such a refutation is called the **Unit Linear Refutability Problem (ULRP)**. In the remainder of this paper, we use the terms unit refutation and unit linear refutation interchangeably.

It is important to note that, unlike the ADD rule, the UADD rule is incomplete. This means that there are HCSs with no linear solutions that do not have a refutation using only the UADD rule. Note that, when applied to Horn constraints, the UADD rule will always derive a Horn constraint.

Definition 3. *The Unit Linear Refutability Problem (ULRP) for HCSs is defined as follows: Given an HCS* **H**, *does* **H** *have a unit linear refutation?*

The principal contribution of this paper is an $O((m+n) \cdot m \cdot n^2)$ time certifying algorithm for ULRP in HCSs (see Sect. 4).

3   Motivation and Related Work

This paper discusses algorithms for determining unit refutations in Horn constraint systems. Horn constraint systems are used widely in program verification applications such as abstract interpretation [1,14] and array bound checking [11]. These constraints are also used in SMT solvers [2,5], infinite state systems, and test-case generation [3]. Therefore, the study of refutation systems for HCSs is well-motivated.

Refutations are negative certificates in that they provide an explanation as to why a constraint system is unsatisfiable. Any problem in the class **NP ∩ coNP** has both "positive" and "negative" certificates, which are succinct (short, polynomial in the size of the input constraint system) [6]. Problems which are **NP-complete** have short positive certificates, but negative certificates must be superpolynomial unless **coNP = NP** [8]. It follows that even if a negative certificate for an **NP-complete** problem is obtained, it is not possible to *verify* its correctness in time polynomial in the size of the input constraint system. One technique for overcoming this issue is to study **incomplete** (weak) but sound refutation systems for constraint systems. The idea underlying incomplete proof systems is to sacrifice completeness for efficient decidability. One such refutation system called *read-once resolution* refutation was introduced in [9]. In this paper, we explore another incomplete refutation system called unit refutation.

Unit resolution has been studied extensively in the construction of refutation systems for Boolean formulas in CNF, especially in Horn clauses. It is well-known that unit resolution is both sound and complete as a refutation procedure for

Horn clauses [13]. In [12], it is shown that the unit refutation of a Horn formula can be extracted in linear time. To the best of our knowledge, the algorithms in this paper are the first algorithms for finding unit refutations of HCSs.

4   Algorithms for ULRP in HCSs

In this section, we show that certifying if an HCS has a unit refutation can be done in polynomial time. Algorithms 4.1 through 4.5 represent our approach.

Let $\mathbf{H}$ be an HCS with m constraints over n variables. For the algorithms in this section, and the proofs in subsequent sections, we use the following notation: 1. $N(\mathbf{H})$ is the set of all-negative constraints in $\mathbf{H}$. 2. $P(\mathbf{H})$ is the set of positive absolute constraints in $\mathbf{H}$. 3. For a constraint $l_k \in \mathbf{H}$, P_k is the set of variables with positive coefficient and N_k is the set of variables with negative coefficient. Note that $|P_k| \leq 1$.

We first provide a series of algorithms (Algorithms 4.1 through 4.3) for reducing an HCS $\mathbf{H}$ down to an HCS $\mathbf{H}'$ by removing variables that cannot be part of any unit refutation.

Algorithm 4.1 maintains a set T of variables which can be used to derive lower bounds on variables appearing in all-negative constraints. For each variable $x_i \in T$, a constraint of the form $x_i \geq b_i$ can be used to help derive a lower bound on a variable in an all negative constraint. As will be shown, any constraint used in a unit refutation of $\mathbf{H}$ must use only variables in T.

Algorithm 4.1. Algorithm for obtaining T

    Input: HCS $\mathbf{H}$
    Output: Set of variables T

1: **procedure** GET-T($\mathbf{H}$)
2:     Let $T := \bigcup_{l_k \in N(\mathbf{H})} N_k$, be a set of variables.
3:     Let $T_{temp} := \emptyset$.
4:     **while** $(T_{temp} \neq T)$ **do**
5:        $T_{temp} := T$.
6:        **for** (each $l_k \in \mathbf{H} \setminus N(\mathbf{H})$) **do**
7:          **if** $P_k \subseteq T$ **then**
8:            $T := T \cup N_k$.
9:     **return** T.

Algorithm 4.2 maintains a set S of variables for which it is possible to derive lower bounds. Each variable $x_i \in S$ is such that it is possible to derive a constraint of the form $x_i \geq b$ from $\mathbf{H}$ through unit derivation. As will be shown, any constraint used in a unit refutation of $\mathbf{H}$ must use only variables in S.

Algorithm 4.3 uses Algorithm 4.1 and Algorithm 4.2 to remove constraints from $\mathbf{H}$ that cannot be used in any unit refutation of $\mathbf{H}$. As stated previously, any constraint used in a unit refutation of $\mathbf{H}$ must use variables in

---

Algorithm 4.2. Algorithm for obtaining S

    Input: HCS **H**
    Output: Set of variables S

1: **procedure** GET-S(**H**)
2:    Let $S := \bigcup_{l_k \in P(\mathbf{H})} P_k$, be a set of variables.
3:    Let $S_{temp} := \emptyset$.
4:    **while** ($S_{temp} \neq S$) **do**
5:        $S_{temp} := S$.
6:        **for** (each $l_k \in \mathbf{H} \setminus P(\mathbf{H})$) **do**
7:            **if** $N_k \subseteq S$ **then**
8:                $S := S \cup P_k$.
9:    **return** S.

---

both the set $T = \text{GET-T}(\mathbf{H})$ and in the set $S = \text{GET-S}(\mathbf{H})$. Thus, any constraint using variables outside of $T \cap S$ can be removed from **H** without invalidating any unit refutation of **H**. Note that removing constraints from **H** may result in variables being removed from T or S. Thus, Algorithm 4.3, continues the constraint removal process until no additional constraints can be removed.

Algorithm 4.3. Algorithm for obtaining reduced system $\mathbf{H}'$

    Input: HCS **H**
    Output: Reduced HCS $\mathbf{H}'$

1: **procedure** REDUCE-HCS(**H**)
2:    Let $\mathbf{H}' := \mathbf{H}$.
3:    Let $\mathbf{H}_{temp} := \emptyset$.
4:    **while** ($\mathbf{H}_{temp} \neq \mathbf{H}'$) **do**
5:        $\mathbf{H}_{temp} := \mathbf{H}'$.
6:        Let $T := \text{GET-T}(\mathbf{H}')$.
7:        Let $S := \text{GET-S}(\mathbf{H}')$.
8:        **for** (each variable x_i in $\mathbf{H}'$) **do**
9:            **if** ($x_i \notin T$ or $x_i \notin S$) **then**
10:               Remove all constraints using x_i from $\mathbf{H}'$.
11:   **return** $\mathbf{H}'$.

Example 1. Let **H** be the **HCS** in System (3).

$$-x_1 - x_2 \geq 0 \quad x_2 - x_3 \geq 4 \tag{3}$$
$$x_1 \geq 1 \quad x_2 - x_1 \geq 0 \quad x_4 \geq 2$$

The only all-negative constraint in System (3) is the constraint: $-x_1 - x_2 \geq 0$. Thus, in GET-T(**H**) we initially have $T = \{x_1, x_2\}$. The constraint: $x_2 - x_3 \geq 4$

is used to add x_3 to T and there is no constraint in System (3) that can be used to add x_4 to T. Thus, it is easy to see that GET-T($\mathbf{H}$) returns the set $\{x_1, x_2, x_3\}$.

The only positive absolute constraints in System (3) are $x_1 \geq 1$ and $x_4 \geq 2$. Thus, in GET-S($\mathbf{H}$) we initially have $S = \{x_1, x_4\}$. The constraint: $x_2 - x_1 \geq 0$ is used to add x_2 to S and there is no constraint in System (3) that can be used to add x_3 to S. Thus, it is easy to see that GET-S($\mathbf{H}$) returns the set $\{x_1, x_2, x_4\}$.

The only variables in both GET-T($\mathbf{H}$) and GET-S($\mathbf{H}$) are x_1 and x_2. Thus, the variables x_3 and x_4 and all constraints using those variables will be removed from $\mathbf{H}$. This results in the HCS in System (4).

$$-x_1 - x_2 \geq 0 \quad x_1 \geq 1 \quad x_2 - x_1 \geq 0 \tag{4}$$

When run on System (4), both Algorithm 4.1 and Algorithm 4.2 return the set $\{x_1, x_2\}$. Thus, no further variables will be removed from the system. Thus, the HCS in System (4) is the HCS $\mathbf{H}'$ returned by REDUCE-HCS($\mathbf{H}$).

For the remaining algorithms, we use an $n \times (m+1)$ matrix $\mathbf{D}$. This matrix represents the positive absolute constraints derivable from $\mathbf{H}$. In Sect. 6 we will show that $\mathbf{D}$ has the following property: 1. For each variable x_i such that $\mathbf{d}_i \neq \mathbf{0}$, the constraint: $x_i \geq d_{i,0}$ is derivable from $\mathbf{H}$ using the UADD rule. 2. For each variable x_i such that $\mathbf{d}_i \neq \mathbf{0}$, the sum $\sum_{k=1}^{m} d_{i,k} \cdot l_k$ results in the constraint: $x_i \geq d_{i,0}$.

Algorithm 4.4 uses the derived absolute constraints represented by the matrix $\mathbf{D}$ and the constraints in $\mathbf{H}$, to derive a new set of positive absolute constraints using the UADD rule.

Algorithm 4.4. Algorithm for updating absolute constraints derived from $\mathbf{H}$

    Input: HCS $\mathbf{H}$, and matrix $\mathbf{D}$.
    Output: Updated matrix $\mathbf{D}'$

1: **procedure** UPDATE-ABSOLUTE($\mathbf{H}$, $\mathbf{D}$)
2:     Let $\mathbf{D}' := \mathbf{D}$.
3:     **for** (each constraint $l_k \in \mathbf{H} \setminus N(\mathbf{H})$) **do**
4:         **if** ($\mathbf{d}_j \neq \mathbf{0}$ for each $x_j \in N_k$) **then**
5:             Let x_i be the variable in P_k.
6:             **if** ($d'_{i,0} < b_k + \sum_{x_j \in N_k} d_{j,0}$) **then**
7:                 $d'_{i,0} := b_k + \sum_{x_j \in N_k} d_{j,0}$.
8:                 **for** ($k' := 1$ **to** m) **do**
9:                     $d'_{i,k'} := \sum_{x_j \in N_k} d_{j,k'}$.
10:                 $d'_{i,k} := d'_{i,k} + 1$.
11:     **return** $\mathbf{D}'$.

Algorithm 4.5 uses Algorithm 4.3 and Algorithm 4.4 to determine whether an HCS **H** has a unit refutation. First, Algorithm 4.5 uses Algorithm 4.3 to remove constraints that cannot be part of any unit refutation of **H**. Then, Algorithm 4.5 repeatedly calls Algorithm 4.4 to use the UADD rule to derive new positive absolute constraints. If **H** has a unit refutation, then after $(n + 1)$ calls to Algorithm 4.4, either the absolute constraints in **D** can be used to derive a contradiction using a single all-negative constraint in **H**, or there is a variable x_i such that the constraint $x_i \geq b_i$ in **D** has been derived using an absolute constraint of the form $x_i \geq b_i'$ that was previously derived from **H** using the UADD rule. In this case, the lower bound on x_i can be increased boundlessly. This can be used to increase the lower bound on other variables in a similar fashion. Thus, at some point an all-negative constraint will be violated.

Algorithm 4.5. Algorithm for Horn ULRP

    Input: HCS **H**
    Output: true if **H** has a unit linear refutation.

 1: **procedure** UNIT-HORN(**H**)
 2:    **H**′ := REDUCE-HCS(**H**).
 3:    Let $\mathbf{D}^0$ be an $n \times (m + 1)$ matrix.
 4:    Initialize every element of **D** to 0.
 5:    **for** (each constraint $l_k \in P(\mathbf{H}')$) **do**
 6:        Let x_i be the variable in P_k.
 7:        $d_{i,0}^0 := b_k$.
 8:        $d_{i,k}^0 := 1$.
 9:    **for** $r = 1$ **to** $(n + 1)$ **do**
10:        $\mathbf{D}^r := $ UPDATE-ABSOLUTE(**H**′, $\mathbf{D}^{(r-1)}$).
11:        **for** (each variable x_i in **H**′) **do**
12:            **if** $(\sum_{l_k : x_i \in P_k} d_{i,k}^r > 1)$ **then**
13:               **return true.**
14:        **for** (each $l_k \in N(\mathbf{H}')$ such that $\mathbf{d}_i^r \neq \mathbf{0}$ for each $x_i \in N_k$) **do**
15:           **if** $(b_k + \sum_{x_i \in N_k} d_{i,0}^r > 0)$ **then**
16:               **return true.**
17:    **return false.**

---

5 Resource Analysis

In this section, we analyze the resources taken by Algorithms 4.1 through 4.5.

Lemma 1. *Let* **H** *be an HCS with m constraints over n variables. Algorithm* GET-T(**H**) *runs in time $O(m \cdot n^2)$.*

Proof. Consider the **while** loop on line 4 of Algorithm 4.1. This loop iterates as long as variables are being added to the set T. Since $|T| \leq n$, this **while** loop iterates at most n times.

Now let us consider a single iteration of the **while** loop. In each iteration, we check each constraint l_k in $\mathbf{H} \setminus N(\mathbf{H})$ to see if $P_k \subseteq T$. This check takes $O(1)$ time, since for each variable x_i we can use an indicator variable to determine if that variable is in T or not. Additionally, we add up to n variables to T. As a result, processing constraint l_k takes $O(n)$ time. Thus, each iteration of the **while** loop takes $O(m \cdot n)$ time. Consequently, Algorithm 4.1 takes $O(m \cdot n^2)$ time. $\qquad \square$

Lemma 2. *Let* $\mathbf{H}$ *be an HCS with* m *constraints over* n *variables. Algorithm* GET-S($\mathbf{H}$) *runs in time* $O(m \cdot n^2)$.

Proof. Consider the **while** loop on line 4 of Algorithm 4.2. This loop iterates as long as variables are being added to the set S. Since $|S| \leq n$, this **while** loop iterates at most n times.

Now let us consider a single iteration of the **while** loop. In each iteration, we check each constraint l_k in $\mathbf{H} \setminus P(\mathbf{H})$ to see if $N_k \subseteq S$. This check takes $O(n)$ time. Additionally, we add at most one variable to S for each constraint. Thus, each iteration of the **while** loop takes $O(m \cdot n)$ time. Consequently, Algorithm 4.2 takes $O(m \cdot n^2)$ time. $\qquad \square$

Lemma 3. *Let* $\mathbf{H}$ *be an HCS with* m *constraints over* n *variables. Algorithm* REDUCE-HCS($\mathbf{H}$) *runs in time* $O(m \cdot n^3)$.

Proof. Consider the **while** loop on line 4 of Algorithm 4.3. This loop iterates as long as variables are being removed from $\mathbf{H}'$. Since $\mathbf{H}'$ starts with n variables, this **while** loop iterates at most n times.

Now let us consider a single iteration of the **while** loop. In each iteration, we call GET-T($\mathbf{H}'$) and GET-S($\mathbf{H}'$). By Lemmas 1 and 2, this takes $O(m \cdot n^2)$ time. Next, for each variable x_i in $\mathbf{H}'$ we check to see if x_i belongs to the sets S and T, then possibly remove up to k constraints from $\mathbf{H}'$. This takes a total of $O(m \cdot n^2)$ time. Thus, each iteration of the **while** loop takes $O(m \cdot n^2)$ time. Consequently, Algorithm 4.3 takes $O(m \cdot n^3)$ time. $\qquad \square$

Lemma 4. *Let* $\mathbf{H}$ *be an HCS with* m *constraints over* n *variables and let* $\mathbf{D}$ *be an* $n \times (m + 1)$ *matrix. Algorithm* UPDATE-ABSOLUTE($\mathbf{H}$, $\mathbf{D}$) *runs in time* $O(m^2 \cdot n)$.

Proof. Consider the **for** loop on line 3 of Algorithm 4.4. This loop iterates at most m times. In each iteration, we update the values in $\mathbf{d}'_i$ for some variable x_i. Note that $\mathbf{d}'_i$ has $(m + 1)$ values and updating each value takes $O(n)$ time. Thus, each iteration of the **for** loop takes $O(m \cdot n)$ time. Thus, Algorithm 4.4 takes $O(m^2 \cdot n)$ time. $\qquad \square$

Theorem 1. *Let* $\mathbf{H}$ *be an infeasible HCS with* m *constraints over* n *variables. Algorithm* UNIT-HORN($\mathbf{H}$) *runs in* $O((m + n) \cdot m \cdot n^2)$ *time.*

Proof. Initially, we set $\mathbf{H}'$ to be REDUCE-HCS($\mathbf{H}$). From Lemma 3, this takes $O(m \cdot n^3)$ time. We then construct and initialize, the matrix $\mathbf{D}^0$. This takes

$O(m \cdot n)$ time. Now let us examine the **for** loop on line 9 of Algorithm 4.5. This loop iterates at most $(n + 1)$ times. In each iteration, we call Algorithm 4.4. From Lemma 4, this takes $O(m^2 \cdot n)$ time. We then perform a $O(m)$ check for each variable in $\mathbf{H}'$ and a $O(n)$ check for each constraint in $N(\mathbf{H}')$. These checks take a total of $O(m \cdot n)$ time. Thus, each iteration of the **for** loop takes $O(m^2 \cdot n)$ time. Consequently, Algorithm 4.5 runs in $O((m + n) \cdot m \cdot n^2)$ time.   □

6   Proof of Correctness

In this section, we will show that UNIT-HORN($\mathbf{H}$) returns **true** if and only if the HCS $\mathbf{H}$ has a unit refutation. First, we will prove several useful properties of HCSs and of Algorithms 4.1 through 4.5.

Lemma 5. *Let* $\mathbf{H}$ *be an HCS. Any unit refutation of* $\mathbf{H}$ *must use an all-negative constraint.*

Proof. Suppose otherwise. Thus, there is a unit refutation R of $\mathbf{H}$ that does not use any all-negative constraints. Thus, each application of the UADD rule must be to constraints of the form: $x_i \geq b_k$ and $x_j - \sum_{x \in N_{k'}} x \geq b_{k'}$. Note that the variable x_j occurs with positive coefficient in the resultant constraint. Thus, this constraint cannot be a contradiction of the form: $0 \geq b$ where $b > 0$. Consequently, R is not a unit refutation of $\mathbf{H}$.   □

Lemma 6. *Let* $\mathbf{H}$ *be an HCS with variable set* X. *Let* T *denote the set of variables returned by* GET-T($\mathbf{H}$). *No unit refutation of* $\mathbf{H}$ *can use any variables in* $X - T$.

Proof. Let x_i be a variable in $\mathbf{H}$ that is not in GET-T($\mathbf{H}$). Suppose that there is a unit refutation R that uses x_i. Thus, there is a constraint $l_k \in \mathbf{H}$ such that $x_i \in N_k$. If $P_k = \emptyset$, then l_k is an all-negative constraint. Thus, x_i would have been added to T by Algorithm 4.1. Consequently, $x_i \in$ GET-T($\mathbf{H}$). This is a contradiction. Thus, $P_k \neq \emptyset$. Let x_j be the variable in P_k. If $x_j \in$ GET-T($\mathbf{H}$), then x_i would have been added to T on line 8 of Algorithm 4.1. Consequently, $x_j \notin$ GET-T($\mathbf{H}$).

This means that any constraint derived from a constraint using x_i, must use a variable x_j not in GET-T($\mathbf{H}$). Note that such a constraint cannot be a contradiction of the form: $0 \geq b$ where $b > 0$. Thus, x_i cannot be part of a unit refutation of $\mathbf{H}$.   □

Lemma 7. *Let* $\mathbf{H}$ *be an HCS and let* $x_i \in$ GET-T($\mathbf{H}$). *There is a sequence of constraints* $l_{k_r}, \ldots l_{k_0}$ *such that* $x_i \in N_{k_r}$, $P_{k_j} \subseteq N_{k_{j-1}}$ *for* $j = 1 \ldots r$, *and* $l_{k_0} \in N(\mathbf{H})$.

Proof. Let T_r, denote the set T after r iterations of the **while** loop on line 4 of Algorithm 4.1.

If $x_i \in T_0$, then there is a constraint l_{k_0} such that $x_i \in N_{k_0}$. The only constraints used to add variables to T before the **while** loop are all-negative constraints. Thus, $l_{k_0} \in N(\mathbf{H})$.

Now assume that we can derive the desired sequence of constraints for all variables in T_{r-1}. If $x_i \in T_r \setminus T_{r-1}$, then there is a constraint l_{k_r} such that $x_i \in N_{k_r}$ and $P_{k_r} \subseteq T_{r-1}$.

Let $x_{i'}$ be the variable in P_{k_r}. We have that $x_{i'} \in T_{r-1}$. By the inductive hypothesis, we can derive the desired sequence for $x_{i'}$. Adding l_{k_r} to this sequence, results in the desired sequence for x_i. Thus, the desired sequence exists for all $x_i \in$ GET-T($\mathbf{H}$). $\square$

Lemma 8. *Let $\mathbf{H}$ be an HCS with variable set X. Let S denote the set of variables returned by* GET-S($\mathbf{H}$). *No unit refutation of $\mathbf{H}$ can use any variables in $X - S$.*

Proof. Let x_i be a variable in $\mathbf{H}$ that is not in GET-S($\mathbf{H}$). Suppose that there is a unit refutation R that uses x_i. Thus, there is a constraint $l_k \in \mathbf{H}$ such that $x_i \in P_k$. If $N_k = \emptyset$, then l_k is a positive absolute constraint. Thus, x_i would have been added to S by Algorithm 4.2. Consequently, $x_i \in$ GET-S($\mathbf{H}$). This is a contradiction. Thus, $N_k \neq \emptyset$. If each $x_j \in N_k$ is in GET-S($\mathbf{H}$), then x_i would have been added to S on line 8 of Algorithm 4.2. Consequently, there is an $x_j \in N_k$ such that $x_j \notin$ GET-S($\mathbf{H}$).

This means that any constraint derived from a constraint using x_i, must use a variable x_j not in GET-S($\mathbf{H}$). Note that such a constraint cannot be a contradiction of the form: $0 \geq b$ where $b > 0$. Thus, x_i cannot be part of a unit refutation of $\mathbf{H}$. $\square$

Lemma 9. *Let $\mathbf{H}$ be an HCS and let $x_i \in$ GET-S($\mathbf{H}$). For some integer b, the constraint: $x_i \geq b$ can be derived from $\mathbf{H}$ by the UADD rule.*

Proof. Let S_r, denote the set S after r iterations of the **while** loop on line 4 of Algorithm 4.2.

If $x_i \in S_0$, then there is a constraint $l_k \in \mathbf{H}$ such that $x_i \in P_k$. The only constraints used to add variables to S before the **while** loop are positive unit constraints. Thus, l_k is a constraint of the form: $x_i \geq b_k$. The constraint: $x_i \geq b_k$ can be trivially derived from $\mathbf{H}$.

Now assume that we can use the UADD rule to derive the constraint: $x_j \geq b_j$ for all $x_j \in S_{r-1}$. If $x_i \in S_r \setminus S_{r-1}$, then there is a constraint l_k such that $x_i \in P_k$ and $N_k \subseteq S_{r-1}$.

By the inductive hypothesis, we can derive the constraint: $x_j \geq b_j$ for each $x_j \in N_k$ using the UADD rule. Note that the constraint l_k is of the form: $x_i - \sum_{x_j \in N_k} x_j \geq b_k$. Thus, using a sequence of unit derivations, we can derive the constraint: $x_i \geq b_k + \sum_{x_j \in N_k} b_j$. Consequently, we can derive a constraint of the form: $x_i \geq b$ for each $x_i \in$ GET-S($\mathbf{H}$). $\square$

Lemma 10. *Let $\mathbf{H}$ be an HCS. $\mathbf{H}$ has a unit refutation if and only if* REDUCE-HCS($\mathbf{H}$) *has a unit refutation.*

Proof. Let $\mathbf{H}'$ be REDUCE-HCS($\mathbf{H}$). First, assume that $\mathbf{H}'$ has a unit refutation R. Note that $\mathbf{H}'$ is constructed by removing constraints from $\mathbf{H}$. Thus, R is a unit refutation of $\mathbf{H}$.

Next, assume that $\mathbf{H}$ has a unit refutation R. Let $\mathbf{H}_1$ be the HCS constructed on line 10 during the first iteration of the **while** loop on line 4 of Algorithm 4.3. Note that $\mathbf{H}_1$ is constructed by removing all constraints using the variables not in the sets GET-T($\mathbf{H}$) and GET-S($\mathbf{H}$) from $\mathbf{H}$. By Lemmas 1 and 2, none of these constraints can be part of any unit refutation of $\mathbf{H}$. Thus, R is a unit refutation of $\mathbf{H}_1$.

Let $\mathbf{H}_i$, $i \geq 2$, be the HCS constructed on line 10 during the ith iteration of the **while** loop on line 4 of Algorithm 4.3. Assume that R is a unit refutation of $\mathbf{H}_{i-1}$. Note that $\mathbf{H}_i$ is constructed by removing all constraints using the variables not in the sets GET-T($\mathbf{H}_{i-1}$) and GET-S($\mathbf{H}_{i-1}$) from $\mathbf{H}_{i-1}$. By Lemmas 1 and 2, none of these constraints can be part of any unit refutation of $\mathbf{H}_{i-1}$. Thus, R is a unit refutation of $\mathbf{H}_i$.

Note that $\mathbf{H}'$ is the HCS constructed during the last iteration of the **while** loop. Thus, by induction, R is a unit refutation of $\mathbf{H}'$. $\qquad\square$

We now look at the matrix $\mathbf{D}$ used to represent positive absolute constraints derivable from $\mathbf{H}$.

Lemma 11. *Let $\mathbf{H}'$ be an HCS with m constraints over n variables and let $\mathbf{D}^r$, $r = 0 \ldots (n+1)$ be the $n \times (m+1)$ matrices in Algorithm 4.5. For $r = 0 \ldots (n+1)$ and for each x_i such that $\mathbf{d}_i^r \neq \mathbf{0}$, the constraint: $x_i \geq d_{i,0}^r$ is derivable from $\mathbf{H}'$ by the UADD rule.*

Proof. We will prove this by induction on r. The matrix $\mathbf{D}^0$ is the matrix constructed on line 9 of Algorithm 4.5. Let x_i be a variable such that $\mathbf{d}_i^0 \neq \mathbf{0}$. By construction, we have that $d_{i,0}^0 = b_k$ where $x_i \geq b_k$ is an absolute constraint in $\mathbf{H}'$. Thus, the constraint: $x_i \geq d_{i,0}^0$ is trivially derivable from $\mathbf{H}'$ by using the UADD rule.

Now assume that the desired property holds for $\mathbf{D}^{r-1}$. Note that $\mathbf{D}^r$ is constructed by calling UPDATE-ABSOLUTE($\mathbf{H}'$, $\mathbf{D}^{r-1}$). Let, x_i be a variable such that $\mathbf{d}_i^r \neq \mathbf{0}$. There are two cases we need to consider, either $\mathbf{d}_i^r = \mathbf{d}_i^{r-1}$ or $\mathbf{d}_i^r \neq \mathbf{0}$ was updated on line 7 of Algorithm 4.4.

In the first case, $d_{i,0}^r = d_{i,0}^{r-1}$. By the inductive hypothesis $x_i \geq d_{i,0}^{r-1}$ is derivable from $\mathbf{H}'$ by the UADD rule. Thus, $x_i \geq d_{i,0}^r$ is derivable from $\mathbf{H}'$ by the UADD rule as desired.

In the second case, there exists a constraint l_k of the form: $x_i - \sum_{x_j \in N_k} x_j \geq b_k$ such that $d_{i,0}^r = b_k + \sum_{x_j \in N_k} d_{j,0}^{r-1}$ and $\mathbf{d}_j^{r-1} \neq \mathbf{0}$ for each $x_j \in N_k$. By the inductive hypothesis, the constraint: $x_j \geq d_{j,0}^{r-1}$ is derivable from $\mathbf{H}'$ by the UADD rule. Applying the unit ADD rule to these constraints and the constraint l_k, we derive the constraint: $x_i \geq b_k + \sum_{x_j \in N_k} d_{j,0}^{r-1}$. Thus, the constraint: $x_i \geq d_{i,0}^r$ is derivable from $\mathbf{H}'$ using the UADD rule as desired. $\qquad\square$

Lemma 12. *Let* $\mathbf{H}'$ *be an HCS with* m *constraints over* n *variables and let* $\mathbf{D}^r$, $r = 0 \ldots (n+1)$ *be the* $n \times (m+1)$ *matrices in Algorithm 4.5. For* $r = 0 \ldots (n+1)$ *and for each* x_i *such that* $\mathbf{d}_i^r \neq \mathbf{0}$, *the sum* $\sum_{k=1}^{m} d_{i,k}^r \cdot l_k$ *results in the constraint:* $x_i \geq d_{i,0}^r$.

Proof. We will prove this by induction on r. The matrix $\mathbf{D}^0$ is the matrix constructed on line 5 of Algorithm 4.5. Let x_i be a variable such that $\mathbf{d}_i^0 \neq \mathbf{0}$. by construction, we have that $d_{i,0}^0 = b_k$ where $x_i \geq b_k$ is an absolute constraint in $\mathbf{H}'$. Additionally, $d_{i,k}^0 = 1$ and all other values in $\mathbf{d}_i^0$ are 0. Thus $\sum_{k=1}^{m} d_{i,k}^0 \cdot l_k$ results in the constraint: $x_i \geq d_{i,0}^0$.

Now assume that the desired property holds for $\mathbf{D}^{r-1}$. Note that $\mathbf{D}^r$ is constructed by calling Update-Absolute($\mathbf{H}'$, $\mathbf{D}^{r-1}$). Let x_i be a variable such that $\mathbf{d}_i^r \neq \mathbf{0}$. There are two cases we need to consider, either $\mathbf{d}_i^r = \mathbf{d}_i^{r-1}$ or $\mathbf{d}_i^r \neq \mathbf{0}$ was updated on line 7 of Algorithm 4.4.

In the first case, $\mathbf{d}^r = \mathbf{d}^{r-1}$. By the inductive hypothesis $\sum_{k=1}^{m} d_{i,k}^{r-1} \cdot l_k$ results in the constraint: $x_i \geq d_{i,0}^{r-1}$. Thus, $\sum_{k=1}^{m} d_{i,k}^r \cdot l_k$ results in the constraint: $x_i \geq d_{i,0}^r$ as desired.

In the second case, there exists a constraint l_k of the form: $x_i - \sum_{x_j \in N_k} x_j \geq b_k$ such that $d_{i,k'}^r = \sum_{x_j \in N_k} d_{j,k'}^{r-1}$ for each $k' \neq k$, $d_{i,k}^r = 1 + \sum_{x_j \in N_k} d_{j,k}^{r-1}$, and $\mathbf{d}_j^{r-1} \neq \mathbf{0}$ for each $x_j \in N_k$. By the inductive hypothesis, for each $x_j \in N_k$, $\sum_{k'=1}^{m} d_{j,k'}^{r-1} \cdot l_{k'}$ results in the constraint: $x_j \geq d_{j,0}^{r-1}$. Thus, $\sum_{k'=1}^{m} \sum_{x_j \in N_k} d_{j,k'}^{r-1} \cdot l_{k'}$ results in the constraint: $\sum_{x_j \in N_k} x_j \geq \sum_{x_j \in N_k} d_{j,0}^{r-1}$. Adding the constraint l_k to this summation results in the constraint: $x_i \geq b_k + \sum_{x_j \in N_k} d_{j,0}^{r-1} = d_{i,0}^r$ as desired. ☐

Lemma 13. *Let* $\mathbf{H}$ *be an HCS. If* Unit-Horn($\mathbf{H}$) *returns* **true** *on line 13, then* $\mathbf{H}$ *has a unit refutation.*

Proof. Since Unit-Horn($\mathbf{H}$) returned **true** on line 13, there must be a variable x_i and $1 \leq r \leq n + 1$, such that $\sum_{l_k : x_i \in P_k} d_{i,k}^r > 1$. Thus, by Lemma 12, more than one constraint with the term x_i was used to derive the constraint: $x_i \geq d_{i,0}^r$. Let $l_{k'}$ be the constraint used to obtain the current value of $\mathbf{d}_i^r$. It follows that, for some $x_{j_{r-1}} \in N_{k'}$, we must have that $\sum_{l_k : x_i \in P_k} d_{j_{r-1},k}^{r-1} \geq 1$. Otherwise,

$$\sum_{l_k : x_i \in P_k} d_{i,k}^r = 1 + \sum_{x_j \in N_{k'}} \sum_{l_k : x_i \in P_k} d_{j,k}^{r-1} = 1.$$

If $x_{j_{r-1}} \neq x_i$, let $l_{k'}$ be the constraint used to obtain the current value of $\mathbf{d}_{j_1}^{r-1}$. Thus, for some $x_{j_{r-2}} \in N_{k'}$, we must have that $\sum_{l_k : x_i \in P_k} d_{j_{r-2},k}^{r-2} \geq 1$. We can continue this process until for some r', we get $x_{j_{r'}} = x_i$.

From Lemma 11, the constraint: $x_i \geq d_{i,0}^{r'}$ is derivable from $\mathbf{H}$ by the UADD rule. Additionally, this derivation is used as part of the unit derivation of the constraint: $x_i \geq d_{i,0}^r$. Note that $d_{i,0}^r > d_{i,0}^{r'}$.

Let R_i be the unit add derivation of $x_i \geq d_{i,0}^r$ and let R_i' be the unit derivation of $x_i \geq d_{i,0}^{r'}$. Replacing R_i' in R_i with a copy of R_i results in a unit derivation of the constraint: $x_i \geq d^r_{i,0} + (d^r_{i,0} - d^{r'}_{i,0}) > d^r_{i,0}$. Thus, for any positive integer z, we can derive the constraint: $x_i \geq d^r_{i,0} + z \cdot (d^r_{i,0} - d^{r'}_{i,0})$.

Observe that, $x_i \in$ GET-T($\mathbf{H}$). By Lemma 7, x_i is used in the unit derivation of some x_j in an all negative constraint l_k whose variables are all variables of $\mathbf{H}'$. Each variable in l_k is in GET-S($\mathbf{H}$). By Lemma 9, a lower bound $x_j \geq b_j$ can be derived for each of these variables through the UADD rule. Thus, we can derive the constraint: $0 \geq b_k + \sum_{x_j \in N_k} b_j$. In particular, we can assume without loss of generality that this derivation uses the variable x_i. If $b_k + \sum_{x_j \in N_k} b_j > 0$, then this is a unit refutation of $\mathbf{H}$. Let $x_i \geq b_i$ be the unit constraint involving x_i used in this derivation. If we replace this constraint with $x_i \geq d^r_{i,0} + z \cdot (d^r_{i,0} - d^{r'}_{i,0})$ such that $z > \frac{b_i - \sum_{x_j \in N_k} b_j - b_k - d^r_{i,0}}{d^r_{i,0} - d^{r'}_{i,0}}$, then we have that the new constraint derived is

$$0 \geq b_k + \sum_{x_j \in N_k} b_j - b_i + d^r_{i,0} + z \cdot (d^r_{i,0} - d^{r'}_{i,0}).$$

Since $z > \frac{b_i - \sum_{x_j \in N_k} b_j - b_k - d^r_{i,0}}{d^r_{i,0} - d^{r'}_{i,0}}$ we have that

$$b_k + \sum_{x_j \in N_k} b_j - b_i + d^r_{i,0} + z \cdot (d^r_{i,0} - d^{r'}_{i,0}) > 0.$$

Thus, $\mathbf{H}$ has a unit refutation. $\square$

Lemma 14. *Let $\mathbf{H}$ be an HCS. If* UNIT-HORN($\mathbf{H}$) *returns* **true** *on line 16, then $\mathbf{H}$ has a unit refutation.*

Proof. Since UNIT-HORN($\mathbf{H}$) returned **true** on line 16, there must be an all-negative constraint l_k of the form: $-\sum_{x_i \in N_k} x_i \geq b_k$ such that, for some r, $\mathbf{d}^r_i \neq \mathbf{0}$ for each $x_i \in N_k$ and $b_k + \sum_{x_i \in N_k} d^r_{i,0} > 0$.

From Lemma 11, each constraint: $x_i \geq d^r_{i,0}$ is derivable from $\mathbf{H}$ by the UADD rule. Together with the constraint: $-\sum_{x_i \in N_k} x_i \geq b_k$, we obtain a unit derivation of the constraint: $0 \geq b_k + \sum_{x_i \in N_k} d^r_{i,0}$. Since $b_k + \sum_{x_i \in N_k} d^r_{i,0} > 0$, this is a unit refutation of $\mathbf{H}$. $\square$

Lemma 15. *Let $\mathbf{H}$ be an HCS. If* UNIT-HORN($\mathbf{H}$) *returns* **false**, *then $\mathbf{H}$ has no unit refutation.*

Proof. Let us assume that for some x_{j_1}, $\mathbf{d}^{n+1}_{j_1} \neq \mathbf{d}^n_{j_1}$. Thus, $d^{n+1}_{j_1,0} > d^n_{j_1,0}$. Let l_k be the constraint used to update $\mathbf{d}^{n+1}_{j_1}$. If for each $x_j \in N_k$, $\mathbf{d}^n_j = \mathbf{d}^{n-1}_j$, then

$$b_k + \sum_{x_j \in N_k} d^{n-1}_{j,0} = b_k + \sum_{x_j \in N_k} d^n_{j,0} = d^{n+1}_{j_1,0} > d^n_{j_1,0}.$$

Thus, Algorithm 4.4, would have set $d^n_{j_1,0} = d^{n+1}_{j_1,0}$. This contradicts our assumption. Thus, for some $x_{j_2} \in N_k$, we must have that $\mathbf{d}^n_{j_2} \neq \mathbf{d}^{n-1}_{j_2}$. Since the constraint: $x_{j_2} \geq d_{j_2,0}^n$ was used to derive the constraint: $x_{j_1} \geq d_{j_1,0}^{n+1}$, $\sum_{l_k : x_{j_2} \in P_k} d_{j_2,k}^{n+1} \geq 1$.

Now, let $l_{k'}$ be the constraint used to update $\mathbf{d}_{j_2}^n$. As before, there must exist a variable $x_{j_3} \in N_{k'}$ such that $\mathbf{d}_{j_3}^{n-1} \neq \mathbf{d}_{j_3}^{n-2}$. Additionally, $\sum_{l_k : x_{j_3} \in P_k} d_{j_1,k}^{n+1} \geq 1$ and $\sum_{l_k : x_{j_3} \in P_k} d_{j_2,k}^n \geq 1$.

Consider the sequence $j_1, j_2, j_3, \ldots, j_{n+1}$ constructed as above. Note that for each $p > q$, $\sum_{l_k : x_{j_p} \in P_k} d_{j_q,k}^{n+1-q} \geq 1$. Since each element of this sequence belongs to the set $1 \ldots n$, we must have that for some $p > q$, $j_p = j_q$. Thus, $\sum_{l_k : x_{j_q} \in P_k} d_{j_q,k}^{n-q} > 1$, since we need to account for the constraint used to derive $x_{j_q} \geq d_{j_q,0}^{n-q}$ and the constraint used to derive $x_{j_q} \geq d_{j_q,0}^{n-p}$. In this case, Algorithm 4.5 would have returned **true** on line 13. Thus, by Lemma 13, $\mathbf{H}$ has a unit refutation.

Thus, for each x_i, $\mathbf{d}_i^{n+1} = \mathbf{d}_i^n$. Consider a constraint $l_k \in \mathbf{H}' \setminus N(\mathbf{H}')$. Let x_i be the variable in P_k. We have that $d_{i,0}^{n+1} \geq b_k + \sum_{x_j \in N_k} d_{j,0}^n$. Thus, $d_{i,0}^{n+1} - \sum_{x_j \in N_k} d_{j,0}^{n+1} \geq b_k$. This means that setting $x_i = d_{i,0}^{n+1}$ satisfies every constraint in $\mathbf{H}' \setminus N(\mathbf{H}')$.

Consider a constraint $l_k \in N(\mathbf{H}')$. Since Algorithm 4.5 returned **false**, then it could not have returned **true** on line 16. Thus, $b_k + \sum_{x_i \in N_k} d_{i,0}^{n+1} \leq 0$. This means that $-\sum_{x_i \in N_k} d_{i,0}^{n+1} \geq b_k$. Thus, setting $x_i = d_{i,0}^{n+1}$ satisfies every constraint in $N(\mathbf{H}')$.

Consequently, setting $x_i = d_{i,0}^{n+1}$ satisfies every constraint in $\mathbf{H}'$. Thus, $\mathbf{H}'$ is feasible and does *not* have a unit refutation. By Lemma 10, $\mathbf{H}$ has no unit refutation. □

Theorem 2. *Algorithm 4.5 solves ULRP for Horn constraints.*

Proof. From Lemmas 13 and 14, we know that if Algorithm 4.5 returns **true**, then the HCS $\mathbf{H}$ has a unit refutation. From Lemma 15, we know that if Algorithm 4.5 returns **false**, then the HCS $\mathbf{H}$ has no unit refutation. Thus, Algorithm 4.5 solves ULRP for Horn constraints. □

Note that in the above proofs, we establish that if $\mathbf{H}$ has no unit-refutation, then $\mathbf{H}'$ is feasible. This gives us the following corollary.

Corollary 1. *Let $\mathbf{H}$ be an HCS. $\mathbf{H}$ has a unit refutation if and only if* REDUCE-HCS*($\mathbf{H}$) is infeasible.*

Proof. First, assume that $\mathbf{H}$ has a unit refutation. From Lemma 10, REDUCE-HCS($\mathbf{H}$) has a unit refutation. Thus, REDUCE-HCS($\mathbf{H}$) is infeasible. Now, assume that REDUCE-HCS($\mathbf{H}$) is infeasible. From the proof of Lemma 15, this means that UNIT-HORN($\mathbf{H}$) does not return **false**. Thus, UNIT-HORN($\mathbf{H}$) returns **true**. By Theorem 2, $\mathbf{H}$ has a unit refutation. □

7   Certificates for Unit Refutation in HCSs

Let $\mathbf{H}$ be an HCS. Note that the certificate returned by Algorithm 4.5 depends on the result of Algorithm 4.5. If Algorithm 4.5 returned **true** on line 13, then the certificate that $\mathbf{H}$ has a unit refutation is constructed as follows:

Let x_i be the variable described in Lemma 13. By Lemma 7, x_i is used in the unit derivation of some x_j in an all negative constraint l_k. By Lemma 9, a lower bound $x_j \geq b_j$ can be derived for each variable in l_k. This derivation uses the variable x_i.

Let R_i be the unit add derivation of $x_i \geq d_{i,0}^r$ and let R_i' be the unit derivation of $x_i \geq d_{i,0}^{r'}$ from $x_i \geq d_{i,0}^r$. By repeating R_i', we can then derive the constraint $x_i \geq d_{i,0}^{r'} + (d_{i,0}^{r'} - d_{i,0}^r)$. If R_i' is repeated l times, we derive the constraint $x_i \geq d_{i,0}^{r'} + l \cdot (d_{i,0}^{r'} - d_{i,0}^r)$.

The unit refutation of $\mathbf{H}$ includes z repetitions of R_i' where $z > \frac{b_i - \sum_{x_j \in N_k} b_j - b_k - d_{i,0}^r}{d_{i,0}^r - d_{i,0}^{r'}}$. Instead of including all z of these repetitions, we only include the initial use of R_i' and indicate that it should be repeated z times. Let b_{max} be the largest absolute value of any element in $\mathbf{b}$, and let d_{max}^r be the largest absolute value of any element in $\mathbf{D}^r$. By construction $d_{max}^0 = b_{max}$. Additionally, $d_{max}^r \leq b_{max} + n \cdot d_{max}^{r-1}$. Thus, $d_{max}^r \in O(n^r \cdot b_{max})$. Since $r \leq n+1$, each element in $\mathbf{D}^r$ can be represented using polynomially many bits. Note that we choose z such that $z > \frac{b_i - \sum_{x_j \in N_k} b_j - b_k - d_{i,0}^r}{d_{i,0}^r - d_{i,0}^{r'}}$. Thus, we can choose a value of z that has polynomially many bits. Note that each portion of the refutation, apart from the repetition of R_i' is polynomial in the size of $\mathbf{H}$. Thus, this certificate is polynomially sized.

If Algorithm 4.5 returned **true** on line 16, then the certificate that $\mathbf{H}$ has a unit refutation is constructed as follows:

Let l_k : $-\sum_{x_i \in N_k} x_i \geq b_k$ be the constraint described in Lemma 14. By Lemma 11, the lower bound $x_i \geq d_{i,0}^r$ can be derived for each variable in l_k by the UADD rule. Let R_i be the derivation of $x_i \geq d_{i,0}^r$. Together with the constraint: $-\sum_{x_i \in N_k} x_i \geq b_k$, we obtain a unit derivation of the constraint: $0 \geq b_k + \sum_{x_i \in N_k} d_{i,0}^r$. Since $b_k + \sum_{x_i \in N_k} d_{i,0}^r > 0$, this is a unit refutation of $\mathbf{H}$.

If Algorithm 4.5 returned **false**, then the certificate that $\mathbf{H}$ has no unit refutation is constructed as follows: 1. Let $\mathbf{D}^{n+1}$ be the last matrix constructed by Algorithm 4.5. 2. Let $\mathbf{x}'$ be the vector such that $x_i' = d_{i,0}^{n+1}$ for $i = 1 \ldots n$. From Lemma 15, $\mathbf{x}'$ is a satisfying assignment to $\mathbf{H}' = \text{REDUCEHCS}(\mathbf{H})$. Thus, $\mathbf{H}'$ has no unit refutation. From Lemma 10, $\mathbf{H}$ does not have a unit refutation. Thus, $\mathbf{x}'$ certifies that $\mathbf{H}$ has no unit refutation.

8   Conclusion

In this paper, we discussed the unit refutation problem in Horn constraint systems. We showed that the existence of a unit refutation can be determined in polynomial time. Recall that unit refutations are domain specific refutations.

That is, unit refutations prove that the current variable domains (absolute constraints) result in a contradiction.

References

1. Cousot, P., Cousot, R.: Abstract interpretation: a unified lattice model for static analysis of programs by construction or approximation of fixpoints. In: POPL, pp. 238–252 (1977)
2. de Moura, L., Owre, S., Ruess, H., Rushby, J.M., Shankar, N.: The ICS decision procedures for embedded deduction. In: IJCAR, pp. 218–222 (2004)
3. Dutertre, B., de Moura, L.: The YICES SMT solver. Technical report, SRI International (2006)
4. Farkas, G.: Über die Theorie der Einfachen Ungleichungen. Journal für die Reine und Angewandte Mathematik **124**(124), 1–27 (1902)
5. Ford, J., Shankar, N.: Formal verification of a combination decision procedure. In: CADE, pp. 347–362 (2002)
6. Garey, M.R., Johnson, D.S.: Computers and Intractability: A Guide to the Theory of NP-Completeness. W. H. Freeman Company, San Francisco (1979)
7. Gupta, G.: Horn logic denotations and their applications. In: Apt, K.R., Marek, V.W., Truszczynski, M., Warren, D.S. (eds.) The Logic Programming Paradigm - A 25-Year Perspective, Artificial Intelligence, pp. 127–159. Springer, Cham (1999)
8. Haken, A.: The intractability of resolution. Theor. Comput. Sci. **39**(2–3), 297–308 (1985)
9. Iwama, K., Miyano, E.: Intractability of read-once resolution. In: Proceedings of the 10th Annual Conference on Structure in Complexity Theory (SCTC 1995), Los Alamitos, CA, USA, pp. 29–36, June 1995. IEEE Computer Society Press (1995)
10. Büning, H.K., Zhao, X.: The complexity of read-once resolution. Ann. Math. Artif. Intell. **36**(4), 419–435 (2002)
11. Lahiri, S.K., Musuvathi, M.: An efficient decision procedure for UTVPI constraints. In: Proceedings of the 5th International Workshop on the Frontiers of Combining Systems, Vienna, Austria, 19–21 September, pp. 168–183. Springer, New York (2005)
12. Minoux, M.: LTUR: a simplified linear-time unit resolution algorithm for Horn formulae and computer implementation. Inf. Process. Lett. **29**(1), 1–12 (1988)
13. Neiman, V.S.: Refutation search for Horn sets by a subgoal-extraction method. J. Log. Program. **9**(2&3), 267–284 (1990)
14. Simon, A., King, A., Howe, J.M.: Two variables per linear inequality as an abstract domain. In: Logic Based Program Synthesis and Transformation, 12th International Workshop, LOPSTR 2002, Madrid, Spain, 17–20 September 2002, Revised Selected Papers, pp. 71–89 (2002)

On Exact Learning of d-Monotone Functions

Nader H. Bshouty$^{(\boxtimes)}$

Technion, Haifa, Israel
bshouty@cs.technion.ac.il

Abstract. In this paper, we study the learnability of the Boolean class of d-monotone functions $f : \mathcal{X} \to \{0,1\}$ from membership and equivalence queries, where $(\mathcal{X}, \leq)$ is a finite lattice. We show that the class of d-monotone functions that are represented in the form $f = F(g_1, g_2, \ldots, g_d)$, where F is any Boolean function $F : \{0,1\}^d \to \{0,1\}$ and $g_1, \ldots, g_d : \mathcal{X} \to \{0,1\}$ are any monotone functions, is learnable in time $\sigma(\mathcal{X}) \cdot (\text{size}(f)/d + 1)^d$ where $\sigma(\mathcal{X})$ is the maximum sum of the number of immediate predecessors in a chain from the largest element to the smallest element in the lattice $\mathcal{X}$ and $\text{size}(f) = \text{size}(g_1) + \cdots + \text{size}(g_d)$, where $\text{size}(g_i)$ is the number of minimal elements in $g_i^{-1}(1)$.

For the Boolean function $f : \{0,1\}^n \to \{0,1\}$, the class of d-monotone functions that are represented in the form $f = F(g_1, g_2, \ldots, g_d)$, where F is any Boolean function and $g_1, \ldots, g_d$ are any monotone DNF, is learnable in time $O(n^2) \cdot (\text{size}(f)/d + 1)^d$ where $\text{size}(f) = \text{size}(g_1) + \cdots + \text{size}(g_d)$.

In particular, this class is learnable in polynomial time when d is constant. Additionally, this class is learnable in polynomial time when $\text{size}(g_i)$ is constant for all i and $d = O(\log n)$.

Keywords: Exact learning · Membership queries · Equivalence queries · d-monotone function

1 Introduction

Let $\mathcal{P} = (\mathcal{X}, \leq)$ be a lattice. A Boolean function $f : \mathcal{X} \to \{0,1\}$ is d-*monotone* if, for any chain $x_1 < x_2 < \cdots < x_t$ in $\mathcal{X}$, the sequence $0, f(x_1), f(x_2), \ldots, f(x_t)$ changes its value at most d times. If $d = 1$, we say that f is a *monotone* function.

In this paper, we study the learnability of d-monotone functions. The first fact that motivates the study of this class is that every Boolean function is d-monotone for some $d \leq n$. The second is Markov's result [15], which states: The minimum number of negation gates in an AND-OR-NOT circuit that computes f is $\log d + O(1)$ if and only if f is an $O(d)$-monotone function. Therefore, learning d-monotone functions can be seen as similar to learning functions with few negations [4].

© The Author(s), under exclusive license to Springer Nature Switzerland AG 2025
I. Finocchi and L. Georgiadis (Eds.): CIAC 2025, LNCS 15679, pp. 34–50, 2025.
https://doi.org/10.1007/978-3-031-92932-8_3

When $\mathcal{X} = \{0,1\}^n$, the problem of learning monotone and d-monotone Boolean functions has been extensively studied in the literature. See [1–5,7,9,10,12] [13,14,16–19].

In the PAC learning without membership queries under the uniform distribution, Bshouty-Tamon [9] and Lange et al. [13,14] proved that monotone functions can be learned in time $\exp(\sqrt{n}/\epsilon)$. Blais et al. [4] extended the result to d-monotone functions. They provided an algorithm that runs in time $\exp(d\sqrt{n}/\epsilon)$ and showed that this algorithm is optimal. See also [3].

In the exact learning with membership and equivalence queries, Angluin [2] proved that any monotone DNF f can be learned in polynomial time (poly(n, size(f))) with size(f) equivalence queries and $n \cdot$ size(f) membership queries, where size(f) is the number of monotone terms (minterms) in f. One possible representation of d-monotone function introduced by Blais et al. [4] uses the fact that every d-monotone function can be expressed as $g_1 \oplus g_2 \oplus \cdots \oplus g_d$, where each g_i is a monotone DNF, and $\oplus$ denotes the exclusive OR (XOR) operation. Takimoto et al. [19] show that if $g_d \Rightarrow g_{d-1} \Rightarrow \cdots \Rightarrow g_1$ and for every $i \le d-1$, there is no term that appears in both[1] g_i and g_{i+1}, then f is learnable from at most $n \prod_i \text{size}(g_i) \le n(\text{size}(f)/d+1)^d$ equivalence queries and $n^3 \prod_i \text{size}(g_i) \le n^3(\text{size}(f)/d+1)^d$ membership queries, where size(f) = size(g_1) + $\cdots$ + size(g_d).

This paper studies the learnability of the d-monotone function in a very general representation. We study the class of d-monotone functions represented in the form $F(g_1, g_2, \ldots, g_d)$ where F is *any* Boolean function $F : \{0,1\}^d \to \{0,1\}$ and each g_i is *any* monotone DNF.

We first state the result in the general setting when $g_i : \mathcal{X} \to \{0,1\}$ where $\mathcal{X}$ is any lattice.

Theorem 1. *Let $(\mathcal{X}, \le)$ be a finite lattice. The class of d-monotone functions $f : \mathcal{X} \to \{0,1\}$, that are represented in the form $f = F(g_1, g_2, \ldots, g_d)$, where F is any Boolean function $F : \{0,1\}^d \to \{0,1\}$ and $g_1, \ldots, g_d : \mathcal{X} \to \{0,1\}$ are any monotone functions, is learnable in time $\sigma(\mathcal{X}) \cdot (\text{size}(f)/d+1)^d$ where $\sigma(\mathcal{X})$ is the maximum sum of the number of immediate predecessors in a chain from the largest element to the smallest element in the lattice $\mathcal{X}$ and $\text{size}(f) = \text{size}(g_1) + \cdots + \text{size}(g_d)$, where $\text{size}(g_i)$ is the number of minimal elements in $g_i^{-1}(1)$.*

The algorithm asks at most $(\text{size}(f)/d+1)^d$ equivalence queries and $\sigma(\mathcal{X}) \cdot (\text{size}(f)/d+1)^d$ membership queries.

For the lattice $\{0,1\}^n$ with the standard $\le$, we have $\sigma(\{0,1\}^n) = n(n+1)/2 = O(n^2)$ and therefore,

Corollary 1. *The class of d-monotone functions $f : \{0,1\}^n \to \{0,1\}$ that are represented in the form $f = F(g_1, g_2, \ldots, g_d)$, where F is any Boolean function*

---

[1] Takimoto et al. claim that their result applies for any g_i that satisfies $g_d \Rightarrow g_{d-1} \Rightarrow \cdots \Rightarrow g_1$ and for every $i \le d-1$, $g_i \ne g_{i+1}$. In this paper, we show that this claim is not entirely accurate. For their algorithm to be valid, it is necessary that for every $i \le d-1$, no term appears in both g_i and g_{i+1}. See also [11] page 560.

and $g_1, \ldots, g_d$ are any monotone DNF, is learnable in time $O(n^2) \cdot (\text{size}(f)/d + 1)^d$, where $\text{size}(f) = \text{size}(g_1) + \cdots + \text{size}(g_d)$.

The algorithm asks at most $(\text{size}(f)/d + 1)^d$ equivalence queries and $n^2 \cdot (\text{size}(f)/d + 1)^d$ membership queries.

In particular, the following classes are learnable in polynomial time (poly(size(f), n)):

1. The class of d-monotone functions where d is constant.
2. The class of $O(\log n)$-monotone functions of size $\text{size}(f) = O(\log n)$.

To compare our result with Takimoto et al. [19], we prove that there is a function f that can be represented as $f = F(g_1, \ldots, g_d)$ and has size s, where the representation $f = G_1 \oplus G_2 \oplus \cdots \oplus G_d$ of Takimoto et al. is of size at least $O(s^d)$. This, by their analysis, implies that for f, their algorithm asks $O(ns^{d^2})$ equivalence queries and $O(n^3 s^{d^2})$ membership queries, while our algorithm asks at most $O(s^d)$ equivalence queries and $O(n^2 s^d)$ membership queries.

2   Definitions and Preliminary Results

Let $\mathcal{X}$ be a finite set. Let $\mathcal{P} = (\mathcal{X}, \leq)$ be a lattice. We say that b is an *immediate predecessor* of a if $b < a$ and there is no c such that $b < c < a$. We say that $a, b \in \mathcal{X}$ are *incomparable* if neither $a \leq b$ nor $b \leq a$ holds. Otherwise, they are *comparable*. The[2] *join* $a \vee b$ of a and b is the smallest element in $\mathcal{X}$ that is greater than or equal to both a and b. For two sets $X_1, X_2 \subseteq \mathcal{X}$, we define the *join* of X_1 and X_2 as $X_1 \vee X_2 = \{x_1 \vee x_2 \mid x_1 \in X_1, x_2 \in X_2\}$. We say that a is a *minimal element* in $\mathcal{S} \subset \mathcal{X}$ if no element in $\mathcal{S}$ is smaller than a. We denote by $\text{Min}(\mathcal{S})$ the set of all minimal elements in $\mathcal{S}$. A *chain* is a totally ordered subset of $\mathcal{X}$. That is, $C \subset \mathcal{X}$ is a chain if every pair of elements in C is comparable.

We define *the maximal predecessor sum* $\sigma(\mathcal{X})$ as the maximum sum of the number of immediate predecessors in a chain from the largest element to the smallest element in a lattice $\mathcal{X}$. Formally, let m be the largest element of $(\mathcal{X}, \leq)$, and let $X = \{x_1, \ldots, x_r\}$ be its set of immediate predecessors. Define the sublattice $(\mathcal{X}_i, \leq)$, where $\mathcal{X}_i = \{x \in \mathcal{X} \mid x \leq x_i\}$, with x_i as the largest element. Then,

$$\sigma(\mathcal{X}) = |X| + \max_{i \in [r]} \sigma(\mathcal{X}_i)$$

where σ of a singleton set is defined as 0.

We will add to the lattice $\mathcal{P}$ a minimum element $\perp \notin \mathcal{X}$ such that $\perp < x$ for all $x \in \mathcal{X}$. This will ease the analysis and the proofs, which are all true without this element.

When $\mathcal{X} = \{0,1\}^n$, for two elements $x, y \in \{0,1\}^n$, we define $x \leq y$ if and only if $x_i \leq y_i$ for all $i \in [n]$. The join $x \vee y$ of x and y is the bitwise OR of x and y. It is easy to see that $(\{0,1\}^n, \leq)$ is a lattice.

---

[2] In a lattice, the join exists, and it is unique.

2.1   The Model

The learning criterion we consider is *exact learning model*. There is a function $f : \mathcal{X} \to \{0,1\}$, called the *target function*, which belongs to a class of functions C. The goal of the learning algorithm is to halt and output a formula h that is logically equivalent to f.

In a *membership query*, the learning algorithm supplies an assignment $a \in \mathcal{X}$ as input to a membership oracle and receives in return the value of $f(a)$. In an *equivalence query*, the learning algorithm supplies any function $h : \mathcal{X} \to \{0,1\}$ as input to an equivalence oracle, and the oracle's response is either "yes" indicating that h is equivalent to f, or a *counterexample*, which is an assignment b such that $h(b) \neq f(b)$.

2.2   Monotone Functions

In this section, we define the concept of monotone functions and give some results.

Let $a \in \mathcal{X}$ and $M_a : \mathcal{X} \to \{0,1\}$ be the function defined by $M_a(x) = 1$ if and only if $x \geq a$. We call M_a a *monotone term* that is generated by a. A *monotone function* f is a disjunction of monotone terms. If f is a monotone function, then f is the disjunction of the monotone terms generated by the elements of $\mathrm{Min}(f^{-1}(1))$. Thus,

$$f = \bigvee_{a \in \mathrm{Min}(f^{-1}(1))} M_a.$$

We will denote $\mathrm{Min}(f) = \mathrm{Min}(f^{-1}(1))$.

The following is a well-known result.

Lemma 1. *The function $f : \mathcal{X} \to \{0,1\}$ is monotone if and only if for every $x \geq y$, we have $f(x) \geq f(y)$.*

The *size* of the monotone function $\mathrm{size}(f)$ is defined as $|\mathrm{Min}(f)| = |\mathrm{Min}(f^{-1}(1))|$. The elements of $\mathrm{Min}(f)$ are called the *minimal elements* of f, and M_a, $a \in \mathrm{Min}(f)$, are called the *minterms* of f. It is easy to see that the minimal elements of a monotone function are incomparable.

For any Boolean function $f : \mathcal{X} \to \{0,1\}$, we define $f(\perp) = 0$.

The following result is easy to prove.

Lemma 2. *Let $f : \mathcal{X} \to \{0,1\}$ be a monotone function. The element a is a minimal element of f if and only if $f(a) = 1$, and for every immediate predecessor b in $\mathcal{X} \cup \{\perp\}$ of a, we have $f(b) = 0$.*

In the full paper [8], we now prove

Lemma 3. *For any two monotone functions g and h, we have:*

1. $\mathrm{Min}(g \vee h) \subseteq \mathrm{Min}(g) \cup \mathrm{Min}(h)$.
2. $\mathrm{Min}(g \wedge h) \subseteq \mathrm{Min}(g) \vee \mathrm{Min}(h)$.
3. $u = v \vee w$ *if and only if* $M_u = M_v \wedge M_w$.

2.3 d-Monotone Functions

This section defines the concept of d-monotone functions and proves some results. Recall that[3] $f(\perp) = 0$.

Definition 1. *Let $f : \mathcal{X} \to \{0,1\}$ be a Boolean function. We say that f is d-monotone if, along any chain $\perp < x_1 < x_2 < \cdots < x_t$ in $\mathcal{X} \cup \{\perp\}$, the function changes its value at most d times.*

It is easy to see that f is monotone if and only if it is 1-monotone or 0-monotone ($f = 0$).

In the full paper [8], we now prove,

Lemma 4. *Let $g_1, \ldots, g_d : \mathcal{X} \to \{0,1\}$ be non-constant monotone Boolean functions and $F : \{0,1\}^d \to \{0,1\}$ be any Boolean function. Then[4] $f = F(g_1, \ldots, g_d)$ is $(d+1)$-monotone.*

If $F(0^d) = 0$, then f is d-monotone.

We note here that for the purpose of learning, we can assume that $F(0^d) = 0$. This is because, if $F(0^d) = 1$, then we can learn $F' = F \oplus 1$ which satisfies $F'(0^d) = 0$, and then recover F as $F = F' \oplus 1$.

2.4   Minimal Elements of a Function

In this section, we extend the definition of minimal element to any Boolean function. Since Lemma 2 is not necessarily true for non-monotone functions, we must define two types of minimal elements: local and global.

For any Boolean function $f : \mathcal{X} \to \{0,1\}$, we say that a is a *local minimal element* of f if $f(a) = 1$ and for every immediate predecessor b of a, $f(b) = 0$. We denote by $\min(f)$ the set of all local minimal elements of f. We say that a is a *global minimal element* of f if $f(a) = 1$ and for every $b < a$ we have $f(b) = 0$. We denote by $\text{Min}(f)$ the set of all global minimal elements of f. Obviously, every global minimal element of f is also a local minimal element of f, and therefore

$$\text{Min}(f) \subseteq \min(f).$$

When the function f is monotone, by Lemma 1 and Lemma 2, $\text{Min}(f) = \min(f)$.

We now prove

Lemma 5. *Let $F : \{0,1\}^d \to \{0,1\}$ where $F(0^d) = 0$. Let $f = F(g_1, g_2, \ldots, g_d)$ where $g_1, g_2, \ldots, g_d$ are monotone functions. Then*

$$\min(f) \subseteq \bigcup_{I \subseteq [d]} \left(\text{Min} \left(\bigwedge_{i \in I} g_i \right) \right) \subseteq \bigcup_{I \subseteq [d]} \left(\bigvee_{i \in I} \text{Min}(g_i) \right).$$

---

[3] This definition is for any Boolean function f. So, $\overline{f}(\perp) = 0$, where $\overline{f}$ denotes the negation of f.

[4] Note here that $f(\perp) = 0$ and may not necessarily be equal to $F(g_1(\perp), \ldots, g_d(\perp)) = F(0, 0, \ldots, 0)$.

If $g_d \Rightarrow g_{d-1} \Rightarrow \cdots \Rightarrow g_1$ then

$$\min(f) \subseteq \bigcup_{i=1}^{d} \text{Min}(g_i).$$

Proof. Let a be a local minimal element of f. Then $f(a) = 1$ and for every immediate predecessor b of a, we have $f(b) = 0$. If $g_i(a) = 0$ for all $i \in [d]$, then $f(a) = F(0^d) = 0$. Therefore, there is some i such that $g_i(a) = 1$.

Let $I \subseteq [d]$ be such that $g_i(a) = 1$ for all $i \in I$ and $g_i(a) = 0$ for all $i \notin I$. Let $h = \wedge_{i \in I} g_i$. Then $h(a) = 1$. Let b be any immediate predecessor of a. Since $b < a$, and g_i are monotone, $g_i(b) = 0$ for every $i \notin I$. Since $f(a) = 1 \neq 0 = f(b)$, we must have $g_i(b) = 0$ for some $i \in I$. Therefore, $h(b) = 0$. Thus, a is a minimal element of $h = \wedge_{i \in I} g_i$, and by Lemma 3, $a \in \vee_{i \in I} \text{Min}(g_i)$.

If $g_d \Rightarrow g_{d-1} \Rightarrow \cdots \Rightarrow g_1$, then $h = \wedge_{i \in I} g_i = g_j$ for $j = \max I$, and then $a \in \text{Min}(g_j)$. $\square$

2.5   The Minimum Monotone Closure of a Function

In this section, we introduce the minimum monotone closure of a function as defined in [6] and the strict monotone representation of a Boolean function as defined in [19], and show how to use them for d-monotone functions.

Let $f : \mathcal{X} \to \{0, 1\}$ be any function. We define the *minimum monotone closure* of f (or simply the *monotone function* of f), $\mathcal{M}(f) : \mathcal{X} \to \{0, 1\}$ to be the function that satisfies $\mathcal{M}(f)(x) = 1$ if there is $y \leq x$ such that $f(y) = 1$. The following is trivial; see, for example, [6].

Lemma 6. *We have*

1. $\mathcal{M}(f)$ *is the minimum monotone function*[5] *that satisfies $f \Rightarrow \mathcal{M}(f)$. In particular,*
2. *If $f(a) = 1$, then $\mathcal{M}(f)(a) = 1$, and if $\mathcal{M}(f)(b) = 0$, then $f(b) = 0$.*
3. $\text{Min}(\mathcal{M}(f)) = \text{Min}(f)$.

The following lemma is proved in [19] for any Boolean function when $d = n$. For d-monotone functions, we prove:

Lemma 7. *Let f be a d-monotone function. Define $f_{i+1} = f_i \oplus \mathcal{M}(f_i) = \overline{f_i} \wedge \mathcal{M}(f_i)$, where $f_1 = f$. Then*

$$f = \mathcal{M}(f_1) \oplus \mathcal{M}(f_2) \oplus \cdots \oplus \mathcal{M}(f_d).$$

Proof. We prove the result by proving the following items:

1. $\mathcal{M}(f_{i+1}) \Rightarrow \mathcal{M}(f_i)$.
2. If $z \in \text{Min}(\mathcal{M}(f_i))$, then $\mathcal{M}(f_i)(z) = 1$ and $\mathcal{M}(f_{i+1})(z) = 0$. In particular, $\text{Min}(\mathcal{M}(f_i)) \cap \text{Min}(\mathcal{M}(f_{i+1})) = \emptyset$.

---

[5] Here, "minimum" means that for any other monotone function g, if $f \Rightarrow g$, then $\mathcal{M}(f) \Rightarrow g$.

3. There exists m such that $\mathcal{M}(f_i)(x) = 0$ for all $i > m$ and all x.
4. Let $g = \mathcal{M}(f_1) \oplus \mathcal{M}(f_2) \oplus \cdots \oplus \mathcal{M}(f_m)$. If $z \in \mathrm{Min}(\mathcal{M}(f_j)) = \mathrm{Min}(f_j)$, then $g(z) = (j \mod 2)$.
5. Let $g = \mathcal{M}(f_1) \oplus \mathcal{M}(f_2) \oplus \cdots \oplus \mathcal{M}(f_m)$. Then $f = g$.
6. If f is d-monotone, then $g(x) = \mathcal{M}(f_1) \oplus \mathcal{M}(f_2) \oplus \cdots \oplus \mathcal{M}(f_d)$.

We prove item 1. If $\mathcal{M}(f_{i+1}) = 0$, the result follows. If $\mathcal{M}(f_{i+1}) \neq 0$, then let z be any element in $\mathcal{X}$ such that $\mathcal{M}(f_{i+1})(z) = 1$. Thus, there exist $y \leq z$ such that $f_{i+1}(y) = 1$. Since $1 = f_{i+1}(y) = \overline{f_i}(y) \wedge \mathcal{M}(f_i)(y)$, we have $\mathcal{M}(f_i)(y) = 1$. Since $\mathcal{M}(f_i)$ is monotone and $z \geq y$, we also have $\mathcal{M}(f_i)(z) = 1$. Therefore, $\mathcal{M}(f_{i+1}) \Rightarrow \mathcal{M}(f_i)$.

We now prove item 2. Let $z \in \mathrm{Min}(\mathcal{M}(f_i)) = \mathrm{Min}(f_i)$. Then $f_i(z) = 1$ and $\mathcal{M}(f_i)(z) = 1$. Thus, $f_{i+1}(z) = f_i(z) \oplus \mathcal{M}(f_i)(z) = 0$. Since $z \in \mathrm{Min}(\mathcal{M}(f_i)) = \mathrm{Min}(f_i)$, for every $y < z$ we have $f_i(y) = 0$ and $\mathcal{M}(f_i)(y) = 0$, and therefore for every $y \leq z$ we have $f_{i+1}(y) = f_i(y) \oplus \mathcal{M}(f_i)(y) = 0$. Therefore, $\mathcal{M}(f_{i+1})(z) = 0$.

Items 1 and 2 imply that $\mathcal{M}(f_{i+1}) \Rightarrow \mathcal{M}(f_i)$ and $\mathcal{M}(f_{i+1}) \neq \mathcal{M}(f_i)$. This implies item 3.

We now show item 4. Let $z \in \mathrm{Min}(\mathcal{M}(f_j))$. By item 2, we have $\mathcal{M}(f_j)(z) = 1$ and $\mathcal{M}(f_{j+1})(z) = 0$. Therefore, by item 1, $\mathcal{M}(f_i)(z) = 0$ for all $i \geq j + 1$ and $\mathcal{M}(f_i)(z) = 1$ for all $i \leq j$. This implies the result.

We now prove item 5. Let $g = \mathcal{M}(f_1) \oplus \mathcal{M}(f_2) \oplus \cdots \oplus \mathcal{M}(f_m)$. Let $x \in \mathcal{X}$. If $\mathcal{M}(f_1)(x) = 0$, then $f(x) = f_1(x) = 0$, and by item 1, $\mathcal{M}(f_i)(x) = 0$ for all i, and therefore $f(x) = g(x)$. If $\mathcal{M}(f_j)(x) = 1$ and $\mathcal{M}(f_{j+1})(x) = 0$, then by item 1, $\mathcal{M}(f_i)(x) = 1$ for all $i \leq j$ and $\mathcal{M}(f_i)(x) = 0$ for all $i > j$. Therefore, $g(x) = (j \mod 2)$. Since $\mathcal{M}(f_{j+1})(x) = 0$, we have $f_{j+1}(x) = 0$. Since for $i \leq j$, $f_{i+1}(x) = f_i(x) \oplus \mathcal{M}(f_i)(x) = f_i(x) \oplus 1$, we have $f_i(x) = f_{i+1}(x) \oplus 1$. Now, since $f_{j+1}(x) = 0$, we get $f(x) = f_1(x) = (j \mod 2)$. Therefore $f(x) = g(x)$.

To prove item 6, it is enough to show that $\mathcal{M}(f_{d+1}) = 0$. Assume to the contrary $\mathcal{M}(f_{d+1}) \neq 0$. We construct a chain of $d + 2$ elements in $\mathcal{X} \cup \{\bot\}$ with alternating values in f and get a contradiction. We start from x_{d+1} a minimal element of $\mathcal{M}(f_{d+1})$. By items 4 and 5, $f(x_{d+1}) = g(x_{d+1}) = (d+1 \mod 2)$. By item 2, $x_{d+1} \notin \mathrm{Min}(\mathcal{M}(f_d))$ and since $\mathcal{M}(f_{d+1}) \Rightarrow \mathcal{M}(f_d)$, $\mathcal{M}(f_d)(x_{d+1}) = 1$ and therefore there is a minimal element $x_d < x_{d+1}$ of $\mathcal{M}(f_d)$. By items 4 and 5, $f(x_d) = g(x_d) = (d \mod 2) \neq f(x_{d+1})$, and so on.

This constructs a chain $x_1 < x_2 < \cdots < x_{d+1}$ with alternating values in f. Since $x_1 \in \mathrm{Min}(\mathcal{M}(f_1)) = \mathrm{Min}(f_1)$, we have $f(x) = f_1(x) = 1$. We now add $\bot$ at the beginning of the chain and get a chain where, along this chain, the value of f is changed $d + 1$ times. Therefore, $\mathcal{M}(f_{d+1}) = 0$. □

Obviously, this representation is unique. We call such representation the *strict monotone representation* of f.

The following lemma presents some properties of this representation.

Lemma 8. *Let f be d-monotone function and let $f = \mathcal{M}(f_1) \oplus \cdots \oplus \mathcal{M}(f_d)$ be the strict monotone representation of f. Then*

1. $\mathcal{M}(f_d) \Rightarrow \mathcal{M}(f_{d-1}) \Rightarrow \cdots \Rightarrow \mathcal{M}(f_1)$.

2. $f_i = \mathcal{M}(f_i) \oplus \mathcal{M}(f_{i+1}) \oplus \cdots \oplus \mathcal{M}(f_d)$.
3. For $j > i$, we have $\text{Min}(\mathcal{M}(f_i)) \cap \text{Min}(\mathcal{M}(f_j)) = \emptyset$.

Proof. Item 1 is item 1 in the proof of Lemma 7.

The proof of item 2 is by induction. First, by Lemma 7, we have $f_1 = f = \mathcal{M}(f_1) \oplus \cdots \oplus \mathcal{M}(f_d)$. Then, by the induction hypothesis, we have

$$f_{i+1} = f_i \oplus \mathcal{M}(f_i) = \mathcal{M}(f_i) \oplus \mathcal{M}(f_{i+1}) \oplus \cdots \oplus \mathcal{M}(f_d) \oplus \mathcal{M}(f_i)$$

$$= \mathcal{M}(f_{i+1}) \oplus \cdots \oplus \mathcal{M}(f_d).$$

To prove item 3, suppose to the contrary $a \in \text{Min}(\mathcal{M}(f_i)) \cap \text{Min}(\mathcal{M}(f_j))$. Since $\mathcal{M}(f_j) \Rightarrow \mathcal{M}(f_{i+1}) \Rightarrow \mathcal{M}(f_i)$, it follows that $a \in \text{Min}(\mathcal{M}(f_{i+1}))$. This contradicts item 2 in the proof of Lemma 7. $\qquad\square$

3   The Algorithm

In this section, we first provide a procedure that builds the hypothesis to the equivalent query. Then we present the algorithm that learns any d-monotone function of the form $F(g_1, \ldots, g_d)$, where $F : \{0,1\}^d \to \{0,1\}$ and each $g_i : \mathcal{X} \to \{0,1\}$ is any monotone Boolean function.

Finally, we establish the following result.

Theorem 1. *Let $(\mathcal{X}, \leq)$ be a finite lattice. The class of d-monotone functions $f : \mathcal{X} \to \{0,1\}$, that are represented in the form $f = F(g_1, g_2, \ldots, g_d)$, where F is any Boolean function $F : \{0,1\}^d \to \{0,1\}$ and $g_1, \ldots, g_d : \mathcal{X} \to \{0,1\}$ are any monotone functions, is learnable in time $\sigma(\mathcal{X}) \cdot (\text{size}(f)/d + 1)^d$ where $\sigma(\mathcal{X})$ is the maximum sum of the number of immediate predecessors in a chain from the largest element to the smallest element in the lattice $\mathcal{X}$ and $\text{size}(f) = \text{size}(g_1) + \cdots + \text{size}(g_d)$, where $\text{size}(g_i)$ is the number of minimal elements in $g_i^{-1}(1)$.*
The algorithm asks at most $(\text{size}(f)/d + 1)^d$ equivalence queries and $\sigma(\mathcal{X}) \cdot (\text{size}(f)/d + 1)^d$ membership queries.

For the lattice $\{0,1\}^n$ with the standard $\leq$, we have

Corollary 1. *The class of d-monotone functions $f : \{0,1\}^n \to \{0,1\}$ that are represented in the form $f = F(g_1, g_2, \ldots, g_d)$, where F is any Boolean function and $g_1, \ldots, g_d$ are any monotone DNF, is learnable in time $O(n^2) \cdot (\text{size}(f)/d + 1)^d$, where $\text{size}(f) = \text{size}(g_1) + \cdots + \text{size}(g_d)$.*
The algorithm asks at most $(\text{size}(f)/d + 1)^d$ equivalence queries and $n^2 \cdot (\text{size}(f)/d + 1)^d$ membership queries.

3.1   Consistent Hypothesis

In this section, we give a procedure **Consistent** that receives d and $\mathcal{X}_0, \mathcal{X}_1 \subseteq \mathcal{X}$ such that there is a d-monotone function f that satisfies $f(x) = 0$ for all $x \in \mathcal{X}_0$ and $f(x) = 1$ for all $x \in \mathcal{X}_1$. The procedure returns a hypothesis h that is a d-monotone function consistent with f on $\mathcal{X}_0 \cup \mathcal{X}_1$. That is, $h(x) = f(x)$ for all $x \in \mathcal{X}_0 \cup \mathcal{X}_1$.

To establish the correctness and analyze the algorithm's complexity, we first prove two lemmas.

Lemma 9. *Let $\mathcal{X}_0, \mathcal{X}_1 \subseteq \mathcal{X}$. Suppose there exists a d-monotone function f such that $f(x) = 0$ for all $x \in \mathcal{X}_0$ and $f(x) = 1$ for all $x \in \mathcal{X}_1$.* **Consistent**$(d, \mathcal{X}_0, \mathcal{X}_1)$ *runs in polynomial time and constructs a d-monotone function h of size $O(|\mathcal{X}_0| + |\mathcal{X}_1|)$ that is consistent with f on $\mathcal{X}_0 \cup \mathcal{X}_1$.*

Proof. Consider the algorithm **Consistent** in Algorithm 1. We prove the correctness by induction on d.

For $d = 1$, the function f is monotone. Suppose there is a monotone function such that $f(x) = 0$ for $x \in \mathcal{X}_0$ and $f(x) = 1$ for $x \in \mathcal{X}_1$. Then, there is no $z \in \mathcal{X}_0$ and $y \in \mathcal{X}_1$ such that $z > y$.

In the first iteration, the procedure defines $F_1 = \vee_{a \in \text{Min}(\mathcal{X}_1)} M_a$ and outputs $h = F_1$. If $z \in \mathcal{X}_1$, then there is $a \leq z$ such that $a \in \text{Min}(\mathcal{X}_1)$. Thus, $M_a(z) = 1$ and consequently $h(z) = 1$. If $z \in \mathcal{X}_0$, there is no $y \in \mathcal{X}_1$ such that $z > y$. Therefore, $M_a(z) = 0$ for all $a \in \text{Min}(\mathcal{X}_1)$, and consequently $h(z) = 0$.

Assume the statement is true for $(d - 1)$-monotone functions. We now prove it for d-monotone functions. Let f be a d-monotone function. In the first iteration of the procedure, it defines $\mathcal{S}_0 = \mathcal{X}_0$, $\mathcal{S}_1 = \mathcal{X}_1$, $W_1 = \text{Min}(\mathcal{S}_1)$, $F_1(x) = \vee_{a \in W_1} M_a(x)$, and $W_0 = \{x \in \mathcal{S}_0 | F_1(x) = 0\}$. After the first iteration, it runs with the new points $\mathcal{S}_1' := \mathcal{S}_0 \backslash W_0$ and $\mathcal{S}_0' := \mathcal{S}_1 \cup W_0$.

We first show that there is a $(d-1)$-monotone function g such that $g(x) = 0$ for all $x \in \mathcal{S}_0' = \mathcal{S}_1 \cup W_0$ and $g(x) = 1$ for all $x \in \mathcal{S}_1' = \mathcal{S}_0 \backslash W_0$.

Assume to the contrary that any function g that is 0 in $\mathcal{S}_0' = \mathcal{S}_1 \cup W_0$ and 1 in $\mathcal{S}_1' = \mathcal{S}_0 \backslash W_0$ is d'-monotone for some $d' \geq d$, and is not $(d-1)$-monotone. Let $\perp < x_1 < x_2 < \cdots < x_t$ be any chain where the function g changes its value d times. Suppose the changes happen in $x_{i_1} < x_{i_2} < \cdots < x_{i_d}$. Since $g(\perp) = 0$, we have $g(x_{i_1}) = 1$ and $g(x_{i_j}) = (j \mod 2)$. Since $g(x_{i_1}) = 1$, we have $x_{i_1} \in \mathcal{S}_1' = \mathcal{S}_0 \backslash W_0$. Therefore $f(x_{i_1}) = 0$. Since $x_{i_1} \in \mathcal{S}_0$ and $x_{i_1} \notin W_0$, we have $F_1(x_{i_1}) = 1$, and therefore, there is $x_0 \leq x_{i_1}$ such that $x_0 \in W_1 = \text{Min}(\mathcal{S}_1)$. In particular, $f(x_0) = 1$. Since $f(x_0) = 1$ and $f(x_{i_1}) = 0$, we have $x_0 \neq x_{i_1}$ and therefore $x_0 < x_{i_1}$.

Let $j \geq 2$. Since $x_{i_j} > x_{i_1} > x_0$, we have $F_1(x_{i_j}) = 1$ and therefore $x_{i_j} \notin W_0$. Thus, $g(x_{i_j}) = \overline{f(x_{i_j})}$ and $f(x_{i_j}) = \overline{g(x_{i_j})} = (j - 1 \mod 2)$ for all $j \geq 2$. Hence, $\perp < x_0 < x_{i_1} < x_{i_2} < \cdots < x_{i_d}$ is a chain for which f changes its value along it $(d + 1)$ times. This implies that f is d''-monotone for some $d'' \geq d + 1$, which is a contradiction.

Now, by the induction hypothesis, $g = F_2 \oplus F_3 \oplus \cdots \oplus F_d$ satisfies $g(x) = 0$ for every $x \in \mathcal{S}_1 \cup W_0$ and $g(x) = 1$ for every $x \in \mathcal{S}_0 \backslash W_0$. We now show that $h = F_1 \oplus g$ is the desired hypothesis. By the definition of W_0, if $x \in \mathcal{S}_0 \backslash W_0$, then $F_1(x) = 1$ and $g(x) = 1$, and therefore $h(x) = 0$. If $x \in W_0$, then $F_1(x) = 0$ and $g(x) = 0$, and therefore $h(x) = 0$. If $x \in \mathcal{S}_1$, then $F_1(x) = 1$ and $g(x) = 0$, and therefore $h(x) = 1$. $\square$

In [19] (page 16), Takimoto et al. claim that if $f = g_1 \oplus g_2 \oplus \cdots \oplus g_d$, where g_i is monotone for every $i \leq d$, $g_{i+1} \neq g_i$, and $g_{i+1} \Rightarrow g_i$ for every $i \leq d-1$, then $g_i = \mathcal{M}(f_i)$. In the appendix of the full paper [8], we show that this claim is not entirely accurate. The following lemma outlines the conditions under which this statement holds.

Algorithm 1. Consistent$(d, \mathcal{X}_0, \mathcal{X}_1)$

1: Let $\mathcal{S}_0 = \mathcal{X}_0; \mathcal{S}_1 = \mathcal{X}_1$.
2: **for** $i = 1$ to d **do**
3:     Let $W_1 \leftarrow \text{Min}(\mathcal{S}_1)$.
4:     Define $F_i = \bigvee_{a \in W_1} M_a$ *If $W_1 = \emptyset$ then $F_i = 0$
5:     $W_0 \leftarrow \{x \in \mathcal{S}_0 \mid F_i(x) = 0\}$
6:     $\mathcal{S}_1 \leftarrow (\mathcal{S}_0 \backslash W_0)$.
7:     $\mathcal{S}_0 \leftarrow \mathcal{S}_1 \cup W_0$.
8: **end for**
9: Output $h = F_1 \oplus F_2 \oplus \cdots \oplus F_d$.

Lemma 10. *If $f = g_1 \oplus \cdots \oplus g_d$, where g_i is monotone function for every $i \leq d$, $g_{i+1} \Rightarrow g_i$ and $\text{Min}(g_{i+1}) \cap \text{Min}(g_i) = \emptyset$ for every $i \leq d-1$, then $\mathcal{M}(f_i) = g_i$.*

Proof. It is enough to prove that $\mathcal{M}(f_1) = g_1$. This is because if we prove that $\mathcal{M}(f_1) = g_1$, then

$$f_2 = f_1 \oplus \mathcal{M}(f_1) = f \oplus \mathcal{M}(f_1) = (g_1 \oplus g_2 \oplus \cdots \oplus g_d) \oplus g_1 = g_2 \oplus \cdots \oplus g_d,$$

and therefore $\mathcal{M}(f_2) = g_2$. Then, by induction, the result follows.

Recall that $f_1 = f$. We first prove that $\mathcal{M}(f) \Rightarrow g_1$. We show that $\text{Min}(\mathcal{M}(f)) \subset \text{Min}(g_1)$. Let $a \in \text{Min}(\mathcal{M}(f)) = \text{Min}(f)$. Then $f(a) = 1$ and for every $b < a$, we have $f(b) = 0$. We now show that $g_1(a) = 1$ and $g_i(a) = 0$ for all $i > 1$. If $g_i(a) = 0$ for all i, then $f(a) = 0$, and we get a contradiction.

If $g_i(a) = 1$ for some $i > 1$, then $g_2(a) = 1$ and there is $a' \in \text{Min}(g_2)$, $a' \leq a$, such that $g_2(a') = 1$. Then $g_1(a') = 1$, and since $\text{Min}(g_1) \cap \text{Min}(g_2) = \emptyset$, there is a $a'' \in \text{Min}(g_1)$ such that $a'' < a'$ and $g_1(a'') = 1$. Since $a'' < a'$ and $a' \in \text{Min}(g_2)$, we have $g_2(a'') = 0$ and therefore $g_i(a'') = 0$ for all $i > 1$. Therefore, $f(a'') = g_1(a'') = 1$. Since $a'' < a' \leq a \in \text{Min}(f)$, we have $f(a'') = 0$, which is a contradiction. Therefore $g_1(a) = 1$ and $g_i(a) = 0$ for all $i > 1$. Since for every $b < a$, $f(b) = 0$, we have for every $b < a$, $g_i(b) = 0$ for all i. This implies that $a \in \text{Min}(g_1)$.

We now prove that $g_1 \Rightarrow \mathcal{M}(f)$. Let $a \in \text{Min}(g_1)$. Then $g_1(a) = 1$ and for every $b < a$, we have $g_1(b) = 0$. Therefore, for every $b < a$ and every $i > 1$, we have $g_i(b) = 0$. If $g_i(a) = 1$ for some $i > 1$, then $g_2(a) = 1$. Then $a \in \text{Min}(g_2)$, and since $\text{Min}(g_1) \cap \text{Min}(g_2) = \emptyset$, we get a contradiction. Therefore, $g_1(a) = 1$, $g_i(a) = 0$ for all $i > 1$ and for every $b < a$, $g_j(b) = 0$ for all $j \geq 1$. Therefore, $f(a) = 1$ and for every $b < a$, $f(b) = 0$. Thus, $a \in \text{Min}(f) = \text{Min}(\mathcal{M}(f))$.   □

The following lemma proves that the output $F_1 \oplus \cdots \oplus F_d$ of the procedure **Consistent** is the strict monotone representation of h.

Lemma 11. *Let $\mathcal{X}_0, \mathcal{X}_1 \subseteq \mathcal{X}$. Suppose there is a d-monotone function f such that $f(x) = 0$ for all $x \in \mathcal{X}_0$ and $f(x) = 1$ for all $x \in \mathcal{X}_1$. Let $h = F_1 \oplus \cdots \oplus F_d$ be the output of **Consistent**$(d, \mathcal{X}_0, \mathcal{X}_1)$. Then $F_i = \mathcal{M}(h_i)$.*

Proof. We use Lemma 10. By step 4 in the procedure **Consistent**, we have that each F_i is a monotone function. Now, it is enough to prove that $\text{Min}(F_1) \cap \text{Min}(F_2) = \emptyset$ and $F_2 \Rightarrow F_1$. Then, the result follows by induction.

Since $\text{Min}(F_1) = \text{Min}(\mathcal{S}_1) = \text{Min}(\mathcal{X}_1) \subseteq \mathcal{X}_1$ and $\text{Min}(F_2) = \text{Min}(\mathcal{S}_0 \backslash W_0) \subseteq \mathcal{X}_0$, we have $\text{Min}(F_1) \cap \text{Min}(F_2) = \emptyset$.

Now if $F_2(z) = 1$, then since $F_2 = \vee_{a \in \text{Min}(\mathcal{S}_0 \backslash W_0)} M_a$ and $\text{Min}(\mathcal{S}_0 \backslash W_0) = \text{Min}(\mathcal{X}_0 \backslash \{x \in \mathcal{X}_0 | F_1(x) = 0\})$, there is an $a \in \mathcal{X}_0 \backslash \{x \in \mathcal{X}_0 | F_1(x) = 0\}$ such that $a \le z$. Then $F_1(a) = 1$ and since F_1 monotone and $z \ge a$, we have $F_1(z) = 1$. Therefore, $F_2 \Rightarrow F_1$. □

3.2   The Main Algorithm

In this section, we present the algorithm and prove Theorem 1 and Corollary 1.

We first prove two lemmas needed to establish the correctness and determine the complexity of the algorithm. The first is:

Lemma 12. *Let $g_1, \ldots, g_d$ be monotone functions. Let h be a monotone function such that*

$$\text{Min}(h) \subseteq \bigcup_{J \subseteq [d]} \bigvee_{i \in J} \text{Min}(g_i). \tag{1}$$

For any $I \subseteq [d]$, we have

$$\text{Min}\left(h \wedge \bigwedge_{i \in I} g_i\right) \subseteq \bigcup_{J \subseteq [d]} \bigvee_{i \in J} \text{Min}(g_i).$$

Proof. Let $a \in \mathcal{X}$. Recall that $M_a : \mathcal{X} \to \{0, 1\}$ is the function that $M_a(x) = 1$ if and only if $x \ge a$.

Let a be a minimal element of $h \wedge \wedge_{i \in I} g_i$. Let $\text{Min}(h) = \{u_1, \ldots, u_t\}$. Then, $h = M_{u_1} \vee M_{u_2} \vee \cdots \vee M_{u_t}$ and

$$h \wedge \wedge_{i \in I} g_i = (M_{u_1} \wedge \wedge_{i \in I} g_i) \vee (M_{u_2} \wedge \wedge_{i \in I} g_i) \vee \cdots \vee (M_{u_t} \wedge \wedge_{i \in I} g_j).$$

By item 1 Lemma 3, a is a minimal element of some $M_{u_\ell} \wedge \wedge_{i \in I} g_i$.

Now, by (1), there is $J_\ell \subseteq [d]$ such that $u_\ell = \vee_{j \in J_\ell} u_{\ell,j}$ where $u_{\ell,j} \in \text{Min}(g_j)$. Therefore, by item 3 in Lemma 3, $M_{u_\ell} = \wedge_{j \in J_\ell} M_{u_{\ell,j}}$ where $M_{u_{\ell,j}}$ is a minterm in g_j. Since $M_{u_{\ell,j}} \Rightarrow g_j$, $M_{u_{\ell,j}} \wedge g_j = M_{u_{\ell,j}}$. Therefore, $M_{u_\ell} \wedge \wedge_{i \in I} g_i = \wedge_{j \in J_\ell} M_{u_{\ell,j}} \wedge \wedge_{i \in I \Delta J_\ell} g_i$.

Thus, by item 2 in Lemma 3,

$$a \in \text{Min}(\wedge_{j \in J_\ell} M_{u_{\ell,j}} \wedge \wedge_{i \in I \Delta J_\ell} g_i) \subseteq \bigvee_{j \in I \cup J_\ell} \text{Min}(g_j).$$

□

The second lemma is given below.

Lemma 13. *Let $f = F(g_1, \ldots, g_d)$ where $F : \{0,1\}^d \to \{0,1\}$ and $g_1, \ldots, g_d$ are monotone functions. Let h be a d-monotone function such that*

$$\bigcup_{i=1}^{d} \mathrm{Min}(\mathcal{M}(h_i)) \subseteq \bigcup_{J \subseteq [d]} \bigvee_{j \in J} \mathrm{Min}(g_j). \tag{2}$$

Then

$$\min(f \oplus h) \subseteq \left(\bigcup_{J \subseteq [d]} \bigvee_{j \in J} \mathrm{Min}(g_j) \right)$$

Proof. Consider

$$G = f \oplus h = F(g_1, \ldots, g_d) \oplus \mathcal{M}(h_1) \oplus \cdots \oplus \mathcal{M}(h_d).$$

Let $a \in \min(G)$ be a local minimal element of G. Then $G(a) = 1$ and for every immediate predecessor b of a we have $G(b) = 0$. Suppose $g_i(a) = 1$ for all $i \in I$, $g_i(a) = 0$ for all $i \notin I$, $\mathcal{M}(h_1)(a) = \cdots = \mathcal{M}(h_\ell)(a) = 1$, and $\mathcal{M}(h_{\ell+1})(a) = \cdots = \mathcal{M}(h_d)(a) = 0$. Since g_i and $\mathcal{M}(h_j)$ are monotone functions, for every immediate predecessor b of a we have $g_i(b) = 0$ for all $i \notin I$ and $\mathcal{M}(h_{\ell+1})(b) = \cdots = \mathcal{M}(h_d)(b) = 0$. Since $f(a) \neq f(b)$, either $\mathcal{M}(h_\ell)(b) = 0$ or $g_i(b) = 0$ for some $i \in I$. Therefore, a is a local minimal element of $H := \mathcal{M}(h_\ell) \wedge \wedge_{i \in I} g_i$ for some $\ell \in [d]$ and $I \subseteq [d]$. Since H is monotone, $\min(H) = \mathrm{Min}(H)$ and therefore

$$a \in \mathrm{Min}\left(\mathcal{M}(h_\ell) \wedge \bigwedge_{i \in I} g_i \right). \tag{3}$$

By (2), (3) and Lemma 12.

$$a \in \bigcup_{J \subseteq [d]} \bigvee_{j \in J} \mathrm{Min}(g_j).$$

$\square$

We now give the proof of the main Theorem. Consider the algorithm **Learn d-Monotone** in Algorithm 2. The following proves Theorem 1.

Theorem 2. *Algorithm* **Learn d-Monotone** *learns d-monotone functions f with at most $R(f)$ equivalence queries and $R(f)\sigma(\mathcal{X})$ membership queries, where*

$$R(f) = \left| \bigcup_{I \subseteq [d]} \bigvee_{i \in I} \mathrm{Min}(g_i) \right| \leq \left(\frac{\mathrm{size}(f)}{d} + 1 \right)^d.$$

Proof. Let $f = F(g_1, g_2, \ldots, g_d)$ be the target function. We will show by induction that at the end of iteration t, the sets $\mathcal{X}_0$, $\mathcal{X}_1$ and the hypothesis h satisfy:

$$\mathcal{X}_0 \cup \mathcal{X}_1 \subseteq \bigcup_{I \subseteq [d]} \bigvee_{i \in I} \mathrm{Min}(g_i) \tag{4}$$

Algorithm 2. Learn d-Monotone

1: $\mathcal{X}_0 = \mathcal{X}_1 = \emptyset$
2: $h \leftarrow 0$
3: **while** $\mathrm{EQ}(h) \neq \mathrm{YES}$ **do**
4:      Let a be a counterexample
5:      **while** there is an immediate predecessor b of a such that $h(b) \neq f(b)$ **do**
6:          $a \leftarrow b$
7:      **end while**
8:      **if** $f(a) = 1$ **then**
9:          $\mathcal{X}_1 \leftarrow \mathcal{X}_1 \cup \{a\}$
10:     **else**
11:         $\mathcal{X}_0 \leftarrow \mathcal{X}_0 \cup \{a\}$
12:     **end if**
13:     $h \leftarrow \mathbf{Consistent}(d, \mathcal{X}_0, \mathcal{X}_1)$
14: **end while**
15: Output h

For every $u \in \mathcal{X}_0 \cup \mathcal{X}_1$ we have $f(u) = h(u)$ $\qquad\qquad$ (5)

and

$$|\mathcal{X}_0 \cup \mathcal{X}_1| = t. \qquad\qquad (6)$$

At the first iteration, we have $h = 0$. The equivalence query returns a' such that $f(a') = 1$. Then, the algorithm in step 5 finds a local minimal element a of f and adds it to $\mathcal{X}_0$ or $\mathcal{X}_1$. Therefore, at the end of the first iteration, by Lemma 5, (4) holds. By Lemma 9, (5) holds. Also, (6) holds since $|\mathcal{X}_0 \cup \mathcal{X}_1| = |\{a\}| = 1$.

Now suppose (4)-(6) hold at the end of iteration t. We prove that they hold at the end of iteration $t + 1$.

At iteration $t + 1$, if $\mathrm{EQ}(h)$ returns a counterexample a', then $f(a') \neq h(a')$ and therefore $f(a') \oplus h(a') = 1$. In step 5 of the algorithm, it continues to go down in the lattice until it finds an a such that $f(a) \oplus h(a) = 1$ and for every immediate predecessor b of a, $f(b) \oplus h(b) = 0$. Such an a exists because $f(\bot) \oplus h(\bot) = 0$. Therefore, $a \in \min(f \oplus h)$. By Lemma 13, we have,

$$a \in \left(\bigcup_{I \subseteq [d]} \bigvee_{i \in I} \mathrm{Min}(g_i) \right).$$

By the induction hypothesis (5), $f(u) = h(u)$ for all $u \in \mathcal{X}_0 \cup \mathcal{X}_1$. Since $f(a) \neq h(a)$, we have $a \notin \mathcal{X}_0 \cup \mathcal{X}_1$ and since a is added either to $\mathcal{X}_0$ or $\mathcal{X}_1$, at iteration $t + 1$, (4) holds and (6) holds at the end of iteration $t + 1$. Now, (5) also holds because a is added to $\mathcal{X}_1$ if $f(a) = 1$ and to $\mathcal{X}_0$ if $f(a) = 0$ and by Lemma 9, $f(u) = h(u)$ for all $u \in \mathcal{X}_0 \cup \mathcal{X}_1 \cup \{a\}$.

This completes the proof of (4)-(6).

Since $\text{size}(f) = \text{size}(g_1) + \cdots + \text{size}(g_d)$, and after each equivalence query, the algorithm adds an element either to $\mathcal{X}_0$ or $\mathcal{X}_1$, and by (4), the number of equivalence queries is at most

$$\left| \bigcup_{I \subseteq [d]} \bigvee_{i \in I} \text{Min}\,(g_i) \right| \leq \prod_{i=1}^{d}(\text{size}(g_i) + 1) - 1$$

$$\leq \left(\frac{\text{size}(f)}{d} + 1 \right)^d = R(f). \qquad \text{AM-GM Inequality}$$

After each equivalence query, the algorithm asks membership queries to go down in the lattice. The worst-case number of membership queries after each equivalence query is $\sigma(\mathcal{X})$. Therefore, the number of membership queries that the algorithm asks is at most $\sigma(\mathcal{X})R(f)$. $\qquad\square$

4   Strict Monotone Representation Size

In this section, we compare the size of the strict monotone representation of f with the size of f using the representation presented in this paper. We show that there exists a d-monotone Boolean function f with $\text{size}(f) = s$ that has size $\Omega((s/d)^d)$ in the strict monotone representation. We also show that this is a tight bound.

Throughout this section, the lattice is $\{0,1\}^n$ with the standard $\leq$.

First, by Lemma 4 and Lemma 7, we have the following:

Lemma 14. $f : \mathcal{X} \to \{0,1\}$ is d-monotone if and only if $f = \mathcal{M}(f_1) \oplus \mathcal{M}(f_2) \oplus \cdots \oplus \mathcal{M}(f_d)$.

We now define two classes of d-monotone functions.

1. The class d-M is the class of d-monotone functions f that are represented as $f = F(g_1, \ldots, g_d)$ where $F : \{0,1\}^d \to \{0,1\}$ is any Boolean function such that $F(0^d) = 0$ and $g_1, g_2, \ldots, g_d : \mathcal{X} \to \{0,1\}$ are any monotone functions.
2. The class d-M$(\oplus\mathcal{M})$ is the class of d-monotone functions f represented in the strict monotone representation $f = \mathcal{M}(f_1) \oplus \mathcal{M}(f_2) \oplus \cdots \oplus \mathcal{M}(f_d)$.

We define $\text{size}(f)$ to be the minimum possible $\text{size}(g_1) + \cdots + \text{size}(g_d)$ of representations of $f = F(g_1, \ldots, g_d)$ in d-M. We define $\text{size}_{\oplus\mathcal{M}}(f) = \text{size}(\mathcal{M}(f_1)) + \cdots + \text{size}(\mathcal{M}(f_d))$.

Before proving the relationship between $\text{size}(f)$ and $\text{size}_{\oplus\mathcal{M}}(f)$, we present two lemmas that will be used to establish this relationship.

Lemma 15. Let $f = F(g_1, \ldots, g_d)$, where g_i are monotone functions and $F(0^d) = 0$. Then, for every k, $\bigcup_{k=1}^{d} \text{Min}(\mathcal{M}(f_k)) \subseteq \bigcup_{I \subseteq [d]} \bigvee_{i \in I} \text{Min}\,(g_i)$.

Proof. By Lemma 11, every hypothesis $h = F_1 \oplus \cdots \oplus F_d$ in the algorithm **Learn** d-**Monotone** 2 satisfies $F_i = \mathcal{M}(h_i)$. Since the final hypothesis of the algorithm is f, the final output of the algorithm is $F_1' \oplus F_2' \oplus \cdots \oplus F_d'$ where $F_i' = \mathcal{M}(f_i)$. In the procedure **Consistent** 1, the minimal elements of all $F_i' = \mathcal{M}(f_i)$ are from $\mathcal{X}_0 \cup \mathcal{X}_1$, and by (4), we have $\mathcal{X}_0 \cup \mathcal{X}_1 \subseteq \bigcup_{I \subseteq [d]} \bigvee_{i \in I} \mathrm{Min}(g_i)$. □

We now prove

Lemma 16. *We have* $\mathrm{size}_{\oplus \mathcal{M}}(f) \leq \left(\frac{\mathrm{size}(f)}{d} + 1 \right)^d - 1$.

Proof. Let $f = F(g_1, \ldots, g_d)$ where $F : \{0,1\}^d \to \{0,1\}$ and $F(0^d) = 0$. Suppose $s_i = \mathrm{size}(g_i)$. By Lemma 15, we have

$$\bigcup_{k=1}^{d} \mathrm{Min}(\mathcal{M}(f_k)) \subseteq \bigcup_{I \subseteq [d]} \bigvee_{i \in I} \mathrm{Min}(g_i).$$

Therefore, by item 3 in Lemma 8 and the AM-GM inequality,

$$\begin{aligned}
\mathrm{size}_{\oplus \mathcal{M}}(f) &= \sum_{k=1}^{d} \mathrm{size}(\mathcal{M}(f_k)) = \sum_{k=1}^{d} |\mathrm{Min}(\mathcal{M}(f_k))| \\
&= \left| \bigcup_{k=1}^{d} \mathrm{Min}(\mathcal{M}(f_k)) \right| \leq \left| \bigcup_{I \subseteq [d]} \bigvee_{i \in I} \mathrm{Min}(g_i) \right| \\
&\leq \prod_{i=1}^{d} (\mathrm{size}(g_i) + 1) - 1 \leq \left(\frac{\mathrm{size}(f)}{d} + 1 \right)^d - 1.
\end{aligned}$$

□

We now show that this bound is tight.

Lemma 17. *There is a d-monotone function f such that*

$$\mathrm{size}_{\oplus \mathcal{M}}(f) = \left(\frac{\mathrm{size}(f)}{d} + 1 \right)^d - 1.$$

Proof. Consider the function $f = y_1 \oplus \cdots \oplus y_d$ where $y_i = x_{i,1} \vee \cdots \vee x_{i,t}$ where $t = n/d$. The size of f is $d(n/d) = n$.

First $y_1 \oplus \cdots \oplus y_d = G_1 \oplus G_2 \oplus \cdots \oplus G_d$ where

$$G_k = \bigvee_{1 \leq i_1 < i_2 < \cdots < i_k \leq d} \left(\bigwedge_{j=1}^{k} y_{i_j} \right).$$

This is because if ℓ of the functions y_i are equal to 1 then $G_1 = G_2 = \cdots = G_\ell = 1$ and $G_{\ell+1} = \cdots = G_d = 0$.

Since $G_d \Rightarrow G_{d-1} \Rightarrow \cdots \Rightarrow G_1$ and $\mathrm{Min}(G_i) \cap \mathrm{Min}(G_{i+1}) = \emptyset$, by Lemma 10, we have $G_i = \mathcal{M}(f_i)$. Now

$$\mathrm{size}_{\oplus \mathcal{M}}(f) = \mathrm{size}(G_1) + \mathrm{size}(G_2) + \cdots + \mathrm{size}(G_d)$$

$$= dt + \binom{d}{2}t^2 + \cdots + \binom{d}{d}t^d$$

$$= (t+1)^d - 1 = \left(\frac{\mathrm{size}(f)}{d} + 1\right)^d - 1.$$

$\square$

References

1. Amano, K., Maruoka, A.: On learning monotone Boolean functions under the uniform distribution. Theor. Comput. Sci. **350**(1), 3–12 (2006). https://doi.org/10.1016/J.TCS.2005.10.012
2. Angluin, D.: Queries and concept learning. Mach. Learn. **2**(4), 319–342 (1987). https://doi.org/10.1007/BF00116828
3. Black, H.: Nearly optimal bounds for sample-based testing and learning of k-monotone functions. CoRR abs/2310.12375 (2023). https://doi.org/10.48550/ARXIV.2310.12375
4. Blais, E., Canonne, C.L., Oliveira, I.C., Servedio, R.A., Tan, L.: Learning circuits with few negations. In: Garg, N., Jansen, K., Rao, A., Rolim, J.D.P. (eds.) Approximation, Randomization, and Combinatorial Optimization. Algorithms and Techniques, APPROX/RANDOM 2015, Princeton, NJ, USA, 24–26 August 2015. LIPIcs, vol. 40, pp. 512–527. Schloss Dagstuhl - Leibniz-Zentrum für Informatik (2015). https://doi.org/10.4230/LIPICS.APPROX-RANDOM.2015.512
5. Blum, A., Burch, C., Langford, J.: On learning monotone Boolean functions. In: 39th Annual Symposium on Foundations of Computer Science, FOCS 1998, Palo Alto, California, USA, 8–11 November 1998, pp. 408–415. IEEE Computer Society (1998). https://doi.org/10.1109/SFCS.1998.743491
6. Bshouty, N.H.: Exact learning Boolean function via the monotone theory. Inf. Comput. **123**(1), 146–153 (1995). https://doi.org/10.1006/INCO.1995.1164
7. Bshouty, N.H.: Simple learning algorithms using divide and conquer. Comput. Complex. **6**(2), 174–194 (1997). https://doi.org/10.1007/BF01262930
8. Bshouty, N.H.: On exact learning of d-monotone functions (2025). https://arxiv.org/abs/2502.01265
9. Bshouty, N.H., Tamon, C.: On the Fourier spectrum of monotone functions. J. ACM **43**(4), 747–770 (1996). https://doi.org/10.1145/234533.234564
10. Goldreich, O., Goldwasser, S., Lehman, E., Ron, D., Samorodnitsky, A.: Testing monotonicity. Combinatorica **20**(3), 301–337 (2000). https://doi.org/10.1007/S004930070011
11. Guijarro, D., Lavín, V., Raghavan, V.: Monotone term decision lists. Theor. Comput. Sci. **259**(1–2), 549–575 (2001). https://doi.org/10.1016/S0304-3975(00)00043-8

12. Harms, N., Yoshida, Y.: Downsampling for testing and learning in product distributions. In: Bojanczyk, M., Merelli, E., Woodruff, D.P. (eds.) 49th International Colloquium on Automata, Languages, and Programming, ICALP 2022, Paris, France, 4–8 July 2022. LIPIcs, vol. 229, pp. 71:1–71:19. Schloss Dagstuhl - Leibniz-Zentrum für Informatik (2022). https://doi.org/10.4230/LIPICS.ICALP.2022.71

13. Lange, J., Rubinfeld, R., Vasilyan, A.: Properly learning monotone functions via local correction. In: 63rd IEEE Annual Symposium on Foundations of Computer Science, FOCS 2022, Denver, CO, USA, 31 October–3 November 2022, pp. 75–86. IEEE (2022). https://doi.org/10.1109/FOCS54457.2022.00015

14. Lange, J., Vasilyan, A.: Agnostic proper learning of monotone functions: beyond the black-box correction barrier. In: 64th IEEE Annual Symposium on Foundations of Computer Science, FOCS 2023, Santa Cruz, CA, USA, 6–9 November 2023, pp. 1149–1170. IEEE (2023). https://doi.org/10.1109/FOCS57990.2023.00068

15. Markov, A.A.: On the inversion complexity of a system of functions. J. ACM **5**(4), 331–334 (1958). https://doi.org/10.1145/320941.320945

16. O'Donnell, R., Servedio, R.A.: Learning monotone decision trees in polynomial time. SIAM J. Comput. **37**(3), 827–844 (2007). https://doi.org/10.1137/060669309

17. O'Donnell, R., Wimmer, K.: KKL, Kruskal-Katona, and monotone nets. In: 50th Annual IEEE Symposium on Foundations of Computer Science, FOCS 2009, Atlanta, Georgia, USA, 25–27 October 2009, pp. 725–734. IEEE Computer Society (2009). https://doi.org/10.1109/FOCS.2009.78

18. Servedio, R.A.: On learning monotone DNF under product distributions. Inf. Comput. **193**(1), 57–74 (2004). https://doi.org/10.1016/J.IC.2004.04.003

19. Takimoto, E., Sakai, Y., Maruoka, A.: The learnability of exclusive-or expansions based on monotone DNF formulas. Theor. Comput. Sci. **241**(1–2), 37–50 (2000). https://doi.org/10.1016/S0304-3975(99)00265-0

Computational Complexity of Combinatorial Distance Matrix Realisation

David L. Fairbairn[1], George B. Mertzios[2]([✉]), and Norbert Peyerimhoff[1]

[1] Department of Mathematical Sciences, Durham University, Durham, UK
[2] Department of Computer Science, Durham University, Durham, UK
george.mertzios@durham.ac.uk

Abstract. The k-CombDMR problem is that of determining whether an $n \times n$ distance matrix can be realised as a sub-matrix by n vertices in some unweighted undirected graph with $n + k$ vertices. This problem has a simple solution in the case $k = 0$. In this paper we show that this problem is polynomial-time solvable for $k = 1$ and $k = 2$, and we provide algorithms to construct such graph realisations by solving appropriate 2-SAT instances. For the case where $k \geq 3$, we prove that the problem becomes NP-complete. We show this by a reduction from the k-colourability problem, where $k \geq 3$. Finally, we present how the simpler problem of tree realisability can be solved in polynomial time for all $k \geq 0$.

Keywords: Distance matrix · Graph realisation · NP-completeness · Polynomial-time algorithm · Graph colourability

1 Introduction

The ability to realise graphs from a partially known distance function (typically given as a distance matrix) is a fundamental problem with wide-ranging practical applications, including network tomography, phylogenetics, and computational network design. Network tomography, for example, relies on understanding the internal structure of a network based on incomplete information about its distances or connectivity [4,10]. Phylogenetics, on the other hand, uses distance matrix realisation to infer evolutionary relationships among species, often by reconstructing evolutionary trees from genetic data. A tree realisation with the fewest vertices often corresponds to the most parsimonious evolutionary scenario [16]. Generating graph realisations is crucial for both synthetic data generation and data analysis (for example clustering), allowing the inference of possible

D. L. Fairbairn—Supported by Tharsus Ltd.
G. B. Mertzios—Supported by the Engineering & Physical Sciences Research Council (EPSRC grant EP/P020372/1).

I. Finocchi and L. Georgiadis (Eds.): CIAC 2025, LNCS 15679, pp. 51–67, 2025.
https://doi.org/10.1007/978-3-031-92932-8_4

structures of underlying systems from observed distances [2,4,7]. In computational network design, minimising the number of vertices (e.g. servers) can significantly reduce network cost and complexity where the cost of edges (e.g. server connections) is negligible compared to the cost of vertices [8,13].

Various graph realisation problems have been studied in the literature, most of them are concerned with weighted graphs with *minimising the sum of the edge weights* as its optimisation criterion [9]. Amongst the results of Hakimi and Yau are a set of necessary and sufficient conditions for graph realisability of a given matrix and a proof of uniqueness of shortest length tree realisations. Dress showed the existence of a weighted graph realisation with minimum total edge weight for any given distance matrix [6]. Moreover, he proved the existence of an optimum solution with at most n^4 vertices for any $n \times n$ distance matrix. In his result, the vertices are only the branch points (vertices with ≥ 3 incident edges) or leaves (vertices with just one incident edges), since all vertices with precisely 2 incident edges can be condensed. Such vertices are called *essential* vertices. Finding weighted graph realisations having the smallest sum of edge weights is NP-hard. More specifically, Althöfer proved that this problem remains NP-hard even in the case where the input distance matrix has integer values (while the edge weights are still real valued) [1]. Althöfer also showed that in the case of integer valued distance matrices, there is always an optimum realisation with rational edge weights. Chung, Garrett and Graham considered a weak version of the weighted graph realisation problem, namely, finding optimum graph realisations for which the distance matrix provides a lower bound on the distances of the corresponding n vertices [4]. They showed that even this weak version of the problem is NP-hard.

In contrast, this paper's focus is the problem of finding combinatorial graph realisations for a prescribed integer valued distance matrix with a prescribed number of additional vertices. Our motivation stems from the field of Multi-Agent Path-Finding (MAPF) [3,12,17], where the problem is to find a set of collision-free paths for a group of agents from their unique start locations to their respective unique goal locations within some graph, or within some temporal graph (i.e. a graph whose structure changes over time). Klobas et al. 2022 [12] studied the computational complexity of the problem of finding temporally disjoint paths and walks in temporal graphs, i.e. paths and walks which do never visit the same vertex at the same time. The recent paper by Atzmon et al. 2023 [3] studied the problem of computing solutions to the MAPF problem, only by utilising the pairwise distances among specific vertices (the "terminals"), while the computed paths are allowed to use any number of non-terminal vertices of the graph. Similarly to [3], in this paper we are again given the pairwise distances among terminal vertices, but we are not given the input graph, and the goal is to generate a graph that respects the given terminal distances by adding the smallest number of additional, non-terminal, vertices.

Throughout this paper we use the notation $[n] = \{1, 2, \ldots, n\}$ for any $n \in \mathbb{N}$ and denote the set of all non-negative integers by $\mathbb{N}_0$ (that is $\mathbb{N}_0 = \mathbb{N} \cup \{0\}$). First we introduce the following problem for every integer $k \in \mathbb{N}_0$.

Problem 1. k-COMBINATORIAL DISTANCE MATRIX REALISATION PROBLEM (k-COMBDMR)

Input: An $n \times n$ matrix D with non-negative integer values.
Question: Does there exist a simple (unweighted) graph $G = (V, E)$ with $|V| \leq n + k$ and an injective mapping $\Phi : [n] \to V$ such that the shortest-path distance function d in G satisfies

$$d(\Phi(i), \Phi(j)) = D_{ij}$$

for all $i, j \in [n]$?

We call such a pair (G, Φ) a *graph realisation* of D. Given an $n \times n$ matrix D, any graph realisation (G, Φ) which has the smallest number of vertices is called a *minimum graph realisation* of D.

Example 2. Two possible graph realisations of the following matrix D, are given in Fig. 1 with $n = 3$, while $k = 3$ and $k = 1$, respectively.

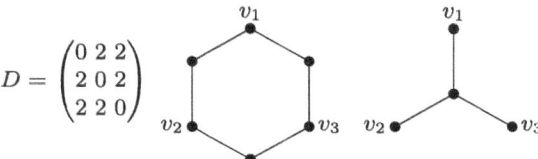

$$D = \begin{pmatrix} 0 & 2 & 2 \\ 2 & 0 & 2 \\ 2 & 2 & 0 \end{pmatrix}$$

Fig. 1. Two graph realisations of the above matrix D, where $\Phi(i) = v_i$ for $i \in [3]$, while $k = 3$ and $k = 1$ in the left and the right realisation, respectively. The right realisation is *minimum*.

It is important to note that k-COMBDMR is distinct from the weighted graph realisation problem. Within the weighted graph realisation problem the edges are equipped with positive real valued weights (their lengths) with the aim to minimise the sum of the edge weights of the graph realisation, whereas in k-COMBDMR we are only concerned with minimising the number of vertices in the graph realisation. Take for instance the distance matrix D below and optimum solutions within these two problems as shown in Fig. 2.

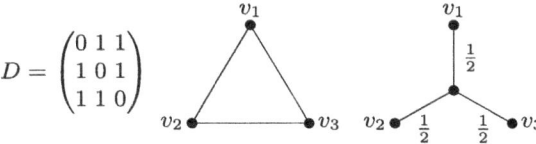

$$D = \begin{pmatrix} 0 & 1 & 1 \\ 1 & 0 & 1 \\ 1 & 1 & 0 \end{pmatrix}$$

Fig. 2. Minimum graph realisations of D above, for k-COMBDMR (Left) and for the weighted graph realisation problem (Right).

Figure 2 shows that optimum solutions for k-COMBDMR and for the weighted graph realisation problem can differ significantly. An optimum solution for k-COMBDMR may not be trivially transformed into an optimum solution for the weighted graph realisation problem, and vice versa. The reader may ask what happens if we consider the weighted graph realisation problem with the additional constraint that the weights must be integers – could we transform any optimum solution for this problem into an optimum solution for k-COMBDMR by replacing weighted edges with paths of length equal to the weight? Figure 3 shows that this is not always the case, as the weighted graph realisation problem with integer weights may have multiple solutions, some of which do not have a corresponding optimum solution for k-COMBDMR under this transformation. Therefore, k-COMBDMR is also a distinct problem from the weighted graph realisation problem with integer weights.

$$D = \begin{pmatrix} 0 & 2 & 2 & 2 & 1 & 3 & 3 & 1 \\ 2 & 0 & 2 & 2 & 3 & 1 & 3 & 1 \\ 2 & 2 & 0 & 2 & 3 & 1 & 1 & 3 \\ 2 & 2 & 2 & 0 & 1 & 3 & 1 & 3 \\ 1 & 3 & 3 & 1 & 0 & 4 & 2 & 2 \\ 3 & 1 & 1 & 3 & 4 & 0 & 2 & 2 \\ 3 & 3 & 1 & 1 & 2 & 2 & 0 & 4 \\ 1 & 1 & 3 & 3 & 2 & 2 & 4 & 0 \end{pmatrix} \tag{1}$$

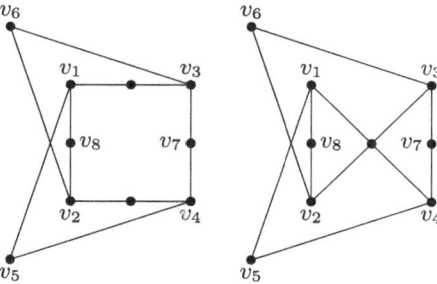

Fig. 3. Optimum graph realisations of D as in (1), with $\Phi(i) = v_i, i \in [8]$ for the weighted graph realisation problem with integer weights $w(e) = 1$ (Left and Right). Only the right graph realisation is optimum for the combinatorial distance realisation problem.

Our Results. In this paper we prove that k-COMBDMR can be solved in polynomial time for $k \leq 2$, while it is NP-complete for $k \geq 3$. In Sect. 3 we provide a polynomial-time algorithm for $k \leq 2$ which solves appropriate 2-SAT instances (see Algorithm 20 and Theorem 21). In Sect. 4 we provide, as our main result, an NP-hardness reduction from the k-colourability problem for every fixed $k \geq 3$

(see Theorem 27). In Sect. 5, we consider the simpler polynomial-time solvable problem of tree realisability for a given distance matrix (see Corollary 30).

In Sect. 2 we introduce our main notions and foundational results. In particular, we introduce the notion of the *q-skeleton* and its associated distance matrix (see Definition 6), which are crucial for the polynomial-time results in Sect. 3. In a nutshell, given the input $n \times n$ distance matrix D, the q-skeleton is the weighted graph of n vertices, where every edge connects vertices of distance at most q in D, and the weight of such an edge is equal to the corresponding distance in D. Although the notion of a q-skeleton is simple and natural, it turns out to be quite powerful, as it allows us to deduce useful upper and lower bounds for the additional vertices needed in a minimum graph realisation of D (see Propositions 8 and 10).

2 Notions and Foundational Results

We begin by identifying the necessary and sufficient conditions for an input matrix D to admit at least one graph realisation.

Definition 3 (Distance matrix). *Let D be an $n \times n$ matrix with non-negative integer valued entries. We call D a* distance matrix *if it satisfies the following properties:*

 (i) All diagonal entries of D are zero and all non-diagonal entries are strictly positive.
 (ii) D is a symmetric matrix.
 (iii) For all $i, j, w \in [n]$, we have

$$D_{iw} + D_{wj} \geq D_{ij}.$$

This definition gives rise to the following result.

Proposition 4. *Let D be an $n \times n$ matrix with non-negative integer valued entries. D is a distance matrix if and only if D admits at least one graph realisation $(G = (V, E), \Phi)$. Furthermore, a graph realisation is obtained by connecting vertices v_i, v_j by a path of length D_{ij} for all $i < j$ such that no two such paths have common interior vertices and $\Phi(i) = v_i$ for all $i \in [n]$. We call such paths* elementary paths *of the graph G.*

Therefore, we will assume that D is a distance matrix with integer valued entries for all instances of k-CombDMR. As Proposition 4 shows, we can always find a graph realisation of a distance matrix D with some number of additional vertices. Note that, in the weighted case, a graph realisation without any additional vertices can be constructed by replacing the elementary paths in Proposition 4 by single edges with appropriate weights. Another immediate consequence of the above construction is the following upper bound on the number of vertices for the existence of a graph realisation of D.

Proposition 5. *Let D be an $n \times n$ distance matrix. Then there exists a graph realisation $(G = (V, E), \Phi)$ of D with*

$$|V| \leq n + \sum_{1 \leq i < j \leq n} (D_{ij} - 1).$$

We now seek to improve the result of Proposition 5, and in doing so, we introduce the following weighted graph, whose distance matrix will be of fundamental importance.

Definition 6 (q-skeleton). *Let D be an $n \times n$ distance matrix and $q \in \mathbb{N}$. The q-skeleton of D is the weighted graph $G^q = (V^q, E^q, w)$ with vertices $V^q = [n]$ and edges*

$$E^q = \{\{i, j\} \in [n] \times [n] \mid (i < j) \wedge (D_{ij} \leq q)\},$$

that is, G^q has an edge between i and j if and only if $D_{ij} \leq q$. Additionally, let the edge weights $w : E^q \to \mathbb{N}$ be given by

$$w(i, j) = D_{ij}, \quad \{i, j\} \in E^q.$$

Let $d_{G^q} : V \times V \to \mathbb{N}_0 \cup \{\infty\}$ be the associated distance function of G^q, that is, $d_{G^q}(i, j)$ is the length of the shortest path between i and j in G^q and equal to ∞ if no such path exists. The $n \times n$ matrix $D^{(q)}$, given by $D_{ij}^{(q)} = d_{G^q}(i, j)$, is called the distance matrix of the q-skeleton of D.

Of particular relevance is the following fact:

$$D_{ij}^{(q)} \begin{cases} = D_{ij} & \text{if } D_{ij} \leq q, \\ \geq D_{ij} & \text{if } D_{ij} > q. \end{cases}$$

Notice, when we have $D^{(q)} = D$ for some q, then we can replace each edge $\{i, j\} \in E^q$ with an elementary path of length D_{ij} in G^q to obtain a graph realisation of D. Furthermore, $D^{(q)}$ can be computed by any polynomial-time weighted all-pairs shortest-paths (APSP) algorithm. In fact, we have the following ordering of the matrices $D^{(q)}$.

Lemma 7. *Let D be an $n \times n$ distance matrix and $m = \max\{D_{ij} : 1 \leq i < j \leq n\}$. Let $D^{(q)}$ be the distance matrix of the q-skeleton of D. Then we have,*

$$D_{ij}^{(1)} \geq D_{ij}^{(2)} \geq \cdots \geq D_{ij}^{(m)} = D_{ij} \quad \text{for all } i, j \in [n].$$

Now let q_0 denote the smallest $q \in \mathbb{N}$ such that $D^{(q)} = D$. Then we have the following improvement on Proposition 5, utilising the q-skeleton of D.

Proposition 8. *Let D be an $n \times n$ distance matrix and $q_0 \in \mathbb{N}$ be the smallest $q \in \mathbb{N}$ such that $D^{(q)} = D$, where $D^{(q)}$ the distance matrix of the q-skeleton of D. Then there exists a graph realisation $(G = (V, E), \Phi)$ of D with*

$$|V| \leq n + \sum_{(1 \leq i < j \leq n) \wedge (2 \leq D_{ij} \leq q_0)} (D_{ij} - 1). \tag{2}$$

Using again the q-skeleton of D we can now provide a lower bound on the number of vertices required for a graph realisation of D (see Proposition 10). To do so, we generalise the earlier notion of elementary paths as follows: for a graph realisation (G, Φ) of D, an *elementary path* is a path of length D_{ij} between vertices v_i and v_j with no interior vertices in $\Phi([n])$. Proposition 10 then follows from the following result.

Proposition 9. *Let $s \in \mathbb{N}$, D be an $n \times n$ distance matrix, $D^{(q)}$ be the distance matrix of the q-skeleton of D and (G, Φ) be a graph realisation of D. If there are no elementary paths of length greater than s in G, then $D^{(s)} = D$.*

Proposition 10. *Let D be an $n \times n$ distance matrix and $q_0 \in \mathbb{N}$ be the smallest $q \in \mathbb{N}$ such that $D^{(q)} = D$, with $D^{(q)}$ the distance matrix of the q-skeleton of D. Any graph realisation $(G = (V, E), \Phi)$ of D must satisfy $|V| \geq n + (q_0 - 1)$.*

Combining Proposition 8 and Proposition 10 we obtain the following Proposition.

Proposition 11. *Let D be an $n \times n$ distance matrix and $D^{(q)}$ be the distance matrix of the q-skeleton of D. Then D has a graph realisation (G, Φ) with $|V| = n$ if and only if $D^{(1)} = D$.*

Due to its general significance throughout this paper, we introduce for any distance matrix D the unweighted graph G_D, which is simply the 1-skeleton G^1 of D as in Definition 6 without the edge weights. Note that Proposition 11 is equivalent to the following theorem.

Theorem 12 ([9]). *For any $n \times n$ distance matrix D, the following statements are equivalent:*

- *A graph realisation $(G = (V, E), \Phi)$ of D with $|V| = n$ exists.*
- *A graph realisation of D is $(G_D = (V_D, E_D), \Phi)$ with $V_D = v_1, \ldots, v_n$ and $\Phi(i) = v_i$ for $i \in [n]$.*

The graph realisation of a distance matrix D with $|V| = n$ is also unique up to isomorphism (See [9]).

0-CombDMR is solved by Theorem 12 as follows:

1. Construct the graph G_D.
2. Check if the distance function of G_D coincides with D.

If D has a graph realisation with $|V| = n$, then we call D a *self-realising distance matrix*. Note then the importance of the graph G_D in the following proposition.

Proposition 13. *Let D be an $n \times n$ distance matrix and $G_D = (V_D, E_D)$ be the associated unweighted graph, with $V_D = \{v_1, \ldots, v_n\}$. Then any graph realisation (G, Φ) of D with $\Phi(i) = v_i$ for $i \in [n]$ has G_D as the induced subgraph on the vertices $v_1, \ldots, v_n$.*

3   Polynomial Solutions of 1-COMBDMR and 2-COMBDMR

In this section, we consider the cases where $k = 1$ and $k = 2$. If D is not self-realising, we know that $|V| \geq n + 1$ for any graph realisation $(G = (V, E), \Phi)$ of D. Specifically for $k = 1$ and $k = 2$ there are three possibilities for a graph realisation, namely, the graph has a single additional vertex ($\bullet$), two additional vertices which are not adjacent ($\vdots$), or two additional vertices which are adjacent ($\vdots$). We now aim to develop polynomial-time algorithms for each of these cases ($\bullet$, $\vdots$, $\vdots$) separately. Our algorithms will also provide such graph realisations if they exists.

The approach we take to solve these problems involves constructing a particular 2-Satisfiability (2-SAT) instance [15] for each case, which can be solved in polynomial-time. We seek to show that a satisfying assignment of the 2-SAT instance gives rise to a graph with each respective property which can subsequently be checked if it is a realisation of D, and if not then we will prove that no graph realisation of D with the respective property exists.

Formally, a 2-SAT instance is expressed as a 2-CNF formula ϕ, which is the conjunction of a set of clauses, that is

$$\phi = \bigwedge_{i=1}^{m} c_i = c_1 \wedge \ldots \wedge c_m$$

for some finite $m \in \mathbb{N}_0$, where each clause c_i is a disjunction of two literals:

$$c_i = (\ell_i^{(1)} \vee \ell_i^{(2)}).$$

Here, each of $\ell_i^{(1)}, \ell_i^{(2)}$ is a literal, i.e., either a variable x or its negation $\bar{x}$.

Within any graph realisation $(G = (V, E), \Phi)$ for these cases with $\Phi(i) = v_i$ we know that the induced subgraph on vertices $v_1, \ldots, v_n$ agrees with G_D by Proposition 13. Therefore, each case must start with the construction of the graph G_D as in Proposition 13.

We now construct the three 2-SAT instances ϕ_1, ϕ_2 and ϕ_3 for the cases ($\bullet$, $\vdots$, $\vdots$), respectively. In the below construction each step will be suffixed with the cases it is relevant to and ϕ represents ϕ_1, ϕ_2 or ϕ_3, respectively. Furthermore, in the case ($\bullet$) let $K = \{n + 1\}$, and in the cases ($\vdots$, $\vdots$) let $K = \{n + 1, n + 2\}$.

1. [$\bullet$, $\vdots$, $\vdots$] Let $G_D = (V_D, E_D)$ be the graph as described in Proposition 13 with vertices $V_D = \{v_1, \ldots, v_n\}$ and $\Phi(i) = v_i$ for $i \in [n]$.
2. [$\bullet$, $\vdots$, $\vdots$] Let $\{x_{i,k} : i \in [n], k \in K\}$ be the set of boolean variables representing the existence of an additional edge $\{v_i, v_k\}$ to those already in G_D. That is, $x_{i,k}$ is true if and only if v_i is adjacent to v_k in G.
3. [$\bullet$, $\vdots$, $\vdots$] For all $i, j \in [n], k \in K$ with $D_{ij} > 2$, we know that the vertices v_i and v_k must not both be adjacent to v_j in any graph realisation of D, as otherwise this would result in a distance of 2 between v_i and v_j. This condition is equivalent to the following clause being satisfied:

$$(\bar{x}_{i,k} \vee \bar{x}_{j,k}). \tag{3}$$

We therefore add the clause (3) to ϕ for all $i, j \in [n], k \in K$ with $D_{ij} > 2$.

4. [●] For all $i, j \in [n]$, with, $D_{ij} = 2$ and $d_{G_D}(v_i, v_j) > 2$, the following boolean expression must be satisfied:

$$(x_{i,n+1} \wedge x_{j,n+1}), \tag{4}$$

meaning that the distance between v_i and v_j must be 2 and realised by a path of length 2 via v_{n+1}. The boolean expression (4) is equivalent to the following two clauses both being satisfied:

$$(x_{i,n+1} \vee x_{i,n+1}), (x_{j,n+1} \vee x_{j,n+1}). \tag{5}$$

Therefore, we add the clauses (5) to ϕ for all $i, j \in [n]$ with $D_{ij} = 2$ and $d_{G_D}(v_i, v_j) > 2$ as there is no other way to realise a distance of 2 between v_i and v_j in G.

5. [❧,❧] For all $i, j \in [n]$, such that, $D_{ij} = 2$ and $d_{G_D}(v_i, v_j) > 2$, we know the following boolean expression must be satisfied:

$$(x_{i,n+1} \wedge x_{j,n+1}) \vee (x_{i,n+2} \wedge x_{j,n+2}), \tag{6}$$

meaning that the distance between v_i and v_j must be 2 and realised via a path of length 2 via v_{n+1} or v_{n+2}. The boolean expression (6), by distributivity, is equivalent to the following 4 clauses being satisfied:

$$(x_{i,n+1} \vee x_{i,n+2}), (x_{j,n+1} \vee x_{j,n+2}),$$
$$(x_{i,n+1} \vee x_{j,n+2}), (x_{j,n+1} \vee x_{i,n+2}). \tag{7}$$

Therefore, we add the clauses (7) to ϕ for all $i, j \in [n]$ such that $D_{ij} = 2$ and $d_{G_D}(v_i, v_j) > 2$.

6. [❧] Compute $D^{(2)}$ the distance matrix of the 2-skeleton of D.

7. [❧] For all $i, j \in [n]$ with $D_{ij} > 3$, we know that, if v_i is adjacent to v_{n+1} then v_j cannot be adjacent to v_{n+2}, as otherwise this would result in a distance of 3 between v_i and v_j. Similarly, if v_i is adjacent to v_{n+2} then v_j cannot be adjacent to v_{n+1}. This condition is equivalent to the following clauses both being satisfied:

$$(\bar{x}_{i,n+1} \vee \bar{x}_{j,n+2}), (\bar{x}_{i,n+2} \vee \bar{x}_{j,n+1}). \tag{8}$$

Therefore, add the clauses (8) to ϕ for all $i, j \in [n]$ with $D_{ij} > 3$.

8. [❧] For all $i, j \in [n]$ with $D_{ij} = 3$ and $D_{ij}^{(2)} > 3$, the following boolean expression must be satisfied (by Lemma 14 below):

$$(x_{i,n+1} \wedge x_{j,n+2}) \vee (x_{i,n+2} \wedge x_{j,n+1}), \tag{9}$$

meaning that the distance between v_i and v_j must be 3 and it must be realised via a path of length 3 through v_{n+1} and v_{n+2}. The boolean expression (9), is equivalent to the following four clauses being satisfied:

$$(x_{i,n+1} \vee x_{i,n+2}), (x_{j,n+1} \vee x_{j,n+2}),$$
$$(x_{i,n+1} \vee x_{j,n+1}), (x_{i,n+2} \vee x_{j,n+2}). \tag{10}$$

Therefore, we add the clauses (10) to ϕ for all $i, j \in [n]$ with $D_{ij} = 3$ and $D_{ij}^{(2)} > 3$.

This concludes the construction of the 2-SAT instances ϕ_1, ϕ_2 and ϕ_3 for the cases ($\bullet$,$\substack{\bullet\\\bullet}$,$\substack{\bullet\\\bullet}$), respectively. As noted in the construction we have the following lemma in the case ($\substack{\bullet\\\bullet}$).

Lemma 14. *Let D be an $n \times n$ distance matrix and $i,j \in [n]$. Assume $D_{ij} = 3$ and $D_{ij}^{(2)} > 3$, where $D^{(2)}$ is the distance matrix of the 2-skeleton of D. In any graph realisation $(G = (V,E), \Phi)$ of D with $V = \{v_i = \Phi(i) : i \in [n]\} \cup \{v_{n+1}, v_{n+2}\}$, with v_{n+1} adjacent to v_{n+2}, any shortest path from v_i to v_j in G must be of the following form:*

$$v_i \to v_{n+1} \to v_{n+2} \to v_j \quad \text{or} \quad v_i \to v_{n+2} \to v_{n+1} \to v_j.$$

Proof. Let $(G = (V,E), \Phi)$ be a graph realisation of D with $V = \{v_i = \Phi(i) : i \in [n]\} \cup \{v_{n+1}, v_{n+2}\}$. Since $D_{ij} = 3$, there exists a shortest path of the form $v_i \to v_s \to v_t \to v_j$ for some $s, t \in [n+2]$. If $\{s,t\} \neq \{n+1, n+2\}$ then this shortest path is a concatenation of elementary paths of length 1 or 2 and therefore $D_{ij} = D_{ij}^{(2)}$, which is a contradiction to the assumption that $D_{ij}^{(2)} > 3$. $\blacksquare$

We now introduce the following definition. A truth assignment of boolean variables $\{x_{i,k}\}$ is said to be *consistent* with a graph $G = (V,E)$ with $V \supset \{v_1, \ldots, v_m\}$ when

$$x_{i,k} = \begin{cases} \text{True} & \text{if } \{v_i, v_k\} \in E, \\ \text{False} & \text{if } \{v_i, v_k\} \notin E. \end{cases}$$

Since all clauses in ϕ_1, ϕ_2 and ϕ_3 are nessessary conditions for a graph realisation of D, we have the following observation.

Observation 15. *Let D be an $n \times n$ distance matrix. If ϕ_1, ϕ_2 or ϕ_3 is not satisfiable then no graph realisation $(G = (V,E), \Phi)$ of D in the respective case ($\bullet$,$\substack{\bullet\\\bullet}$,$\substack{\bullet\\\bullet}$) exists.*

Note that any graph realisation $(G = (V,E), \Phi)$ of D for the cases ($\bullet$,$\substack{\bullet\\\bullet}$,$\substack{\bullet\\\bullet}$) gives rise to a satisfying assignment $\mathbf{X}$ of ϕ_1, ϕ_2 or ϕ_3, respectively, which is consistent with G.

We now seek to show that we require only a single satisfying assignment to determine whether such a graph realisation exists. Let $G_{\phi_i, \mathbf{X}}$ be the unique graph with vertex set $V = \{v_1, \ldots, v_n\}$, and additional vertices v_{n+1}, v_{n+2} (if required), having G_D as the induced subgraph on the vertices $v_1, \ldots, v_n$ and consistent with a satisfying assignment $\mathbf{X}$ of ϕ_i. For such a graph, let $D(\phi_i, \mathbf{X})$ denote the $n \times n$ distance matrix of $G_{\phi_i, \mathbf{X}}$ over the vertices $\{v_1, \ldots, v_n\}$. Computing $D(\phi_i, \mathbf{X})$ can be done in polynomial-time via an all-pairs shortest-paths (APSP) algorithm.

Lemma 16. *Let $\mathbf{X}$ be a satisfying assignment of $\phi \in \{\phi_1, \phi_2\}$ and $D^{(2)}$ be the distance matrix of the 2-skeleton of D. Then*

$$D(\phi, \mathbf{X}) = D^{(2)}.$$

Lemma 17. *Let* $\mathbf{X}$ *be a satisfying assignment of* ϕ_3 *and* $D^{(3)}$ *be the distance matrix of the 3-skeleton of* D. *Then*

$$D(\phi_3, \mathbf{X}) = D^{(3)}.$$

An immediate consequence of Lemma 16 and Lemma 17 is the following corollary.

Corollary 18. *Let* $\mathbf{X}$ *and* $\mathbf{X}'$ *be two distinct satisfying assignments of* $\phi \in \{\phi_1, \phi_2, \phi_3\}$. *Then*

$$D(\phi, \mathbf{X}) = D(\phi, \mathbf{X}').$$

Corollary 18 tells us that the distance matrix $D(\phi_i, \mathbf{X})$ of $G_{\phi_i, \mathbf{X}}$ is invariant over all satisfying assignments of ϕ_i for $i \in \{1, 2, 3\}$. Therefore, if we can find a single satisfying assignment $\mathbf{X}$ of ϕ_i, then we can construct $G_{\phi_i, \mathbf{X}}$, and if $D(\phi_i, \mathbf{X}) = D$ then we have found a graph realisation $(G = (V, E), \Phi)$ of D for the respective case $(\bullet, \mathbf{\vdots}, \mathbf{\vdots})$. It remains to show that, in the case $D(\phi_i, \mathbf{X}) \neq D$, no graph realisation of D with the respective property exists.

Proposition 19. *Let* D *be an* $n \times n$ *distance matrix. If* $D(\phi_i, \mathbf{X}) \neq D$ *for some satisfying assignment* $\mathbf{X}$ *of* ϕ_i, $i \in \{1, 2, 3\}$, *then no graph realisation* $(G = (V, E), \Phi)$ *of* D *with the respective property* $(\bullet, \mathbf{\vdots}, \mathbf{\vdots})$ *exists.*

Proof. Assume that such a graph realisation $(G = (V, E), \Phi)$ of D exists with the respective property $(\bullet, \mathbf{\vdots}, \mathbf{\vdots})$. By Observation 15, we know that there exists a satisfying assignment $\mathbf{X}'$ of ϕ_i, consistent with the graph G, such that $D(\phi_i, \mathbf{X}') = D$. By Corollary 18, we know that $D(\phi_i, \mathbf{X}) = D(\phi_i, \mathbf{X}')$ which is a contradiction to $D(\phi_i, \mathbf{X}) \neq D$.

Therefore, we have the following polynomial-time algorithm to determine whether there exists a graph realisation $(G = (V, E), \Phi)$ of D with the respective property $(\bullet, \mathbf{\vdots}, \mathbf{\vdots})$. Moreover, the algorithm produces such a graph realisation if it exists.

Algorithm 20 (Solving 1-CombDMR and 2-CombDMR).
Input: *An* $n \times n$ *distance matrix* D *and property* $(\bullet, \mathbf{\vdots}, \mathbf{\vdots})$.
Output: *A graph realisation* $(G = (V, E), \Phi)$ *of* D *with the respective property* $(\bullet, \mathbf{\vdots}, \mathbf{\vdots})$ *if it exists, or a statement that no such graph realisation exists.*

1. *Let* $\Phi(i) = v_i$ *for* $i \in [n]$.
2. *Construct the 2-CNF formula* ϕ, *corresponding to the input property.*
3. *Compute a satisfying assignment* $\mathbf{X}$ *of* ϕ, *if it exists.*
4. *If* ϕ *is not satisfiable, then no graph realisation* $(G = (V, E), \Phi)$ *of* D *with the respective property* $(\bullet, \mathbf{\vdots}, \mathbf{\vdots})$ *exists. (By Observation 15)*
5. *If* ϕ *is satisfiable, then construct the graph* $G_{\phi, \mathbf{X}}$ *consistent with the satisfying assignment* $\mathbf{X}$.
6. *Compute the distance matrix* $D(\phi, \mathbf{X})$ *of* $G_{\phi, \mathbf{X}}$ *(using any APSP algorithm).*

7. If $D(\phi, \mathbf{X}) = D$ then $(G_{\phi, \mathbf{X}}, \Phi)$ is a graph realisation of D with the respective property ($\bullet, \natural, \natural$).
   Otherwise, if $D(\phi, \mathbf{X}) \neq D$ then no graph realisation of D with the respective property ($\bullet, \natural, \natural$) exists. (By Proposition 19)

The following theorem summarises of our results in this section.

Theorem 21. *The 1-*CombDMR *and 2-*CombDMR *problems are polynomial-time solvable.*

4 k-CombDMR is NP-Complete for $k \geq 3$

In this section we prove that k-CombDMR is NP-complete for every $k \geq 3$, via a reduction from k-colourability, which is known to be NP-complete [11,14,18]. For the readers' convenience we restate the k-colourability problem as follows:

Problem 22 (**k-colourability**). Given a graph $G = (V, E)$. Does there exist a function $\chi : V \to [k]$ such that for all $\{i, j\} \in E$ we have $\chi(i) \neq \chi(j)$?

As we will prove, k-Colourability can be reduced to the k-CombDMR problem by the following reduction algorithm.

Algorithm 23 (Reduction of k-Colourability to k-CombDMR).
Input: *A connected simple undirected graph $G_c = (V_c, E_c)$ for which we want to determine if it is k-colourable.*
Output: *A distance matrix D such that G_c is k-colourable if and only if k-*CombDMR *for D is a YES-instance.*

1. *Enumerate the vertices of G_c such that $V_c = \{v_1, \ldots, v_{n_c}\}$ where $n_c = |V_c|$.*
2. *Construct the gadget graph $G_g = (V_g, E_g)$, with $V_c \subseteq V_g$, as follows. We subdivide each edge in E_c twice, i.e., we replace each edge by a path of length 3 (containing two new vertices). For every pair of non-adjacent vertices in G_c, we add a path of length 2 between them (containing one new vertex). We enumerate the vertices of G_g such that $V_g = \{v_1, \ldots, v_{n_c}, v_{n_c+1} \ldots, v_{n_g}\}$ where $v_{n_c+1}, \ldots, v_{n_g}$ are the new vertices and $n_g = |V_g|$, see Fig. 4.*
3. *Let d_{G_g} denote the shortest path distance function of G_g. Construct the $n \times n$ distance matrix D where $n = n_g + 1$, with entries,*

$$
\begin{aligned}
D_{ij} &= d_{G_g}(v_i, v_j) && \text{for } i, j \in [n_g], \\
D_{i'n} &= D_{ni'} = 2 && \text{for } i' \in [n_c], \\
D_{i''n} &= D_{ni''} = 3 && \text{for } i'' \in [n_g] \setminus [n_c], \\
D_{nn} &= 0.
\end{aligned}
$$

This will result in a distance matrix of the form:

$$D = \begin{bmatrix} & & \begin{matrix} 2 \\ \vdots \\ 2 \\ 3 \\ \vdots \\ 3 \end{matrix} \\ \begin{matrix} D_{ij} = d_{G_g}(v_i, v_j) \\ i, j \in [n_g] \end{matrix} & \\ \hline 2 \cdots \cdots 2 \; 3 \cdots \cdots 3 & 0 \end{bmatrix}$$

An example of the construction of a gadget graph as in Algorithm 23 is illustrated in Fig. 4. Figure 5 illustrates an example graph realisation of D, where the nth row and column correspond to the vertex v_n.

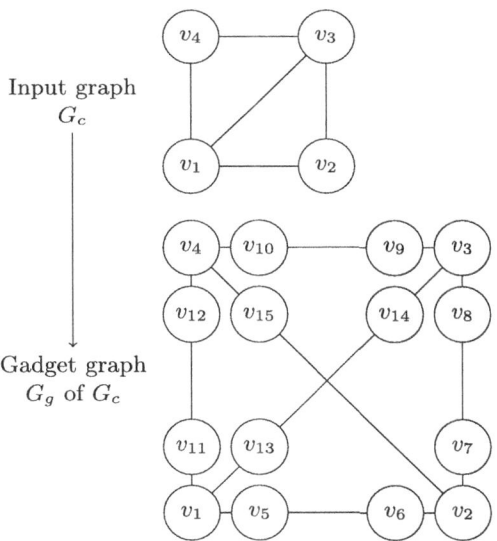

Fig. 4. Example of construction of a gadget graph G_g from an input graph G_c as in Algorithm 23 with old vertices $v_1, \ldots, v_4$ (i.e., $n_c = 4$) and new vertices $v_5, \ldots, v_{15}$ (i.e., $n_g = 15$).

Proposition 24. *The constructed matrix D of Algorithm 23 satisfies the conditions of Definition 3 and is therefore a distance matrix.*

Now that we have established that the constructed matrix D is a valid distance matrix, our next aim is to prove that, if D is a YES-instance of k-COMBDMR then G_c is k-colourable (Proposition 26 below). We start with the following useful lemma.

Lemma 25. *Given an input graph $G_c = (V_c, E_c)$ and $k \in \mathbb{N}$. Let $(G = (V, E), \Phi)$ be any graph realisation of the constructed $n \times n$ distance matrix D by Algorithm 23, with $|V| = n + k$ vertices. Let $v_{n+1}, \ldots, v_{n+k} \in V \setminus \Phi([n])$. Any two vertices v_i and v_j adjacent in the input graph G_c cannot both be adjacent to the same vertex $v \in \{v_{n+1}, \ldots, v_{n+k}\}$ in G.*

Proof. Given a graph realisation $(G = (V, E), \Phi)$ of D with $|V| = n + k$ vertices and $\Phi(i) = v_i$ for $i \in [n]$, with D constructed by Algorithm 23 and let d_G denote the shortest path distance function of G. As v_i and v_j are adjacent in G_c, by construction $D_{ij} = 3$ and $d_G(v_i, v_j) = 3$. If v_i and v_j were both adjacent to the same vertex $v \in \{v_{n+1}, \ldots, v_{n+k}\}$ in G then $d_G(v_i, v_j) \leq 2$, which would be a contradiction to G being a graph realisation of D. □

Proposition 26. *Given an input graph $G_c = (V_c, E_c)$ and D the $n \times n$ matrix as constructed in Algorithm 23. If D is a YES-instance of k-COMBDMR then G_c is k-colourable.*

Proof. Assume a graph realisation $(G = (V, E), \Phi)$ of D as constructed by Algorithm 23 with $|V| = n + k$ exists. Without loss of generality let $\Phi(i) = v_i$ for $i \in [n]$ and let $\{v_{n+1}, \ldots, v_{n+k}\} = V \setminus \Phi([n])$. Then, by Lemma 25 we know that any two adjacent vertices in G_c cannot both be adjacent to the same vertex $v \in \{v_{n+1}, \ldots, v_{n+k}\}$ in G. Furthermore, we know that each vertex v_i for $i \in [n_c]$ must be adjacent to at least one of the vertices $v \in \{v_{n+1}, \ldots, v_{n+k}\}$ in G to realise the $D_{in} = D_{ni} = 2$ distances in D. We construct a colouring of the vertices of G_c by assigning a colour to each of the vertices $v_{n+1}, \ldots, v_{n+k}$ and then assign the same colour to any vertex v_i for $i \in [n_c]$ which is adjacent to that vertex (with arbitrary choice in the case of multiple adjacent vertices $v_{n+1}, \ldots, v_{n+k}$). This is a valid k-colouring due to Lemma 25. □

The colour assignment in the above proof is illustrated in Fig. 5 as a continuation of the example in Fig. 4.

The following theorem states that the implication in Proposition 26 is, in fact, an equivalence.

Theorem 27. *Let $k \in \mathbb{N}$, G_c be an input graph for Algorithm 23 and D be the constructed distance matrix. Then G_c is k-colourable if and only if D is a YES-instance of k-COMBDMR.*

Proof. The forward direction is given by Proposition 26. It remains to prove that if G_c is k-colourable then D is a YES-instance of k-COMBDMR. Let G_c have a k-colouring $\chi : V_c \to [k]$. We begin by constructing a graph realisation $(G = (V, E), \Phi)$ with $V = \{v_1, \ldots, v_{n+k}\}$ of the $n \times n$ distance matrix D. Let $\Phi(i) = v_i$ for $i \in [n]$. The edge set E of G is determined by the following requirements:

- The induced subgraph of G on the vertices $\{v_1, \ldots, v_{n_g}\}$ coincides with the gadget graph G_g.
- v_n is not adjacent to any of the vertices of the gadget graph G_g.

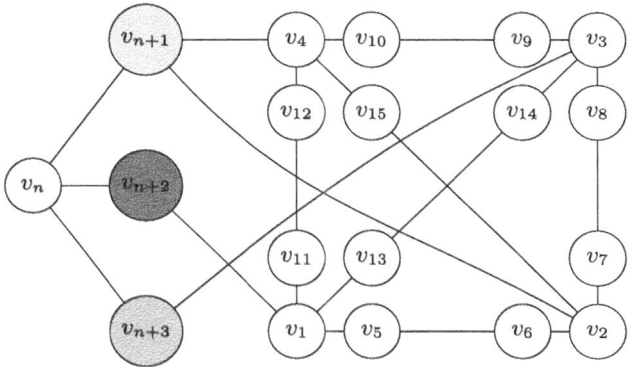

Fig. 5. A graph realisation of D constructed from the example in Fig. 4 with $k = 3$. In accordance with the proof of Proposition 26, the vertex v_1 inherits the colour of vertex v_{n+2}, the vertices v_2 and v_4 inherit the colour of vertex v_{n+2}, and the vertex v_3 inherits the colour of vertex v_{n+3}.

– For $j \in [k]$ the neighbours of v_{n+j} are precisely the following: v_n and all vertices v_i in $\{v_1, \ldots, v_{n_c}\}$ whose colour is j, that is, $\chi(v_i) = j$.

We now show that (G, Φ) is indeed a graph realisation of D. As each vertex v_i for $i \in [n_g] \setminus [n_c]$ is adjacent to some v_j for $j \in [n_c]$ in G and not adjacent to any vertex in $\{v_n, \ldots, v_{n+k}\}$, it suffices to verify the following equalities:

$$d_G(v_n, v_i) = 2 \qquad \text{for } i \in [n_c], \tag{11}$$
$$d_G(v_i, v_j) = D_{ij} \qquad \text{for } i, j \in [n_c]. \tag{12}$$

Let $i \in [n_c]$. By construction, v_i is adjacent to v_{n+j} with $j = \chi(v_i)$ and v_{n+j} is adjacent to v_n, therefore $d_G(v_n, v_i) \leq 2$ and (11) follows from the fact that v_n is not adjacent to v_i. For (12), we distinguish between two cases: if v_i and v_j are adjacent in G_c then $D_{ij} = 3$ and $d_{G_g}(v_i, v_j) = 3$ by construction. Moreover, $\chi(v_i) \neq \chi(v_j)$ implies that v_i and v_j are not adjacent to the same vertex in $\{v_{n+1}, \ldots, v_{n+k}\}$ in G. Therefore, there is no shortest path of length smaller than 3 between v_i and v_j in G. If v_i and v_j are not adjacent in G_c then $D_{ij} = 2$ and $d_{G_g}(v_i, v_j) = 2$ by construction. We do not add an edge between v_i and v_j in G, and therefore $d_G(v_i, v_j) = 2 = D_{ij}$. Hence, (G, Φ) is a graph realisation of D.

Note that k-COMBDMR $\in$ NP since any distance matrix of a finite graph can be computed in polynomial-time. Therefore, Theorem 27 implies the next theorem.

Theorem 28. k-COMBDMR *is NP-complete for all $k \in \mathbb{N}, k \geq 3$.*

5   Tree Realisations

In this section, we discuss the restricted case of combinatorial tree realisations of distance matrices. For a given $n \times n$ distance matrix D, we call a graph realisation (G, Φ) of D a *tree realisation* of D if G is a tree. In contrast to Proposition 5 for the general graph realisation problem, it is no longer true that any distance matrix always admits a combinatorial tree realisation with sufficiently many vertices. To see this, observe that the distance matrix of C_3 in Fig. 2 can only be represented by a graph that contains a triangle, and thus is not a tree. While C_3 admits a tree realisation in the weighted case, the distance matrix of C_4 does not have a weighted tree realisation.

Zareckiĭ shows that the tree realisation problem can be solved in $O(n^4)$ time [19] and provides a set of necessary and sufficient conditions for a distance matrix to have a tree realisation:

Theorem 29 ([19]). *Let D be an $n \times n$ matrix. Then D is a distance matrix and there exists a unique minimal combinatorial tree realisation $(T = (V, E), \Phi)$ of D if and only if*

(a) For all $i, j \in [n]$: $D_{ij} \in \mathbb{Z}$, $D_{ij} = D_{ji} > 0$ for all $i \neq j$, $D_{ii} = 0$.
(b) For all $i, j, k \in [n]$: $D_{ij} + D_{jk} - D_{ik}$ is even.
(c) For all $i, j, k, l \in [n]$: At least two of $D_{ij} + D_{kl}, D_{ik} + D_{jl}, D_{il} + D_{jk}$ are equal and at least the third.

For clarity, condition (c) of Theorem 29 is equivalent to

$$D_{ij} + D_{kl} \leq \max(D_{ik} + D_{jl}, D_{il} + D_{jk}),$$

for all $i, j, k, l \in [n]$, where minimal is with respect to the number of vertices in the tree.

Zareckiĭ also provides an $O(n^4)$ algorithm to construct the tree realisation of a distance matrix if it exists [19].

There exists an $O(n^2)$ algorithm to solve the minimum *weighted* tree realisation problem [5]. Provided the input distance matrix is integer valued, their algorithm can be adapted to solve the *combinatorial* tree realisation problem for integer valued distance matrices in $O(n^2)$ time.

Corollary 30. *If there is a minimal weighted tree realisation of D (containing only essential vertices) which has exclusively integer edge weights, then each edge can be replaced by an elementary path of the same length to obtain a combinatorial tree realisation of the original distance matrix. Otherwise, if this minimal weighted tree realisation (which is necessarily unique) requires a non-integer edge weight, then the distance matrix does not admit a combinatorial tree realisation.*

References

1. Althöfer, I.: On optimal realizations of finite metric spaces by graphs. Discret. Comput. Geom. **3**(2), 103–122 (1988)
2. Anirudh Sabnis, A., Sitaraman, R.K., Towsley, D.: Occam: an optimization based approach to network inference. SIGMETRICS Perform. Eval. Rev. **46**(2), 36–38 (2019)
3. Atzmon, D., Bernardini, S., Fagnani, F., Fairbairn, D.: Exploiting geometric constraints in multi-agent pathfinding. In: Proceedings of the International Conference on Automated Planning and Scheduling, vol. 33, no. 1, pp. 17–25 (2023)
4. Chung, F., Garrett, M., Graham, R., Shallcross, D.: Distance realization problems with applications to internet tomography. J. Comput. Syst. Sci. **63**(3), 432–448 (2001)
5. Culberson, J.C., Rudnicki, P.: A fast algorithm for constructing trees from distance matrices. Inf. Process. Lett. **30**(4), 215–220 (1989)
6. Dress, A.: Trees, tight extensions of metric spaces, and the cohomological dimension of certain groups: a note on combinatorial properties of metric spaces. Adv. Math. **53**(3), 321–402 (1984)
7. Díaz, J., Diner, Ö.Y., Serna, M., Serra, O.: The multicolored graph realization problem. Discret. Appl. Math. **354**, 146–159 (2024). 18th Cologne-Twente Workshop on Graphs and Combinatorial Optimization (CTW 2020)
8. Feremans, C., Labbé, M., Laporte, G.: Generalized network design problems. Eur. J. Oper. Res. **148**(1), 1–13 (2003)
9. Hakimi, S.L., Yau, S.S.: Distance matrix of a graph and its realizability. Q. Appl. Math. **22**, 305–317 (1965)
10. Herman, G.T., Kuba, A.: Discrete Tomography: Foundations, Algorithms, and Applications. Springer, New York City (2012)
11. Karp, R.M.: Reducibility among combinatorial problems. In: Complexity of Computer Computations (Proceedings of the Symposium, IBM Thomas J. Watson Res. Center, Yorktown Heights, N.Y., 1972), pp. 85–103. The IBM Research Symposia Series, Plenum, New York-London (1972)
12. Klobas, N., Mertzios, G.B., Molter, H., Niedermeier, R., Zschoche, P.: Interference-free walks in time: temporally disjoint paths. Auton. Agent. Multi-Agent Syst. **37**(1), 1 (2023)
13. Kortsarz, G., Nutov, Z.: Approximating some network design problems with node costs. Theor. Comput. Sci. **412**(35), 4482–4492 (2011)
14. Lovász, L.: Coverings and coloring of hypergraphs. In: Proceedings of the Fourth Southeastern Conference on Combinatorics, Graph Theory and Computing (Florida Atlantic Univ., Boca Raton, Fla., 1973). Congress. Numer., vol. VIII, pp. 3–12. Utilitas Math., Winnipeg, MB, USA (1973)
15. Papadimitriou, C.H.: Computational Complexity. Addison-Wesley, Reading (1994)
16. Semple, C., Steel, M.: Phylogenetics, Oxford Lecture Series in Mathematics and its Applications, vol. 24. Oxford University Press, Oxford (2003)
17. Stern, R., et al.: Multi-agent pathfinding: definitions, variants, and benchmarks. In: Proceedings of the Symposium on Combinatorial Search (SoCS) (2019)
18. Stockmeyer, L.: Planar 3-colorability is polynomial complete. SIGACT News **5**(3), 19–25 (1973)
19. Zareckiĭ, K.A.: Constructing a tree on the basis of a set of distances between the hanging vertices. Uspehi Mat. Nauk **20**(6), 90–92 (1965)

Online Range Assignment Problems

Paz Carmi, Matthew J. Katz$^{(\boxtimes)}$, and Idan Tomer

Ben Gurion University of the Negev, Beer Sheva, Israel
{carmip,matya}@bgu.ac.il, idantom@post.bgu.ac.il

Abstract. We study two online range assignment problems. In the first problem, introduced by de Berg et al. [4], P is a growing set of transceivers, initially consisting of a single transceiver p_1. Upon the arrival of a new transceiver p_i, one needs to (re)assign a range to at most one of the 'previous' transceivers, so that p_i is covered by at least one of them. We adopt the common rule that a transceiver's range (which is initially 0) can never decrease over time. The cost of such an online range assignment is $\sum_{i=1}^{n} \rho(p_i)$, where $\rho(p_i)$ is the final range assigned to p_i, and we wish to compare it with the cost of an optimal offline assignment which satisfies the intermediate coverage requirements. We present several results for this problem, which improve some of the previous results of de Berg et al. [4].

We also introduce a new problem, in which we have a set T of stationary transmitters and an initially-empty set S of mobile receivers. Upon the arrival of a new receiver s, it must first specify its intended route, after which one needs to (re)assign a range to at most one of the transmitters in T, so that s's route is fully covered by at least one of them. We consider the case where the routes of the receivers are unit line segments and prove a constant factor upper bound on the competitive ratio of our range assignment algorithm.

Keywords: online algorithms · competitive analysis · range assignment · broadcast

1 Introduction

Let $P = \{p_1, \ldots, p_n\}$ be a set of points in the plane, representing the set of transceivers of some wireless network. A range assignment to P assigns a transmission range r_p to each of the transceivers $p \in P$. A point p_j *receives* or is *covered* by point p_i, if $r_{p_i} \geq |p_i p_j|$, where $|p_i p_j|$ denotes the Euclidean distance between p_i and p_j. The *communication graph* induced by a range assignment to P is the directed graph over P, in which there is an edge from p_i to p_j if and only if p_j receives p_i. The *cost* of a range assignment is $\sum_{i=1}^{n} r_{p_i}^{\alpha}$ for some $\alpha \geq 2$. In the *broadcast range assignment* problem, the goal is to assign ranges to the points in P, such that (i) the induced communication graph contains a path from p_1 to each of the points in P, and (ii) the cost of the assignment is

I. Finocchi and L. Georgiadis (Eds.): CIAC 2025, LNCS 15679, pp. 68–82, 2025.
https://doi.org/10.1007/978-3-031-92932-8_5

minimum. This problem (sometimes with additional constraints) has received considerable attention; see e.g. [1–3].

de Berg et al. [4] introduced the online version of the broadcast range assignment problem. In this version, only p_1 is given in advance, and the rest of the points arrive one by one, such that p_i arrives before p_{i+1}, for $2 \leq i \leq n-1$. Upon the arrival of p_i, $i \geq 2$, we may (re)assign a range to exactly one of the previous points $p_1, \ldots, p_{i-1}$. However, when reassigning a range to a point, the new range cannot be smaller than the current one. More precisely, if p_i is not yet covered by one of the previous points, then we must (re)assign a range to one of them, so that p_i becomes covered, and otherwise, we may or may not (re)assign a range to one of the previous points. As usual, a solution to the online version is evaluated with respect to the optimal solution to the offline version, where $p_1, \ldots, p_n$ are given in advance (i.e., the *competitive ratio*). However, we still require that, for any $1 \leq i \leq n - 1$, the offline solution restricted to $P_i = \{p_1, \ldots, p_i\}$ contains a broadcast tree for P_i rooted at p_1, that is, it contains a path from p_1 to each of the points in P_i.

Let $nn(p_j)$ be the nearest neighbor of p_j among the points in P_{j-1}. de Berg et al. proposed the following algorithm (among other algorithms) for the online version. Upon the arrival of p_i, if p_i is not yet covered, then assign the range $C \cdot |nn(p_i)p_i|$ to the point $nn(p_i)$, where C is some constant. They analyzed the cases $C = 1$ and $C = 2$ (referred to as NN and 2-NN) and obtained both upper and lower bounds on the competitive ratio in these cases. More precisely, for $C = 1$ and $\alpha = 2$, they obtained an upper bound of 322 and a lower bound of approximately 7.61 on the competitive ratio, and for $C = 2$ and $\alpha = 2$ they obtained an upper bound of 36.

In a subsequent paper, which is less relevant to our work, de Berg et al. [5] studied the dynamic version of the broadcast range assignment problem, where points can be inserted into and deleted from P (except for p_1). They introduce an additional parameter, namely k, which is the number of range modifications (in both directions) allowed per an update and analyze the cost of the broadcast tree maintained by their algorithms with respect to k.

In this paper, we study the online version of the broadcast range assignment problem, as defined by de Berg et al. [4] (see above). In addition, we define and study the following online range assignment problem with mobile receivers. In this problem, we have a set T of stationary transmitters and an initially empty set S of mobile receivers. The receivers arrive one at a time. When a receiver s arrives, it is required to specify its intended route, after which we must make sure that there exists at least one transmitter $t \in T$ that covers s. We say that transmitter t *covers* receiver s if $r_t \geq |tp|$, for any point p on s's route. More precisely, upon the arrival of s, we may (re)assign a range to at most one of the transmitters in T, so that s is covered by a transmitter in T. Moreover, when resetting the range of a transmitter, we may only increase it. The cost of a solution is $\sum_{t \in T} r_t^\alpha$ and, as in the online broadcast problem, we compare it to the cost of an optimal offline solution (that also satisfies the coverage requirements that occur along the way).

1.1 Our Contribution

We restrict our attention to the most common case where $\alpha = 2$. We first study the online broadcast range assignment problem, for which we show that the optimal value of C (for the analysis method of de Berg et al.) is $1 + \frac{1}{\sqrt{2}}$. The upper bound on the competitive ratio that we obtain for this value of C is 34. Moreover, we present a corresponding lower bound of 14.74. We also obtain an improved upper bound of 85 on the competitive ratio for $C = 1$.

Next, we study the online range assignment problem with mobile receivers, in which the routes of the receivers are unit line segments. We prove an upper bound of 392 on the competitive ratio of C-NN with $C = 3 + \sqrt{6}$.

2   Static Receivers

The *2-NN* algorithm introduced by de Berg et al. [4] receives a sequence of points $P = (p_1, ..., p_n)$, arriving one at a time. Upon the arrival of point p_i, $i > 1$, whose current range is 0, if p_i is not already covered by a point p_ℓ, for some $1 \leq \ell < i$, then let p_j be the nearest neighbor of p_i (i.e., $p_j = nn(p_i)$) among the points $p_1, \ldots, p_{i-1}$. The algorithm assigns p_j the range $2 \cdot |p_i p_j|$. The *C-NN* algorithm is defined similarly, except that it assigns p_j the range $C \cdot |p_i p_j|$ instead of $2 \cdot |p_i p_j|$.

Let $\rho(p_i)$ denote the final range of p_i, i.e., after the arrival of p_n. Then the cost of this assignment of ranges $\rho = (\rho(p_1), \ldots, \rho(p_n))$ to the points of P is

$$\text{cost}(\rho) = \sum_{i=1}^{n} \rho(p_i)^2.$$

We show that the best value for C is $1 + \frac{1}{\sqrt{2}}$ (see clarification below). In particular, by setting $C = 1 + \frac{1}{\sqrt{2}}$, we obtain an upper bound smaller than 34 on the competitive ratio of the resulting range assignment, thus improving the previous upper bound of 36 obtained for $C = 2$ by de Berg et al. [4]. Our proof of the upper bound is essentially the same as the proof of de Berg et al. (with some adaptations), and it is for this proof that we show that the value $C = 1 + \frac{1}{\sqrt{2}}$ is optimum.

Consider the assignment of radii $\rho = (R'_1, \ldots, R'_n)$ to the points of P, produced by algorithm C-NN. For the purpose of the analysis, we define *charging disks* as follows: For each point p_j with $R'_j > 0$, let p_i be the point that determined R'_j, thus $R'_j = C \cdot |p_i p_j|$. We associate with p_i the *charging disk* d_i, which is the disk centered at p_i with radius $r_i = k|p_i p_j|$, where $k = \frac{C-1}{C}$. Notice that a point p_i can have at most one charging disk.

Let OPT be an optimal range assignment (obeying all intermediate requirements). We wish to bound the ratio between $\text{cost}(\rho)$ and the cost of OPT, i.e. the competitive ratio. To do so, we prove the following two statements.

- The charging disks are pairwise disjoint.
- The sum of their areas is bounded by some constant times the sum of the areas of the disks defined by OPT.

We then use the fact that $R'_j = \frac{C}{k}r_i$, where p_i is the point that determined R'_j, to bound the competitive ratio of the C-NN algorithm.

2.1   The Charging Disks Are Pairwise Disjoint

For two points $p, p' \in P$, we write that $p \prec p'$ if and only if p arrives before p', i.e., there exist indices i and j, such that $p = p_i$, $p' = p_j$, and $i < j$. Let $p \prec v$ be two points with charging disks d_p and d_v, respectively. We prove that $d_p \cap d_v = \emptyset$. Let $q \prec p$ be the point whose final range was determined by p, i.e. $R'_q = C \cdot |pq|$. The radius of d_p is therefore $k|pq|$. Let $w \prec v$ be the point whose final range was determined by v, i.e. $R'_w = C \cdot |vw|$. The radius of d_v is therefore $k|vw|$, see Fig. 1.

We know that

(1) $|qv| > R'_q = C \cdot |pq|$, since otherwise v is already within q's range at the time of its arrival.

(2) If $w \neq p$, then $|wv| \leq |pv|$, since otherwise v would not determine the final range of w.

(3) $|pv| \geq |qv| - |pq|$, since by the triangle inequality $|qv| \leq |pq| + |pv|$.

(4) $k = \frac{C-1}{C} \Rightarrow (1-k)C - 1 = 0$.

We now use the above observations to prove that $d_p \cap d_v = \emptyset$, or in other words, that $|pv| - k|pq| - k|vw| > 0$. (If $w = p$, then the radius of d_v is $k|pv|$, and we begin the sequence below from the second expression.) Indeed

$$|pv| - k|pq| - k|vw| \overset{(2)}{\geq} |pv| - k|pq| - k|pv| =$$

$$(1-k)|pv| - k|pq| \overset{(3)}{\geq} (1-k)(|qv| - |pq|) - k|pq| =$$

$$(1-k)|qv| - |pq| \overset{(1)}{>} (1-k)C \cdot |pq| - |pq| = ((1-k)C - 1)|pq| \overset{(4)}{=} 0.$$

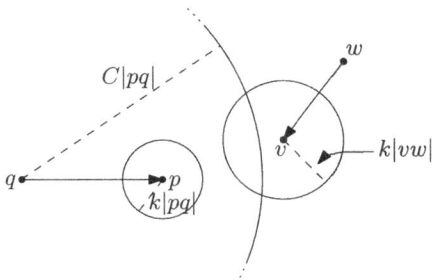

Fig. 1. $d_p \cap d_v = \emptyset$.

2.2   Areas Analysis

Let s be an arbitrary point in P whose assigned range is greater than 0, and assume w.l.o.g. that the range assigned to s by OPT is 1. Let P_s be the subset of P consisting of all points p_i, such that (i) $s \prec p_i$, (ii) p_i is covered by s (i.e., it is in the disk of radius 1 around s), and (iii) p_i is not covered by any point that precedes s. Now, consider our solution, and let P'_s be all points in P_s with an associated charging disk. For each $p_i \in P'_s$, let q_i be the point whose final range was determined by p_i, and let $Q'_s = \{q_i \mid p_i \in P'_s\}$. Since p_i lies within distance 1 from s, we have $|p_i q_i| \leq 1$. We wish to bound the total area of the charging disks associated with the points of P'_s. The radius of each such disk d_{p_i} is $k|p_i q_i| \leq k$ and it is centered at p_i (whose distance from s is at most 1), so d_{p_i} is contained in a disk D of radius $1 + k$ centered at s, see Fig. 2. We conclude that the total area of the charging disks associated with points in P'_s is at most the area of D. That is, $\sum_{p_i \in P'_s} r_i^2 \leq (1+k)^2$.

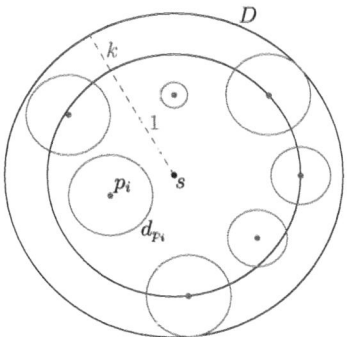

Fig. 2. The charging disks associated with the points of P'_s are contained in D.

Bounding the Competitive Ratio. From the inequality $\sum_{p_i \in P'_s} r_i^2 \leq (1+k)^2$, we can obtain an upper bound on the sum $\mathrm{cost}(Q'_s) = \sum_{q_i \in Q'_s} (R'_{q_i})^2$, as follows. Recall that $r_i = k|p_i q_i| = \frac{k}{C} R'_{q_i}$, so we have

$$\sum_{p_i \in P'_s} r_i^2 = \sum_{p_i \in P'_s} \left(\frac{k}{C}\right)^2 (R'_{q_i})^2 = \left(\frac{k}{C}\right)^2 \sum_{q_i \in Q'_s} (R'_{q_i})^2 \leq (1+k)^2, \text{ or}$$

$$\sum_{q_i \in Q'_s} (R'_{q_i})^2 \leq \left(\frac{C}{k}\right)^2 (1+k)^2 = \left(\frac{C}{\frac{C-1}{C}}\right)^2 \cdot \left(1 + \frac{C-1}{C}\right)^2 = \frac{(2C-1)^2 \cdot C^2}{(C-1)^2}.$$

It is easy to verify that the value of C that minimizes the latter expression (and is not smaller than 1) is $1 + \frac{1}{\sqrt{2}}$, and substituting C with this value we get

$$\text{cost}(Q'_s) = \sum_{q_i \in Q'_s} (R'_{q_i})^2 \leq 17 + 12\sqrt{2} \approx 33.971 < 34.$$

We charge $\text{cost}(Q'_s)$ to the squared of the range of s, i.e., 1. Finally, since each range R'_j corresponds to exactly one of the disks of OPT, we get that $\text{cost}(\rho)/\text{cost}(\hat{O}PT) < 34$.

The following theorem summarizes the main result up to now.

Theorem 1. *The competitive ratio of the C-NN algorithm, for $C = 1 + \frac{1}{\sqrt{2}}$, is slightly less than 34, and for any $C \neq 1 + \frac{1}{\sqrt{2}}$, the competitive ratio of the C-NN algorithm, assuming the charging-disks-based analysis, is worse.*

2.3   Lower Bound

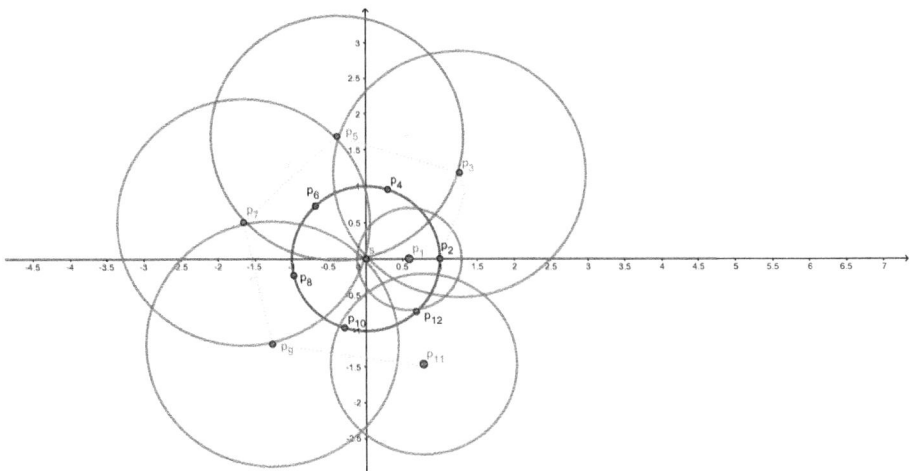

Fig. 3. A lower bound of 14.74 on the competitive ratio of the C-NN algorithm for $C = 1 + \frac{1}{\sqrt{2}}$.

Let $\varepsilon \ll 1$ and consider the following scenario; see Fig. 3.

- $s = (0,0)$
- $p_1 = (\frac{1}{C} - \varepsilon, 0)$
- $p_2 = (1,0)$
- p_4 – the top intersection point of the circle of radius 1 centered at s and the bisector of s and p_1.

- p_{2i}, for $3 \leq i \leq 6$ – the point on the circle of radius 1 centered at s, such that $\angle p_{2i-2}sp_{2i} = \frac{\pi}{3}$.
- p_{2i-1}, for $2 \leq i \leq 5$ – the point beyond p_{2i} on the ray from p_{2i+2} passing through p_{2i}, such that $|p_{2i-1}p_{2i}| = 1 - \varepsilon$.
- p_{11} – the point beyond p_{12} on the ray from p_1 passing through p_{12}, such that $|p_{11}p_{12}| = |p_1p_{12}| - \varepsilon$.
- Finally, p_{2i-1}, for $2 \leq i \leq 6$, is connected by a chain of points, in which the distance between consecutive points is at most ε, beginning at a point within distance ε of p_{2i-3} and ending at a point within distance ε of p_{2i-1}. (The chain to p_3 begins at a point within distance ε of p_2.) The points arrive between the arrival of p_{2i-2} and p_{2i-1}, and are located such that their distance to p_{2i} always exceeds the distance $|p_{2i-1}p_{2i}|$. (Hence the chain to p_3, for example, does not follow the straight line between p_2 and p_3.)

We first observe that the optimal assignment in this scenario is to assign range 1 to s, so that it covers p_1 and p_{2i}, for $1 \leq i \leq 6$. The remaining points p_{2i-1}, for $2 \leq i \leq 6$, are covered via the point chains. For example, consider the chain between p_3 and p_5. Then, p_3 is assigned range $C\varepsilon$ to cover the first point of the chain, and the first point is assigned range $C\varepsilon$ to cover the second point, etc. Finally, the last point of the chain is assigned range $C\varepsilon$ to cover p_5. Since the contribution of the ranges $C\varepsilon$ to the total cost is negligible, assuming ε is sufficiently small, we get that the cost of the optimal assignment is essentially 1.

We now compute the cost of the assignment obtained by the C-NN algorithm.

- Since the distance between s and p_1 is $\frac{1}{C} - \varepsilon$, s is assigned range $1 - C\varepsilon \approx 1$.
- p_1 is the closest to p_2, so p_1 is assigned range $C|p_1p_2| = C - 1 + C\varepsilon \approx C - 1$.
- p_3 is connected via the chain of points between p_2 and p_3, with no significant contribution to the total cost.
- Since the distance between p_{2i-1} and p_{2i}, for $2 \leq i \leq 5$ is $1 - \varepsilon$, p_{2i-1} is the closest point to p_{2i} upon p_{2i}'s arrival, so p_{2i-1} is assigned range $C(1-\varepsilon) \approx C$.
- The point p_{11} is the closest to p_{12} and is thus assigned range $C|p_{11}p_{12}| \approx C \cdot 0.74 = 1.26$. (We compute $|p_{11}p_{12}|$ by computing $|p_1p_{12}|$, where the latter computation is done by observing that $\triangle sp_4p_1$ is an isosceles triangle with $|sp_4| = |p_1p_4| = 1$ and $|sp_1| = 1/C$ and that $\angle p_4sp_{12} = \frac{2\pi}{3}$, and using the law of cosines.)

Summing up all the costs, ignoring the ranges $C\varepsilon$ whose contribution is negligible, we get

$$1 + (C - 1)^2 + 4 \cdot C^2 + 1.26^2 \approx 14.74.$$

We conclude that 14.74 is a lower bound on the competitive ratio of the C-NN algorithm. Finally, by slightly adapting this construction, we get a lower bound of 20.13 for the 2-NN algorithm. Indeed, $1 + 1 + 4 \cdot 2^2 + (2 \cdot 0.73)^2 \approx 20.13$.

3   Mobile Receivers

In this section, T is a set of stationary transmitters and P is an initially empty set of mobile receivers. The receivers arrive one at a time. When a receiver p

requests to join the network, it must specify its intended route, after which we need to make sure that there exists at least one transmitter $t \in T$, whose range is long enough to cover p. The initial range of the transmitters in T is 0. We say that transmitter t *covers* receiver p if $R_t \geq |tq|$, for any point q on p's route. More precisely, upon the arrival of p, we may (re)assign a range to at most one of the transmitters in T, so that p is covered by a transmitter in T. Moreover, when resetting the range of a transmitter, we may only increase it. The cost of a solution is $\sum_{t \in T} R_t^2$ and, as in the online broadcast problem, we compare it to the cost of an optimal offline solution (that also satisfies the coverage requirements that occur along the way).

Recall that we are assuming that the routes of the receivers are directed line segments of length 1.

Notation. Let p be a receiver and let p_s and p_e be its starting and ending points, i.e., p moves along the segment from p_s to p_e. Then, the *distance* between a transmitter t and p is $d(p,t) = \max\{|p_s t|, |p_e t|\}$, and we say that t covers p if $R_t \geq d(p,t)$. We also write $t = nn(p)$ if $d(p,t) \leq d(p,t')$, for any $t' \in T$.

The Algorithm. When a new receiver p arrives, if it is not already covered by one of the transmitters, then assign t the range $C \cdot d(p,t)$, where $t = nn(p)$ and $C \geq 1$ is a constant to be determined later.

Analysis. We use charging disks as in the previous section, however, the different setting raises some issues which must be addressed. Let t be a transmitter with final range $R_t > 0$, and let p be the receiver that determined R_t, i.e., $R_t = C \cdot d(p,t)$. Recall that $d(p,t) = \max\{|p_s t|, |p_e t|\}$. We place the charging disk of p at the endpoint of p's route that determines $d(p,t)$. That is, if for example $d(p,t) = |p_e t|$, then the charging disk of p is the disk d_{p_e} centered at p_e with radius $k|p_e t|$, where now $k = \frac{C-3}{C+1}$.

The Charging Disks are Pairwise Disjoint. Let t and t' be two transmitters with positive final ranges R_t and $R_{t'}$, respectively. Let p and p' be the receivers that determine R_t and $R_{t'}$, respectively, and assume that p arrives before p'. Then, upon the arrival of p, the range of t is (re)set to cover p, i.e., t is assigned range $C \cdot d(p,t) = Cx$ (where $d(p,t) = x$), and upon the arrival of p', the range of t' is (re)set to cover p', i.e., t' is assigned range $C \cdot d(p',t') = Cy$ (where $d(p',t') = y$). Since p' determines $R_{t'}$, we know that (i) $d(p',t) > C \cdot d(p,t)$, and (ii) $d(p',t') \leq d(p',t)$; see Fig. 4.

Assume, w.l.o.g., that $x = d(p,t) = |p_e t|$ and $y = d(p',t') = |p'_s t'|$, and let d_{p_e} and $d_{p'_s}$ be the charging disks of p and p', respectively. We prove that $d_{p_e} \cap d_{p'_s} = \emptyset$, by proving that $|p_e p'_s| - kx - ky \geq 0$. Since at least one of the endpoints p'_s, p'_t is out of t's range, i.e., outside the disk of radius Cx centered at t, there exists $z > 0$, such that $d(p',t) = Cx + z$. Then, $y = |p'_s t'| \leq Cx + z$. Moreover, by the triangle inequality, $|tp'_s| \leq |tp_e| + |p_e p'_s| = x + |p_e p'_s|$, so $|p_e p'_s| \geq |tp'_s| - x \geq Cx + z - 1 - x$. Finally, we notice that $x \geq \frac{1}{2}$, as $|p_s p_e| = 1$ and

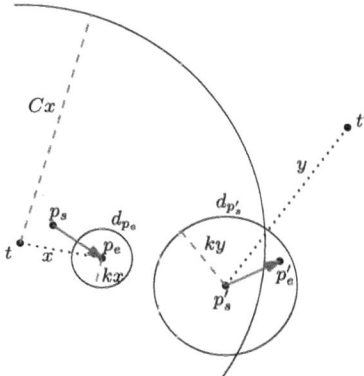

Fig. 4. The charging disks d_{p_e} and $d_{p'_s}$ are disjoint.

x is the longer of the two distances $|p_s t|$ and $|p_e t|$. Using all these observations we get

$$
\begin{aligned}
|p_e p'_s| - kx - ky &\geq Cx + z - 1 - x - kx - k(Cx + z) \\
&= (1 - k)(Cx + z) - (1 + k)x - 1 \\
&\geq (1 - k)Cx - (1 + k)x - 1 \\
&= \frac{4Cx}{C + 1} - \frac{(2C - 2)x}{C + 1} - 1 = 2x - 1 \geq 0.
\end{aligned}
$$

Areas Analysis. Let t be an arbitrary transmitter in T whose assigned range is greater than 0, and assume w.l.o.g. that the range assigned to t by OPT is 1. As in the static version, we obtain the inequality

$$
\sum (k \cdot d(p_i, t_i))^2 \leq (1 + k)^2
$$

where the sum is over all the receivers p_i that are served by t in OPT, but determine the ranges of transmitters t_i in our solution.

Bounding the Competitive Ratio. Again, as in the static version, we have

$$
\sum (R_{t_i})^2 \leq \left(\frac{C}{k}\right)^2 (1 + k)^2 = \left(\frac{C(C + 1)}{C - 3}\right)^2 \cdot \left(\frac{2C - 2}{C + 1}\right)^2 = \frac{4(C - 1)^2 C^2}{(C - 3)^2}.
$$

It is easy to verify that the value of C that minimizes the latter expression (and is not smaller than 1) is $3 + \sqrt{6}$, and substituting C with this value we get that $\sum (R_{t_i})^2 = 196 + 80\sqrt{6} < 392$. We conclude that the competitive ratio of our solution is less than 392.

The following theorem summarizes the main result of this section.

Theorem 2. *The competitive ratio of our algorithm for the mobile version, for $C = 3 + \sqrt{6}$, is slightly less than 392.*

4   Improved Upper Bound for NN

In this section, we obtain an improved upper bound for the NN algorithm. Suppose that the NN algorithm increases the range of $nn(p_i)$ upon the insertion of p_i, so that its new and final range is $|nn(p_i)p_i|$. Let m_i be the midpoint between $nn(p_i)$ and p_i. As in [4], we associate with p_i a charging disk d_i centered at m_i. However, the radius of our charging disk is $r_i = \gamma \cdot |nn(p_i)p_i|$, where $\gamma = \frac{\sqrt{3}-1}{4}$, which is much larger than the radius used in [4]. This difference enables us to obtain a much better upper bound, however, the proof becomes more involved.

Let p_i be the point determining the final range of q_i, and let p_j be the point determining the final range of q_j. Moreover, assume that $p_i \prec p_j$. Recall that in order to bound the competitive ratio from above, we need to show that the associated charging disks d_i and d_j are disjoint. Assume w.l.o.g. (by scaling and rotation) that

$$q_i = (0,0),\ \ p_i = (1,0),\ \ \text{and}\ p_j = (x_{p_j}, y_{p_j}).$$

Then, the center of d_i is $m_i = (0.5, 0)$ and its radius is $r_i = \gamma \cdot |q_i p_i| = \gamma$. We denote the center of d_j, which is the midpoint between p_j and q_j, by m_j and its radius by $r_j = \gamma \cdot |q_j p_j|$.

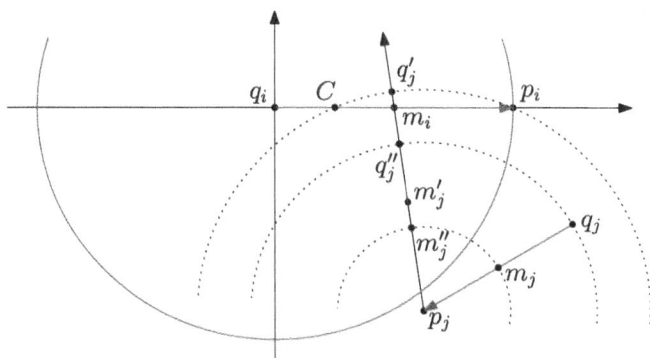

Fig. 5. Improved upper bound setup.

We define the function

$$f(p_j, q_j) = |m_i m_j| - (r_i + r_j). \tag{1}$$

We will assume w.l.o.g. that $y_{p_j} \le 0$. We will also assume that $x_{p_j} \ge 0.5$ and justify this assumption towards the end of the section. Let q'_j be the point at which the ray $\overrightarrow{p_j m_i}$ intersects the circle centered at p_j of radius $|p_j p_i|$, and let m'_j be the midpoint between p_j and q'_j; see Fig. 5 for an illustration. Let q''_j be the point on the ray $\overrightarrow{p_j m_i}$, such that $|p_j q_j| = |p_j q''_j|$, and let m_j'' be the midpoint between p_j and q_j''. Thus, $|p_j m_j| = |p_j m_j''|$, see Fig. 5.

To show that the charging disks are disjoint we need to show that $f(p_j, q_j) \geq 0$. We do this by first proving that $f(p_j, q'_j) \leq f(p_j, q_j)$, and then proving that $f(p_j, q'_j) \geq 0$.

We start by showing that m'_j is between p_j and m_i. This is clearly true if q'_j is between p_j and m_i, so assume that it is not, i.e., m_i is between p_j and q'_j. We use the following simple geometric observation.

Observation 4.1. *Let $\triangle ABC$ be an isosceles triangle, such that $|AB| = |AC|$ and $\angle BAC < 120°$. Let r be a ray emanating from A that intersects the edge BC, and let m be a point on r, such that $2|Am| = |AB|$. Then m is inside the triangle $\triangle ABC$.*

We apply the observation as follows: p_j and p_i play the roles of A and B, respectively, C is the intersection point of the segment $q_i p_i$ with the circle centered at p_j of radius $|p_j p_i|$, and r is the ray passing through q'_j. Since the angle $\angle q_i p_j p_i < 90°$ (p_j is outside the disk with center at q_i and radius $|q_i p_i|$), we have that $m = m'_j$ is inside triangle $\triangle C p_i p_j$. Thus,

Conclusion 4.1. *(i) m'_j is between p_j and m_i, and (ii) m''_j is between p_j and m'_j (and therefore also between p_j and m_i), since $|p_j q''_j| = |p_j q_j| \leq |p_j p_i| = |p_j q'_j|$.*

Claim 4.1. *If $f(p_j, q'_j) \geq 0$, then $f(p_j, q_j) \geq 0$.*

Proof. By the triangle inequality, we have that

$$|m_i p_j| \leq |m_i m_j| + |m_j p_j| \quad \overset{|m_j p_j| = |m''_j p_j|}{\implies}$$

$$|m_i p_j| \leq |m_i m_j| + |m''_j p_j| \quad \overset{Conclusion\ 4.1(ii)}{\implies}$$

$$|m_i m''_j| + |m''_j p_j| \leq |m_i m_j| + |m''_j p_j| \quad \implies$$

$$|m_i m''_j| \leq |m_i m_j| \quad \overset{Conclusion\ 4.1(ii)}{\implies}$$

$$|m_i m'_j| \leq |m_i m_j|$$

And since $|p_j q'_j| \geq |p_j q_j|$, we have that $r'_j \geq r_j$ and thus $f(p_j, q'_j) \leq f(p_j, q_j)$.

It remains to show that $f(p_j, q'_j) \geq 0$. We distinguish between the cases $x_{p_j} \geq 1$ (Claim 4.2) and $0.5 \leq x_{p_j} < 1$ (Claim 4.3 and Claim 4.4).

Claim 4.2. *If $x_{p_j} \geq 1$, then $f(p_j, q'_j) \geq 0$.*

Proof. For $x_{p_j} = 1$, we have that $y_{p_j} < 0$ (otherwise, $p_i = p_j$, and p_j is already covered by q_i), thus $\triangle m_i p_i p_j$ is a right triangle, and $|m_i p_j|^2 = |p_i m_i|^2 + |p_i p_j|^2$, see Fig. 6. By the choice of q'_j, we have that $|p_i p_j| = |q'_j p_j|$. Recall that from Conclusion 4.1(i) m'_j is between m_i and p_j (actually, even q'_j is between m_i and p_j), thus $|m_i p_j| = |m_i m'_j| + |m'_j p_j|$.

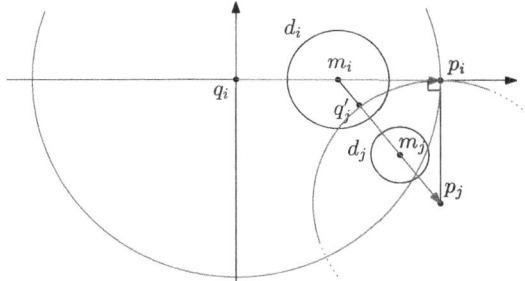

Fig. 6. Proof of Claim 4.2. For $x_{p_j} = 1$, $\triangle p_i m_i p_j$ is a right-angled triangle. For $x_{p_j} > 1$, $|m_i p_j|^2 > |p_i m_i|^2 + |p_i p_j|^2$.

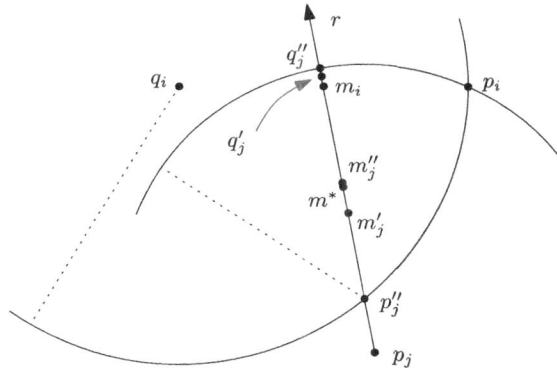

Fig. 7. Illustration of the Proof of Claim 4.3.

We wish to show that $f(p_j, q'_j) = |m_i m'_j| - (\gamma + \gamma |q'_j p_j|) \geq 0$. We notice that

$$|m_i m'_j| - (\gamma + \gamma |q'_j p_j|) \geq 0 \qquad \Leftrightarrow$$
$$\gamma + \gamma |q'_j p_j| \leq |m_i m'_j| \qquad \Leftrightarrow$$
$$\gamma + (\gamma + 0.5)|q'_j p_j| \leq |m_i m'_j| + 0.5|q'_j p_j| = |m_i m'_j| + |m'_j p_j| = |m_i p_j| \qquad \Leftrightarrow$$
$$(\gamma + (\gamma + 0.5)|q'_j p_j|)^2 \leq |m_i p_j|^2 = |p_i p_j|^2 + |p_i m_i|^2 = |q'_j p_j|^2 + 0.25$$

and a simple calculation shows that the last inequality is always true. For $x_{p_j} > 1$, we have that $|m_i p_j|^2 > |p_i m_i|^2 + |p_i p_j|^2$ and the proof holds in the same manner.

From now on we assume that $0.5 \leq x_{p_j} \leq 1$.

Claim 4.3. *Let p''_j be the intersection of the disk of radius 1 centered at q_i and the segment $\overline{p_j m_i}$. Let q''_j be the point on the ray emanating from p''_j passing through m_i at distance $|p''_j p_i|$ from p''_j. If $f(p''_j, q''_j) \geq 0$ then $f(p_j, q'_j) \geq 0$.*

Proof. Let m'_j be the midpoint between p_j and q'_j, m''_j be the midpoint between p''_j and q''_j and $m^\star$ be the midpoint between p''_j and q'_j; see Fig. 7. Let r be the ray emanating from p_j and passing through m_i. For any two points a, b on r, we write $a \ll_r b$, if $|p_j a| < |p_j b|$; see Fig. 8. We have

$$|p_j q''_j| = |p_j p''_j| + |p''_j q''_j| = |p_j p''_j| + |p''_j p_i| \geq |p_j p_i| = |p_j q'_j|, \tag{2}$$

where the inequality follows from the triangle inequality. Thus, $q'_j \ll_r q''_j$. Observe that

$$m'_j \ll_r m^\star \ll_r m''_j \ll_r m_i, \tag{3}$$

where the last relation follows from Observation 4.1 (with $A = p''_j$, $B = p_i$, and C is the point at distance $|p''_j p_i|$ from p''_j to the left of the ray r such that $\angle p_i p''_j C = 90°$). Observe also that

$$|m'_j m^\star| = 0.5|p_j p''_j| \implies |m^\star m_i| + 0.5|p_j p''_j| = |m'_j m_i|.$$
$$|m^\star m''_j| = 0.5|q'_j q''_j| \implies |m''_j m_i| + 0.5|q'_j q''_j| = |m^\star m_i|.$$

And therefore,

$$|m''_j m_i| = |m'_j m_i| - (|m'_j m^\star| + |m^\star m''_j|) = |m'_j m_i| - 0.5(|p_j p''_j| + |q'_j q''_j|). \tag{4}$$

$$
\begin{aligned}
f(p_j, q'_j) &- f(p''_j, q''_j) = \\
&|m_i m'_j| - \gamma - \gamma|p_j q'_j| - (|m_i m''_j| - \gamma - \gamma|p''_j q''_j|) \overset{Eq.~(4)}{=} \\
&|m_i m'_j| - \gamma|p_j q'_j| - (|m_i m'_j| - 0.5(|p_j p''_j| + |q'_j q''_j|) - \gamma|p''_j q''_j|) = \\
&- \gamma|p_j q'_j| - (-0.5(|p_j p''_j| + |q'_j q''_j|) - \gamma(|p_j q'_j| + |q'_j q''_j| - |p_j p''_j|)) = \\
&- \gamma|p_j q'_j| + 0.5(|p_j p''_j| + |q'_j q''_j|) + \gamma(|p_j q'_j| + |q'_j q''_j| - |p_j p''_j|) = \\
&0.5(|p_j p''_j| + |q'_j q''_j|) + \gamma(|q'_j q''_j| - |p_j p''_j|) = \\
&(0.5 - \gamma)|p_j p''_j| + (0.5 + \gamma)|q'_j q''_j| \geq 0, \quad \text{for } \gamma < 0.5.
\end{aligned}
$$

We showed that $f(p_j, q'_j) - f(p''_j, q''_j) \geq 0$. By Claim 4.3, it remains to show that $f(p''_j, q''_j) \geq 0$. To simplify the notation, we replace $x_{p''_j}$ by x and $f(p''_j, q''_j)$ by $f(x, q''_j)$.

Claim 4.4. $f(x, q''_j) \geq 0$.

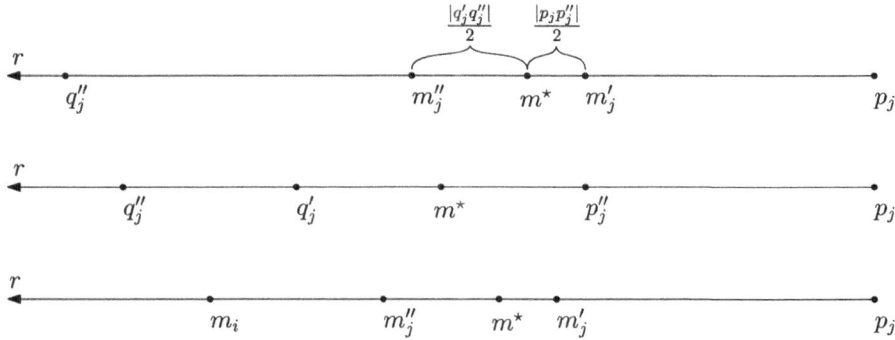

Fig. 8. Proof of Claim 4.3. The relative order of the points along the ray r.

Proof. Recall that we are assuming that p_j'''s y-coordinate is negative, so $p_j'' = (x, -\sqrt{1 - x^2})$.

$$|q_j'' p_j''| = |p_i p_j''| = \sqrt{(1 - x)^2 + 1 - x^2} = \sqrt{2}\sqrt{1 - x}$$

$$|m_i p_j''| = \sqrt{(0.5 - x)^2 + 1 - x^2} = \sqrt{1.25 - x}$$

$$r_{p_j''} = \gamma |q_j'' p_j''| = \gamma |p_i p_j''| = \gamma \cdot \sqrt{2}\sqrt{1 - x}$$

$$|m_j'' p_j''| = \frac{|p_i p_j''|}{2} = \frac{\sqrt{1 - x}}{\sqrt{2}}$$

$$|m_i m_j''| = |m_i p_j''| - |m_j'' p_j''| = \sqrt{1.25 - x} - \frac{\sqrt{1 - x}}{\sqrt{2}}$$

$$f(x, q_j'') = |m_i m_j''| - (r_{p_i} + r_{p_j''}) = \sqrt{1.25 - x} - \frac{\sqrt{1 - x}}{\sqrt{2}} - \gamma(|q_j'' p_j''| + |q_i p_i|)$$

$$f(x, q_j'') = \sqrt{1.25 - x} - \frac{\sqrt{1 - x}}{\sqrt{2}} - \frac{\sqrt{3} - 1}{4}\left(\sqrt{2}\sqrt{1 - x} + 1\right)$$

It is easy to verify that $f(x, q_j'')$ is monotonically increasing, for $x \geq 0$, and that $f(0.5, q_j'') = 0$. Thus, $f(x, q_j'') \geq 0$, for $0.5 \leq x \leq 1$, as claimed.

We have shown that the charging disks are pairwise disjoint assuming $x_{p_j} \geq 0.5$. If $x_{p_j} < 0.5$, let $\hat{p}_j$ and $\hat{q}_j$ be the reflections of p_j and q_j about the line $x = 0.5$. By the proof above the charging disks of p_i and $\hat{p}_j$ are disjoint, and therefore so are the charging disks of p_i and p_j.

Now, let OPT be an optimal range assignment and let $\mathcal{D}_{\text{OPT}}$ be the set of disks used by OPT. To obtain the upper bound for the NN algorithm, we notice that $\sum_{i=1}^{n-1} r_i^2 \leq \sum_{d \in \mathcal{D}_{\text{OPT}}} ((1.5 + \gamma) \cdot r(d))^2 = (1.5 + \gamma)^2 \cdot \text{cost}(OPT)$ (which is essentially as we did for the C-NN algorithm). So the cost of the our solution, which is $\frac{1}{\gamma^2}\sum_{i=1}^{n-1} r_i^2$, is at most $\frac{1}{\gamma^2} \cdot (1.5 + \gamma)^2 \cdot \text{cost}(OPT)$. Finally, recalling that $\gamma = \frac{\sqrt{3} - 1}{4}$, we get that the cost of our solution is less than $85 \cdot \text{cost}(OPT)$.

The following theorem summarizes the main result of this section.

Theorem 3. *The competitive ratio of the NN algorithm is slightly less than 85.*

References

1. Caragiannis, I., Kaklamanis, C., Kanellopoulos, P.: A logarithmic approximation algorithm for the minimum energy consumption broadcast subgraph problem. Inf. Process. Lett. **86**(3), 149–154 (2003)
2. Clementi, A., Penna, P., Silvestri, R.: On the power assignment problem in radio networks. Mob. Netw. Appl. **9**(2), 125–140 (2004)
3. Das, G.K., Das, S., Nandy, S.C.: Range assignment problem in wireless network. In: Algorithms, Architectures and Information Systems Security. Statistical Science and Interdiciplinary Research, vol. 3, pp. 195–224. World Scientific (2008)
4. de Berg, M., Markovic, A., Umboh, S.W.: The online broadcast range-assignment problem. Algorithmica **85**(12), 3928–3956 (2023)
5. de Berg, M., Sadhukhan, A., Spieksma, F.: Stable approximation algorithms for the dynamic broadcast range-assignment problem. SIAM J. Discret. Math. **38**(1), 790–827 (2024)

General Position Subset Selection in Line Arrangements

Adrian Dumitrescu$^{(\boxtimes)}$ [ORCID]

Algoresearch L.L.C., Milwaukee, WI 53217, USA
ad.dumitrescu@algoresearch.org

Abstract. Given a set of points in the plane, the GENERAL POSITION SUBSET SELECTION problem is that of finding a maximum-size subset of points in general position, i.e., with no three points collinear. The problem is known to be NP-complete and APX-hard, and the best approximation ratio known is $\Omega\left(\mathsf{OPT}^{-1/2}\right) = \Omega(n^{-1/2})$. Here we obtain better approximations in two special cases:

(I) A constant factor approximation for the case where the input set consists of lattice points and is *dense*, which means that the ratio between the maximum and the minimum distances in P is of the order of $\Theta(\sqrt{n})$.

(II) An $\Omega\left((\log n)^{-1/2}\right)$-approximation for the case where the input set is the set of vertices of a *generic n-line arrangement*, i.e., one with $\Omega(n^2)$ vertices.

The scenario in (I) is a special case of that in (II). Our approximations rely on probabilistic methods and results from incidence geometry.

1 Introduction

A set of points in the plane is said to be in *general position* if no 3 points are collinear. The problem of selecting a large subset of points from the $k \times k$ grid of integer points with no three points collinear (i.e., in general position) goes back more than 100 years, see for example [8] and [6, Ch. 10]. In particular, it is not known if one can always select $2k$ points from the $k \times k$ grid so that no 3 points are collinear [8].

A key quantity in the process of selecting a large subset from an input set of points is the number of *collinear triples* in the set. Payne and Wood [18] obtained the following upper bound on this number. The proof relies on the classical points and lines incidence bound due to Szemerédi and Trotter [21]; see also [17, Chap. 10] and [20] for a modern approach of this topic. (The special case $\ell = \sqrt{n}$ can also be found in [13] or in [22, p. 313].)

Lemma 1 [18]. *Let P be a set of n points in the plane with at most ℓ collinear. Then the number of collinear triples in P is $T = O(n^2 \log \ell + \ell^2 n)$.*

By applying the above lemma together with a lower bound on the independence number of a hypergraph due to Spencer [19], Payne and Wood [18] obtained the following result:

© The Author(s), under exclusive license to Springer Nature Switzerland AG 2025
I. Finocchi and L. Georgiadis (Eds.): CIAC 2025, LNCS 15679, pp. 83–90, 2025.
https://doi.org/10.1007/978-3-031-92932-8_6

Theorem 1 [18]. *Let P be a set of n points in the plane with at most ℓ collinear. Then P contains a subset of $\Omega\left(n/\sqrt{n \log \ell + \ell^2}\right)$ points in general position. In particular, if $\ell = O(\sqrt{n})$, then P contains a subset of $\Omega\left((n/\log \ell)^{1/2}\right)$ points in general position.*

A cursory examination of the above lower bound shows that if the input set has a large subset of points in general position (of, say, linear, or nearly linear size), then the above guarantee is nearly $\sqrt{n} \ll n$, i.e., roughly a factor of $\sqrt{n}$ smaller than the truth.

Given a set P of points in the plane, the GENERAL POSITION SUBSET SELECTION problem (GPSS, for short), is that of finding a maximum-size subset of points in general position. The current best approximation for the problem is a greedy algorithm due to Cao [7, Chap. 3], achieving a ratio of $\Omega\left(\mathsf{OPT}^{-1/2}\right) = \Omega(n^{-1/2})$; here OPT is the value of an optimum solution to GPSS. See also [9, Chap. 9] and [11] for further aspects of this problem.

Preliminaries. For a set P of n points in the plane, consider the ratio (sometimes also called *spread*)

$$D(P) = \frac{\max\{|ab| : a, b \in P, a \neq b\}}{\min\{\ ab| : a, b \in P, a \neq b\}},$$

where $|ab|$ is the Euclidean distance between points a and b. We may assume without loss of generality that $\min\{|ab| : a, b \in P, a \neq b\} = 1$. In this case $D(P) = \max\{|ab| : a, b \in P, a \neq b\}$ is the diameter of P. A standard disk packing argument shows that if $|P| = n$, then $D(P) \geq \alpha_0\, n^{1/2}$, where

$$\alpha_0 := 2^{1/2} 3^{1/4} \pi^{-1/2} \approx 1.05, \tag{1}$$

provided that n is large enough; see [23, Prop. 4.10]. On the other hand, a $\sqrt{n} \times \sqrt{n}$ section of the integer lattice shows that this bound is tight up to a constant factor. An n-element point set P satisfying the condition $D(P) \leq \alpha\, n^{1/2}$, for some constant $\alpha \geq \alpha_0$, is said to be α-*dense*; see for instance [12; 24] and Fig. 1 for an illustration.

According to a classical result of Beck [4], if P is a set of n points in the plane where at most ℓ points of P are collinear, then P determines at least $\Omega(n(n - \ell))$ distinct lines. In particular, if $\ell \leq cn$, where $c < 1$ is a constant, then P determines $\Omega(n^2)$ distinct lines. By duality, see e.g., [14], for a set L of n lines in the plane, where at most ℓ lines of L are concurrent, the number of vertices of the corresponding arrangement is $\Omega(n(n - \ell))$. We say that an arrangement of n lines is *generic* if it has $\Omega(n^2)$ vertices.

As in [25], here we follow the convention that the approximation ratio of an algorithm for a maximization problem is less than 1. Throughout this paper all logarithms are in base 2.

Our Results. First, we obtain a constant factor approximation for the GENERAL POSITION SUBSET SELECTION problem in dense lattice point sets; its approximation ratio depends on the spread ratio of the input.

Theorem 2. *Given an α-dense set of n lattice points, a constant factor approximation for the* GENERAL POSITION SUBSET SELECTION *problem can be computed in polynomial time. If $c = c(\alpha)$ denotes its approximation ratio, then $c(\alpha) = \Omega\left(\alpha^{-2}\right)$.*

Second, we obtain an $\Omega((\log n)^{-1/2})$-approximation for the special case where the input set is the set of vertices of an n-line arrangement under the relatively mild genericity assumption.

Theorem 3. *Given n lines in the plane that make a generic line arrangement, an $\Omega\left((\log n)^{-1/2}\right)$-approximation for the* GENERAL POSITION SUBSET SELECTION *problem for the set of vertices of the corresponding line arrangement, can be computed by a randomized algorithm in expected polynomial time.*

Essentially the same proof (of this theorem in Sect. 3) yields the following:

Corollary 1. *Given n lines in the plane, let V denote the vertex set of the corresponding line arrangement. If $V' \subseteq V$ is a subset of vertices with $|V'| = \Omega(n^2)$, an $\Omega\left((\log n)^{-1/2}\right)$-approximation for the* GENERAL POSITION SUBSET SELECTION *problem in V' can be computed by a randomized algorithm in expected polynomial time.*

It is worth noting that the scenario in Theorem 2 is a special case of the scenario in Theorem 3. Indeed any finite set P of lattice points in the integer grid is a subset of the vertices of an arrangement $\mathcal{A}^+$ of axis-parallel lines, induced by H and V, the sets of horizontal and vertical lines incident to points in P, respectively. Moreover, if P is a dense set, then $\mathcal{A}^+$ is generic, since

$$|H|, |V| = \Omega(|P|/(\alpha\sqrt{n})) = \Omega(\sqrt{n}),$$

thus $\mathcal{A}^+$ has $|H| \cdot |V| = \Omega(n)$ vertices. Observe that *not* all vertices in the arrangement $\mathcal{A}^+$ are necessarily in P.

2   Subset Selection in Dense Lattice Point Sets

In this section we prove Theorem 2. First, we introduce the setup.

For a positive integer m, the $m \times m$ grid is the set of points in the plane $G_m = \{(x, y) : x, y \in \{0, 1, \ldots, m - 1\}\}$. We have $|G_m| = m^2$. Following [26], let $k(m)$ denote the minimum number of colors in a coloring of the points in G_m such that no three collinear points are monochromatic. Since no three points in a single row or column can receive the same color, we have $k(m) \geq m/2$. It was shown by Wood [26] that for any $\varepsilon > 0$, $k(m) \leq (2 + \varepsilon)m$ for every $m \geq M(\varepsilon)$. Among others, this is based on the fact that for any $\varepsilon > 0$, there exists a prime

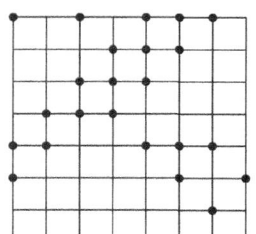 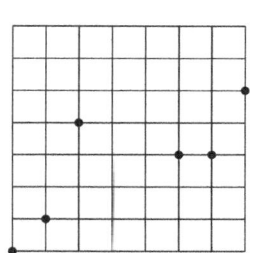 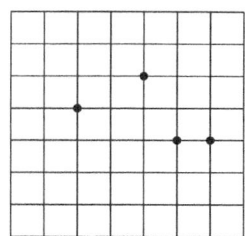

Fig. 1. Left: a 2-dense set of $n = 25$ points in the 8×8 grid ($m = 8$). Center: V_0; $|V_0| = m = 8$ and $p = 11$; points in V_i may lie outside the grid, e.g., $(3,9)$ lies 2 units above it. Right: The approximation algorithm returns $P \cap V_0$ (or $P \cap V_3$); here $|P \cap V_0| = |P \cap V_3| = 4$.

between m and $(1 + \varepsilon)m$, for every $m \geq M(\varepsilon)$. For instance, a nonasymptotic result, more convenient for small m, is that there exists a prime between m and $6m/5$, for every $m \geq 25$ [16]. (And by the well-known Bertrand-Chebyshev theorem, there exists a prime between m and $2m$, for every m.)

We briefly recall Wood's argument [26] relying on a construction of Erdős [10]. Let $p \geq m$ be a prime; in practice we may pick the smallest prime at least m or a prime at least m that is close to m. For any integer i, let

$$V_i = \{(x, (x^2 \bmod p) + i) \colon x = 0, 1, \ldots, m - 1\}.$$

Note that $|V_i| = m$, and that V_i may be *not* contained in G_m. Refer to Fig. 1. A Vandermonde determinant calculation shows that V_i is in general position for every $i \in \mathbb{Z}$. Moreover, $V_i \cap V_j = \emptyset$ whenever $i \neq j$. Each point $(x, y) \in G_m$ is in V_i where $i = y - (x^2 \bmod p)$. Since

$$1 - p = 0 - (p - 1) \leq y - (x^2 \bmod p) \leq m - 1,$$

the union of the $m+p-1$ sets V_i over $i = 1-p, \ldots, m-1$ consists of $m+p-1$ color classes where each class is in general position. That is, G_m can be covered by $m+p-1$ m-element point sets in general position. As mentioned in the previous paragraph, $m + p - 1 \leq (2 + \varepsilon)m$ for every $m \geq M(\varepsilon)$, or $m + p - 1 \leq 11m/5$ for every $m \geq 25$, as needed.

Proof of Theorem 2. Since P is α-dense, we may assume that $P \subseteq G_m$, where $m = \lceil \alpha\sqrt{n} \rceil$, after a suitable translation by an integer vector. The algorithm chooses a prime $p \geq m$ close to m. By the Prime Number Theorem, this involves *primality testing* [15, Ch. 14.6] of no more than $O(m^{0.525}) = O(n^{0.263})$ odd integers in the interval $[m, m + 2m^{0.525}]$ [2]. As argued above, G_m is covered by $m+p-1 \leq (2+\varepsilon)m$ m-element point sets V_i in general position. By monotonicity (i.e., every subset of a set in general position is likewise in general position), the sets $P \cap V_i$ are in general position, and the algorithm outputs the largest one, with at least $\frac{n}{(2+\varepsilon)m}$ elements.

Since each of the m rows of G_m contains at most two elements of an optimal solution, we have $\mathsf{OPT} \leq 2m$. Consequently, the ratio $\mathsf{ALG/OPT}$ is bounded from below as

$$\frac{\mathsf{ALG}}{\mathsf{OPT}} \geq \frac{n}{2 \cdot (2 + \varepsilon)m^2} = \Omega\left(\frac{1}{\alpha^2}\right).$$

Clearly the resulting algorithm runs in polynomial time; we give a few details below.

Choosing the prime p and generating the sets V_i takes $O(n)$ time. Selecting one V_i with a maximal size $|P \cap V_i|$ can be done withing the same time, and so the overall running time is $O(n)$. □

Remark. Using a deterministic algorithm for primality testing makes the approximation algorithm deterministic as well.

3   Subset Selection in Generic Line Arrangements

In this section we prove Theorem 3. Recall that an arrangement of n lines is generic if it has $\Omega(n^2)$ vertices. See Fig. 2.

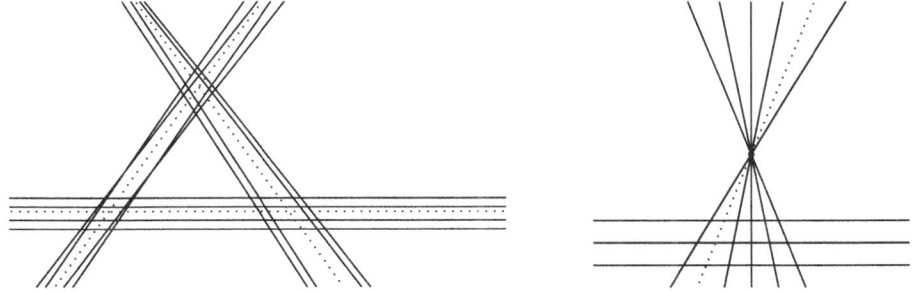

Fig. 2. Left: a generic line arrangement consisting of three bundles of $n/3$ nearly parallel lines each. Right: a non-generic line arrangement consisting of 3 parallel lines and $n - 3$ concurrent lines.

Next, we introduce the setup. Given a set P of points in the plane, let $\ell(P)$ denote the largest number of points in P that is incident to a line determined by P. We have $2 \leq \ell(P) \leq |P|$, with $\ell(P) = 2$ if and only if P is in general position and $\ell(P) = |P|$ if and only if P is collinear.

Lemma 2. *Let V be the set of vertices of an n line arrangement $\mathcal{A} = \mathcal{A}(L)$, where $L = \{\ell_1, \ldots, \ell_n\}$. Then $\ell(V) \leq n - 1$.*

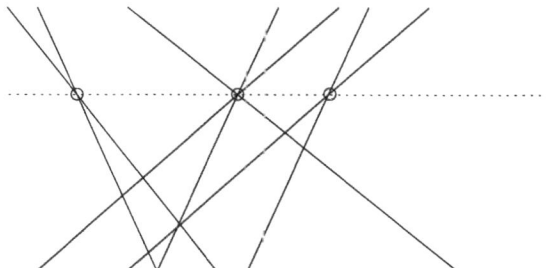

Fig. 3. Collinear vertices in a line arrangement –incident to an induced line.

Proof. We distinguish two cases. Let $V' \subseteq V$ be a set of collinear vertices, say on a line h. If $h \in L$, then obviously $|V'| \leq n - 1$. If $h \notin L$, scan h, say, from left to right, and observe that each vertex in V' uses up at least two "new" lines from L, thus $|V'| \leq n/2$, see Fig. 3. □

Let V be the set of vertices of a generic n line arrangement $\mathcal{A} = \mathcal{A}(L)$. Let $N = |V| \leq \binom{n}{2}$; we also have $N \geq cn^2$ by the assumption. By Lemma 2, $\ell := \ell(V) \leq n - 1$, thus by Lemma 1 the number of collinear triples is bounded as:

$$T = O(N^2 \log \ell + \ell^2 N) = O(N^2 \log n + n^2 N). \tag{2}$$

Proof of Theorem 3. The subset selection algorithm consists of two steps: (i) random sampling from V, and (ii) application of the deletion method. See, e.g., [1, Chap. 3]; and a similar application in [27]. In the first step, a random subset $X \subset V$ is chosen by selecting points independently with probability $p = k/N$, for a suitable k to be determined. Note that $E[\|X\|] = N \cdot k/N = k$.

To select a subset in general position, one needs to avoid one type of obstacles, collinear triples. One obstacle can be eliminated by deleting one point from the subset in the second step. In particular, it suffices to choose k so that

$$E[Tp^3] \leq k/2,$$

since this implies that the expected number of remaining points is at least $k - k/2 = k/2$. Let c_1 denote the constant hidden in (2). Using the previous upper bound on T, and recalling that $N \geq cn^2$, it suffices to choose k so that

$$2c_1 k^2 (N \log n + n^2) \leq N^2. \tag{3}$$

Setting $k = c' \frac{n}{\sqrt{\log n}}$ for a sufficiently small $c' > 0$ satisfies the inequality, proving the lower bound on the size of the output. In addition, this setting is valid for ensuring that $p = k/N < 1$.

In summary, the expected number of points in general position found by the algorithm is $\Omega\left(\frac{n}{\sqrt{\log n}}\right)$. To analyze the approximation ratio it suffices to notice that $\mathsf{OPT} \le 2n$. Indeed, for each line $\ell_i \in L$, $i = 1, \ldots, n$, at most two vertices in V can appear in any solution, thus $\mathsf{OPT} \le 2n$. It follows that the approximation ratio is

$$\frac{\mathsf{ALG}}{\mathsf{OPT}} = \frac{1}{2n} \cdot \Omega\left(\frac{n}{\sqrt{\log n}}\right) = \Omega\left(\frac{1}{\sqrt{\log n}}\right),$$

as claimed.

Since repeated random samplings are independent, the probability of error for the resulting randomized algorithm of not finding the required number of points can be made arbitrarily small by repetition –in a standard fashion. Clearly the resulting algorithm runs in expected polynomial time; we give a few details below.

Computing the set of vertices V and performing the random sampling takes $O(n^2)$ time. The expected sample size is $O(k) = O(n)$. Using the point-line duality, the algorithm computes the line arrangement dual to the point sample in $O(n^2)$ time [5, Ch. 8]. It repeatedly removes lines passing through vertices incident to at least three lines, each in $O(n)$ time, and updates the line arrangement. This step is repeated until the arrangement is simple and thus, the points left in the sample are in general position. The expected running time is $O(n^2)$. □

Proof of Corollary 1. We perform random sampling from V', let $N = |V'|$, and proceed as in the proof of Theorem 3. Inequality (3) is again satisfied due to the assumption $|V'| = \Omega(n^2)$. □

4 Conclusion

We conclude with some problems for further investigation:

1. Can the approximation factor in Theorem 3 be improved? Perhaps using ideas from [3, Sec. 2]?
2. Can a constant factor approximation be obtained for the problem of finding a largest subset in general position among the vertices of an n-line arrangement?
3. Can better approximation factors for the general position subset selection problem be obtained in other interesting scenarios?

References

1. Alon, N., Spencer, J.: The Probabilistic Method, 4th edn. Wiley, New York (2016)
2. Baker, R., Harman, G., Pintz, J.: The difference between consecutive primes, II. Proc. Lond. Math. Soc. **83**(3), 532–562 (2001)
3. Balogh, J., Clemen, F.C., Dumitrescu, A., Liu, D.: Subset selection problems in planar point sets. Preprint arXiv:2412.14287 (2024)
4. Beck, J.: On the lattice property of the plane and some problems of Dirac, Motzkin and Erdős in combinatorial geometry. Combinatorica **3**, 281–297 (1983)
5. de Berg, M., Cheong, O., van Kreveld, M., Overmars, M.: Computational Geometry, Algorithms and Applications, 3rd edn. Springer, New York (2008)

6. Braß, P., Moser, W., Pach, J.: Research Problems in Discrete Geometry. Springer, New York (2005)
7. Cao, C.: Study on two optimization problems: line cover and maximum genus embedding, Master's thesis, Texas A&M University (2012)
8. Dudeney, H.: A puzzle with pawns. In: Amusements in Mathematics, Nelson, Edinburgh, p. 94 (1917)
9. Eppstein, D.: Forbidden Configurations in Discrete Geometry. Cambridge University Press, Cambridge (2018)
10. Erdős, P., Roth, K.F.: On a problem of Heilbronn. J. Lond. Math. Soc. 1(3), 198–204 (1951)
11. Froese, V., Kanj, I., Nichterlein, A., Niedermeier, R.: Finding points in general position. Int. J. Comput. Geom. Appl. 27(4), 277–296 (2017)
12. Kovács, I., Tóth, G.: Dense point sets with many halving lines. Disc. Comput. Geom. 64(3), 965–984 (2020)
13. Lefmann, H.: Distributions of points in the unit square and large k-gons. Eur. J. Comb. 29(4), 946–965 (2008)
14. Matoušek, J.: Lectures on Discrete Geometry. Springer, New York (2002)
15. Motwani, R., Raghavan, P.: Randomized Algorithms. Cambridge University Press, Cambridge (1995)
16. Nagura, J.: On the interval containing at least one prime number. Proc. Jpn. Acad. 28(4), 177–181 (1952)
17. Pach, J., Agarwal, P.: Combinatorial Geometry. Wiley-Interscience, New York (1995)
18. Payne, M., Wood, D.: On the general position subset selection problem. SIAM J. Disc. Math. 27(4), 1727–1733 (2013)
19. Spencer, J.: Turán's theorem for k-graphs. Disc. Math. 2(2), 183–186 (1972)
20. Székely, L.: Crossing numbers and hard Erdős problems in discrete geometry. Comb. Probab. Comput. 6, 353–358 (1997)
21. Szemerédi, E., Trotter, W.T.: Extremal problems in discrete geometry. Combinatorica 3, 381–392 (1983)
22. Tao, T., Vu, V.: Additive Combinatorics, Cambridge Studies in Advanced Mathematics, p. 105. Cambridge University Press, Cambridge, United Kingdom (2006)
23. Valtr, P.: Convex independent sets and 7-holes in restricted planar point sets. Disc. Comput. Geom. 7(2), 135–152 (1992). https://doi.org/10.1007/BF02187831
24. Valtr, P.: Planar point sets with bounded ratios of distances. PhD Thesis, Freie Universität Berlin (1994)
25. Williamson, D., Shmoys, D.: The Design of Approximation Algorithms. Cambridge University Press, Cambridge (2011)
26. Wood, D.: A note on colouring the plane grid. Geombinatorics 13(4), 193–196 (2004)
27. Zhang, Z.: A note on arrays of dots with distinct slopes. Combinatorica 13(1), 127–128 (1993)

Branching Programs with Extended Memory: New Insights

Suryajith Chillara[1(✉)] and Nithish Raja[2]

[1] Center for Security, Theory and Algorithmic Research (CSTAR),
International Institute of Information Technology - Hyderabad (IIIT-H),
Hyderabad, India
suryajith.chillara@iiit.ac.in
[2] Department of Mathematics and Computer Science,
Technische Universiteit Eindhoven, Eindhoven, The Netherlands
n.r.raja@tue.nl

Abstract. One of the central themes of *Algebraic Complexity Theory* is to understand the relative computation power of algebraic complexity classes VP and VNP. VP is a class of polynomial families which can be *computed efficiently* by algebraic circuits. VNP is a class of polynomials which can be *efficiently expressed* through polynomial families in VP, but we do not know if they can be computed efficiently. It is a long standing open problem in this area to show that VP is a strict subset of VNP. At this juncture, it is fair to believe that newer characterization of these complexity classes could help us understand these models better (and possibly help us in resolving this question in restricted settings like non-commutativity at the very least).

It is well known that Algebraic Branching Programs (ABPs) characterize the class VBP (or equivalently VP_skew). Recently, Mengel [MFCS, 2013] showed that ABPs can be extended with memory and the computational models thus obtained can be used to characterize VP and VNP. In particular, Mengel showed that ABPs with a single stack characterize VP, and branching programs with random-access memory characterize VNP.

In this work we show that algebraic branching programs with just 2 stacks efficiently simulates the polynomial families in VNP, and two stack branching programs are VNP-complete.

Further, we observe that k-stack branching programs (for all polynomially bounded k) are no more powerful than 2-stack branching programs (upto a polynomial blowup in size).

This strengthens the work of Mengel in the following way – VBP vs. VP vs. VNP are characterized by algebraic branching programs with 0, 1, and 2 stacks, respectively, or no-memory, a stack as memory, and a queue as memory respectively, which hold even in the non-commutative setting.

We also refine Mengel's characterization of VP through stack-height characterization.

N. Raja—This work was done while this author was a masters student at IIIT Hyderabad, India.

I. Finocchi and L. Georgiadis (Eds.): CIAC 2025, LNCS 15679, pp. 91–104, 2025.
https://doi.org/10.1007/978-3-031-92932-8_7

1　Introduction

Algebraic Models of Computation. An algebraic circuit is the most commonly encountered computational model in algebraic complexity. It is a DAG where the leaf nodes are labelled with variables or constants and the intermediate nodes are labelled with operations such as $+, \times$. Another important computational model is the algebraic branching program (ABP). ABPs are source-to-sink DAGs with the edges labelled with variables or constants. The polynomial computed by an ABP is given by sum of weights of all source-to-sink path where weight of a path is defined as product of weight of edges in it (refer to [7,8] for a detailed discussion on these computational models).

VP, VBP, VNP, *Reductions and Completeness.* The class VP contains all polynomial families that have polynomially-bounded degree, and are computed by polynomial-sized circuits. Similarly, the class VBP contains those polynomial families in VP which can also be computed by polynomial-sized ABPs. Another important class in algebraic complexity is VNP. It contains all polynomial families that can be expressed as a boolean sum of a VP polynomial family. That is, $(p_n)_{n \in \mathbb{N}} \in$ VNP if $p_n(\bar{x}) = \sum_{\bar{e} \in \{0,1\}^{m-n}} q_m(\bar{x}, \bar{e})$ where $m \in \mathrm{poly}(n)$ and $(g_m)_{m \in \mathbb{N}} \in$ VP.

Definition 1 (p-projection). *Let $p_n(x_1, \ldots, x_n)$ and $q_m(y_1, \ldots, y_m)$ be two polynomials. Then p_n is said to be a p-projection of q_m if there exists a simple map $\sigma : \{x_i\}_{i \in [n]} \rightarrow \{y_i\}_{i \in [m]} \cup \mathbb{F}$ and $m \in \mathrm{poly}(n)$ s.t. $p_n(x_1, \ldots, x_n) = q_m(\sigma(x_1), \ldots, \sigma(x_n))$.*

Valiant [9] showed that polynomial families (Perm_n) and (HC_n) are complete for VNP under p-projections. In other words, a polynomial family is in VNP if and only if it can be expressed as a p-projection of either Permanent or Hamiltonian cycle families. Valiant further showed that a polynomial family is in VNP if its coefficients can be computed in #P/poly (Valiant's criterion). We refer the readers to [5,7,9] for finer details.

1.1　Branching Programs with Auxiliary Memory

Let S be a stack, and Σ be its alphabet. For each letter $s \in \Sigma$, we can define stack operations $\mathsf{push}(s)$ and $\mathsf{pop}(s)$. We use nop to denote no operation on the stack. A sequence of operations on the stack S is a sequence $\mathsf{op}_1, \mathsf{op}_2, \ldots, \mathsf{op}_r$ where (for all $i \in [r]$) either op_i is of the form $\mathsf{push}(s)$ or $\mathsf{pop}(s)$, for some letter $s \in \Sigma$, or of the form nop. Realizable sequences of stack operations are defined inductively.

1. Empty sequence is realizable.
2. A realizable sequence starts with an empty stack and ends with an empty stack.
3. If P is a sequence of realizable stack operations, then for all $s \in \Sigma$, $\mathsf{push}(s) \cdot P \cdot \mathsf{pop}(s)$ is also a realizable sequence. Further, $\mathsf{nop} \cdot P$ and $P \cdot \mathsf{nop}$ are also realizable sequences.

4. If P and Q are two realizable sequences of stack operations, then $P \cdot Q$ is also a realizable sequence of stack operations.

Definition 2 (Definition 3.1, [6]). *Let S be a stack over the alphabet Σ. A stack branching program (SBP) G is an algebraic branching program with an additional edge labeling σ where $\sigma : E \mapsto \{\mathrm{op}(s) \mid \mathrm{op} \in \{\mathrm{push}, \mathrm{pop}\}, s \in \Sigma\} \cup \{\mathrm{nop}\}$. A source-to-sink path $P = e_1, e_2, \ldots, e_r$ has the sequence of stack operations $\sigma(P) = \sigma(e_1), \sigma(e_2), \ldots, \sigma(e_r)$. We call P a stack realizable path if $\sigma(P)$ is a stack realizable sequence. Polynomial $f(x_1, \ldots, x_n)$ thus computed is given by the sum of weights of all stack realizable source-to-sink paths.*

$$f(x_1, \ldots, x_n) = \sum_{P : P \in s \rightsquigarrow t \ \mathit{stack\ realizable\ paths}} \mathrm{wt}(P).$$

Using this model of computation, Mengel proved the following theorem.

Theorem 1 (Theorem 3.3, [6]). *A polynomial family $\{f_n\}_{n \in \mathbb{N}}$ is in VP if and only if a family of polynomial-sized stack branching programs computes it.*

We make this theorem of [6] a bit more precise by adapting the proof of depth reduction of [11] and show the following.

Theorem 2. *A polynomial family $\{f_n\}_{n \in \mathbb{N}}$ is in VP if and only if a family of polynomial-sized stack branching programs of stack-height at most $O(\log^2 n)$ computes it.*

This additional characterization (along with Lemma 2) gives us an alternative proof to the fact that polynomial families in VP can be simulated by quasi-polynomial sized algebraic branching programs.

k-stack Branching Programs. In this work we also introduce k-stack branching programs, a generalization of stack branching programs obtained by allowing the number of stacks to be a parameter.

Definition 3. *Let $S_1, \ldots, S_k$ be k many stacks over the alphabet Σ. A k-stack branching program G is an algebraic branching program with an additional edge labeling σ where*

$$\sigma : E \mapsto \{(\mathrm{op}_1, \ldots, \mathrm{op}_k) \mid \forall \, j \in [k], \mathrm{op}_j \in \{\mathrm{push}_j(s_j), \mathrm{pop}_j(s'_j), \mathrm{nop}\} \ \mathit{for} \ s_j, s'_j \in \Sigma\}$$

and, for each $j \in [k]$, push_j and pop_j are push and pop operations on stack S_j, respectively. For each $i \in [k]$, let π_i be the projection of a k-tuple to its ith element. A source-to-sink path $P = e_1 e_2 \ldots e_r$ has the sequence of stack operations $\sigma(P) = \sigma(e_1), \sigma(e_2), \ldots, \sigma(e_r)$. For each $i \in [k]$, let $\sigma_i(P) = \pi_i(\sigma(e_1)), \pi_i(\sigma(e_2)), \ldots, \pi_i(\sigma(e_r))$.

P is a k-stack realizable path if for each $i \in [k]$, $\sigma_i(P)$ is a stack realizable sequence. Polynomial $f(x_1, \ldots, x_n)$ thus computed is given by the sum of weights of all k-stack-realizable $s \rightsquigarrow t$ paths.

$$f(x_1, \ldots, x_n) = \sum_{\pi : \pi \in s \rightsquigarrow t \ \mathit{k-stack-realizable paths}} \mathrm{wt}(\pi).$$

Using this definition, we prove that k-stack branching programs are VNP-complete for $2 \le k \le \mathrm{poly}(n)$.

Theorem 3. *Let $n, k = k(n)$ be natural numbers such that n is large, and $2 \le k \le \mathrm{poly}(n)$. A polynomial family $\{f_n\}_{n \in \mathbb{N}} \in$ VNP if and only it can be computed by a family of polynomial-sized k-stack branching programs.*

A generic approach to prove Theorem 3 is the following.

1. Show that Hamiltonian cycle polynomial family (or Permanent polynomial family resp.) can be computed by polynomial-sized 2-stack branching programs. This can be done via gadget constructions (see Appendix B (Appendix A resp.) in the full version [4]). That is, for all $n \in \mathbb{N}$, there is a polynomial-sized 2-stack branching program that computes HC_n (Perm_n resp.), and
2. Show that for any natural number k that is polynomially-bounded, families of polynomials computed by k-stack branching programs are in VNP (see Lemma 1).

However for the sake of this article, we would like to replace the approach in Item 1 with the following (more) insightful theorem.

Theorem 4. *Given a k-stack branching program ($k \ge 1$) of size s computing the polynomial $q_n(x_1, \ldots, x_n)$, there exists a $(k+1)$-stack branching program of size $\mathrm{poly}(s, n)$ computing*

$$ p_n(x_1, \ldots, x_n) = \sum_{(e_1, \ldots, e_m) \in \{0,1\}^m} q_{n+m}(x_1, \ldots, x_n, e_1, \ldots, e_m). \quad (m \in \mathrm{poly}(n)) $$

We also provide efficient constructions of 2-stack branching program computing the Perm_n and HC_n polynomials (see Appendices A and B in the full version of this paper [4]).

By putting Lemma 1, the fact that HC_n can be efficiently computed by 2-stack branching programs, and Theorem 4 together, we also get alternate proofs for the following well-known statements.

1. Families of polynomials that are obtained through *Exponential sum* of families of polynomials from either VBP or VP, are in VNP, and
2. VNP is closed under exponential summation (cf. [2, Theorem 2.19]).

2   Improved Characterization of VP

It was shown by Mengel [6] that a polynomial p_n computed by a multiplicatively disjoint circuit (see [5] for a detailed discussion on multiplicatively disjoint circuits) of size s can also be computed by a stack branching program of size $\mathrm{poly}(s)$ and any polynomial p_n computed by a stack branching program of size s' can also be computed by a circuit of size $\mathrm{poly}(s')$. These two statements combined gives us a characterization of VP i.e., a polynomial family $(p_n)_{n \in \mathbb{N}}$ is in VP if and only if there exists a poly(n)-sized stack branching program computing it. In this section, we improve upon this characterization by showing that any degree d polynomial computed by a circuit of size s can also be computed by a stack branching program of size poly(n, s, d) and stack height $O(\log(s)\log(d))$. Setting $s, d \in$ poly(n), we get that any polynomial family $(p_n)_{n \in \mathbb{N}}$ is in VP if and only if there exists a poly(n)-sized stack branching program computing it and the stack height of such a stack branching program is bound by $O(\log^2(n))$. This can be seen as a stack height reduction result for stack branching program analogous to the depth reduction result for circuits shown in [11].

Before we describe the technical details, we first borrow some notation from [11]. For a gate w in a given circuit C, we use $f(w)$ to denote the polynomial computed at w, and $d(w)$ to denote degree of polynomial computed at w. Further, we have the following.

Definition 4. *Let C be a circuit and v, w are gates in the circuit. $f(v)$, $f(w)$ denote the polynomial computed at gates v, w respectively and $d(v)$, $d(w)$ denote the degree of the polynomial computed at gates v, w respectively. Then, the function $f(v; w)$ is defined as shown below and $d(f(v; w)) = d(w) - d(v)$.*

$$f(v; w) = \begin{cases} 1 & \text{if } v = w, \\ 0 & \text{if } v \neq w \text{ and } f(w) \in \mathbb{F} \cup \{x_i\}_{i=1}^n \text{ i.e., } w \text{ is a leaf node,} \\ f(v; w') + f(v; w'') & \text{if } w = w' + w'', \\ f(w'')f(v; w') & \text{if } w = w' \times w'' \text{ s.t. } d(w'') \leq d(w'). \end{cases}$$

Definition 5 (Frontier gates). *Let C be a circuit computing an arbitrary polynomial. For any value of a, the set V_a is defined as follows.*

$$V_a = \{t \mid t \text{ is a gate in } C, d(t) > a, t = t' \times t'', d(t'') \leq d(t') \leq a\}. \tag{1}$$

Theorem 5. *Let P_n be a degree d polynomial computed by a circuit C of size $|C|$. Then, there exists a* poly$(|C|)$*-sized stack branching program with stack height $O(\log|C| \log d)$ computing P_n.*

Proof. We will now adapt and build on the proof of [6, proposition 3.6] and thus we recommend the readers to look at it. Further, the idea is to show the following.

1. Every gate w in C has a corresponding pair of vertices $\{s_w, t_w\}$ in a relaxed stack branching program (see [6] for a detailed discussion on relaxed stack branching program) s.t. $\sum$wt(realizable s_w to t_w walks of length m_w) $= f(w)$, $m_w = O(d(w)^2)$ and stack height at most $O(\log(d(w)))$, and
2. Every pair of gates v, w s.t. $f(v; w)$ is non-zero, there exists a pair of vertices $\{s_{vw}, t_{vw}\}$ in the relaxed stack branching program s.t. $\sum$wt(length m_{vw} realizable s_{vw} to t_{vw} walks) $= f(v; w)$, $m_{vw} = O(d(f(v; w)^2))$ and stack height $O(\log(d(f(v; w))))$.

The above statements are proven by induction on the degree of the polynomial computed at each gate. At any stage i, we consider all gates w and pairs of
gates v, w s.t. $d(w), d(f(v; w)) \in (2^{i-1}, 2^i]$.

Base Case. Consider the gate w individually and then the pair of gates v, w
s.t. $d(w)$ and $d(f(v; w))$ is 1. Clearly, $f(w)$ and $f(v; w)$ compute variables or
constants or linear forms. Edges with weights as variables and constants can be
added to the relaxed stack branching program. Any linear form $\sum_{j=1}^{n} c_j x_j$ can
be computed as shown in Fig. 1. Note that the stack alphabet used here will be
$\{\langle w, i \rangle \mid \forall w \in C, i \in [n]\} \cup \{\langle w, v, i \rangle \mid \forall v, w \in C, i \in [n]\} \cup \{u_k, v_k, u_{k'}, v_{k'} \mid$
$u_k, v_k, u_{k'}, v_{k'} \in C\}$ and the stack symbols used in Fig. 1 will change to $\langle w, v, i \rangle$
if the linear form is computed by $f(v; w)$.

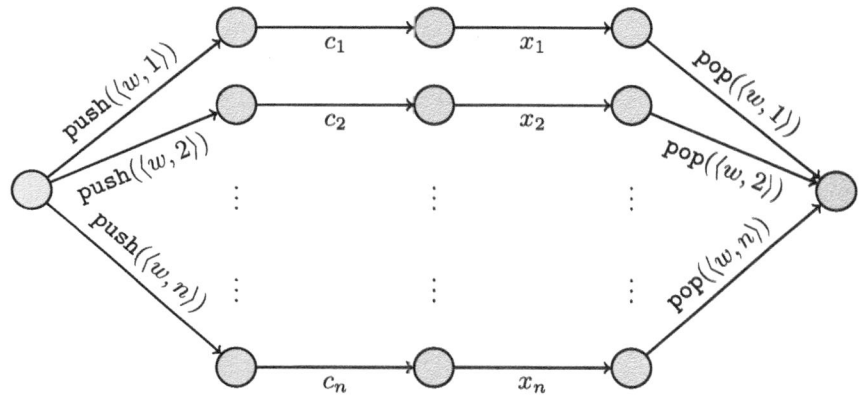

Fig. 1. Gadget computing linear form

Clearly, the stack height is ≤ 1 and path length $\leq 1 + 3^{(0+1)} \leq 4$.

Inductive hypothesis

1. Consider all gates w s.t. $d(w) \in (2^{i-1}, 2^i]$. Assume that there exists pairs of
   vertices $\{s_w, t_w\}$ in the polynomial-sized relaxed stack branching program s.t.
   $\sum \text{wt}(\text{length } m_w \text{realizable } s_w \text{to } t_u \text{ walks }) = f(w)$ satisfying $m_w \leq d(w) +$
   $3^{(i+1)}$ and stack height is $O(i)$.
2. Consider all pairs of gates v, w s.t. $f(v; w)$ is non-zero and
   $d(f(v; w)) \in (2^{i-1}, 2^i]$. Assume that there exists pairs of vertices
   $\{s_{vw}, t_{vw}\}$ in the polynomial-sized relaxed stack branching program s.t.
   $\sum \text{wt}(\text{length } m_{vw} \text{Prealizable } s_{vw} \text{to } t_{vw} \text{walks})$ is equal to $f(v; w)$ satisfying
   $m_{vw} \leq d(f(v; w)) + 3^{(i+1)}$ and stack height is $O(i)$.

Increment step

1. Consider all gates w s.t. $d(w) \in (2^i, 2^{i+1}]$. Due to [11], we write $f(w) =$
   $\sum_{t \in V_a} f(t')f(t'')f(t; w)$ s.t. $d(t'), d(t''), d(f(t; w)) \in (2^{i-1}, 2^i]$ and $a = 2^i$. Due

to the inductive assumption, we get that a relaxed stack branching program exists with pairs of vertices $\{s_{t'}, t_{t'}\}$, $\{s_{t''}, t_{t''}\}$ and $\{s_{tw}, t_{tw}\}$ computing each $f(t')$, $f(t'')$ and $f(t; w)$ respectively.

Next, recall from [6, Proposition 3.6] that we can construct a relaxed stack branching program with source and sink vertices s_w, t_w computing $f(w)$. For each product term we add 3 vertices, for addition we add 2 vertices and at most m_w vertices to make the walk lengths equal. Trivially, $V_a \leq |C|$ and number of gates w is also bounded by $|C|$. Therefore, the total number of vertices added can be at most $(3|C| + m_w + 2)|C|$. Since, we push one stack alphabet for addition and one for multiplication, the overall stack height increases by 2.

We know that $d(t) = d(t') + d(t'')$, $d(f(t; w)) = d(w) - d(t)$ and due to the inductive assumption, we get $m_{t'} \leq d(t') + 3^{(i+1)}$, $m_{t''} \leq d(t'') + 3^{(i+1)}$ and $m_{tw} \leq d(f(t; w)) + 3^{(i+1)}$.

$$m_w \leq \max_t \{m_{t'} + m_{t''} + m_{tw}\} \tag{2}$$

$$\leq \max_t \{d(t') + d(t'') + d(f(t; w)) + 3^{(i+2)}\} \tag{3}$$

$$\leq \max_t \{d(t) + d(w) - d(t) + 3^{(i+2)}\} \tag{4}$$

$$\leq \max_t \{d(w) + 3^{(i+2)}\} \tag{5}$$

$$\leq d(w) + 3^{(i+2)}. \tag{6}$$

2. Consider all pairs of gates v, w s.t. $d(f(v; w)) \in (2^i, 2^{i+1}]$. Due to [11], we write $f(v; w) = \sum_{t \in V_a} f(t'')f(v; t')f(t; w)$ s.t. $d(t'')$, $d(f(v; t'))$, $d(f(t; w)) \leq (2^{i-1}, 2^i]$ and $a = 2^i + d(v)$. Due to the inductive assumption, we get that a relaxed stack branching program exists with pairs of vertices $\{s_{t''}, t_{t''}\}$, $\{s_{vt'}, t_{vt'}\}$ and $\{s_{tw}, t_{tw}\}$ computing $f(t'')$, $f(v; t')$ and $f(t; w)$ respectively.

Once again, by following the steps in [6, Proposition 3.6], we construct a relaxed stack branching program with source and sink vertices s_{vw}, t_{vw} computing $f(v; w)$. Since, we look at pairs of gates, the total number of vertices added can be at most $(3|C| + m_{vw} + 2)|C|^2$. Similar to the previous case, we push one stack alphabet for addition and one for multiplication. This causes the overall stack height to increase by 2.

We know that $d(f(v; t')) = d(t') - d(v)$, $d(f(t; w)) = d(w) - d(t)$ and due to the inductive assumption, we get $m_{t''} \leq d(t'') + 3^{(i+1)}$, $m_{vt'} \leq d(f(v; t')) + 3^{(i+1)}$ and $m_{tw} \leq d(f(t; w)) + 3^{(i+1)}$.

$$m_{vw} \leq \max_t \{m_{t''} + m_{vt'} + m_{tw}\} \tag{7}$$

$$\leq \max_t \{d(t'') + d(t') - d(v) + d(w) - d(t) + 3^{(i+2)}\} \tag{8}$$

$$\leq \max_t \{d(w) - d(v) + 3^{(i+2)}\} \tag{9}$$

$$\leq d(f(v; w)) + 3^{(i+2)}. \tag{10}$$

The maximum walk length becomes $d + 3^{\lceil \log(d) \rceil + 1} \in O(d^2)$. Now, we can convert the relaxed stack branching program to a normal stack branching program as shown in [6, lemma 3.5] and the size of the stack branching program remain polynomial.

The overall stack height is atmost $2 \log(d)$. However, every alphabet in the stack is from the set $\{\langle w, i \rangle \mid \forall w \in C, i \in [n]\} \cup \{\langle w, v, i \rangle \mid \forall v, w \in C, i \in [n]\} \cup \{u_k, v_k, u_{k'}, v_{k'} \mid u_k, v_k, u_{k'}, v_{k'} \in C\}$. Converting the stack alphabets to binary, the overall stack height becomes $O((\log |C|)(\log d))$.   □

3   Computational Power of k-Stack Branching Programs

In this section, we first provide a proof for Theorem 4 and through this, we explore the computational power of k-stack branching programs.

Proof of Theorem 4. In the figures below (and henceforth in this document), for the sake brevity, edge weights, and stack operations on edges are only mentioned when they are different from 1, and when they are different from nop respectively. The default edge weight is 1, and the default operation is nop.

Let G_Q be the k-stack branching program that computes the polynomial $Q_{n+m}(x_1, \ldots, x_n, y_1, \ldots, y_m)$, and $|G_Q| = \text{poly}(n)$.

Let the graph gadgets $G_{\text{push}(0,1)}$, $G_{\text{pop}(0,1)}$, G_{init} and G_{reset} be constructed as shown in Figs. 2a, 2b, 3 and 4 respectively.

Let the helper gadgets $G_{k+1 \to 1}$, $G_{1 \to k+1}$, $G_{(i)}$ and $G'_{(i)}$ be as described in Figs. 5a, 5b, 6 and 7. Note that these gadgets make use of the $(k+1)$th stack.

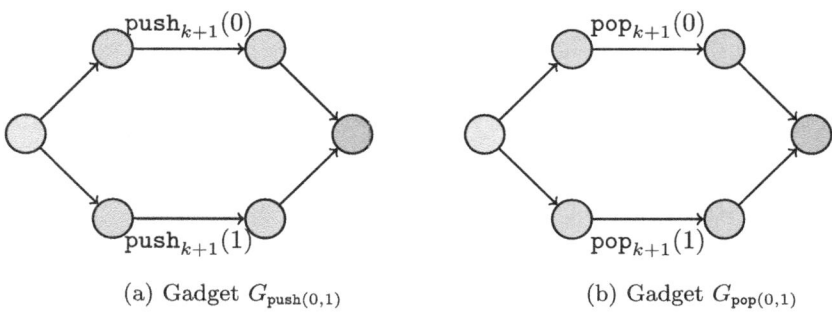

(a) Gadget $G_{\text{push}(0,1)}$ (b) Gadget $G_{\text{pop}(0,1)}$

Fig. 2. Gadget $G_{\text{push}(0,1)}$ and $G_{\text{pop}(0,1)}$

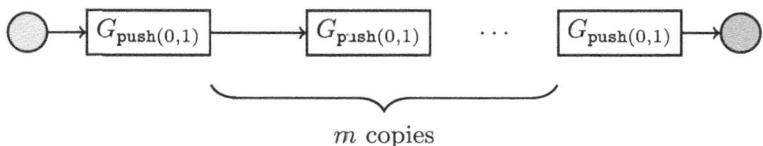

m copies

Fig. 3. Gadget G_{init}

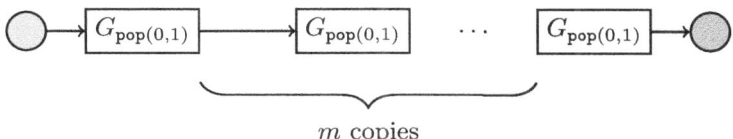

Fig. 4. Gadget G_{reset}

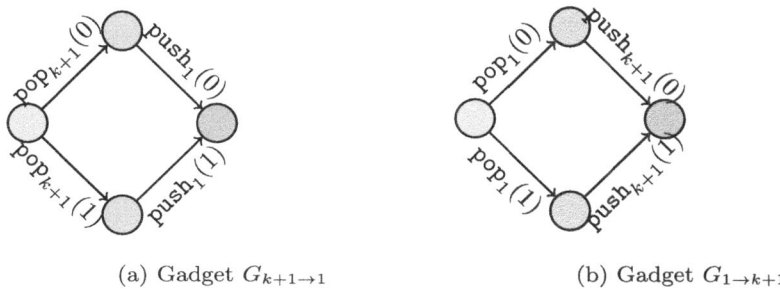

(a) Gadget $G_{k+1\to 1}$ (b) Gadget $G_{1\to k+1}$

Fig. 5. Gadgets $G_{k+1\to 1}$ and $G_{1\to k+1}$

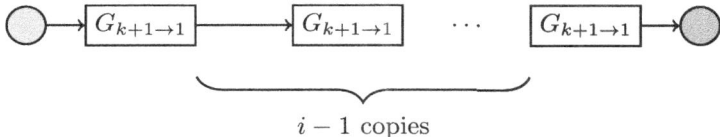

Fig. 6. Gadget $G_{(i)}$

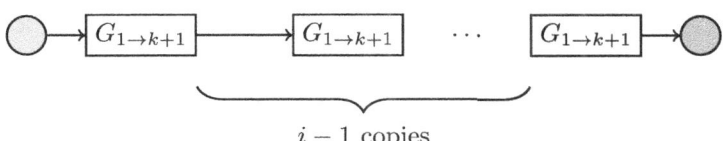

Fig. 7. Gadget $G'_{(i)}$

Any edge in G_Q with edge label y_i ($\forall\ i \in [m]$) is now replaced by a corresponding graph gadget H_i (shown in Fig. 8).

We now construct a $(k+1)$-stack branching program G_P by putting together G_{init}, modified G_Q and G_{reset} as shown in Fig. 9.

Note that each source to sink path in the gadget G_{init} fills m many $\{0, 1\}$ values into the $(k+1)^{\text{th}}$ stack. We now associate i^{th} entry of this stack with the assignment of the variable y_i. The top of the stack corresponds to the assignment for y_1 and the bottom most element, to y_m. Thus, a path from source to sink in G_{init} fixes the variables $y_1, \ldots, y_m$.

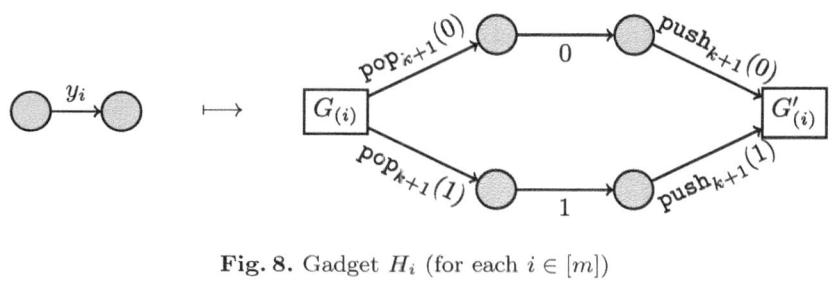

Fig. 8. Gadget H_i (for each $i \in [m]$)

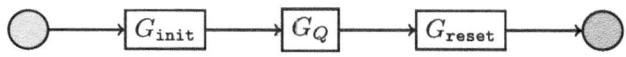

Fig. 9. $(k+1)$-stack branching program G_P

To access the assignment for a variable y_i, we need to access the ith element from the top, in the $(k+1)$th stack. Towards that we use the helper gadgets $G_{(i)}$ and $G'_{(i)}$. $G_{(i)}$ moves the $i-1$ elements at the top of the $(k+1)^{\text{th}}$ stack to the first stack. Next, if a 0 is popped, then we add an edge with weight 0. If a 1 is popped, an edge with weight 1 is added. Finally, $G'_{(i)}$ moves back all the elements moved by $G_{(i)}$ to $(k+1)^{\text{th}}$ stack.

It is easy to see that in any source to sink path of G_P, its subsection corresponding to G_{init} determines the path taken in each instance of graph gadgets H_i ($i \in [m]$).

Putting it all together, we get that G_P computes the polynomial $P(x_1, \ldots, x_n)$.

$$P(x_1, \ldots, x_n) = \sum_{(e_1, \ldots, e_m) \in \{0,1\}^m} Q(x_1, \ldots, x_n, e_1, \ldots, e_m)$$

$$= \sum_{\mathbf{e} = (e_1, \ldots, e_m) \in \{0,1\}^m} \left(\sum_{\tau : \pi \in s \leadsto t \ k-stack-realizablepathsinG_Q} \text{wt}(\pi|_{\mathbf{e}}) \right)$$

$$= \sum_{\pi : \pi \in s \leadsto t \ k-stack-realizablepathsinG_Q} \left(\sum_{\mathbf{e} = (e_1, \ldots, e_m) \in \{0,1\}^m} \text{wt}(\pi|_{\mathbf{e}}) \right)$$

$$= \sum_{\pi' : \pi' \in s \leadsto t \ (k+1)-stack-realizablepathsinG_P} \text{wt}(\pi').$$

In the above math block, the last equality follows from the fact that each k-stack realizable path is associated with each of the 2^m many paths of G_{init} which in turn correspond to the assignment $\mathbf{y} \leftarrow \mathbf{e}$, and $\text{wt}(\pi|_{\mathbf{e}})$ corresponds to the weight of the path π after substituting for $y_1, \ldots, y_m$ by $e_1, \ldots, e_m$.   □

Corollary 1. *Every polynomial family in* VNP *is computed by a polynomial-sized 2-stack branching programs.*

Proof. By definition, we know that every polynomial family $(p_n)_{n \in \mathbb{N}}$ in VNP can be expressed as $\sum_{\bar{e} \in \{0,1\}^{m-n}} g_m(\bar{x}, \bar{e})$ where the polynomial family $(g_m)_{m \in \mathbb{N}}$ is in VP and $m = \text{poly}(n)$. Due to Theorem 5, we know that every polynomial family in VP has an efficient stack branching program computing it. Combining it with the above theorem, we see that every polynomial family in VNP has a 2-stack branching program computing it.   □

The above corollary tells us that every polynomial family in VNP can be computed efficiently by 2-stack branching programs. However, is every polynomial family computed by a 2-stack branching program inside VNP? We give an affirmative answer to this question using the following lemma. In fact, the lemma proves a much stronger result. We prove that statement for k-stack branching program for every $k \in [1, \text{poly}(n)]$.

Lemma 1. *Let $(p_n)_{n \in \mathbb{N}}$ be a polynomial family such that for each n, p_n is computed by a polynomial-sized k-stack branching program where $0 \leq k \leq \text{poly}(n)$. Then $(p_n)_{n \in \mathbb{N}}$ is in VNP.*

Proof. Note that each k-stack realizable, source-to-sink path computes a monomial. Multiple such paths may compute the same monomial and together they would contribute to the coefficient of the monomial. Given a monomial, we can compute all the 2-stack realizable paths contributing to this in $\#P/\text{poly}$ – over a sequence of all non-deterministic choices at each node to go from the source to sink, we can accumulate the edge weights (including the stack operations) of all the edges along the traversal and check if – the product of weights of edges equals the given monomial, and if the sequence of stack operations are 2-stack realizable. Putting this together with Valiant's criterion completes the proof.   □

Combining Theorem 4 with Lemma 1, we get the promised characterization 2-stack branching program of VNP. However, notice that we have proven Theorem 4 and Lemma 1 for an arbitrary k. Consider the following scenario, a polynomial p_n, in VNP, is computed by a 2-stack branching program and the boolean sum of p_n is computed by a 3-stack branching program (given by Theorem 4). However, Lemma 1 tells us that any polynomial computed by 3-stack branching program is also inside VNP. Therefore, we have also given a stack branching program based proof showing the closure of VNP under boolean sum (this was originally proven by Valiant [10, Theorem 5]). Another detail to note from the proof of Theorem 4 is that it can be modified slightly and used when p_n is computed by an ABP instead. To be more precise, using the proof of Theorem 4 with 2 additional stacks instead of 1, we see that if a polynomial is computed by a size s ABP, then its boolean sum is computed by 2-stack branching program i.e., the boolean sum of any polynomial computed by an efficient ABP lies inside VNP. This gives an alternate proof for boolean sum of polynomial families in VBP lying inside VNP (originally obtained as a consequence of [10, proposition 1] and [10, proposition 2]).

4   Relevance of Stack Height

Mengel proves that any 1-stack branching program with a larger alphabet size ($|\Sigma| > 2$) can be simulated by a 1-stack branching program (over binary alphabet) with poly-logarithmic blow-up in length and polynomial blow-up in size. Thus, it is sufficient to work with binary alphabet. This observation further implies that stack branching programs with constant-sized alphabet, logarithmic stack-height, and of size s can be simulated by algebraic branching programs of size poly(s). We extend this line of thought and state the following.

Lemma 2. *Let $P(x_1, \ldots, x_n)$ be a polynomial computed by a k-stack branching program (over binary alphabet) of size s, and stack-height at most H (for every stack). Then $P(x_1, \ldots, x_n)$ can also be computed by an algebraic branching program of size at most $O(2^{k(H+1)} \cdot s)$.*

Proof. Let G be the k-stack branching program. Let V and E be the vertex and edge sets of G. Let $\mathcal{S}$ be the set of all possible heights of stack configurations.

$$\mathcal{S} = \{(h_1, \ldots, h_k) \mid \forall\, i \in [k], 0 \leq h_i \leq H\}.$$

For a stack-height vector $(h_1, \ldots, h_k)$ in $\mathcal{S}$, there are $\prod_{i=1}^{k} 2^{h_i}$ many different stack configurations. In total, there are at most

$$\sum_{0 \leq h_1, \ldots, h_k \leq H} \left(\prod_{i=1}^{k} 2^{h_i}\right) \leq \left(\sum_{j=0}^{H} 2^j\right)^k \leq 2^{(H+1)k}.$$

Let V' and E' be initialized to empty sets. Let us use t to denote the quantity $2^{(H+1)k}$. For each vertex $v \in V$, we make t many copies of this vertex and add them to set V'. v_C be the copy of v that corresponds to the configuration C of k-stacks. For each edge $e = (u, v) \in E$ with weight wt(e) and the associated stack operations $(\mathsf{op}_1, \ldots, \mathsf{op}_k)$, we add an edge between u_{C_1} and v_{C_2} (with edge weight wt(e)), for all stack configurations C_1 and C_2 such that C_2 is obtained from C_1 through the stack operations $(\mathsf{op}_1, \ldots, \mathsf{op}_k)$. Let G' be the algebraic branching program obtained at the end of the above process with V' and E' as its vertex and edge sets.

It is now straightforward to form a bijection between realizable source-to-sink paths in the G and source-to-sink paths in G'.   □

Observe that we can trade-off H for k and vice versa and get the following statements. The set of families of polynomials computed by polynomial-sized k-stack branching programs with $k = O(1)$ and H logarithmic, is equivalent to VBP.

5   Discussion

5.1   What About VF

In this paper we mainly talk about branching programs with memory and their characterizations of VP and VNP in general and non-commutative settings. So far we did not make any observations about the formulas. Ben-Or and Cleve [1] showed that algebraic formulas are computationally equivalent to width-3 branching programs. Mengel [6, Lemma 3.9] showed that any polynomial family computed by polynomial-sized algebraic circuits can be computed with a family of width-2 stack branching programs over the binary alphabet. By analysing [6, Lemma 3.9] carefully, we can get that any polynomial family that is computed by polynomial-sized algebraic branching programs can also be computed by polynomial-sized width-2 stack branching programs over the binary alphabet, with stack-height at most logarithmic. In other words, adding a logarithmically height-bounded stack to a constant-width algebraic branching program adds computational power (assuming VF $\neq$ VBP) and this is in contrast to the fact that general 1-stack branching programs with logarithmic stack-height are computationally equivalent to algebraic branching programs.

5.2   Future Directions

- In this work, we showed that boolean summation of a polynomial can be computed using exactly one additional stack Theorem 4. To be precise, given a k-stack branching program that efficiently computes polynomial $q(\bar{x})$, we can construct a $(k+1)$-stack branching program that computes the polynomial $p = \sum_{e \in \{0,1\}^{|x|}} q(\bar{x}, \bar{e})$. The next step would be to perform the reverse. Given a k-stack branching program computing the polynomial $p(\bar{x})$ and the guarantee that polynomial p is of the form $p = \sum_{e \in \{0,1\}^{|x|}} q(\bar{x}, \bar{e})$, can we construct a $(k-1)$-stack branching program or k-stack branching program that computes $q(\bar{x})$?
- Due to [6], we know that VP is characterized by stack branching programs. Does this hold in the monotone setting as well? We answer this for mVNP and 2-stack branching programs by showing that the permanent polynomial family, which is known to be not in mVNP, can be computed efficiently by a monotone 2-stack branching program (??).
- Are k-stack branching programs strictly more powerful than monotone k-stack branching programs? In all the well known computational models, the only way to perform cancellations is via negation. However, k-stack branching programs have an alternate way of performing negations (by making the corresponding paths non-realizable). Therefore, it is not immediately clear if negations grant any additional computational power. This is not known even with the restriction of $k = 1$.
- Since monotone 2-stack branching programs are able to compute polynomial families that are outside mVNP, it would be interesting to figure out the exact computational power of monotone 2-stack branching programs. Do they

characterize any of the known classes beyond monotone VNP (see [3] for details on classes beyond mVNP).

– Can we use the tools from weighted pushdown automata theory to answer the VP vs VNP question in the non-commutative setting.

References

1. Ben-Or, M., Cleve, R.: Computing algebraic formulas using a constant number of registers. SIAM J. Comput. **21**(1), 54–58 (1992)
2. Bürgisser, P.: Completeness and Reduction in Algebraic Complexity Theory, Algorithms and computation in mathematics, vol. 7. Springer, Heidelberg (2000)
3. Chatterjee, P., Gajjar, K., Tengse, A.: Monotone classes beyond VNP. In: 43rd IARCS Annual Conference on Foundations of Software Technology and Theoretical Computer Science (FSTTCS 2023). Leibniz International Proceedings in Informatics (LIPIcs), vol. 284. Schloss Dagstuhl – Leibniz-Zentrum für Informatik, Dagstuhl, Germany (2023). https://doi.org/10.4230/LIPIcs.FSTTCS.2023. 11. https://drops.dagstuhl.de/entities/document/10.4230/LIPIcs.FSTTCS.2023. 11
4. Chillara, S., Raja, N.: Branching Programs with Extended Memory: New Insights (2024). https://hal.science/hal-04737571
5. Malod, G., Portier, N.: Characterizing valiant's algebraic complexity classes. J. Complex. **24**(1), 16–38 (2008)
6. Mengel, S.: Arithmetic branching programs with memory. In: Mathematical foundations of computer science 2013, Lecture Notes in Computer Science, vol. 8087, pp. 667–678. Springer, Heidelberg (2013). https://doi.org/10.1007/978-3-642-40313-259
7. Saptharishi, R.: A survey of lower bounds in arithmetic circuit complexity, version 9.0.3 (2021). https://github.com/dasarpmar/lowerbounds-survey/releases/tag/v9. 0.3
8. Shpilka, A., Yehudayoff, A.: Arithmetic circuits: a survey of recent results and open questions. Found. Trends Theor. Comput. Sci. **5**(3-4), 207–388 (2010). https://doi. org/10.1561/0400000039
9. Valiant, L.G.: Completeness classes in algebra. In: Conference Record of the Eleventh Annual ACM Symposium on Theory of Computing (Atlanta, Ga., 1979), pp. 249–261. ACM, New York (1979)
10. Valiant, L.G.: Reducibility by algebraic projections. Enseign. Math. (2) **28**(3-4), 253–268 (1982)
11. Valiant, L.G., Skyum, S., Berkowitz, S., Rackoff, C.: Fast parallel computation of polynomials using few processors. SIAM J. Comput. **12**(4), 641–644 (1983)

Tatami Printer: Physical ZKPs for Tatami Puzzles

Suthee Ruangwises[✉][iD]

Department of Computer Engineering, Chulalongkorn University, Bangkok, Thailand
suthee@cp.eng.chula.ac.th

Abstract. Tatami puzzles are pencil puzzles with an objective to partition a rectangular grid into rectangular regions such that no four regions share a corner point, as well as satisfying other constraints. In this paper, we develop a physical card-based protocol called *Tatami printer* that can help verify solutions of Tatami puzzles. We then use the Tatami printer to construct zero-knowledge proof protocols for two such puzzles: Tatamibari and Square Jam. These protocols enable a prover to show a verifier the existence of the puzzles' solutions without revealing them.

Keywords: zero-knowledge proof · card-based cryptography · Tatamibari · Square Jam · puzzle

1 Introduction

Pencil puzzles are puzzles written on paper that can be solved using logical reasoning. Examples of them include Sudoku, Numberlink, and Nonogram. Pencil puzzles are classified into several categories based on their main theme. Puzzles involving partitioning a rectangular grid into multiple regions to satisfy certain rules are often called *decomposition puzzles*.

Many decomposition puzzles, especially the ones originated in Japan, specify that the regions must be rectangular, and also have a common *corner constraint*: no four regions can share a corner point. This rule came from the arrangement of Japanese Tatami mats. Hence, these puzzles can be collectively called *Tatami puzzles*, a subcategory of decomposition puzzles [11].

1.1 Tatamibari

Tatamibari is a Tatami puzzle developed by a Japanese publisher Nikoli. The puzzle consists of an $m \times n$ rectangular grid, with some cells containing a +, |, or − symbol. The objective of this puzzle is to partition the grid into rectangles, with each one containing exactly one symbol.

- If the symbol is a +, that rectangle must be a square.
- If the symbol is a |, that rectangle must have a greater height than width.
- If the symbol is a −, that rectangle must have a greater width than height.

© The Author(s), under exclusive license to Springer Nature Switzerland AG 2025
I. Finocchi and L. Georgiadis (Eds.): CIAC 2025, LNCS 15679, pp. 105–118, 2025.
https://doi.org/10.1007/978-3-031-92932-8_8

In addition, no four rectangles can share a corner point [6]. See Fig. 1.

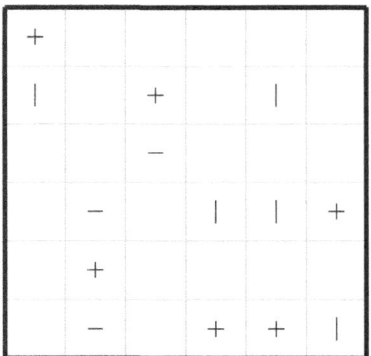 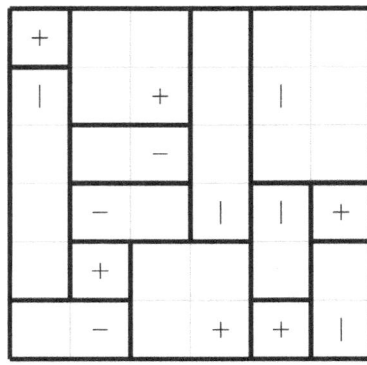

Fig. 1. An example of a 6 × 6 Tatamibari puzzle (left) and its solution (right)

Deciding whether a given Tatamibari puzzle has a solution has recently been proved to be NP-complete [1].

1.2 Square Jam

Square Jam is another Tatami puzzle developed by Eric Fox. The puzzle consists of an $n \times n$ square grid, with some cells containing a positive integer. The objective of this puzzle is to partition the grid into squares; each square may contain any number (including zero) of cells with an integer, but the integers in those cells must all be equal, and also equal to the side length of that square. In addition, no four squares can share a corner point [5]. See Fig. 2.

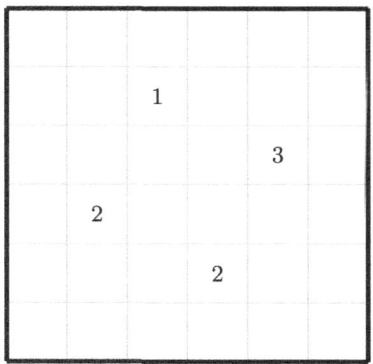 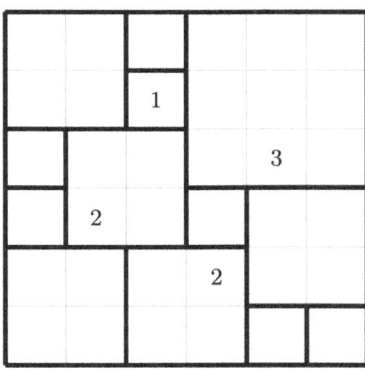

Fig. 2. An example of a 6 × 6 Square Jam puzzle (left) and its solution (right)

1.3   Zero-Knowledge Proof

Suppose Agnes designed a difficult Tatami puzzle and challenged her friend Brian to solve it. After a while, he could not solve the puzzle and began to doubt whether it actually has a solution. Agnes needs to convince him that her puzzle has a solution *without* revealing it (which would render the challenge pointless). A *zero-knowledge proof (ZKP)* makes this difficult task possible.

Introduced by Goldwasser et al. [3] in 1989, a ZKP is an interactive protocol between a prover P and a verifier V. Both of them are given a computational problem x, but its solution w is known to only P. A ZKP with perfect completeness and soundness must satisfy the following properties.

1. **Perfect Completeness:** If P knows w, then V always accepts.
2. **Perfect Soundness:** If P does not know w, then V always rejects.
3. **Zero-knowledge:** V obtains no information about w, i.e. there is a probabilistic polynomial time algorithm S (called a *simulator*), not knowing w but having an access to V, such that the outputs of S and of the actual protocol follow exactly the same probability distribution.

In 1991, Goldreich et al. [2] proved that a ZKP exists for every NP problem, implying one can construct a computational ZKP for any NP pencil puzzle via a reduction. Such construction, however, is neither practical nor intuitive as it requires cryptographic primitives. Instead, many researchers focused on developing physical ZKPs using objects found in everyday life such as a deck of playing cards. These card-based protocols have the benefit that they do not require computers and also allow external observers to verify that the prover truthfully executes them (which is often a challenging task for digital protocols). They also have didactic values and can be used to teach the concept of ZKP to non-experts.

1.4   Card-Based Zero-Knowledge Proof Protocols

There is a line of work dedicated to design physical card-based ZKP protocols for specific pencil puzzles, including ABC End View [14], Goishi Hiroi [14], Heyawake [12], Hitori [12], Juosan [9], Kakuro [10], Masyu [8], Nonogram [13], Numberlink [16], Nurikabe [12], Ripple Effect [17], Shikaku [15], Slitherlink [8], Sudoku [19], Sumplete [4], Takuzu [9], and Toichika [14].

In 2024, Ruangwises and Iwamoto [18] proposed a more generic protocol called *printing protocol*, which can help verify solutions of several decomposition puzzles such as Five Cells and Meadows. Their protocol, however, has a limitation that it cannot check constraints involving relationships between multiple regions, and thus cannot be used to verify solutions of Tatami puzzles.

1.5   Our Contribution

In this paper, we propose a card-based protocol called *Tatami printer*, which is modified from the printing protocol of Ruangwises and Iwamoto [18].

The Tatami printer can print numbers or symbols from a template onto a target area from the puzzle grid, constituting a new rectangular region in the grid. It also verifies that the printed region does not overlap with existing regions, and that no four regions share a corner point.

We then use the Tatami printer to construct ZKP protocols for two Tatami puzzles: Tatamibari and Square Jam.

2 Preliminaries

2.1 Cards

In our protocol, each card may have a number or a symbol written on the front side (e.g. $\boxed{1}$, $\boxed{+}$, $\boxed{\heartsuit}$), or may be a blank card (denoted by $\boxed{}$). All cards have indistinguishable back sides denoted by $\boxed{?}$.

2.2 Pile-Shifting Shuffle

Given a matrix M of cards, a *pile-shifting shuffle* [20] rearranges the columns of M by a uniformly random cyclic shift. It can be implemented by putting all cards in each column into an envelope, and repeatedly picking some envelopes from the bottom and putting them on the top of the pile.

Each card in M can be replaced by a stack of cards (as long as every stack in the same row has the same number of cards), and the protocol still works in the same way.

2.3 Chosen Cut Protocol

Given a sequence $C = (c_1, c_2, ..., c_q)$ of q face-down cards, a *chosen cut protocol* [7] enables P to select a desired card c_i without revealing i to V. It also reverts C back to its original state after P finishes using c_i.

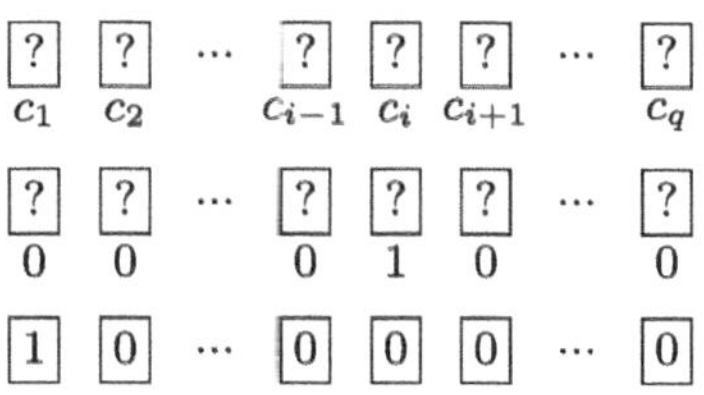

Fig. 3. A $3 \times q$ matrix M constructed in Step 1 of the chosen cut protocol

In the chosen cut protocol, P performs the following steps.

1. Construct the following $3 \times q$ matrix M (see Fig. 3).
   (a) In Row 1, place the sequence C.

(b) In Row 2, place a face-down $\boxed{1}$ at Column i and face-down $\boxed{0}$s at all other columns.

(c) In Row 3, place a face-up $\boxed{1}$ at Column 1 and face-up $\boxed{0}$s at all other columns.

2. Turn all cards face-down. Apply the pile-shifting shuffle to M.

3. Turn over all cards in Row 2 and locate the position of the only $\boxed{1}$. A card in Row 1 directly above this $\boxed{1}$ will be the desired card c_i.

4. After finishing using c_i, place c_i back into M at the same position.

5. Turn all cards face-down. Apply the pile-shifting shuffle to M again.

6. Turn over all cards in Row 3 and locate the position of the only $\boxed{1}$. Shift the columns of M cyclically such that this $\boxed{1}$ moves to Column 1. M is now reverted back to its original state.

Each card in C can be replaced by a stack of cards (as long as every stack has the same number of cards), and the protocol still works in the same way.

2.4   Printing Protocol of Ruangwises and Iwamoto

In a recent work, Ruangwises and Iwamoto [18] developed the following printing protocol.

A $p \times q$ matrix of face-down cards (called a *template*) and another $p \times q$ matrix of face-down cards of an area from the puzzle grid are given (the front side of all cards are known to P but not to V). This protocol verifies that positions in the area corresponding to non-blank cards in the template are initially empty (consisting of all blank cards). Then, it places all non-blank cards from the template at the corresponding positions in the area, replacing the original blank cards (see Fig. 4) without revealing any card to V.

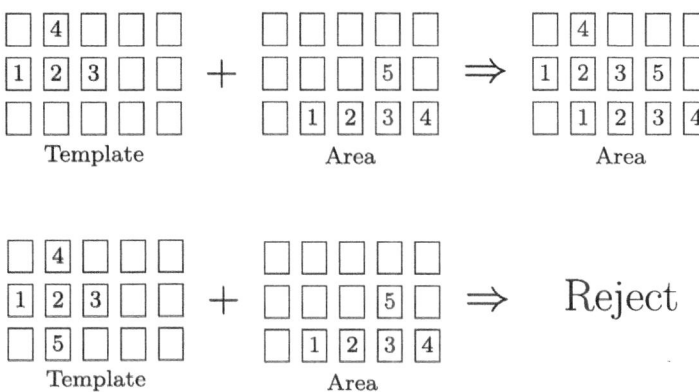

Fig. 4. Examples of a successful print from a 3×5 template onto a 3×5 area (top) and an unsuccessful print due to overlapping non-blank cards (bottom)

In the printing protocol, P performs the following steps.

1. Place each card from the template on top of a corresponding card from the area, creating pq two-card stacks.
2. For each of the pq stacks, perform the following steps.
   (a) Apply the chosen cut protocol to select a blank card. (If the preconditions are met, at least one card must be blank; if both cards are blank, P can select any of them.)
   (b) Reveal that the selected card is blank (otherwise V rejects) and remove it from the stack.

After these steps, all non-blank cards from the template are placed at the corresponding positions in the area, and V is convinced that these positions in the area were initially empty.

For decomposition puzzles with simple constraints, P can repeatedly apply the printing protocol to print each region of the partition according to P's solution onto the puzzle grid, convincing V that these regions do not overlap.

However, a major limitation of this protocol is that each region is separately printed, so it cannot check constraints involving relationships between multiple regions. Therefore, it cannot be used to verify solutions of Tatami puzzles.

3   Our Proposed Protocol: Tatami Printer

Considering the limitation of the printing protocol, we aim to modify the protocol so that it can also check the corner constraint of Tatami puzzles.

First, observe that if no four rectangles share a corner point, each point inside (or on the boundary of) the puzzle grid can be a corner point of at most two rectangles. Threfore, on each cell we add a "corner counter" to keep track of the number of rectangles having the top-left corner of that cell as their corner point.

On top of each card in a template, we place a ♣ if its top-left corner is a corner point of a rectangle the template represents, and a ♡ if its top-left corner is not a corner point. Also, on top of each card in the puzzle grid, we place two cards (initially both ♡s) to keep track of the number of rectangles having its top-left corner as their corner point (which is the number of ♣s among them).

When printing a $p \times q$ template onto a $p \times q$ area from the grid (the front side of all cards are known to P but not to V), P performs the following steps.

1. In the template, separate the top cards of all stacks (called the *corner counter part*) from the bottom cards (called the *main part*) and put them in another $p \times q$ matrix of cards. Also, in the area, separate the top two cards of all stacks (the corner counter part) from the bottom cards (the main part) and put them in another $p \times q$ matrix of two-card stacks (see Fig. 5).
2. For the main part, place each card from the template on top of a corresponding card from the area, creating pq two-card stacks.
3. For each of the pq stacks, perform the following steps.

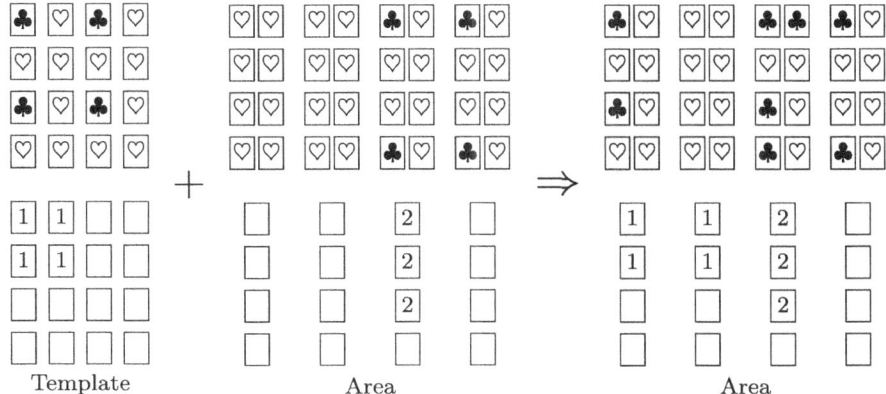

Fig. 5. An example of a successful print from a 4×4 template onto a 4×4 area, with the top matrices being the corner counter part and the bottom matrices being the main part

  (a) Apply the chosen cut protocol to select a blank card. (If the preconditions are met, at least one card must be blank; if both cards are blank, P can select any of them.)
  (b) Reveal that the selected card is blank (otherwise V rejects) and remove it from the stack.
4. For the corner counter part, place each card from the template on top of a corresponding two-card stack from the area, creating pq three-card stacks.
5. For each of the pq stacks, perform the following steps.
  (a) Apply the chosen cut protocol to select a ♡. (If the preconditions are met, at least one card must be a ♡; if both cards are ♡s, P can select any of them.)
  (b) Reveal that the selected card is a ♡ (otherwise V rejects) and remove it from the stack.

After these steps, all non-blank cards from the main part of the template are placed at the corresponding positions in the area. All ♣s from the corner counter part of the template are also added to the corresponding positions in the area. V is convinced that these positions in the area were initially empty, and that each point in the area is a corner point of at most two rectangles.

4   ZKP Protocol for Tatamibari

First, from a solution of Tatamibari, one can fill a symbol on every cell according to the rule of the puzzle (a $+$, $|$, or $-$ depending on the size of a rectangle that cell belongs to). We call this instance an *extended solution* of the puzzle.

The key observation is that in the extended solution of Tatamibari, there are only mn different sizes of rectangles (as there are m possible heights and n possible widths), and the symbol on every cell in a rectangle of each size is fixed. Therefore, only mn different templates are required.

In our protocol, P constructs mn templates, one for each size of rectangle. Each template has size $(m + 1) \times (n + 1)$ (to support a corner counter of the bottom-right corner), and the rectangle is placed at the top-left corner of the template. A cell inside the rectangle is represented by a card with a symbol according to the rule, while a cell outside the rectangle is represented by a blank card (see Fig. 6).

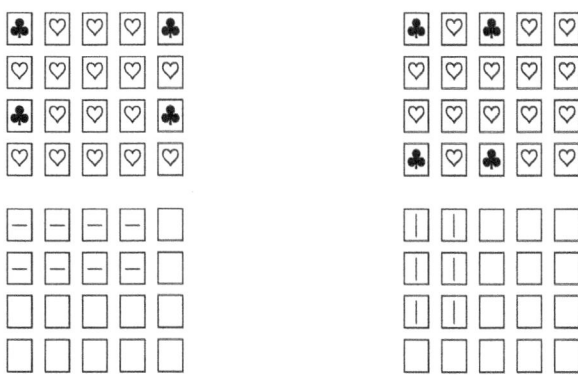

Fig. 6. 4×5 templates of rectangles with sizes 2×4 (left) and 3×2 (right), with the top matrices being the corner counter part and the bottom matrices being the main part

4.1   Main Protocol

Initially, P publicly places two ♡s on top of a blank card on every cell in the grid. To handle edge cases, P publicly appends m rows and n columns of "dummy" stacks to the bottom and to the right of the grid. Then, P turns all cards face-down. We now have an $2m \times 2n$ matrix of three-card stacks.

Observe that if we arrange all $4mn$ stacks in the matrix into a single sequence $A = (a_1, a_2, ..., a_{4mn})$, starting at the top-left corner and going from left to right in Row 1, then from left to right in Row 2, and so on, we can locate exactly where the four neighbors of any given card are. Namely, the cards on the neighbor to the left, right, top, and bottom of a cell containing a_i are a_{i-1}, a_{i+1}, a_{i-2n}, and a_{i+2n}, respectively. Therefore, we can select any area from the grid by applying the chosen cut protocol to select the top-left corner cell of that area, and the rest will follow as the chosen cut protocol preserves the cyclic order.

P also constructs mn templates of all mn sizes of rectangle and lets V verify that all templates are correct (otherwise V rejects).

Suppose that in P's extended solution, the grid is partitioned into k rectangles $B_1, B_2, ..., B_k$. Note that k is public information, which is the number of cells containing a symbol in the original puzzle. For each $i = 1, 2, ..., k$, P performs the following steps.

1. Apply the chosen cut protocol to select an $(m+1) \times (n+1)$ area whose top-left corner is the top-left corner of B_i.
2. Apply the chosen cut protocol to select a template of a rectangle with the same size as B_i.
3. Apply the Tatami printer to the selected template and area.
4. Reconstruct a template that has just been used and replenish the pile of templates with it. Let V verify again that all mn templates are correct (otherwise V rejects). Note that V does not know which template has just been used.

Finally, P reveals all cards from the main part on the cells that contain a symbol in the original puzzle. V verifies that the symbols on the cards match the ones in the cells (otherwise V rejects). P also reveals the main part of all dummy stacks that they are still blank (otherwise V rejects). If all verification steps pass, then V accepts.

This protocol uses $\Theta(m^2 n^2)$ cards and $\Theta(kmn)$ shuffles.

4.2 Proof of Correctness and Security

We will prove the perfect completeness, perfect soundness, and zero-knowledge properties of this protocol.

Lemma 1 (Perfect Completeness). *If P knows a solution of the Tatamibari puzzle, then V always accepts.*

Proof. Suppose P knows an extended solution of the puzzle. Consider each i-th iteration of the main protocol.

- In Step 3, since $B_1, B_2, ..., B_k$ form a partition of the puzzle grid, B_i does not overlap with $B_1, B_2, ..., B_{i-1}$. Also, from the corner constraint, each point in the grid can be a corner point of at most two rectangles. Thus, the Tatami printer will pass.
- In Step 4, since P reconstructs a template that has just been used, all mn templates are correct, and thus this step will pass.

Therefore, every iteration will pass. After k iterations, all symbols in P's extended solution will be printed on the grid, so all symbols in the original puzzle will match the ones on the corresponding cards.

Hence, we can conclude that V always accepts.   □

Lemma 2 (Perfect Soundness). *If P does not know a solution of the Tatamibari puzzle, then V always rejects.*

Proof. We will prove the contrapositive of this statement. Suppose V accepts, which means the verification passes in all steps. Consider the main protocol.

Since Step 4 passes for every iteration, all mn templates are correct after each iteration (and also at the beginning of the protocol), which implies the symbols printed in every iteration form a shape of a rectangle and follow the rule of the puzzle.

In Step 3, since the Tatami printer passes for every iteration, B_i does not overlap with $B_1, B_2, ..., B_{i-1}$ for every i. Also, since the final verification passes, the combined area of $B_1, B_2, ..., B_k$ must cover the whole puzzle grid, i.e. $B_1, B_2, ..., B_k$ form a partition of the grid. Furthermore, each point in the grid can be a corner point of at most two rectangles, so no four rectangles share a corner point.

Since the final verification passes, all symbols in the original puzzle match the ones on the corresponding cards.

Hence, we can conclude that the puzzle grid is partitioned into rectangles according to the rules, which implies P must know a valid solution of the puzzle.   □

Lemma 3 (Zero-Knowledge). *During the verification, V obtains no information about P's solution.*

Proof To prove the zero-knowledge property, we will show that any interaction between P and V can be simulated by a simulator S that does not know P's solution. It is sufficient to prove that all distributions of cards that are turned face-up can be simulated by S.

- In Steps 3 and 6 of the chosen cut protocol in Sect. 2.3, because of the pile-shifting shuffles, the $\boxed{1}$ has probability $1/q$ to be at each of the q columns. Therefore, these two steps can be simulated by S.
- In Steps 3(b) and 5(b) of the Tatami printer in Sect. 3, there is only one deterministic pattern of the cards that are turned face-up. Therefore, these two steps can be simulated by S.
- In Step 4 of the main protocol, there is only one deterministic pattern of the cards that are turned face-up (all correct templates). Therefore, this step can be simulated by S.

Hence, we can conclude that V obtains no information about P's solution. □

5   ZKP Protocol for Square Jam

First, from a solution of Square Jam, one can fill an integer on every cell according to the rule of the puzzle (equal to the side length of a square that cell belongs to). We call this instance an *extended solution* of the puzzle.

The key observation is that in the extended solution of Square Jam, there are only n different sizes of square, and the integer on every cell in a square of each size is fixed. Therefore, only n different templates are required.

However, unlike in Tatamibari, the number of squares in P's solution is not public information, so P cannot reveal that number to V. Therefore, P has to apply the Tatami printer for n^2 times, sometimes printing nothing, as V knows that there are at most n^2 squares.

In our protocol, P constructs n templates, one for each size of square. Each template has size $(n + 1) \times (n + 1)$, and the square is placed at the top-left corner of the template. A cell inside the square is represented by a card with a number equal to the side length of that square, while a cell outside the square is represented by a blank card (see Fig. 7). In addition, P constructs a *blank template* of the same size, with the main part consisting of all blank cards and the corner counter part consisting of all $\heartsuit$s.

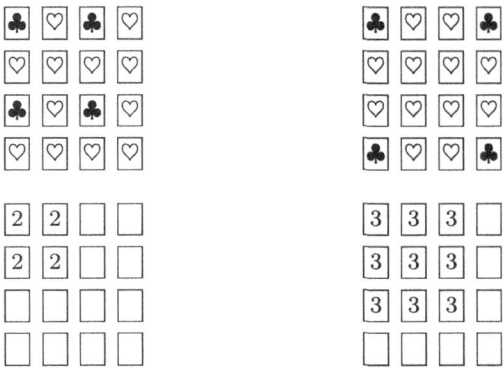

Fig. 7. 4×4 templates of squares with side lengths 2 (left) and 4 (right), with the top matrices being the corner counter part and the bottom matrices being the main part

5.1 Main Protocol

Initially, P publicly places two $\heartsuit$s on top of a blank card on every cell in the grid. To handle edge cases, P publicly appends n rows and n columns of "dummy" stacks to the bottom and to the right of the grid. Then, P turns all cards face-down. We now have a $2n \times 2n$ matrix of three-card stacks.

P also constructs templates of all n sizes of square, plus a blank template, and lets V verify that all $n + 1$ templates are correct (otherwise V rejects).

Suppose that in P's extended solution, the grid is partitioned into k squares $B_1, B_2, ..., B_k$. For each $i = 1, 2, ..., n^2$, P performs the following steps.

1. Apply the chosen cut protocol to select an $(n+1) \times (n+1)$ area whose top-left corner is the top-left corner of B_i (if $i > k$, select any area).
2. Apply the chosen cut protocol to select a template of a square with the same size as B_i (if $i > k$, select a blank template).
3. Apply the Tatami printer to the selected template and area.

4. Reconstruct a template that has just been used and replenish the pile of templates with it. Let V verify again that all $n + 1$ templates are correct (otherwise V rejects).

Finally, P reveals all cards from the main part on the cells that contain an integer in the original puzzle. V verifies that the integers on the cards match the ones in the cells (otherwise V rejects). P also reveals the main part of all dummy stacks that they are still blank (otherwise V rejects). If all verification steps pass, then V accepts.

This protocol uses $\Theta(n^3)$ cards and $\Theta(n^4)$ shuffles.

5.2   Proof of Correctness and Security

We will prove the perfect completeness, perfect soundness, and zero-knowledge properties of this protocol.

Lemma 4 (Perfect Completeness). *If P knows a solution of the Square Jam puzzle, then V always accepts.*

Proof Suppose P knows an extended solution of the puzzle. Consider each i-th iteration of the main protocol ($i \leq k$).

– In Step 3, since $B_1, B_2, ..., B_k$ form a partition of the puzzle grid, B_i does not overlap with $B_1, B_2, ..., B_{i-1}$. Also, from the corner constraint, each point in the grid can be a corner point of at most two rectangles. Thus, the Tatami printer will pass.
– In Step 4, since P reconstructs a template that has just been used, all $n + 1$ templates are correct, and thus this step will pass.

Therefore, the first k iterations will pass. All subsequent iterations just print a blank template (we call them *trivial iterations*), so they will also pass. After the first k iterations, all integers in P's extended solution will be printed on the grid. Subsequent trivial iterations do not modify the grid, so all integers in the original puzzle will match the ones on the corresponding cards.

Hence, we can conclude that V always accepts.   □

Lemma 5 (Perfect Soundness). *If P does not know a solution of the Square Jam puzzle, then V always rejects.*

Proof We will prove the contrapositive of this statement. Suppose V accepts, which means the verification passes in all steps. Consider the main protocol.

Since Step 4 passes for every iteration, all $n + 1$ templates are correct after each iteration (and also at the beginning of the protocol), which implies the integers printed in every nontrivial iteration form a shape of a square and are equal to the side length of that square.

Let $D_1, D_2, ..., D_\ell$ be the squares printed in each nontrivial iteration in this order, for some $\ell \leq n^2$. In Step 3, since the Tatami printer passes for every iteration, D_i does not overlap with $D_1, D_2, ..., D_{i-1}$ for every i. Also, since the final verification passes, the combined area of $D_1, D_2, ..., D_\ell$ must cover the whole puzzle grid, i.e. $D_1, D_2, ..., D_\ell$ form a partition of the grid. Furthermore, each point in the grid can be a corner point of at most two squares, so no four squares share a corner point.

Since the final verification passes, all integers in the original puzzle match the ones on the corresponding cards.

Hence, we can conclude that the puzzle grid is partitioned into squares according to the rules, which implies P must know a valid solution of the puzzle. □

Lemma 6 (Zero-Knowledge). *During the verification, V obtains no information about P's solution.*

Proof To prove the zero-knowledge property, we will show that any interaction between P and V can be simulated by a simulator S that does not know P's solution. It is sufficient to prove that all distributions of cards that are turned face-up can be simulated by S.

The zero-knowledge property of the chosen cut protocol and the Tatami printer has been proved in the proof of Lemma 3, so it is sufficient to consider only the main protocol.

In Step 4 of the main protocol, there is only one deterministic pattern of the cards that are turned face-up (all correct templates). Therefore, this step can be simulated by S.

Hence, we can conclude that V obtains no information about P's solution. □

6   Future Work

We developed the Tatami printer protocol, which can check the corner constraint of Tatami puzzles. However, this protocol still has a limitation that it cannot check other types of constraint involving relationships between multiple regions, e.g. a constraint in *Fillmat* which specifies that horizontally or vertically adjacent rectangles cannot contain the same number. An interesting future work is to develop a technique to check such constraints.

References

1. Adler, A., Bosboom, J., Demaine, E.D., Demaine, M.L., Liu, Q.C., Lynch, J.: Tatamibari Is NP-complete. In: Proceedings of the 10th International Conference on Fun with Algorithms (FUN), pp. 1:1–1:24 (2020)
2. Goldreich, O., Micali, S., Wigderson, A.: Proofs that yield nothing but their validity and a methodology of cryptographic protocol design. J. ACM **38**(3), 691–729 (1991)
3. Goldwasser, S., Micali, S., Rackoff, C.: The knowledge complexity of interactive proof systems. SIAM J. Comput. **18**(1), 186–208 (1989)
4. Hatsugai, K., Ruangwises, S., Asano, K., Abe, Y.: NP-completeness and physical zero-knowledge proofs for sumplete, a puzzle generated by ChatGPT. N. Gener. Comput. **42**(3), 429–448 (2024)

5. Janko, A., Janko, O.: Square Jam. https://www.janko.at/Raetsel/Square-Jam/index.htm
6. Janko, A., Janko, O.: Tatamibari. https://www.janko.at/Raetsel/Tatamibari/index.htm
7. Koch, A., Walzer, S.: Foundations for actively secure card-based cryptography. In: Proceedings of the 10th International Conference on Fun with Algorithms (FUN), pp. 17:1–17:23 (2020)
8. Lafourcade, P., Miyahara, D., Mizuki, T., Robert, L., Sasaki, T., Sone, H.: How to construct physical zero-knowledge proofs for puzzles with a "single loop" condition. Theor. Comput. Sci. **888**, 41–55 (2021)
9. Miyahara, D., et al.: Card-based ZKP protocols for takuzu and juosan. In: Proceedings of the 10th International Conference on Fun with Algorithms (FUN), pp. 20:1–20:21 (2020)
10. Miyahara, D., Sasaki, T., Mizuki, T., Sone, H.: Card-based physical zero-knowledge proof for Kakuro. IEICE Trans. Fund. Electron. Commun. Comput. Sci. **102**(9), 1072–1078 (2019)
11. puzz.link: List of puzzle types. https://puzz.link/list.html
12. Robert, L., Miyahara, D., Lafourcade, P., Mizuki, T.: Card-based ZKP for connectivity: applications to nurikabe, hitori, and heyawake. N. Gener. Comput. **40**(1), 149–171 (2022)
13. Ruangwises, S.: An improved physical ZKP for nonogram and nonogram color. J. Comb. Optim. **45**(5), 122 (2023)
14. Ruangwises, S.: Verifying the first nonzero term: physical ZKPs for ABC end view, Goishi Hiroi, and Toichika. J. Comb. Optim. **47**(4), 69 (2024)
15. Ruangwises, S., Itoh, T.: How to physically verify a rectangle in a grid: a physical ZKP for shikaku. In: Proceedings of the 11th International Conference on Fun with Algorithms (FUN), pp. 24:1–24:12 (2022)
16. Ruangwises, S., Itoh, T.: Physical zero-knowledge proof for numberlink puzzle and k vertex-disjoint paths problem. N. Gener. Comput. **39**(1), 3–17 (2021)
17. Ruangwises, S., Itoh, T.: Physical zero-knowledge proof for ripple effect. Theor. Comput. Sci. **895**, 115–123 (2021)
18. Ruangwises, S., Iwamoto, M.: Printing protocol: physical ZKPs for decomposition puzzles. N. Gener. Comput. **42**(3), 331–343 (2024)
19. Sasaki, T., Miyahara, D., Mizuki, T., Sone, H.: Efficient card-based zero-knowledge proof for Sudoku. Theor. Comput. Sci. **839**, 135–142 (2020)
20. Shinagawa, K., et al.: Card-based protocols using regular polygon cards. IEICE Trans. Fund. Electron. Commun. Comput. Sci. **100**(9), 1900–1909 (2017)

On the Price of Anarchy in Packet Routing Games with FIFO

Daniel Schmand[1]([⊠])(iD), Torben Schürenberg[1]([⊠])(iD), and Martin Strehler[2]([⊠])(iD)

[1] University of Bremen, Bremen, Germany
{schmand,torsch}@uni-bremen.de
[2] Westsächsische Hochschule Zwickau, Zwickau, Germany
martin.strehler@fh-zwickau.de

Abstract. We investigate packet routing games in which network users selfishly route themselves through a network over discrete time, aiming to reach the destination as quickly as possible. Conflicts due to limited capacities are resolved by the first-in, first-out (FIFO) principle. Building upon the line of research on packet routing games initiated by Werth et al. [21], we derive the first non-trivial bounds for packet routing games with FIFO. Specifically, we show that the price of anarchy is at most 2 for the important and well-motivated class of uniformly fastest route equilibria introduced by Scarsini et al. [16] on any linear multigraph. We complement our results with a series of instances on linear multigraphs, where the price of stability converges to at least $\frac{e}{e-1}$. Furthermore, our instances provide a lower bound for the price of anarchy of continuous Nash flows over time on linear multigraphs which establishes the first lower bound of $\frac{e}{e-1}$ on a graph class where the monotonicity conjecture is proven by Correa et al. [5].

Keywords: Competitive Packet Routing · FIFO · Price of Anarchy · Price of Stability · Makespan Objective

1 Introduction

1.1 Motivation

Selfish routing games, also known as competitive routing games, have been the subject of extensive study due to their wide-ranging applications in traffic and communication networks. These games involve multiple agents, such as road users or data packets, competing for limited resources like network paths and bandwidth. When considering road traffic, the idea of modeling with a continuous flow model appears unrealistic. Given that vehicles are indivisible and possess substantial size, the applicability of particles that can be fractionated at will seems dubious with respect to capturing all relevant dynamics. Consequently, there has been a recent surge in the study of atomic dynamic flow models.

Werth et al. [21] considered such a model with deterministic queues, which also ensures the first-in, first-out (FIFO) property, from a game-theoretic perspective. The authors examine two variants, each pertaining to distinct objective

I. Finocchi and L. Georgiadis (Eds.): CIAC 2025, LNCS 15679, pp. 119–135, 2025.
https://doi.org/10.1007/978-3-031-92932-8_9

functions. For narrowest flows, the cost of a user is the most expensive resource on her path and users try to minimize these bottlenecks. It is NP-hard to determine according flows. The price of anarchy (PoA) is tightly bounded by the number of players. Conversely, when dealing with the *makespan* objective function, where every user strives to minimize their arrival time, equilibria are easy to compute in the single commodity case. However, the PoA has been left open. Since the introduction of this model by Werth et al., numerous variants have been studied, but the PoA in the original model is still an unresolved problem.

This paper aims to take a step towards addressing this gap. We find a constant upper bound of 2 for the class of linear multigraphs regarding the makespan objective and provide a sequence of linear multigraphs that converge to a price of stability (PoS) and therefore also a PoA of at least $\frac{e}{e-1}$.

1.2   Informal Model Description

Before delving into a comprehensive description of the model in Sect. 2, we offer an informal overview that is intended to provide a more effective categorization within the multifaceted realm of routing games.

We are examining a dynamic competitive routing game involving atomic players on a directed graph $G = (V, E)$. Each edge $e \in E$ is equipped with an integral transit time $\tau(e) \in \mathbb{N}_{>0}$. Each player routes an unsplittable packet of unit size through this network from origin s to destination d, that is, the players choose an $s - d$ path, start at time zero and try to reach d as quickly as possible, i.e., players aim to minimize the arrival time at d. Furthermore, each edge is equipped with a point queue. At any point in time arbitrarily many packets may enter such a queue, but the queue has a limited outflow rate, that is, at most ν_e packets may leave the queue at a particular time step. These queues operate according to the FIFO principle, ensuring that packets exit in the same sequence they entered. There is no limit to the maximum storage capacity of these queues. Consequently, the total latency on an edge is composed of the transit time and any potential waiting time in the queue.

In particular, we consider so-called uniformly fastest route (UFR) equilibria that have also been analyzed by [16]. Such a UFR equilibrium is achieved if for every player and for every node v on its route, there exists no other route such that this player can arrive earlier at v. In other words, a UFR equilibrium also takes into account arrival time at intermediate nodes. For the social optimum, we consider the makespan of the problem, that is, assuming all players start at time zero, we minimize the arrival time of the last player.

1.3   Related Literature

The diversity of routing games mirrors the breadth of their applications. We will give a brief overview of the origin and development, focusing particularly on the most relevant results for the present work.

A common characteristic across all games is that each player possesses a utility function, typically characterized by parameters such as latency[1] (see, e.g., [1,12,13,16], residence time [3], bottleneck [21], or arrival penalties[2] [9] and chooses a path (strategy) from its origin to its destination. A state in the game is referred to as a (pure) Nash equilibrium when no individual player can decrease the private utility function by unilaterally altering their route.

Equilibrium solutions are often compared to a system optimum with respect to some objective function. Common objectives include minimizing the total latency (sum over all users) [10], the total completion time (i.e. makespan) [21], the (average) delay [16], or the throughput [12]. The ratio between the worst equilibrium and the system optimum is referred to as the price of anarchy (PoA). In addition, we also consider the price of stability (PoS), which refers to the ratio between the best equilibrium and the system's optimal state.

In the categorization of such competitive routing games, it is necessary to make several differentiations. Initially, the games can be classified as either static or dynamic. In a static setup, players choose their path, and interference arises when both paths share the same edge. This interference can be attributed to load-dependent latencies [15], or due to the edge capacities that constrain the total flow on an edge [6]. Pigou's two link network [13] and Braess' Paradox [1] are pioneering examples in the realm of static games.

Conversely, dynamic games incorporate the element of time, with particles navigating the network in a time-sensitive manner. Typically, the subsequent player may experience delays due to the preceding player, as the former must wait until the latter clears the next edge. Furthermore, it is common to require adherence to the FIFO principle. Moreover, the capacity typically constrains the flow rate (measured in flow units per time unit), rather than the aggregate flow passing through an edge. A notable early and often adapted example is Vickrey's queuing model [20], which introduced point queues of no physical space and factored in varying departure times. Since then, the game-theoretic aspects of flows over time have been a topic of extensive discussion within the traffic community. From the perspective of algorithmic game theory, Koch and Skutella [12] have demonstrated that Nash flows over time with the underlying deterministic queuing model can be interpreted as a sequence of specific static flows.

The second differentiation pertains to non-atomic versus atomic games. In non-atomic flows, one can envision an infinite number of players, each controlling an infinitesimally small packet. On the other hand, atomic flows, as introduced by Rosenthal [14], are characterized by each entity controlling a substantial portion of the total flow. Typically, these games involve a finite number of players, each with an unsplittable packet of unit size.

Non-atomic models are relatively well-understood due to their amenability to analytical methods. For example, in non-atomic models, the value of an equi-

---

[1] In the literature latency is sometimes also called travel time.

[2] Deviation from the desired arrival time is often taken into account as an additive penalty in the objective of dynamic traffic assignment models (DTA) used in the traffic community.

librium flow is often unique [4]. On the other hand, the PoA is still an open question in many dynamic scenarios. A significant advancement was made by Correa et al. [5], where they established that if one is permitted to restrict the inflow rate of the equilibrium to the optimum flow's initial inflow rate, the PoA concerning the makespan is bounded by $\frac{e}{e-1}$. Unfortunately, these findings are contingent upon a monotonicity conjecture [5]. Despite its intuitive appeal, this conjecture has only been validated for relatively straightforward graph classes, such as linear multigraphs, to date.

The suitability of non-atomic models, particularly for depicting scenarios such as road traffic, is a subject of ongoing debate, since it is unclear whether particle size is irrelevant in such applications. A game-theoretic approach based on a dynamic, discrete model with point queues was introduced by Werth [21]. Subsequent to this approach, numerous variants of atomic routing games have been explored, given that these models necessitate the incorporation of additional concepts or intricate modeling details. A particular aspect is simultaneity: what happens when two players want to use the same edge at the same time? This necessitates a concept for parallel processing (fair time sharing, see [11]) or tie-breaking rules. For the latter, predefined priorities on players [10,21] or edges [2,17,21] are typically employed.

Atomic models are often more challenging to analyze since they usually do not have unique equilibria, but some results were obtained for special setting. For example, Scarsini et al. [16] studied a game where the inflow occurs in generations of players, also emphasizing the relevance of UFR equilibria. Similarly, Cao et al. [2] studied a game where the inflow rate never exceeds the network capacity, adding the concept of dynamic route choices.

However, the relationship between discrete and continuous routing models remains partially elusive. When considering system optimal solutions for a single source and a single sink, the similarities are pronounced. Already Ford and Fulkerson [8] pioneered the computation of dynamic flows for discrete time based on static flow computations. Fleischer and Tardos [7] successfully addressed this problem for continuous time. A comprehensive summary is available by Skutella [19]. The situation becomes more intricate when examining equilibria. A Nash equilibrium in the continuous case is an equilibrium in the discrete case, if and only if the continuous flow can be interpreted as a flow of discrete packets, which is not possible in general. Conversely, while every state in the discrete setting can be interpreted as a feasible continuous flow [7], the equilibrium property is often lost. That is, a Nash equilibrium in the discrete case does not necessarily translate to an equilibrium in the continuous case.

1.4   Our Contribution

We consider an atomic dynamic routing game on linear multigraphs[3], where we have a linearly ordered node set $V = \{v_1, \ldots, v_m\}$ and the edge set consists of multiple edges (v_j, v_{j+1}). For this graph class, we prove the PoA to be at

---

[3] This graph class is also known in the literature as chain-of-parallel networks [2,16].

most 2, presenting the first result of a constant PoA on a graph class completely independent of the number of players or the network size. Specifically, we do not impose artificial restrictions on the inflow rate of the network. Furthermore, we show that whenever players have the choice between multiple quickest paths, the sum of queues in the network is maximized when players greedily opt for the longest queues available.

Moreover, we show that there are instances of linear multigraphs, where the PoS converges to at least $\frac{e}{e-1}$, indicating that discrete Nash flows even on such a restricted graph class are non-trivial. Furthermore, this result aligns with the upper bound for the PoA in a non-atomic routing game investigated by Correa et al. [5] where the inflow rate is restricted to that of an optimal flow and a certain monotonicity conjecture holds. We show that our instances can be translated to the setting of Correa et al., thereby establishing a lower bound of $e/(e-1)$ on the PoA in their context. With the monotonicity conjecture validated for these instances by the authors of [5], we achieve the first tight bound on the PoA for a non-trivial class of graphs concerning Nash flows over time.

This paper is organized as follows. Upon establishing the required notation, we proceed to examine system optimal flows within our framework in Sect. 2. Subsequently, we demonstrate the unique structure of the least favorable UFR equilibrium and establish an upper bound of 2 on the PoA on linear multigraphs in Sect. 3. We wrap up our findings in Sect. 4 by formally showing a strong connection to the dynamic model and by establishing a lower bound on the PoS of $\frac{e}{e-1}$ for this class of graphs. Please note that, due to space limitations, some proofs have been omitted and are provided in the full version of this paper [18].

2   Preliminaries

We consider the classic packet routing game introduced by [21]. Formally, a packet routing game Γ is given by the tuple $\Gamma = (N, G = (V, E), \tau : E \to \mathbb{N}_{>0}, s \in V, d \in V)$, where $N = [n] := \{1, \ldots, n\}$ with $n \in \mathbb{N}_{>0}$ is the player set, G is a multigraph with nodes V and directed edges E. Additionally, $\tau(e) \in \mathbb{N}_{>0}$ denotes the transit time of an edge $e \in E$. Each player has one packet located at the source node $s \in V$ and aims to send this packet as fast as possible to the destination node d.

In this work, we focus on packet routing games on *linear multigraphs* and show the first non-trivial bounds on the PoA for packet routing games with FIFO policy. A linear multigraph for $s \neq d$ is given by $G = (V, E)$ with $V = \{s = v_0, v_1, \ldots, v_m = d\}$ and $j = j' + 1$ for all edges $(v_{j'}, v_j) \in E$. We call all edges from v_{j-1} to v_j together with the nodes v_{j-1} and v_j the j-th layer of the graph. We assume that the edges $e_1^j, e_2^j, \ldots$ in each layer j are ordered with respect to their transit times $\tau(e_1^j) \leq \tau(e_2^j) \leq \ldots$. We denote the set of all linear multigraphs by $\mathcal{G}$.

Since each player is associated with a single packet, we refer to the packet of a player as the player itself. Given G, s, and d, the strategy space $\mathcal{S} := \mathcal{P}^n$ for the players is given by all simple $s-d$ paths $\mathcal{P}$ in G. A tuple $S = (P_{(1)}, \ldots, P_{(n)}) \in \mathcal{S}$

is called a state of the game. Each state $S \in \mathcal{S}$ induces a network loading in the following sense.

Network Loading: Given a state S of the game, we define the positions of every player at any point in time by the following algorithm: We initialize at time $t = 0$ all queues $(q_e)_{e \in E}$ to be empty and position all packets at s. We add each player to the queue of the first edge in their paths. Players in queues are ordered according to FIFO, i.e., by their arrival time at e, where we break ties according to the player's index in favor of the player with the lower index. We now iterate over t until all packets have arrived at d: For each edge e we add all players i with $e \in P_{(i)}$ who have left the queue of their previous edge e' on their path at time $t - \tau(e')$ to queue q_e. Subsequently, for every edge e with a non-empty queue, we remove the first player from the queue.[4] Hence, the player arrives at $\tau(e) + t$ at the head of e. If a player i is both added to q_e and removed from q_e at the same time t, we say *player i does not queue on e*. Furthermore, when a player leaves the queue of an edge at time t, we refer to this edge as *used (at time t)*. Note that this procedure can be turned into a polynomial time algorithm by keeping track and iterating only over relevant times t, where players leave queues.

We now introduce some additional notation for the network loading: The *waiting time* $w_e^i(S)$ of player i on edge e is defined as the time that player i spends in queue q_e. The *latency* $l_e^i(S)$ of player $i \in N$ on edge $e \in E$ is defined by $l_e^i(S) := \tau(e) + w_e^i(S)$. The *workload* $l_e(S, t)$ on an edge e at time t under the strategy profile S is defined as the latency that a fictional player entering e at time t would experience on that edge if she had the largest index among all players present at q_e at that time. For the chosen $s - d$ path $P_{(i)} = (e_1, \ldots, e_m)$ of player i, the arrival time $a_{v_j}^i(S)$ of i at the head v_j of the edge e_j is given by $a_{v_j}^i(S) = \sum_{r=1}^{j} l_{e_r}^i(S)$. For a fixed strategy profile S, this yields a uniquely defined *arrival pattern* $a_v(S) = (a_v^i(S))_{i \in N}$ for every node $v \in V$, which we interpret as an ordered vector of the arrival times of all players for that node. For $v = s$, we set $a_s(S) = 0^n := (0, \ldots, 0) \in \mathbb{N}_{\geq 0}^n$, since all players start at time 0 at s. We call the number of players that arrive at node v at time t the *inflow of v at time t*. We refer to all players with arrival time t at node v as the *generation t at v*. We define the *completion time* of player $i \in N$ to be $C_i(S) := a_d^i(S)$ and the (total) *completion time* of the state to be $C(S) := \max_{i \in N} C_i(S)$. Since the network loading algorithm uniquely determines the positions of all players at any given time, we can compute all these values through a post-processing step.

Social Objective: The social objective is to minimize the completion time $C(S)$ among all $S \in \mathcal{S}$. A state $S^* \in \mathcal{S}$ is called an *optimal state* for the game if $S^* \in \mathrm{argmin}_{S \in \mathcal{S}} C(S)$.

Example: To illustrate notation, we refer the reader to Fig. 1. For the player set $N = [3]$ and the strategies $S = ((e_1^1, e_1^2), (e_2^1, e_1^2), (e_1^1, e_1^2))$ we obtain the waiting

---

[4] Note that we assume edge capacities to be equal to 1 to simplify notation, but we can extend the model to arbitrary edge capacities as we discuss in the full version.

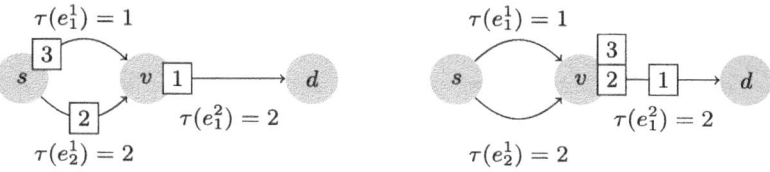

Fig. 1. An example to illustrate the notation. The left figure depicts the positions of players at $t = 1$, the right figure for $t = 2$.

times $w^1_{e^1_1}(S) = w^2_{e^1_1}(S) = 0$, $w^3_{e^1_1}(S) = 1$ in the first layer. The latencies in the first layer are $l^1_{e^1_1}(S) = 1, l^2_{e^1_1}(S) = l^3_{e^1_1}(S) = 2$ concluding in the workloads of $l_{e^1_1}(S, 0) = l_{e^1_2}(S, 0) = 3$ and the arrival pattern $a_v(S) = (1, 2, 2)$ at v. For the second layer, we obtain waiting times $w^1_{e^2_1}(S) = w^2_{e^2_1}(S) = 0$, $w^3_{e^2_1}(S) = 1$, latencies $l^1_{e^2_1}(S) = l^2_{e^2_1}(S) = 2$, $l^3_{e^2_1}(S) = 3$, workloads $l_{e^2_1}(S, 0) = 2, l_{e^2_1}(S, 1) = 3$, $l_{e^2_1}(S, 2) = 4, l_{e^2_1}(S, 3) = 3$ and the arrival pattern at d is $a_d(S) = (3, 4, 5)$ with completion time $C(S) = 5$.

Equilibria: A state $S = (P_{(1)}, \ldots, P_{(n)}) \in \mathcal{S}$ is called a *uniformly fastest route (UFR) equilibrium* if for every player $i \in N$, $P \in \mathcal{P}$ and node $v \in P_{(i)} \cap P$ it holds that $a^i_v(S) \leq a^i_v(S')$ for $S' = (P_{(1)}, \ldots, P_{(i-1)}, P, P_{(i+1)}, \ldots, P_{(n)})$. This means that in a UFR equilibrium, no player can achieve a better arrival time at any node v on her chosen path by unilaterally altering her strategy to another path. Following the work of Scarsini et al. [16], we believe that this definition aptly reflects the self-interested behavior of players in road traffic and we consider only this kind of equilibria, here[5]. The set of all such UFR equilibria is denoted by $\mathcal{S}_{\text{eq}}$. To improve readability, we often drop the term *UFR* and simply call them *equilibria* in the rest of the paper. It should be noted that this set is not empty. Assume all players select their paths sequentially in the order of their indices, starting with the player with the smallest index, and every player chooses a uniform fastest route given the choices of the players with lower indices. Since none of the later players can arrive at an intermediate node before a player with lower index, no player can displace an earlier player and we indeed obtain an equilibrium.

We start by observing that the completion time in any equilibrium of our packet routing game is realized by the player with the highest index.

Lemma 1. *For every packet routing game Γ on a linear multigraph and $S \in \mathcal{S}_{\text{eq}}$ it holds that $C(S) = C_n(S)$.*

Proof. We will show the even stronger statement that $a^i_v(S) \leq a^{i+1}_v(S)$ at every node $v \in V$. Assume this would not hold. Let v_j be the node with smallest index with $a^i_{v_j}(S) > a^{i+1}_{v_j}(S)$. Assume that player i would change her strategy only in this layer by choosing the same edge e as player $i+1$. Since $a^i_{v_{j-1}}(S) \leq a^{i+1}_{v_{j-1}}(S)$

---

[5] Details can be found in full version of this paper.

the player i enters the queue q_e not later than player $i + 1$. Since the state is the same up to node v_{j-1} the time that other players that use edge e arrive at edge e does not change. Hence, player i is removed from the queue not later than player $i + 1$ is removed in state S from this queue and thus arrives strictly earlier at v_j than in S. This contradicts $S \in \mathcal{S}_{\mathrm{eq}}$. $\qquad\square$

In essence, the proof of Lemma 1 also shows that all equilibria share a common characteristic with the specially constructed equilibrium above: all players arrive at every node and in particular at the destination d in the order of their indices.

Price of Anarchy and Price of Stability: In terms of the social objective, our goal within the game Γ is to minimize the completion time $C(S)$ across all $S \in \mathcal{S}$, i.e., the arrival time of the player arriving last at the destination node d. Since the players only optimize their own arrival times and not the social objective function, this equilibrium might not be efficient with respect to this social objective function. Two popular measures of inefficiency are the price of anarchy (PoA) and the price of stability (PoS) which for a given packet routing game Γ and an optimal state S^* are defined by

$$\mathrm{PoA}(\Gamma) := \frac{\max\limits_{S \in \mathcal{S}_{\mathrm{eq}}} C(S)}{\min\limits_{S \in \mathcal{S}} C(S)} = \frac{\max\limits_{S \in \mathcal{S}_{\mathrm{eq}}} C(S)}{C(S^*)}, \quad \mathrm{PoS}(\Gamma) := \frac{\min\limits_{S \in \mathcal{S}_{\mathrm{eq}}} C(S)}{\min\limits_{S \in \mathcal{S}} C(S)} = \frac{\min\limits_{S \in \mathcal{S}_{\mathrm{eq}}} C(S)}{C(S^*)}$$

and for a set $\mathcal{H}$ of graphs as

$$\mathrm{PoA}(\mathcal{H}) := \sup_{\Gamma : G(\Gamma) \in \mathcal{H}} \mathrm{PoA}(\Gamma), \quad \mathrm{PoS}(\mathcal{H}) := \sup_{\Gamma : G(\Gamma) \in \mathcal{H}} \mathrm{PoS}(\Gamma).$$

By definition we have $\mathrm{PoS}(\mathcal{H}) \leq \mathrm{PoA}(\mathcal{H})$. If the underlying game Γ is not immediately apparent from context, it will be indicated through a superscript to all definitions.

Understanding an Optimal State S^*: The first step in bounding the PoA and the PoS is understanding an optimal state. The optimal state for a packet routing problem sends all packets as fast as possible to the destination d. A very related problem is the discrete maximum dynamic flow problem, where as many packets as possible are being sent to d in a given time horizon. Ford and Fulkerson introduced this problem in [8] and showed how to solve it by a single static flow computation. This static flow is then repeated over time to obtain a flow maximizing the number of packets arriving at d within a given time horizon T. We will revisit these classical results to provide the following structure of an optimal state.

Proposition 2. *For a packet routing game Γ on $G \in \mathcal{G}$ let P^1 be a shortest $s - d$ path in G, and let P^j, $j > 1$ be a shortest $s - d$ path in G after the deletion of $P^1, \ldots, P^{j-1}$. Let k be such that $\tau(P^k) := \sum_{e \in P^k} \tau(e) \leq C(S^*)$ and $\tau(P^{k+1}) > C(S^*)$ if P^{k+1} exists. There is an optimal state $S^* \in \mathcal{S}$ with the following properties.*

1. *There are no queues in any layer $\ell \geq 2$.*
2. $C(S^*) + 1 - \tau(P^1)$ *packets use* P^1 *and* $C(S^*) + \delta_j - \tau(P^j)$ *packets use* P^j
   with $\delta_j \in \{0, 1\}$ *for* $2 \leq j \leq k$ *and* $\sum_{j=2}^{k} \delta_j = n - k \cdot C(S^*) - 1 + \sum_{j=1}^{k} \tau(P^j)$.

3   Upper Bound on the Price of Anarchy

In this section, we will show that the PoA for the class of linear multigraphs is bounded above by 2. We first focus on subgraphs consisting of only two consecutive nodes. This approach allows us to derive certain structural properties that will be instrumental in our proof. Furthermore, we extend the model by allowing a distinct at s that differs from the zero vector, i.e., not all players may be present at s at the beginning. Instead, they arrive according to the schedule dictated by the starting pattern. Formally, players will enter the queue of the first edge of their path at their respective starting times.

In linear multigraphs, the sequence of nodes from s to d is uniquely defined. The strategy of a player in an equilibrium always consists of edges with the shortest latency to the next nodes in the sequence, given the strategy of all players of lower index. The only difference between different equilibria is the choice of the edge to use when multiple edges have the same latency for that player. A policy $\phi(\Gamma, a_s)$ is a mapping that maps a game Γ and a starting pattern a_s of the players set at node s to a state of the game that is an equilibrium.

Our attention will be centered on a specific policy, referred to as the *greedy queue policy* $\hat{\phi}(\Gamma, a_s)$. Under this policy, the players sequentially choose their edges and the current player always opts for an edge with smallest latency. In case of a tie, each player decides for one of these edges with the longest queue. If there are still multiple edges with these properties, we break ties in favor of the edge with the lowest index among these options. We denote the unique strategy vector arising from the greedy queue policy by $\hat{S} = \hat{\phi}(\Gamma, a_s)$ for any given packet routing game Γ and starting pattern a_s.

In the subsequent discussion, we aim to demonstrate that $\hat{S}$ simultaneously maximizes the arrival times of all players at d and we will bound the difference to an optimal state. To achieve this goal, we start by showing that the greedy queue policy $\hat{S} = \hat{\phi}(\Gamma, 0^n)$ produces the worst completion time in any linear multigraph setting, i.e., $C(\hat{S}) = \sup_{S \in \mathcal{S}_{eq}} C(S)$ for every Γ (Proposition 6). Afterwards, we will observe that removing edges from a network only enlarges the completion time, when players choose $s - d$ paths according to the greedy queue policy (Lemma 7). This will allow us to bound the PoA by first fixing an optimal state and, second, deleting all edges from G that are not used in the optimum. The equilibrium in the resulting graph will utilize, at most, the edges that are employed in the optimal state, thereby enabling us to establish an upper bound for the arrival times (Theorem 9).

We start by providing three fundamental properties of $\hat{S}$. The first one states that the workload on parallel used edges differs by at most one. This can be used to show the second property that the greedy queue policy maximizes the individual arrival times in a single-layer network. The third lemma guarantees that a pointwise smaller starting pattern will never worsen the completion time of any player with the greedy queue policy.

Lemma 3. *Let Γ be a packet routing game where $G(\Gamma) = (V = \{s, d\}, E)$ represents a linear multigraph with a single layer and let $a_s \in \mathbb{N}_{\geq 0}^n$ be a starting pattern at s. Furthermore, we require the players to adhere to the strategy profile $\hat{S} = \phi(\Gamma, a_s)$. If two players leave the queues on edges e_i^1 and e_j^1, $i < j$, at the same time t, then for the workload it holds that $l_{e_j^1}(\hat{S}, t) \leq l_{e_i^1}(\hat{S}, t) \leq l_{e_j^1}(\hat{S}, t) + 1$.*

Lemma 4. *Let Γ be a packet routing game with $G(\Gamma) = (V = \{s, d\}, E)$ being a linear multigraph with only one layer and $a_s \in \mathbb{N}_{\geq 0}^n$ be a starting pattern at s. If players adhere to the strategy profile $\hat{S} = \phi(\Gamma, a_s)$, then $C_i(\hat{S}) \geq C_i(S)$ for all $S \in \mathcal{S}_{eq}$ and all $i \in N$.*

Lemma 5. *Let Γ be a packet routing game with $G(\Gamma) = (V = \{s, d\}, E)$ and $a, b \in \mathbb{N}_{\geq 0}^n$ be starting patterns at s with $a \leq b$ pointwise. Let $\hat{S}^a = \phi(\Gamma, a)$ and $\hat{S}^b = \phi(\Gamma, b)$ denote the equilibria arising from the policy greedy queue for a and b, respectively. For every $i \in N$ denote by $C_i(\hat{S}^a)$ and $C_i(\hat{S}^b)$ the corresponding completion times of player i. Then, $C_i(\hat{S}^a) \leq C_i(\hat{S}^b)$ for all $i \in N$.*

Although the statement of Lemma 5 appears very natural, the proof is rather technical. The main idea of the proof is to restrict on patterns, which only differ in one position by one. Then, we introduce a dummy player and carefully swap this dummy player with consecutive players to transform one pattern into the other, while keeping track of the changes in the arrival patterns.

Now we are ready to show that the greedy queue policy constructs an equilibrium simultaneously maximizing the arrival times of all players under all equilibria.

Proposition 6. *Let Γ be a packet routing game with $G(\Gamma) \in \mathcal{G}$ (and starting pattern 0^n). The greedy queue policy constructs an equilibrium maximizing the arrival times of all players at d under all equilibria $S \in \mathcal{S}_{eq}$. Specifically, for $\hat{S} = \phi(\Gamma, 0^n)$, it holds that $C(S) \leq C(\hat{S})$.*

Proof. Fix an arbitrary equilibrium $S \in \mathcal{S}_{eq}$. We will show this statement by carefully transforming S to $\hat{S}$ in m steps. In the j-th iteration of the transformation, the players follow the strategy of S until node v_{m-j} and then follow the greedy queue policy. Denote this strategy as S^j. Observe that $S = S^0$ and $\hat{S} = S^m$. For any fixed $j \in \{0, \ldots, m-1\}$ the strategies S^j and S^{j+1} are identical until node v_{m-j-1} and therefore also the arrival patterns at v_{m-j-1}. With Lemma 4, we know that the arrival pattern of S^{j+1} at node v_{m-j} is pointwise at least as big as the arrival pattern of S^j at this node. Using Lemma 5 iteratively yields $C_i(S^j) \leq C_i(S^{j+1})$ for all $j \in \{0, \ldots, m-1\}$ and all $i \in N$. Iterating this argument yields $C_i(S^0) \leq C_i(S^m)$ for all $i \in N$, and in particular $C(S) = C_n(S) = C_n(S^0) \leq C_n(S^m) = C(S^m) = C(\hat{S})$. $\qquad\square$

To effectively compare the social optimum with any greedy Nash Equilibrium two additional properties are required. The following technical lemma demonstrates that, under the greedy queue policy, deleting edges is never beneficial for any player.

Lemma 7. *Let Γ_1 and Γ_2 be packet routing games that differ only in the underlying graphs with $G(\Gamma_1) = (V, E_1) \in \mathcal{G}$ and $G(\Gamma_2) = (V, E_1 \cup E_2) \in \mathcal{G}$ with $\hat{S}_{G_1} = \phi(\Gamma_1, 0^n)$ and $\hat{S}_{G_2} = \phi(\Gamma_2, 0^n)$. It holds that $C(\hat{S}_{G_1}) \geq C(\hat{S}_{G_2})$.*

Finally, we present our last lemma which asserts that the transit time of any player in any equilibrium in any layer with inflow at most k is bounded by the k-th shortest transit time in that layer.

Lemma 8. *Let j be a layer with at least k edges $e_1^j, e_2^j, \ldots$ in a linear multigraph. Let $S \in \mathcal{S}_{eq}$ be an equilibrium for a packet routing game on this graph. If the maximum inflow into v_{j-1} under S is at most k at any point in time, the latency of a player never exceeds the k-th shortest transit time $\tau(e_k^j)$ in that layer.*

Proof. Consider the first time t in an equilibrium where a player i uses an edge with a latency exceeding $\tau(e_k^j)$. Because t was the earliest time with this property, we know that at time $t - 1$ no player experiences a latency that exceeds $\tau(e_k^j)$. As at most k players enter the queues at time t and we remove one player from the queue of each used edge afterwards, either there are k used edges and the latencies do not increase, or there are at most $k - 1$ used edges, i.e., one of the edges $e_1^j, e_2^j, \ldots, e_k^j$ is unused. In the latter case, the latency is trivially bounded by $\tau(e_k^j)$. Thus, there is still an edge with latency $\tau(e_k^j)$, which finishes the proof. □

With Proposition 6 and all the lemmata at hand, we are now able to prove an upper bound of 2 on the PoA for the class of linear multigraphs.

Theorem 9. *For the family $\mathcal{G}$ of linear multigraphs, we have $\mathrm{PoA}(\mathcal{G}) \leq 2$.*

Proof. For a fixed packet routing game Γ, there is a k such that there exists an optimal flow S^* that is essentially temporally repeated and uses k disjoint paths due to Theorem 2. Consider the subgraph $G' = (V, E')$ of $G = (V, E)$, which consists only of all edges used by this optimal flow. Consider the induced packet routing game Γ' from Γ that differs only on the underlying graph, i.e., $G(\Gamma') = G'$ and $G(\Gamma) = G$. Clearly it holds that $C^\Gamma(S^*) = C^{\Gamma'}(S^*)$. Proposition 6 and Lemma 7 show that for the worst equilibria $\hat{S}_G = \phi(\Gamma, 0^n)$ and $\hat{S}_{G'} = \phi(\Gamma', 0^n)$ in the two graphs we have $C^\Gamma(\hat{S}_G) \leq C^{\Gamma'}(\hat{S}_{G'})$ and, hence, $\mathrm{PoA}(\Gamma) \leq \mathrm{PoA}(\Gamma')$. Thus, it suffices to show the claim for the subgraph G'.

If G' has only one layer, the definitions of the equilbria and the social optimum coincide as there are no intermediate nodes. Thus, S^* has the same arrival pattern at d as the equilibrium $\hat{S}_{G'}$ and hence $\mathrm{PoA}(\Gamma') = 1$.

If G' consists of more than one layer, we will separately discuss the latency of the last player from s to v_1 and from v_1 to d, respectively. In any UFR equilibrium S, every player aims to be as fast as possible at node v_1. Thus, for each time t, a player arrives at v_1 via all edges e^1 with $\tau(e^1) \leq t$. It is immediate that for each t, the equilibrium S maximizes the number of players arriving at v_1 by t. Additionally, the players arrive at v_1 in the order of their indices due to Lemma 1. Thus, the n-th player reaches v_1 in S not later than the last player reaches v_1 in S^*, and thus we have in particular $a_{v_1}^n(S) \leq \max_{i \in N} a_{v_1}^i(S^*) \leq C^{\Gamma'}(S^*)$. For the second part from v_1 to d, note that the inflow in each layer except the first is at most k, since G' is constructed by k edge-disjoint $s - d$ paths. Therefore, by Lemma 8, a player in an equilibrium has a maximum latency of $\tau(e_k^j)$ in any layer $j > 1$. We conclude that for each player in an equilibrium, we can bound the total latency from v_1 to d from above by $\sum_{j=2}^m \tau(e_k^j)$. As seen in Theorem 2, we have $\sum_{j=1}^m \tau(e_k^j) \leq C^{\Gamma'}(S^*)$. Therefore, we can further bound the total latency from v_1 to d from above by $\sum_{j=2}^m \tau(e_k^j) \leq C^{\Gamma'}(S^*) - \tau(e_k^1)$.

Adding both parts, we obtain

$$C^{\Gamma'}(\hat{S}_{G'}) \leq a_{v_1}^n(S) + \left(C^{\Gamma'}(S^*) - \tau(e_k^1) \right) \leq 2 \cdot C^{\Gamma'}(S^*).$$

Thus, for all Γ with $G(\Gamma) \in \mathcal{G}$ it follows

$$\text{PoA}(\Gamma) \leq \text{PoA}(\Gamma') = \frac{C^{\Gamma'}(\hat{S}_{G'})}{C^{\Gamma'}(S^*)} \leq \frac{2 \cdot C^{\Gamma'}(S^*)}{C^{\Gamma'}(S^*)} \leq 2 \,.$$

This yields $\text{PoA}(\mathcal{G}) = \sup_{\Gamma : G(\Gamma) \in \mathcal{G}} \text{PoA}(\Gamma) \leq 2$, which finishes the proof. $\square$

4   Lower Bound on the Price of Stability

4.1   Constructing Bad Instances

In this section, we will establish a lower bound on the PoS and therefore also on the PoA. We will construct a sequence of packet routing games $(\Gamma_i)_{i \in \mathbb{N}}$ with $\text{PoA}(\Gamma_i) = \text{PoS}(\Gamma_i)$ and $\lim_{i \to \infty} \text{PoS}(\Gamma_i) \geq \frac{e}{e-1}$, where each $G(\Gamma_i) \in \mathcal{G}$, i.e., each underlying graph is a linear multigraph. This lower bound is slightly above 1.582.

We introduce a subclass $\mathcal{G}_{k,\ell} \subseteq \mathcal{G}$ for $0 \leq \ell < k$, $\ell \in \mathbb{N}_{\geq 0}$ and $k \in \mathbb{N}_{>0}$. A graph $G \in \mathcal{G}_{k,\ell}$ is a linear multigraph that consists of nodes $V = \{s = v_0, v_1, \ldots, v_{k-\ell} = d\}$ and k edges in every layer. In layer j from v_{j-1} to v_j, these k edges consist of $j - 1$ edges with a layer specific transit time of τ_j (called special edges) and $k - (j - 1)$ edges with a transit time of 1 (called standard edges). In Fig. 2, the graphs are visualized for different values of ℓ.

An informal interpretation of the graph class $\mathcal{G}_{k,\ell}$ is at hand. The value k clearly represents the number of edges in each layer of the graph. Furthermore, it also determines the maximum number of nodes in the graph, which is $k + 1$. On the other hand, ℓ signifies the number of layers that are truncated from the graph. Thus, a graph in $\mathcal{G}_{k,\ell}$ has $k - \ell + 1$ nodes. Additionally, the amount of special edges incrementally increases by one as we move from one layer to the next, starting with zero special edges in the first layer.

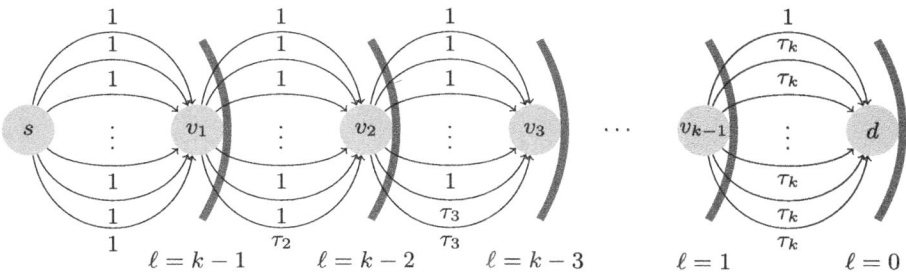

Fig. 2. Visualization of graphs in $\mathcal{G}_{k,\ell}$. For a given ℓ, the graph G consists of all layers left of the red curved line and the last node before the line serves as destination node d. (Color figure online)

Theorem 10. *There exists a sequence of packet routing games* $(\Gamma_i)_{i \in \mathbb{N}_{>0}}$ *on linear multigraphs with* $\lim_{i \to \infty} \mathrm{PoS}(\Gamma_i) \geq \frac{e}{e-1}$ *where each* $G(\Gamma_i) \in \mathcal{G}$*. Thus, we obtain* $\mathrm{PoS}(\mathcal{G}) \geq \frac{e}{e-1} \geq 1.582$*.*

Proof. For each $i \in \mathbb{N}_{>0}$, we choose Γ_i to be the game characterized by $G(\Gamma_i) \in \mathcal{G}_{k,\ell}$ with $\ell = i$ and $k = \lceil ei \rceil$. The player set is $N(\Gamma_i) = [n_i]$ with $n_i = k! = \lceil ei \rceil!$ and the transit time of the special edges in layer $j \geq 2$ is set to

$$\tau_j = \frac{n_i}{k - j + 1} - \frac{n_i}{k - j + 2} + 2.$$

The underlying concept of this construction can be explained as follows. The transit times on the special edges are deliberately set high to ensure that no equilibrium flow uses them. As a result, the network progressively narrows with each layer, leading to inevitable queuing on all standard edges. Note that the transit times of the special edges increase for higher layers. Thus, we additionally truncate the construction at a certain point to prevent the narrow layers from significantly enlarging the social optimum too much. We begin by showing by induction on the arrival times of players at intermediate nodes that no special edges are used in any equilibrium. Specifically, we show that for $j \geq 1$ at each time step $j \leq t \leq \frac{n_i}{k-j+1} + j - 1$ exactly $k - j + 1$ players arrive at v_j. Consider a fixed game Γ_i and an arbitrary equilibrium $S_i \in \mathcal{S}_{\mathrm{eq}}^{\Gamma_i}$. First, as the first layer is solely composed of k standard edges, trivially no player uses a special edge and it is immediate that k players arrive at v_1 at each time $1 \leq t \leq \frac{n_i}{k}$. Suppose by induction hypothesis that this holds up to layer j, i.e., $k - j + 1$ players arrive at time steps $j \leq t \leq \frac{n_i}{k-j+1} + j - 1$ at v_j. We have $k - j$ standard edges in layer $j + 1$, hence, the throughput on standard edges is one smaller than the inflow. Therefore, in each time step, the number of players in queues in this layer increases by 1. After $\frac{n_i}{k-j+1}$ time steps, all n_i players arrived at v_j. At this point in time, there are $\frac{n_i}{k-j+1}$ queued players in $k-j$ many queues, that is, each queue has length exactly $\frac{n_i}{(k-j+1)(k-j)} = \frac{n_i}{k-j} - \frac{n_i}{k-j+1}$. Since τ_{j+1} is strictly larger than this value, the special edges in layer $j + 1$ are indeed never used. Furthermore,

$k - j = k - (j + 1) + 1$ players arrive at each time step $j + 1 \leq t \leq \frac{n_i}{k-j+2} + j$ at v_{j+1}.

As no special edges are used by any equilibrium $S_i \in \mathcal{S}_{eq}$, this implies that every equilibrium has the same completion time. The n_i-th player arrives at node v_j at time $\frac{n_i}{k-j+1} + j - 1$ and thus we have

$$C^{\Gamma_i}(S_i) = (k - \ell) + \frac{n_i}{k - (k - \ell) + 1} - 1 = (k - \ell - 1) + n_i \cdot \left(\frac{1}{\ell + 1}\right). \quad (1)$$

As shown in the full version of this paper [18], we can derive the following upper bound on the completion time of an optimal flow.

$$C^{\Gamma_i}(S_i^*) \leq \left(\frac{3k^2 - 4k\ell - k + \ell^2 + \ell}{2k}\right) + n_i \cdot \left(\frac{k - \ell - 1}{(\ell + 1)k} + \frac{1}{k} - \frac{1}{k}\sum_{j=\ell+2}^{k}\frac{1}{j}\right) \quad (2)$$

For the j-th harmonic number H_j, it is well known that $H_j = \ln(j) + \gamma + \frac{1}{2j} - \varepsilon_j$ with $0 \leq \varepsilon_j \leq \frac{1}{8j^2}$ and the Euler-Mascheroni constant $\gamma \approx 0.577$. Thus,

$$\lim_{\ell \to \infty} \sum_{j=\ell+2}^{\lceil e\ell \rceil}\frac{1}{j} = \lim_{\ell \to \infty} \left(H_{\lceil e\ell \rceil} - H_{\ell+1}\right)$$

$$= \lim_{\ell \to \infty} \left(\ln(\lceil e\ell \rceil) + \frac{1}{2\lceil e\ell \rceil} - \varepsilon_{\lceil e\ell \rceil} - \ln(\ell + 1) - \frac{1}{2(\ell + 1)} + \varepsilon_{\ell+1}\right)$$

$$= \lim_{\ell \to \infty} \left(\underbrace{\ln\left(\frac{\lceil e\ell \rceil}{\ell + 1}\right)}_{\to \ln(e)=1} + \underbrace{\frac{1}{2\lceil e\ell \rceil} - \varepsilon_{\lceil e\ell \rceil} - \frac{1}{2(\ell + 1)} + \varepsilon_{\ell+1}}_{\to 0}\right) = 1. \quad (3)$$

Since the growth of n_i dominates the term, we obtain

$$\text{PoS}(\mathcal{G}) \geq \lim_{i \to \infty} \text{PoS}(\Gamma_i) = \lim_{i \to \infty} \frac{C^{\Gamma_i}(S_i)}{C^{\Gamma_i}(S_i^*)}$$

$$\overset{(1),(2)}{\geq} \lim_{i \to \infty} \frac{(k - \ell - 1) + n_i \cdot \left(\frac{1}{\ell+1}\right)}{\left(\frac{3k^2 - 4k\ell - k + \ell^2 + \ell}{2k}\right) + n_i \cdot \left(\frac{k-\ell-1}{(\ell+1)k} + \frac{1}{k} - \frac{1}{k}\sum_{j=\ell+2}^{k}\frac{1}{j}\right)}.$$

By simplifying this expression, we get

$$\text{PoS}(\mathcal{G}) \geq \lim_{i \to \infty} \frac{\frac{1}{\ell+1}}{\frac{k-\ell-1}{(\ell+1)k} + \frac{1}{k} - \frac{1}{k}\sum_{j=\ell+2}^{k}\frac{1}{j}} = \lim_{i \to \infty} \frac{1}{1 - \frac{\ell+1}{k}\sum_{j=\ell+2}^{k}\frac{1}{j}}$$

$$= \lim_{i \to \infty} \frac{1}{1 - \underbrace{\frac{\ell+1}{\lceil e\ell \rceil}}_{\to \frac{1}{e}} \underbrace{\sum_{j=\ell+2}^{\lceil e\ell \rceil}\frac{1}{j}}_{\to 1, \text{ due to } (3)}} = \frac{1}{1 - \frac{1}{e} \cdot 1} = \frac{1}{\left(\frac{e-1}{e}\right)} = \frac{e}{e-1},$$

which finishes the proof. $\qquad\qquad\qquad\qquad\qquad\qquad\qquad\qquad\qquad\qquad\qquad\square$

4.2   Implications for Flows over Time

Flows over time can essentially be seen as the continuous variant of packet routings. For a comprehensive introduction to the topic, we refer to Skutella's survey article [19].

Correa et al. proved an upper bound on the makespan-PoA of Nash flows over time of $\frac{e}{e-1}$ if a *monotonicity conjecture* holds [5]. In essence, the monotonicity conjecture states that the completion time of a Nash flow over time does not decrease as the inflow rate increases while the total amount of flow and the underlying network remains the same.

Proposition 11 (Correa et al. [5]). *The makespan-PoA is upper bounded by $\frac{e}{e-1}$ if the monotonicity conjecture holds.*

Correa et al. [5] proved the monotonicity conjecture for linear multigraphs. Interestingly, the same authors also provided a lower bound on the PoA of $\frac{e}{e-1}$, by a sequence of instances for which the monotonicity conjecture could not been proven. We tighten the PoA bound on linear multigraphs by observing that the instance used in the proof of Theorem 10 can be extended to Nash flows over time. The proof exploits the fact that the instance is constructed with evenly loaded edges, such that interpreting integral flow values continuously does not allow for any improvement in the travel time for any flow particle.

Proposition 12. *The makespan-PoA is lower bounded by $\frac{e}{e-1}$ even in linear multigraphs.*

5   Conclusion

We provided an upper bound of 2 on the PoA in linear multigraphs for one of the most natural packet routing games. Interestingly, our upper bound is independent of the network size and the number of players which is in stark contrast to other packet routing games. Furthermore, a sequence of linear multigraphs with a PoS that converges to at least $\frac{e}{e-1}$ has been presented. It is evident that our findings extend to serial concatenations of parallel path networks through subdivisions of edges in linear multigraphs and we obtain the same bounds on the PoA in an extended model where we allow for integer-valued edge capacities, as shown in the full version of this paper [18].

A natural next step would be to further tighten the bounds and establish an upper bound on the PoA for more general graph classes. Furthermore, considering the PoA when not restricted to UFR equilibria is an interesting research direction. Future research could explore the PoA for other relevant social objective functions as well.

Acknowledgments. TS acknowledges funding by the Deutsche Forschungsgemeinschaft (DFG) - Project number 281474342.

Disclosure of Interests. The authors have no competing interests to declare that are relevant to the content of this article.

References

1. Braess, D.: Über ein Paradoxon aus der Verkehrsplanung. Unternehmensforschung **12**, 258–268 (1968). https://doi.org/10.1007/BF01918335
2. Cao, Z., Chen, B., Chen, X., Wang, C.: A network game of dynamic traffic. In: Daskalakis, C., Babaioff, M., Moulin, H. (eds.) Proceedings of the 2017 ACM Conference on Economics and Computation, EC '17, Cambridge, MA, USA, 26–30 June 2017, pp. 695–696. ACM (2017). https://doi.org/10.1145/3033274.3085101
3. Cao, Z., Chen, B., Chen, X., Wang, C.: Bounding residence times for atomic dynamic routings. Math. Oper. Res. **47**(4), 3261–3281 (2022). https://doi.org/10.1287/moor.2021.1242
4. Cominetti, R., Correa, J.R., Larré, O.: Existence and uniqueness of equilibria for flows over time. In: Aceto, L., Henzinger, M., Sgall, J. (eds.) ICALP 2011. LNCS, vol. 6756, pp. 552–563. Springer, Heidelberg (2011). https://doi.org/10.1007/978-3-642-22012-8_44
5. Correa, J., Cristi, A., Oosterwijk, T.: On the price of anarchy for flows over time. Math. Oper. Res. **47**(2), 1394–1411 (2022). https://doi.org/10.1287/moor.2021.1173
6. Correa, J.R., Schulz, A.S., Stier-Moses, N.E.: Selfish routing in capacitated networks. Math. Oper. Res. **29**(4), 961–976 (2004). https://doi.org/10.1287/moor.1040.0098
7. Fleischer, L., Tardos, É.: Efficient continuous-time dynamic network flow algorithms. Oper. Res. Lett. **23**(3–5), 71–80 (1998)
8. Ford, L.R., Jr., Fulkerson, D.R.: Constructing maximal dynamic flows from static flows. Oper. Res. **6**(3), 419–433 (1958)
9. Han, K., Friesz, T.L., Yao, T.: Existence of simultaneous route and departure choice dynamic user equilibrium. Transport. Res. Part B: Methodol. **53**, 17–30 (2013). https://doi.org/10.1016/j.trb.2013.01.009
10. Harks, T., Peis, B., Schmand, D., Tauer, B., Vargas Koch, L.: Competitive packet routing with priority lists. ACM Trans. Econ. Comput. (TEAC) **6**(1), 1–26 (2018). https://doi.org/10.1145/3184137

11. Hoefer, M., Mirrokni, V.S., Röglin, H., Teng, S.H.: Competitive routing over time. Theor. Comput. Sci. **412**(39), 5420–5432 (2011). https://doi.org/10.1016/j.tcs.2011.05.055
12. Koch, R., Skutella, M.: Nash equilibria and the price of anarchy for flows over time. Theory Comput. Syst. **49**(1), 71–97 (2011)
13. Pigou, A.: The Economics of Welfare. Macmillan, New York (1920)
14. Rosenthal, R.W.: The network equilibrium problem in integers. Networks **3**(1), 53–59 (1973). https://doi.org/10.1002/net.3230030104
15. Roughgarden, T.: Selfish Routing and The Price of Anarchy. MIT press, Cambridge (2005)
16. Scarsini, M., Schröder, M., Tomala, T.: Dynamic atomic congestion games with seasonal flows. Oper. Res. **66**(2), 327–339 (2018). https://doi.org/10.1287/OPRE.2017.1683
17. Scheffler, R., Strehler, M., Vargas Koch, L.: Routing games with edge priorities. ACM Trans. Econ. Comput. **10**(1), 1–27 (2022). https://doi.org/10.1145/3488268
18. Schmand, D., Schürenberg, T., Strehler, M.: On the price of anarchy in packet routing games with FIFO (2025). https://doi.org/10.48550/arXiv.2502.04811
19. Skutella, M.: An Introduction to Network Flows over Time. In: Cook, W.J., Lovász, L., Vygen, J. (eds.) Research Trends in Combinatorial Optimization, pp. 451–482. Springer, Heidelberg (2009). https://doi.org/10.1007/978-3-540-76796-1_21
20. Vickrey, W.S.: Congestion theory and transport investment. Am. Econ. Rev. **59**(2), 251–260 (1969)
21. Werth, T., Holzhauser, M., Krumke, S.: Atomic routing in a deterministic queuing model. Oper. Res. Perspect. **1**(1), 18–41 (2014). https://doi.org/10.1016/j.orp.2014.05.001

On the Computational Complexity of Graph Reconstruction

Cristina Bazgan[1], Morgan Chopin[2], André Nichterlein[3],
and Camille Richer[1,2(✉)]

[1] Université Paris-Dauphine, PSL Research University, CNRS, UMR 7243,
LAMSADE Paris, France
`cristina.bazgan@lamsade.dauphine.fr`, `camille.richer@dauphine.eu`
[2] Orange Innovation, Châtillon, France
`morgan.chopin@orange.com`
[3] Algorithmics and Computational Complexity, Technische Universität Berlin,
Berlin, Germany
`andre.nichterlein@tu-berlin.de`

Abstract. One can observe a dynamic process like an epidemic or a
rumor spreading by tracking timestamps when persons become infected
or start spreading a rumor. The edges of the underlying network in which
this dynamic process unfolds is often hidden. The goal is to recover the
edges from such timestamp-information of several dynamic processes.
   We study the computational complexity of the corresponding optimization problem. Herein, given such observations from multiple
dynamic processes, the task is to find a graph with a minimum number
of edges that is consistent with all these observations. We differentiate
a highly contagious case where each infected person will infect all its
neighbors in the graph and a resistant case where a person gets infected
only after all its neighbors are infected. While the latter case turns out
to be polynomial-time solvable, the former case is NP-complete but efficiently solvable in restricted cases. Our investigations of these two cases
reveal connections to the problems of reconstructing the edge set of a
graph from vertex covers or from dominating sets, that is, given is set of
subsets where each subset forms a vertex cover or each subset forms a
dominating set in the (unknown) graph.

Keywords: Network Inference · Graph Recovery · Linear Threshold
Model · Fixed-Parameter-Tractability

1 Introduction

Diffusion processes in which an infectious behavior spreads over the vertices of
a graph capture for example diseases spreading through a population, malware
attacks on computer/mobile networks, activation of synapses in the brain, or
videos going viral, among others [11, 19]. We use a basic model where the vertices
can switch from an *inactive* state to an *active* state (modeling e. g. the vertex
getting infected or informed) but not the other way around. The activation
process proceeds in discrete time-steps. At time-step one, a subset of vertices is

I. Finocchi and L. Georgiadis (Eds.): CIAC 2025, LNCS 15679, pp. 136–152, 2025.
https://doi.org/10.1007/978-3-031-92932-8_10

Table 1. Overview over our results. The first row shows the computational complexity for the feasibility problems, the second row for the optimization problems. The number of cascades is ℓ and the maximum number of time-steps in a cascade is τ.

	threshold equal to 1	threshold equal to degree
Feas.	P (Theorem 1)	P (Theorem 1)
Opt.	NP-hard for $\tau = 2$ (Theorems 3 and 7) FPT for $\tau + \ell$ (Theorem 6) P for $\ell = 2$ and on ordered inputs (Theorems 5 and 9)	P (Theorem 2)

active. In subsequent time-steps, if an inactive vertex v has a "sufficient" number of active neighbors, then v becomes active; such propagation models have been studied extensively [1,6,16,21]. A cascade is a record of which vertices become active at each time-step. Given a set of observed cascades on a vertex set V, we aim to find an undirected graph in which all these cascades could occur.

Previous work on this problem and its variants focuses on randomized approaches and studies conditions under which the original graph can be reconstructed (with high probability) [2,8,14,18,22]. We complement this work by investigating the computational complexity of finding a graph consistent with the observed cascades in a deterministic setting; we refer to this as the feasibility problem. As many social networks are sparse, we further study the optimization variant of finding a consistent graph with a minimum number of edges.

We employ the linear threshold model [16] and consider two of its extremes that still allows for propagation: First, the threshold-1-model: an inactive vertex becomes active if at least one neighbor is active. This refers to e. g. a setting where there is essentially no immunity to an infection. In this model, if a cascade activates all vertices in two time-steps, then the vertices active at time-step one form a dominating set. Thus, if all cascades have two time-steps, then we want to reconstruct a graph from a set of dominating sets. The second model is unanimity threshold: an inactive vertex becomes active only after all its neighbors are active. It models e. g. vertices being almost immune to an infection. With unanimity threshold, there are at most two time-steps in a cascade, and the vertices in the second time-step form an independent set while the vertices in the first time-step form a vertex cover. Thus, our problem can be rephrased as reconstructing a graph without isolated vertices from a set of independent sets, respectively.

Our Results. Refer to Table 1 for an overview. We prove that the feasibility problem (see Sect. 2 for a formal definition) is polynomial-time solvable for both threshold models. For unanimity threshold, the optimization problem is polynomial-time solvable too. However, for threshold 1 the optimization problem turns out to be NP-hard, even if all cascades have only two time-steps. That is, reconstructing a graph with a minimum number of edges from dominating sets is NP-hard as well. We identify efficiently solvable special cases: First, when the number of cascades ℓ and the number of time-steps τ are small; here we show fixed-parameter tractability. Second, we obtain a polynomial-time algorithm when the vertices of the input are "ordered" in the sense that in every cascade, the vertex v_i is activated not later than v_j for all $i < j$. Finally, we extend this polynomial-time algorithm to the case in which the vertex pairs violating this property form a matching.

Related Work. In many real-world problems it is impossible to directly observe a network, but one can still observe the nodes becoming infected and hope to recover the topology of the underlying graph and its parameters by observing the "infection times" of the nodes. The problem of recovering the edges of an unknown directed graph from the observations of cascades propagating over it is known in the literature as the *Network Inference Problem* [2,4,8,13,22]. It has been studied in both the discrete [14,18,22] and the continuous setting [2,8,13]. Much of this work is based on the independent cascades model [16], which is probabilistic. The goal therein is to establish necessary and sufficient conditions to recover the directed graph with high probability, with a focus on the number of observations needed to do so [2,14,18]. However little is known about the computational complexity of this problem.

Close combinatorial problems are GRAPH RECONSTRUCTION, where the task is to recover the edges of a hidden undirected graph using a minimum number of distance queries between the vertices [15], and TARGET SET SELECTION, which captures diffusion processes and is well-studied from the viewpoint of parameterized complexity [3,5,7,10.20]. The latter is essentially the reverse of our problem: given a graph and a linear threshold function, the task is to find a minimum number of seed vertices that start the diffusion process and eventually activate all vertices. Note that using threshold 1 makes TARGET SET SELECTION trivial: select one vertex per connected component. In contrast, we show that the problem of recovering the graph under this model is algorithmically challenging.

2   Preliminaries

For $n \in \mathbb{N}$ we set $[n] = \{1, 2, \ldots, n\}$. Let G be a simple, undirected, and unweighted graph. We denote the set of vertices of G by $V(G)$ and the set of edges of G by $E(G)$. We denote the degree of a vertex $v \in V(G)$ by $\deg_G(v)$ and the set of neighbors of v by $N_G(v)$. For two subsets of vertices $W, U \subseteq V(G)$ we denote $E_G(W, U)$ the set of edges with one endpoint in W and another in U. If $U = W$, then we write $E_G(U) = E_G(U, U)$. If the graph is clear from context, then we drop the subscript. We denote with $H \subseteq G$ that H is a subgraph of G. For a subset of vertices $W \subseteq V(G)$, we denote with $G[W]$ the subgraph *induced* by W. We denote by K_n the complete graph on n vertices (also called a clique of order n).

Diffusion Model. Let G be a graph and let $\mathrm{thr} \colon V \to \mathbb{N}$ be a threshold function. We consider the propagation rule where a vertex v becomes active in a time-step if at least $\mathrm{thr}(v)$ vertices from $N_G(v)$ are active in the previous time-step. Formally, let $\mathcal{A}^t_{G,\mathrm{thr}}(S)$ denote the set of vertices of G that are active in the tth time-step when $S \subseteq V(G)$ is the set of initially active vertices, that is, $\mathcal{A}^1_{G,\mathrm{thr}}(S) = S$ and $\mathcal{A}^{t+1}_{G,\mathrm{thr}}(S) = \mathcal{A}^t_{G,\mathrm{thr}}(S) \cup \{v \in V \mid \mathrm{thr}(v) \le |N_G(v) \cap \mathcal{A}^t_{G,\mathrm{thr}}(S)|\}$. The propagation process stops when no further vertex becomes active, that is, when $\mathcal{A}^t_{G,\mathrm{thr}}(S) = \mathcal{A}^{t+1}_{G,\mathrm{thr}}(S)$ for some $t > 0$.

We consider two threshold functions: *unanimity* threshold and constant threshold equal to *one*. In unanimity threshold, we have $\mathrm{thr}(v) = \deg(v)$ for all $v \in V$. Thus, a vertex $v \in V$ is activated at time-step $t \ge 2$ if all its neighbors are activated at time-step $t - 1$. As a consequence, no vertex gets activated at time-steps $t \ge 3$. In threshold one, we have $\mathrm{thr}(v) = 1$ for each vertex $v \in V$; that is, v is activated at time-step $t \ge 2$ if at least one of its neighbors is activated at time-step $t - 1$. We denote the two cases by $\mathrm{thr} \equiv \deg$ and $\mathrm{thr} \equiv 1$ respectively.

Problem Statement. Let V be a set of vertices. A *cascade* $C = (C^1, \ldots, C^\tau)$ of length τ on V is a sequence of τ pairwise disjoint subsets of V; intuitively C is a record of observed vertex activations over time. We say a graph G is *consistent* with cascade C under threshold function thr if activating vertices C^1 at time-step 1 will get the vertices of C^t activated in the tth round of the resulting propagation process. We formally define consistency as follows:

Definition 1. *A graph G is* consistent *with a cascade C of length τ under threshold function* thr *if $C^t \subseteq \mathcal{A}^t_{G,\mathrm{thr}}(C^1) \backslash \mathcal{A}^{t-1}_{G,\mathrm{thr}}(C^1)$ for each $t = 2, \ldots, \tau$.*

By default, we assume that we have *full* cascades, that is, $\bigcup_{t=1}^\tau C^t = V$. If C is a full cascade, then we have $C^t = \mathcal{A}^t_{G,\mathrm{thr}}(C^1) \backslash \mathcal{A}^t_{G,\mathrm{thr}}(C^1)$ as no vertex is unaccounted for and, hence, no vertex could additionally be activated at time step t. If we deviate from this setting by allowing $\bigcup_{t=1}^\tau C^t \subset V$, then we explicitly mention that we allow *partial* cascades.

We observe ℓ cascades $C_1, C_2, \ldots, C_\ell$ over V, where cascade i has τ_i time-steps. Our goal is to find a graph G that is consistent with these ℓ cascades and that has a minimum number of edges. Note that if $\mathrm{thr} \equiv \deg$, then the concrete threshold values depend on G; this is not the case for $\mathrm{thr} \equiv 1$. The decision variant is defined as follows:

GRAPH RECONSTRUCTION FROM CASCADES (GRC)
Input: A set V of n vertices, ℓ full cascades $C_1, \ldots, C_\ell$ of length at most τ, $m \in \mathbb{N}$, and a threshold function $\mathrm{thr} \colon V \to \mathbb{N}$.
Question: Is there a graph $G \subseteq K_n$ over V with at most m edges that is consistent with all cascades under thr?

The input size is $n \cdot \ell$. The *feasibility* problem FGRC associated with GRC asks for a subgraph consistent with all cascades, regardless of the number of edges (i. e. GRC with $m = \binom{n}{2}$). The *optimization* problem MIN GRC asks for a subgraph consistent with all cascades and minimizing the number of edges.

If we allow partial cascades in the input, then we add PARTIAL to the problem name and we use the abbreviation GRPC. We state our hardness results for the decision problem GRC and our algorithms for MIN GRC.

We discover connections to SET COVER and its variants.

d-SET COVER
Input: A universe $U = \{u_1, \ldots, u_\ell\}$, a collection $\mathcal{F} = \{S_1, \ldots, S_n\}$ of n subsets of U of size at most d, and an integer $k \leq n$.
Question: Is there a set cover $\mathcal{F}' \subseteq \mathcal{F}$ with $|\mathcal{F}'| \leq k$ and $\cup_{S \in \mathcal{F}'} S = U$?

Without size constraint for the sets in $\mathcal{F}$, we call the problem SET COVER. Both variants are NP-complete [12]. The optimization version of these two problems (minimize the size of $\mathcal{F}'$) are indicated by the prefix MIN in the respective problem name. MIN 2-SET COVER is closely related to MIN EDGE COVER; both problems are polynomial-time solvable [12].

3   Basic Observations and Feasibility

In this section, we establish basic observations and use them to show that the feasibility problems are polynomial-time solvable. Subsequently, we consider an instance $\mathcal{I} = (V, C_1, \ldots, C_\ell, m, \text{thr})$ of GRC where either $\text{thr} \equiv 1$ or $\text{thr} \equiv \text{deg}$.

Observation 1. *A cascade with a single time-step is consistent with any graph and can thus be ignored. Moreover if* $\text{thr} \equiv \text{deg}$, *then there can be at most two time-steps in each cascade.*

We call a vertex pair u, v a *forbidden edge* if uv is "inconsistent" with at least one cascade: For $\text{thr} \equiv \text{deg}$, this will be vertex pairs that appear together in time-step two in some cascade. For $\text{thr} \equiv 1$, this will be vertex pairs that appear in some cascade in time-steps that are too far apart. We refer to Figs. 1 and 2 for an illustration of forbidden edges (and further notation defined below).

Definition 2. *A vertex pair uv is a* forbidden edge *if there exists a cascade C_i, $i \in [\ell]$ such that:*

- $\text{thr} \equiv \text{deg}$ *and* $u, v \in C_i^2$; *or*
- $\text{thr} \equiv 1$ *and* $u \in C_i^t$, $v \in C_i^{t'}$, *with* $|t - t'| > 1$.

As the name indicates and as we show next, these forbidden edges are exactly the edges that cannot be in any solution.

Lemma 1. *Let G be a graph over V consistent with all cascades. Then there is no forbidden edge in G. Moreover, if $uv \notin E(G)$ is not forbidden, then $G' = (V, E(G) \cup \{uv\})$ is consistent with all cascades.*

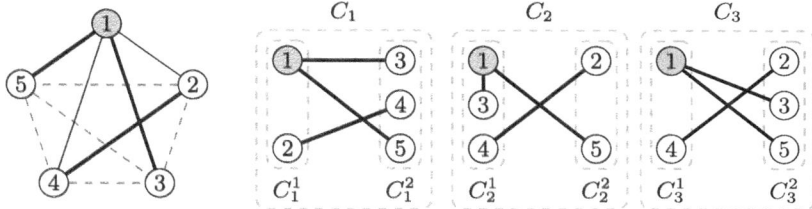

Fig. 1. The three cascades on the right form an instance of MIN GRC with threshold unanimity. The cascades are highlighted by dashed boxes, as well as the time-steps within the cascades (from left to right). On the left is a K_5: dashed edges are forbidden, solid edges form the feasibility graph G_f and bold edges form an optimal solution that is also represented in the cascades. The vertex 1 highlighted in gray forms the set V^1. The associated MIN 2-SET COVER instance (see Theorem 2) is $U = \{2, \ldots, 5\}$ and $\mathcal{F} = \{\{2\}, \{3\}, \{4\}, \{5\}, \{2, 4\}\}$. (Color figure online)

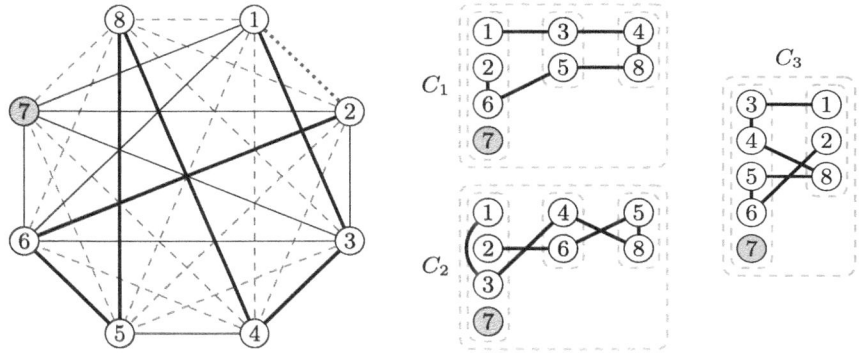

Fig. 2. The three cascades on the right form an instance of MIN GRC with threshold 1. On the left is a K_8: dashed edges are forbidden, the dotted edge is useless (see Definition 3), solid edges form the feasibility graph G_f and bold edges form an optimal solution that is also represented in the cascades (see Theorem 5). The vertex 7 highlighted in gray forms the set V^1 and it is isolated in this optimal solution graph.

Proof. Assume, for sake of contradiction, that G contains a forbidden edge uv. If thr $\equiv 1$, then there is a cascade C_i where u and v are at least two time-steps apart; say $u \in C_i^t$ and $v \in C_i^{t'}$ with $t' > t + 1$. Then u being active at time-step t would imply v becoming active at time-step $t + 1 < t'$, contradicting C_i. If thr $\equiv$ deg, according to Lemma 1 we can assume that each cascade has exactly two time-steps. Then the second time-step of each cascade must be an independent set in the solution.

Now let $uv \notin E(G)$ be a non-forbidden vertex pair. If thr $\equiv 1$, then u can be at most one time-step ahead of v (and vice versa). Thus, neither u nor v is activated earlier in G' because of uv, and adding an edge cannot result in a vertex being activated later. If thr $\equiv$ deg, as uv is not forbidden, it follows that in each cascade at least one of u and v is in the first time-step. Fix an arbitrary cascade C_i and assume wlog. that $u \in C_i^1$ and $v \in C_i^2$. As G is consistent with C_i, it follows that $N_G(v) \subseteq C_i^1$. Thus, adding uv results in $N_{G'}(v) \subseteq C_i^1$ and v is activated at time-step two. Hence, G' is consistent with all cascades.   □

We call a vertex pair u, v a *useless edge* if uv, although not forbidden, can be safely excluded from a solution.

Definition 3. *A vertex pair uv is a* useless edge *if*

- $\mathrm{thr} \equiv \deg$ *and for all $i \in [\ell], u, v \in C_i^1$; or*
- $\mathrm{thr} \equiv 1$ *and for all $i \in [\ell]$, there is $t \in [\tau_i]$ such that $u, v \in C_i^t$.*

Lemma 2. *Let G be a graph over V consistent with all cascades. If uv is a useless edge, then $G' = (V, E(G) \setminus \{uv\})$ is consistent with all cascades.*

Proof. As uv is useless, for each cascade, u and v are activated in the same time-step. If $\mathrm{thr} \equiv 1$, then removing uv does not change the time-steps at which u and v become active; thus $G' = (V, E(G) \setminus \{uv\})$ is consistent with all cascades. If $\mathrm{thr} \equiv \deg$, then G being consistent implies that $u, v \in C_i^1$ for each $i \in [\ell]$. Hence, deleting uv does not change the consistency of the graph with the cascades.   □

Observe that in a feasible solution, for each $t \geq 2$ and $i \in [\ell]$, every $v \in C_i^t$ has at least one neighbor $u \in C_i^{t-1}$. We say that u *covers* v in C_i, or alternatively that uv covers v in C_i. We define

$$V^1 := \bigcap_{i=1}^{\ell} C_i^1$$

the set of vertices that are activated without propagation in every cascade. The complement set $\overline{V^1} = V \setminus V^1$ then corresponds to the vertices that need to be activated by propagation in at least one cascade. Observe that for $\mathrm{thr} \equiv 1$ and $\mathrm{thr} \equiv \deg$ we can remove all but one vertex of V^1 from the instance.

Data Reduction Rule 1. *If $u, v \in V^1$, then delete u.*

Lemma 3 ($\star$). *Data Reduction Rule 1 is correct.*

The proof of Lemma 3 is deferred to a full version. We next show that focusing on non-forbidden, non-useless edges is sufficient for checking feasibility. To this end, we introduce further notation.

Definition 4. *The feasibility graph is $G_f = (V, E_{\mathrm{thr}})$ where E_{thr} is the set of non-forbidden, non-useless vertex pairs under the given threshold function thr.*

We next show that the name feasibility graph is indeed meaningful.

Lemma 4. *The instance $\mathcal{I}$ is feasible if and only if the feasibility graph $G_f = (V, E_{\mathrm{thr}})$ is consistent with all cascades.*

Proof. "⇒:" If $\mathcal{I}$ is feasible, then there exists a graph $H = (V, F)$ that is consistent with all cascades. By Lemmas 1 and 2, we can assume that H does not contain any forbidden or useless edges. Hence, $H \subseteq G_f$ and Lemma 1 implies that G_f is consistent with all cascades.

"⇐:" If G_f is consistent with all cascades, then G_f certifies that $\mathcal{I}$ is indeed feasible.   □

Note that checking whether G_f is consistent with all cascades can be done in polynomial time: For each cascade C_i, simply activate all vertices in C_i^1, simulate the propagation process, and compare the results with C_i. Hence, we obtain:

Theorem 1. *FGRC is polynomial-time solvable for* thr $\equiv 1$ *and* thr $\equiv$ deg.

For unanimity threshold thr $\equiv$ deg, we can also solve the corresponding optimization problem in polynomial time.

Theorem 2. *MIN GRC is polynomial-time solvable for* thr $\equiv$ deg.

Proof. Consider an instance $\mathcal{I} = (V, C_1, \ldots, C_\ell, \text{thr} \equiv \text{deg})$ of MIN GRC. By Theorem 1 we can verify the feasibility in polynomial-time. Assume now that the instance is feasible and let G_f be the feasibility graph.

Note that no vertex in $\overline{V^1}$ can be isolated in G_f, as it would not be activated. Moreover, if no vertex in $\overline{V^1}$ is isolated in a spanning subgraph $G \subseteq G_f$, then G is feasible: For each cascade C_i, each vertex $u \in C_i^2$ needs at least one neighbor in C_i^1 and by definition of G, we have that $N_G(v) \subseteq N_{G_f}(v) \subseteq C_i^1$. Hence, we need to select as few edges as possible that cover all vertices in $\overline{V^1}$. This is essentially a MIN 2-SET COVER-instance $(U, \mathcal{F})$: The universe is $\overline{V^1}$ (a vertex in V^1 does not need to be covered). For each edge $e \in E(G_f)$, add the set $e \setminus V^1$ in $\mathcal{F}$. The sets in an optimum solution of $(U, \mathcal{F})$ directly correspond to the edges of an optimal solution (see Fig. 1 for an example). As MIN 2-SET COVER is polynomial-time solvable [12], the theorem follows.   □

4   Bounded Input Dimension

In this section, we consider all thresholds being equal to one and study the cases when ℓ or τ (or both) are bounded.

Bounded τ. We start with the case in which each cascade has at most two timesteps and state it is NP-complete; the proof is deferred to a full version.

Theorem 3 (⋆). *GRC with* thr $\equiv 1$ *is NP-complete for* $\tau = 2$.

Note that GRC is fixed-parameter tractable with respect to the number of edges in the solution graph: Exhaustively applying Data Reduction Rule 1 results in an instance with $|V^1| \leq 1$. As every vertex in $\overline{V^1}$ requires at least one incident edge, it follows that the number of edges in the solution graph is $\Theta(n)$. Hence, brute-forcing all possible solution graphs and selecting the one with the fewest edges is sufficient to show fixed-parameter tractability. In contrast, we obtain W[2]-hardness when allowing partial cascades in the input.

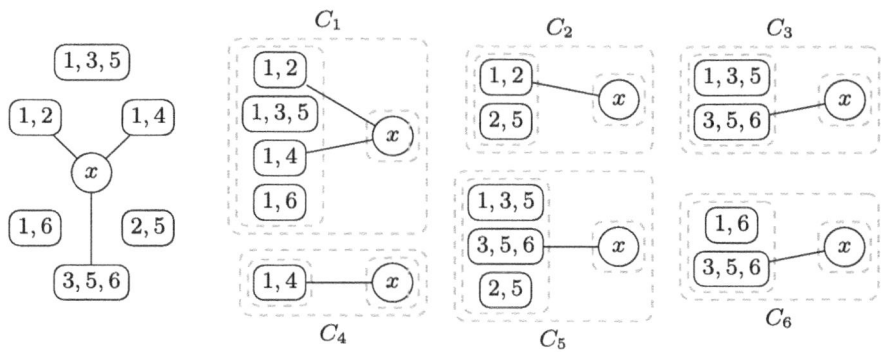

Fig. 3. Illustration of the reduction from SET COVER to GRPC with thr $\equiv 1$: On the right are six cascades constructed from the following instance of SET COVER: $U = \{1,2,3,4,5,6\}$, $\mathcal{F} = \{S_1 = \{1,2\}, S_2 = \{1,3,5\}, S_3 = \{1,4\}, S_4 = \{1,6\}, S_5 = \{2,5\}, S_6 = \{3,5,6\}\}$. On the left a graph consistent with the cascades.

Theorem 4. *GRPC with* thr $\equiv 1$ *and* $\tau = 2$ *is W[2]-hard parameterized by the number of edges in the solution graph.*

Proof. We construct a parameterized reduction from SET COVER, which is W[2]-complete with respect to k [9]. Given an instance $\mathcal{I} = (U, \mathcal{F}, k)$ of SET COVER, we define an instance $\mathcal{I}' = (V, C_1, \ldots, C_\ell, m, \text{thr} \equiv 1)$ of GRPC as follows: Set $V = \mathcal{F} \cup \{x\}$ where x is a dummy vertex. For each $u_i \in U$, create a partial cascade $C_i = (\{S_j \in \mathcal{F} \mid u_i \in S_j\}, \{x\})$ (see Fig. 3 for an illustration). Set $m = k$. We show now that $\mathcal{I}$ has a set cover of size at most k if and only if there is a graph consistent with all cascades in $\mathcal{I}'$ with at most m edges.

"$\Rightarrow$" Let $\mathcal{F}'$, $|\mathcal{F}'| \leq k$, be a solution of $\mathcal{I}$. Then the graph G where x is connected to vertices S_i such that $S_i \in \mathcal{F}'$ is consistent with the given cascades and has at most m edges.

"$\Leftarrow$" Suppose that there is a graph G consistent with the cascades in $\mathcal{I}'$ with at most m edges. For each $i \in [\ell]$, G is consistent with cascade C_i so x must be connected to at least one S_j with $u_i \in S_j$ for each $i \in [\ell]$. Moreover x is incident to at most k edges, so $N_G(x)$ is a cover of U with at most k subsets. $\square$

Bounded ℓ. We next consider the case in which ℓ is small. As a corollary of our next result we obtain a polynomial-time algorithm for $\ell = 2$. However, the complexity for constant $\ell \geq 3$ remains open.

We say that an edge uv is *useful* in cascade C if it is non-forbidden and its endpoints u, v are in distinct time-steps in C (u covers v or v covers u in C). A non-forbidden, non-useless edge must be useful in at least one cascade.

Theorem 5. *MIN GRC with* thr $\equiv 1$ *is polynomial-time solvable when every edge of the feasibility graph is useful in at most 2 cascades.*

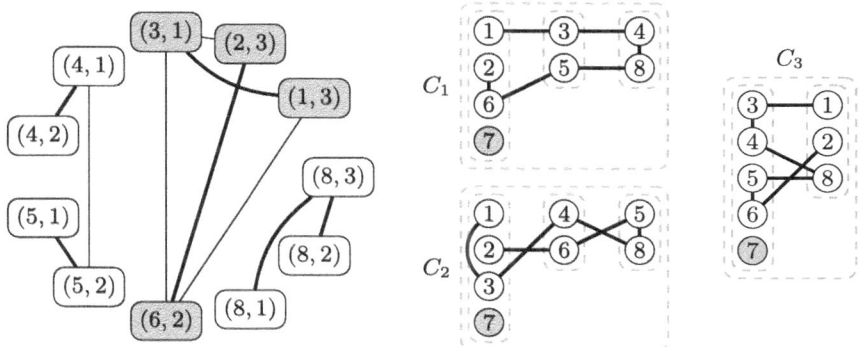

Fig. 4. Illustration of the reduction from MIN GRC with thr $\equiv 1$ to MIN 2-SET COVER. The MIN GRC instance is that of Fig. 2. The reduced instance is represented on the left: U is the set of vertices, $\mathcal{F}$ consists of sets of size 2 (the edges) and singletons (gray vertices). Since every singleton is contained in a set of size 2, we can ignore them. Bold edges represent an optimal solution $S_{13}, S_{26}, S_{34}, S_{48}, S_{56}, S_{58}$ and the corresponding solution to MIN GRC is represented with the cascades on the right.

Proof. We provide a polynomial-time reduction from MIN GRC to MIN SET COVER. Let $\mathcal{I} = (V, C_1, \ldots, C_\ell, \text{thr} \equiv 1)$ be an instance of MIN GRC with thr $\equiv$ 1. We construct in polynomial time an instance $\mathcal{I}' = (U, \mathcal{F})$ of MIN SET COVER (see Fig. 4 for an illustration).

Set the universe U as the set of pairs (v, i) such that the vertex $v \in \overline{V^1}$ must be covered in cascade C_i: $U = \{(v, i) \mid i \in [\ell], v \in V \backslash C_i^1\}$. For each $e \in E_{\text{thr}}$, e is useful in at least one cascade, and in each cascade C_i where e is useful, e covers the endpoint located in the later time-step. Then when e is taken in the solution graph, the elements of U that are covered are $S_e = \{(v, i) \in U \mid \exists t \in \{2, \ldots, \tau_i\}, C_i^t \cap e = \{v\}$ and $C_i^{t-1} \cap e \neq \emptyset\}$. Set $\mathcal{F} = \{S_e \mid e \in E_{\text{thr}}\}$.

Observe that any set cover $\mathcal{F}' \subseteq \mathcal{F}$ corresponds to an edge set E such that (V, E) is consistent with all cascades and vice versa. Hence, this reduction is also approximation-preserving.

The size of S_e is the number of cascades in which e is useful. Thus all subsets in $\mathcal{F}$ have size at most ℓ, and if every edge is useful in at most 2 cascades then every S_e has size at most 2 and we reduce to the polynomial-time solvable MIN 2-SET COVER.   □

Corollary 1. *MIN GRC with* thr $\equiv 1$ *is polynomial-time solvable when* $\ell = 2$.

Bounded τ and ℓ. In the last part of this section we turn to the case in which τ and ℓ are small. We obtain fixed-parameter tractability by constructing an ILP where the number of variables is bounded by a (exponential) function in τ and ℓ. This construction is based on the following notion that classifies the vertices:

Definition 5. *Let $v \in V$. The class γ of v is an ℓ-tuple where the i-th coordinate γ_i is the activation time-step of v in C_i. A class γ is represented in a set of vertices if there exists at least one vertex of class γ in it.*

Observe that an edge is forbidden between two vertices of classes γ, β if $\max_{i \in [\ell]} |\gamma_i - \beta_i| > 1$. An edge is useless between two vertices of the same class.

Theorem 6. *MIN GRC with thr $\equiv 1$ parameterized by the number of cascades ℓ and the maximum number of time-steps in a cascade τ is fixed-parameter tractable.*

Proof. We provide a fixed-parameter tractable algorithm for the optimization problem. The algorithm first computes the number of edges between every two classes in an optimal solution using an ILP. In a second step, it constructs a solution matching the numbers provided by the ILP.

Let $\mathcal{I} = (V, C_1, \ldots, C_\ell, \text{thr} \equiv 1)$ be an instance of MIN GRC. Let Θ be the set of classes represented in V: $|\Theta| \leq \tau^\ell$. For $i \in [\ell]$ and $t \in [\tau_i]$, denote Θ_i^t the classes represented in C_i^t. Given a class $\gamma \in \Theta$, denote $V(\gamma) \subseteq V$ the vertices of that class. We say that a vertex has *class-neighborhood* $X \subseteq \Theta$ in a solution graph if it has at least one neighbor in each $V(\beta), \beta \in X$ and no neighbor of a class not in X.

We now specify our ILP. For each class γ and class-neighborhood $X \subseteq \Theta$, we add a variable $x_{\gamma,X}$ denoting the number of vertices in $V(\gamma)$ that (will) have class-neighborhood X. For each pair of classes γ, β, we add a variable $y_{\gamma,\beta}$ denoting the (futur) number of edges in $E(V(\gamma), V(\beta))$. Whenever $E(V(\gamma), V(\beta))$ should be empty because these edges are forbidden or useless, the corresponding y-variable is set to zero and the x-variables $x_{\gamma,X}$ such that $\beta \in X$ are set to zero too. For a fixed γ, there are at most 3^ℓ classes that can have an edge with γ. Thus there are at most $(3\tau)^\ell$ non-zero y-variables and at most $\tau^\ell \cdot 2^{3^\ell}$ non-zero x-variables. The ILP is as follows:

$$\min \sum_{\gamma,\beta \in \Theta} y_{\gamma,\beta} \tag{1}$$

$$\text{s.t. } y_{\gamma,\beta} \geq \sum_{X \subseteq \Theta: \beta \in X} x_{\gamma,X} \qquad \forall \gamma, \beta \in \Theta \tag{2}$$

$$\sum_{X \subseteq \Theta: X \cap \Theta_i^{t-1} \neq \emptyset} x_{\gamma,X} = |V(\gamma)| \qquad \forall i \in [\ell], 2 \leq t \leq \tau_i : \gamma \in \Theta_i^t \tag{3}$$

$$\sum_{X \subseteq \Theta} x_{\gamma,X} = |V(\gamma)| \qquad \forall \gamma \in \Theta \tag{4}$$

$$y_{\gamma,\beta}, x_{\gamma,X} \in \{0, 1, 2, \ldots\} \qquad \forall \gamma, \beta \in \Theta, X \subseteq \Theta \tag{5}$$

The objective (1) is to minimize the total number of edges in the graph. Constraints (2) monitor the y-variables: $y_{\gamma,\beta}$ is the maximum between the number of vertices in $V(\gamma)$ that have a neighbor in $V(\beta)$, and the number of vertices in $V(\beta)$ that have a neighbor in $V(\gamma)$. Constraints (3) ensure feasibility: for

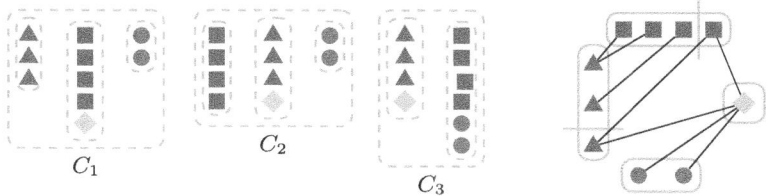

Fig. 5. The three cascades on the left form an instance of MIN GRC with thr ≡ 1 and 4 classes. On the right is an optimal solution constructed from a solution of the ILP. The square and triangle classes are subdivided according to the class-neighborhood: for instance, the square class is divided between those that have a neighbor in the triangle class and those that have one in the triangle class and in the diamond class.

each cascade C_i and each time-step $t \geq 2$, all vertices of class γ represented in C_i^t should have at least one neighbor in a class represented in C_i^{t-1}, partitioning $V(\gamma)$ according to the class-neighborhoods $X \in \Theta : X \cap \Theta_i^{t-1} \neq \emptyset$. Constraints (4) ensure that the partitions defined by (3) are identical for each γ.

To prove correctness it remains to show that the ILP admits a solution with objective value at most m if and only if there exists a graph G consistent with all cascades and with at most m edges.

$\Rightarrow$ Suppose there is a feasible solution to the ILP that achieves an objective value at most m. We next construct a solution to $\mathcal{I}$ with at most m edges.

Start from the empty graph $G = (V, E)$ on V. Partition the vertices according to their class. Then for each class γ, partition $V(\gamma)$ according to the variables $(x_{\gamma,X})_{X \subseteq \Theta}$: for each $x_{\gamma,X} > 0$, there is a subset of $x_{\gamma,X}$ vertices that is *labeled* by X. Each vertex gets exactly one label due to constraint (4). Then add edges in G as follows: Let $\gamma, \beta \in \Theta$. Denote by $V_{\gamma \to \beta}$ the vertices of $V(\gamma)$ that are in a subset of $V(\gamma)$ labeled by a set X containing β. Similarly, denote by $V_{\beta \to \gamma}$ the vertices of $V(\beta)$ that are in a subset of $V(\beta)$ labeled by a set X containing γ. All $y_{\gamma,\beta}$ edges between $V(\gamma)$ and $V(\beta)$ must be between $V_{\gamma \to \beta}$ and $V_{\beta \to \gamma}$. According to constraints (2), $y_{\gamma,\beta} \geq \max(|V_{\gamma \to \beta}|, |V_{\beta \to \gamma}|)$. Add to G a maximum matching between $V_{\gamma \to \beta}$ and $V_{\beta \to \gamma}$. If there are unmatched vertices in the larger of the two sets, connect them to any vertex of the smaller set. This uses exactly $\max(|V_{\gamma \to \beta}|, |V_{\beta \to \gamma}|)$ edges. For each vertex $v \in V(\gamma)$ labeled by X, v has class-neighborhood X in G. For each $\gamma, \beta \in \Theta$, we add at most $y_{\gamma,\beta}$ edges in the solution graph. The total number of edges is thus at most m. See Fig. 5 for an illustration of the solution construction.

The solution graph is consistent with every cascade: Let $i \in [\ell], t \geq 2$ and $v \in C_i^t$ of class $\gamma \in \Theta_i^t$. Because of constraint (3), we know that $V(\gamma)$ is partitioned into subsets labeled by $X : X \cap \Theta_i^{t-1} \neq \emptyset$. The class-neighborhood of v contains at least one class $\beta \in \Theta_i^{t-1}$ and v has at least one neighbor of class β. Thus v has at least one neighbor in C_i^{t-1}.

$\Leftarrow$ Consider a feasible graph $G = (V, E)$ with at most m edges. Partition V according to the classes and partition each class according to the class-neighborhood in G. Set the variables accordingly. The objective is the number of edges in G, at most m.

The ILP has $\tau^\ell \cdot 2^{3^\ell} + (3\tau)^\ell$ variables ignoring those that are set to zero, and it has $2(3\tau)^\ell + \ell \cdot \tau^\ell + \tau^\ell$ constraints. Since an ILP instance $\mathcal{I}'$ on p variables can be solved in time $O(p^{2.5p+o(p)} \cdot |\mathcal{I}'|)$ [17], we obtain a total running time of $\tau^{O(\ell \tau^\ell 2^{3^\ell})} 2^{O((3\tau)^\ell 2^{3^\ell})} n^2$. $\qquad\square$

5   Structural Restrictions in the Feasibility Graph

In this section, we study the impact of structural restrictions of the feasibility graph G_f on the computational complexity of GRC with thr $\equiv 1$. First, we start by considering the case that G_f has small treewidth. Second, we investigate the case that G_f is "essentially directed", that is, the non-forbidden vertex pairs uv appear in all cascades in the same order.

Low Treewidth. We next prove that GRC remains NP-hard, even if the vertex cover number of G_f (i.e. number of vertices to remove to get an edgeless graph) is two. This implies that the problem remains NP-hard if G_f has treewidth two.

Theorem 7. *GRC with thr $\equiv 1$ is NP-complete for $\tau = 3$ even when the vertex cover number of the feasibility graph is 2.*

Proof. We provide a polynomial-time reduction from SET COVER that is similar to one by Gomez-Rodriguez et al. [13]. Given an instance $\mathcal{I} = (U, \mathcal{F}, k)$ of SET COVER, the reduced instance of GRC $\mathcal{I}' = (V, C_0, C_1, \ldots, C_\ell, m, \text{thr} \equiv 1)$, is defined by $V = \mathcal{F} \cup \{x, d\}$ where x, d are new vertices, $C_0 = (\{d\}, \mathcal{F} \cup \{x\})$, for each $i \in [\ell]$, $C_i = (\{S_j : u_i \in S_j\}, \{x, d\}, \{S_j : u_i \notin S_j\})$, and $m = k + n + 1$.

The feasibility graph G_f of $\mathcal{I}'$ is the complete bipartite graph between $\{x, d\}$ and $\mathcal{F}$, plus the edge xd: the edges between the vertices from $\mathcal{F}$ are either forbidden or useless. Thus, $\{x, d\}$ is a vertex cover of G_f.

We show now that $\mathcal{I}$ has a set cover of size at most k if and only if there is a graph consistent with the cascades in $\mathcal{I}'$ with at most m edges.

"$\Rightarrow$" If $\mathcal{I}$ is a yes-instance, then let $\mathcal{F}'$ be a solution of $\mathcal{I}$. Then the graph G where d is connected to all the vertices and x is connected to vertices S_i such that $S_i \in \mathcal{F}'$ is a graph consistent with the given cascades of at most m edges.

"$\Leftarrow$" Suppose that there is a graph consistent with the cascades in $\mathcal{I}'$ with at most m edges. Because of C_0, d is connected to all other vertices, that is $n + 1$ edges. For each $i \in [\ell]$, G is consistent with cascade C_i so x must be connected to at least one S_j with $u_i \in S_j$ for each $i \in [\ell]$. Moreover x is incident to xd and at most $k = m - (n + 1)$ other edges, so it gives a cover of U with at most k subsets. $\qquad\square$

Directed Feasibility Graph. We say that an edge $uv \in E_{\mathrm{thr}}$ is *unidirectional*, denoted by $u \leq v$, if for all cascades in which uv is useful, u is in the time-step before that of v. In other words, u can cover v in some cascades but v can never cover u. Otherwise, a non-forbidden, non-useless edge uv is *bidirectional*.

We can direct each unidirectional edge uv in the feasibility graph from u to v when $u \leq v$. If all edges are unidirectional, then the resulting digraph turns out to be acyclic. This will be the crucial property exploited by our polynomial-time algorithm for this case. Note that this result is in stark contrast to the NP-hardness in Theorem 4 for partial cascades: All edges in the reduction of Theorem 4 are unidirectional.

Theorem 8. MIN *GRC with* thr $\equiv 1$ *and no bidirectional edge in the feasibility graph is polynomial-time solvable.*

Proof. Let $\mathcal{I} = (V, C_1, \ldots, C_\ell, \mathrm{thr} \equiv 1)$ be an instance of MIN GRC without bidirectional edges in G_f. By Theorem 1, the feasibility of $\mathcal{I}$ can be checked in polynomial time. We now assume that $\mathcal{I}$ is feasible and we show how to compute an optimal solution.

Let D be the digraph obtained from the feasibility graph by orienting each edge uv from u to v when $u \leq v$. Then D is a directed acyclic graph: There is no bidirectional edge so no cycle of length 2. Assuming there is a cycle $(v_1, \ldots, v_k)$ of length $k \geq 3$, let C_i be a cascade such that $v_1 \in C_i^t$ and $v_2 \in C_i^{t+1}$. The next $v_3, \ldots, v_k$ must be in C_i^{t+1} or in a later time-step, making the arc $v_k v_1$ impossible. Denote by $N^-(v)$ the in-neighbors of v in D. Let $v \in \overline{V^1}$. We say that $u \in N^-(v)$ is a *best predecessor* of v if u can cover v in all cascades where v is not in the first time-step. We show that there is always a best predecessor of v.

If there is a single $u \in N^-(v)$, then u must cover v everywhere since the instance is feasible. Otherwise, let $u_1, u_2 \in N^-(v)$. Then they are always in the same time-step or in consecutive time-steps. If they are always in the same time-step, they can cover v in exactly the same cascades. If not, then they are connected by a unidirectional edge. Assuming $u_1 \leq u_2$, then u_1 can cover v in a strictly larger set of cascades than u_2.

We build an optimal solution as follows: For every vertex $v \in \overline{V^1}$, find a best predecessor u of v. Such a vertex always exists since the instance is feasible. Add the edge uv in the solution. Every $v \in \overline{V^1}$ is covered in the cascades where it needs to be covered, so the solution is feasible, and it uses exactly $|\overline{V^1}|$ edges which is the minimum number of edges required, thus it is optimal.   □

Without bidirectional edges, we were able to turn the feasibility graph in a directed acyclic graph in the previous theorem. While bidirectional edges destroy this property, we show next that the graph remains "sufficiently acyclic" for a polynomial-time algorithm when the bidirectional edges form a matching.

Theorem 9. MIN *GRC with* thr $\equiv 1$ *and such that bidirectional edges form a matching in the feasibility graph is polynomial-time solvable.*

Proof. Let $\mathcal{I} = (V, C_1, \ldots, C_\ell, \text{thr} \equiv 1)$ be an instance of MIN GRC such that any vertex is incident to at most one bidirectional edge in G_f. By Theorem 1, the feasibility of $\mathcal{I}$ can be checked in polynomial time. We now assume that $\mathcal{I}$ is feasible and we show how to compute an optimal solution.

Let D be the digraph obtained from the feasibility graph by orienting according to $u \leq v$ when the edge is unidirectional and with arcs in both directions when the edge is bidirectional. Then D has no cycle of length 3: Assume there is a cycle (v_1, v_2, v_3). According to the proof of Theorem 8, it must contain at least one bidirectional edge. Since they form a matching, it must contain exactly one. Assume $v_1 v_3$ is the bidirectional edge. Then there is a cascade C_i such that $v_1 \in C_i^t, v_3 \in C_i^{t+1}$, and a cascade C_j such that $v_3 \in C_j^{t'}, v_1 \in C_j^{t'+1}$. Since $v_1 \leq v_2$, v_2 is in $C_j^{t'+1}$ or in a later time-step, but it contradicts $v_2 \leq v_3$.

Denote by $N^-(v)$ the in-neighbors of v in D. Let $v \in \overline{V^1}$ be incident only to unidirectional edges. First we show that if $u_1, u_2 \in N^-(v)$ are connected by a bidirectional edge, then for any other $u_3 \in N^-(v)$, u_3 is connected to u_1 and u_2 with unidirectional edges.

Since u_1 and u_3 are connected to v with unidirectional edges, u_1 and u_3 are always in the same or consecutive time-steps. However u_1 and u_3 cannot be always in the same time-step, because then $u_2 u_3$ would be a second bidirectional edge incident to u_2. Thus, $u_1 u_3 \in E(G_f)$ and $u_1 u_3$ is a unidirectional edge as already one edge incident to u_1 is bidirectional. The same reasoning applies to u_2 and u_3, so $u_2 u_3$ is unidirectional.

Then we show that either $u_3 \leq u_1, u_2$ or $u_1, u_2 \leq u_3$: Assume $u_1 \leq u_3$. Then $u_2 \leq u_3$, otherwise there would be a cycle of length 3 in D. Similarly, if $u_3 \leq u_1$ then $u_3 \leq u_2$ too. So two vertices of $N^-(v)$ connected by a bidirectional edge share the same predecessors and successors in $N^-(v)$.

We now show that if there is no best predecessor of v (see proof of Theorem 8), then at least two edges are required to cover v in all cascades. Assume there is no best predecessor in $N^-(v)$. According to the proof of Theorem 8, and what was said above, it implies that there are two vertices $u_1, u_2 \in N^-(v)$ connected by a bidirectional edge and with no predecessor in $N^-(v)$. Then it is impossible to cover v in all cascades with a single edge and at least two edges are required. The edges $u_1 v, u_2 v$ are sufficient to cover v in all cascades.

We build an optimal solution as follows: For each $v \in \overline{V^1}$ that is not incident to a bidirectional edge, find a single best predecessor u if there is one. Otherwise find two predecessors u_1, u_2 connected by a bidirectional edge covering v in all cascades. Add respectively uv or $u_1 v, u_2 v$ in the solution. For each bidirectional edge uv, compute the number of edges needed to cover u and v when uv is in the solution and when it is not: When uv is not in the solution, apply the previous method to find one or two edges needed to cover u and v independently. When uv is in the solution, u and v are covered by uv in some cascades. Ignoring these cascades, find one or two edges needed to cover u, v in the remaining cascades. Keep the best of the two options.

Vertices that are not incident to a bidirectional edge are covered independently and we use the minimum number of edges to do so. Vertices that are incident to a bidirectional edge are covered independently from any other vertex except the other endpoint of the bidirectional edge. Since we compare the two options, the computed solution is optimal. □

6   Conclusion

We formalized GRAPH RECONSTRUCTION FROM CASCADES as a combinatorial problem and analyzed its computational complexity. While settling several cases, there are a few questions left for future work: First, what is the parameterized complexity of GRC for thr ≡ 1 with respect to the parameter number ℓ of cascades? Second, how well can MIN GRC be approximated? The reduction in Theorem 5 is approximation preserving and transfers the simple ℓ-approximation for ℓ-SET COVER to MIN GRC. However, is there a constant-factor approximation? Third, having only studied the two extremes thr ≡ 1 and thr ≡ deg leaves several well-studied thresholds (thr ≡ 2 or thr ≡ $\lceil \deg /2 \rceil$) unaccounted for. Finally, we only briefly discussed partial cascades, that is, the setting where we have incomplete information. This leaves a lot of open questions in this setting.

Disclosure of Interests. The authors have no competing interests to declare that are relevant to the content of this article.

References

1. Abrahamson, K.A., Downey, R.G., Fellows, M.R.: Fixed-parameter tractability and completeness IV: on completeness for W[P] and PSPACE analogues. Ann. Pure Appl. Logic **73**(3), 235–276 (1995)
2. Abrahao, B.D., Chierichetti, F., Kleinberg, R., Panconesi, A.: Trace complexity of network inference. In: The 19th ACM SIGKDD International Conference on Knowledge Discovery and Data Mining, KDD 2013, pp. 491–499. ACM (2013). https://doi.org/10.1145/2487575.2487664
3. Bazgan, C., Chopin, M., Nichterlein, A., Sikora, F.: Parameterized inapproximability of target set selection and generalizations. Computability **3**(2), 135–145 (2014). https://doi.org/10.3233/COM-140030
4. Beerliova, Z., et al.: Network discovery and verification. IEEE J. Sel. Areas Commun. **24**(12), 2168–2181 (2006)
5. Ben-Zwi, O., Hermelin, D., Lokshtanov, D., Newman, I.: Treewidth governs the complexity of target set selection. Disc. Optim. **8**(1), 87–96 (2011). https://doi.org/10.1016/j.disopt.2010.09.007
6. Chen, N.: On the approximability of influence in social networks. SIAM J. Disc. Math. **23**(3), 1400–1415 (2009)
7. Chopin, M., Nichterlein, A., Niedermeier, R., Weller, M.: Constant thresholds can make target set selection tractable. Theory Comput. Syst. **55**(1), 61–83 (2013). https://doi.org/10.1007/s00224-013-9499-3
8. Daneshmand, H., Gomez-Rodriguez, M., Song, L., Schölkopf, B.: Estimating diffusion network structures: recovery conditions, sample complexity & soft-thresholding algorithm. In: Proceedings of the 31th International Conference

on Machine Learning, ICML 2014. JMLR Workshop and Conference Proceedings, vol. 32, pp. 793–801. JMLR.org (2014). http://proceedings.mlr.press/v32/daneshmand14.html

9. Downey, R.G., Fellows, M.R.: Fundamentals of Parameterized Complexity. Springer, Heidelberg (2013)

10. Dvořák, P., Knop, D., Toufar, T.: Target set selection in dense graph classes. SIAM J. Disc. Math. **36**(1), 536–572 (2022). https://doi.org/10.1137/20m1337624

11. Easley, D.A., Kleinberg, J.M.: Networks, Crowds, and Markets - Reasoning About a Highly Connected World. Cambridge University Press,Cambridge (2010). https://doi.org/10.1017/CBO9780511761942. http://www.cambridge.org/gb/knowledge/isbn/item2705443/?site_locale=en_GB

12. Garey, M.R., Johnson, D.S.: Computers and Intractability: A Guide to the Theory of NP-Completeness. Freeman, New York (1979)

13. Gomez-Rodriguez, M., Leskovec, J., Krause, A.: Inferring networks of diffusion and influence. ACM Trans. Knowl. Disc. Data (TKDD) **5**(4), 1–37 (2012)

14. Hoffmann, J., Basu, S., Goel, S., Caramanis, C.: Learning mixtures of graphs from epidemic cascades. In: III, H.D., Singh, A. (eds.) Proceedings of the 37th International Conference on Machine Learning. Proceedings of Machine Learning Research, vol. 119, pp. 4342–4352. PMLR (2020). https://proceedings.mlr.press/v119/hoffmann20a.html

15. Kannan, S., Mathieu, C., Zhou, H.: Graph reconstruction and verification. ACM Trans. Algor. **14**(4) (2018)

16. Kempe, D., Kleinberg, J.M., Tardos, É.: Maximizing the spread of influence through a social network. Theory Comput. **11**, 105–147 (2015). https://doi.org/10.4086/toc.2015.v011a004

17. Lenstra, H.W., Jr.: Integer programming with a fixed number of variables. Math. Oper. Res. **8**(4), 538–548 (1983)

18. Netrapalli, P., Sanghavi, S.: Learning the graph of epidemic cascades. SIGMETRICS Perform. Eval. Rev. **40**(1), 211–222 (2012). https://doi.org/10.1145/2318857.2254783

19. Newman, M.: Networks: An Introduction. Oxford University Press, Oxford (2010). https://doi.org/10.1093/ACPROF:OSO/9780199206650.001.0001

20. Nichterlein, A., Niedermeier, R., Uhlmann, J., Weller, M.: On tractable cases of target set selection. Soc. Netw. Anal. Min. **3**(2), 233–256 (2013). https://doi.org/10.1007/s13278-012-0067-7

21. Peleg, D.: Local majorities, coalitions and monopolies in graphs: a review. Theor. Comput. Sci. **282**, 231–257 (2002)

22. Pouget-Abadie, J., Horel, T.: Inferring graphs from cascades: a sparse recovery framework. In: Bach, F.R., Blei, D.M. (eds.) Proceedings of the 32nd International Conference on Machine Learning, ICML 2015. JMLR Workshop and Conference Proceedings, vol. 37, pp. 977–986. JMLR.org (2015). http://proceedings.mlr.press/v37/pouget-abadie15.html

Efficient Certifying Algorithms for Linear Classification

Vincenzo Bonifaci$^{(\boxtimes)}$ ⓘ and Sara Galatro ⓘ

Dipartimento di Matematica e Fisica, Università Roma Tre, Rome, Italy
{vincenzo.bonifaci,sara.galatro}@uniroma3.it

Abstract. An efficient certifying algorithm is proposed that, given a set
of n points in $\mathbb{R}^d$ with binary labels, either returns a hyperplane separating the points, or identifies $d + 2$ of the labeled points that cannot
be separated by any hyperplane. The existence of such $d + 2$ points in
the inseparable case is known to be guaranteed by Kirchberger's theorem
in combinatorial geometry; we show how to compute these points efficiently. We then propose a dimension-free and constructive extension of
Kirchberger's theorem, where for any $\varepsilon > 0$ one finds either a separating
hyperplane, or $O(1/\varepsilon^2)$ of the labeled points that cannot be separated
with normalized margin ε by any hyperplane. Our algorithms are based
on solving one primal-dual pair of linear programs with d primal and
n dual variables, and at most $n - d$ linear equation systems with $O(d)$
equations and $O(d)$ unknowns.

Keywords: Certifying algorithm · Binary classification · Linear
duality · Combinatorial convexity · Kirchberger's theorem · Machine
learning

1 Introduction

During the last couple of decades, the notion of *certifying algorithm* has been
proposed to formalize the idea that algorithms should supplement their answers
with certificates, in order to ease the task of checking solution correctness [3,
15,19]. A certifying algorithm is an algorithm that produces, along with each
output, a *certificate* or *witness* that the particular output indeed satisfies the
input-output relation required by the computational task at hand. For instance,
while a traditional algorithm to check whether a graph is bipartite might only
return a yes/no answer, thus requiring its user to blindly trust the computation,
a certifying algorithm could instead return a 2-coloring of the graph when the
graph is bipartite, and an odd cycle subgraph when the graph is not bipartite; in
both cases the end user can easily check the answer's correctness. Additionally,
evidence of an odd cycle in the graph makes it simpler for the user to understand
why the search for a proper bipartition was fruitless.

The notion of certifying algorithm has received attention in several sub-areas
of algorithm design, such as graph recognition, and a wealth of significant algorithms have been designed (or redesigned) with the aim of making them certifying; see for example [10,11,15,20,25] and references therein. In this work, we

I. Finocchi and L. Georgiadis (Eds.): CIAC 2025, LNCS 15679, pp. 153–169, 2025.
https://doi.org/10.1007/978-3-031-92932-8_11

propose to apply the idea of certifying algorithms to machine learning tasks. Namely, we consider binary linear classification, one of the fundamental tasks in supervised machine learning [22]. In binary linear classification, one is provided with a set of n binary-labeled data points in $\mathbb{R}^d$, and the goal is to construct a hyperplane separating the data points into the two labeled classes (if possible). Clearly, if an algorithm provides a normal vector v to a supposed separating hyperplane, separation can be easily checked by computing inner products between the data points and the vector v. However, if the algorithm fails to construct a separating hyperplane, how can the user easily check that, indeed, no separating hyperplane exists? And can one identify some data points that obstruct the separation?

For binary linear classification, a result in combinatorial geometry known as Kirchberger's theorem ensures that, in the inseparable case, $d + 2$ labeled points exist, among the n given, that are inseparable (see Theorem 3 for a formal statement and references). When $d \ll n$, this would provide easily checkable evidence of the impossibility of separation. However, Kirchberger's theorem is existential. In this work, we show how to efficiently *compute* the $d + 2$ witness points guaranteed by the theorem. We then provide a constructive, dimension-free extension of Kirchberger's theorem, where for any $\varepsilon > 0$ one finds either a separating hyperplane, or $O(1/\varepsilon^2)$ of the labeled points that cannot be separated with normalized margin ε by any hyperplane (the *normalized margin* being the minimum distance between the data points and the hyperplane, normalized by the diameter of the point set; see Sect. 4 for a formal definition).

We note that, to some extent, one can see the problem of certifying linear classification as an issue in interpretable machine learning [16,27]. Citing Lipton et al. [16], "to trust an AI model [...] you might care not only about *how often* a model is right, but also *for which examples* it is right." In this sense, identification of the witness points guaranteed by Kirchberger's theorem provides the end user of the binary classification method with concrete, combinatorial counter-evidence to the linear separability hypothesis.

Main Results and Techniques. Our main results are twofold:

-   We show how Kirchberger's theorem can be made constructive and certifying, by designing an algorithm that given n binary-labeled points in $\mathbb{R}^d$, either returns a hyperplane separating the points, or identifies $d + 2$ of the labeled points that cannot be separated by any hyperplane (Theorem 5). The algorithm is based on solving one primal-dual pair of linear programs with d primal and n dual variables, and subsequently solving at most $n - d$ linear systems in $O(d)$ variables and equations each (Algorithm 1).
-   We provide a constructive and dimension-free generalization of Kirchberger's theorem, where for any $\varepsilon > 0$ one finds either a separating hyperplane, or $O(1/\varepsilon^2)$ of the labeled points that cannot be separated with normalized margin ε by any hyperplane (Theorem 6). The algorithm is based on linear programming and random sampling (Algorithm 2).

Our techniques are based on linear duality combined with ideas from combinatorial convexity. In particular, we extend proof ideas from *Carathéodory's theorem* [7] and, for the dimension-free generalization, from the so-called *Approximate Carathéodory's theorem* [1,21,28]. Both are results in combinatorial convexity: indeed, the inseparability certificates constructed by our algorithms are partly combinatorial in nature. The second algorithm is randomized, as the proof of the Approximate Carathéodory's theorem uses a probabilistic method.

Related Work. While the notion of efficiently checkable certificates is pervasive in theoretical computer science, viewing certificates as a pragmatic approach to result checking was a later development [5,19]. According to McConnell et al. [19], the term "certifying algorithm" was first used in [15]. McConnell et al. [19] explicitly define the notion of certifying algorithm and put forward the thesis (that we subscribe to) that "certifying algorithms are much superior to non-certifying algorithms, and that for complex algorithmic tasks, only certifying algorithms are satisfactory." An introduction to certifying algorithms can be found in the survey by Alkassar et al. [3]. Certifying algorithms have been designed for several problems [10,11,15,20,25], including graph connectivity, planarity testing, convex hulls, network flows, and matching; the LEDA algorithmic library implemented many checkers for such problems [20]. As far as we could establish however, certifying algorithms have not been explicitly advocated or studied in the context of machine learning tasks.

Binary linear classification is one of the fundamental problems in supervised machine learning [22]. Linear and convex programming techniques have been used to approach binary classification since at least the 1960s [9,17,24], later giving rise to the idea of Support Vector Machines (SVMs) [6]. In such a context, the usefulness of convex duality theory is well-known [22, Section 5.2].

Kirchberger's theorem [13] was proved by mathematician Paul Kirchberger (a student of David Hilbert) in 1902, in the context of approximation theory. Several proofs of the theorem are known, all relating the theorem to results in combinatorial convexity and discrete geometry, such as Carathéodory's theorem and Helly's theorem [29,30].

Carathéodory's theorem itself has seen several applications in linear and combinatorial optimization [4,21,26]. Closer to our setting, Haghighatkhah et al. [12] have used a constructive version of Carathéodory's theorem for bias removal in a classification context. Although the overarching goal of the algorithms is rather different, we reuse such a subroutine from [12, Section 2] in one of our algorithms. Our second algorithm is instead based on the Approximate Carathéodory theorem, the proof of which applies a probabilistic method [1,28].

Organization. In Sect. 2 we recall the main definitions and results relating to linear separability and combinatorial convexity, and we define the *Certified Linear Classification* problem. In Sect. 3 we present an algorithm for Certified Linear Classification and prove its correctness; this algorithm can be interpreted as a constructive and certifying extension of Kirchberger's theorem. In Sect. 4 we propose a dimension-free extension of Kirchberger's theorem, outline a corresponding algorithm and prove its correctness. We give some concluding remarks in Sect. 5.

2   Problem and Notation

Linear Separability. Let P and Q be two subsets in $\mathbb{R}^d$. Then P and Q are called (strictly) *linearly separable* if there exists a hyperplane H separating the two subsets, that is, if there exist a vector $v \in \mathbb{R}^d$ and constants $b, c \in \mathbb{R}$ ($c \neq 0$) such that

$$v^\top x + b \geq c, \text{ for all } x \in P \quad \text{and} \quad v^\top x + b \leq -c, \text{ for all } x \in Q.$$

Without loss of generality one can take $c = 1$ after rescaling. The above condition is then equivalent to the existence of $v \in \mathbb{R}^{d+1}$ such that

$$yv^\top \begin{pmatrix} x \\ 1 \end{pmatrix} \geq 1, \quad \text{for all } x \in P \cup Q, \tag{1}$$

where $y = 1$ whenever $x \in P$ and $y = -1$ whenever $x \in Q$.

Convexity. We recall some notions from combinatorial convexity. Given a set P, the *convex hull* of $P \subseteq \mathbb{R}^d$ is the smallest convex set that contains P, and it is denoted as $CH(P)$. One can also characterize the convex hull $CH(P)$ as the set of every convex combination of all finite collections of points in P:

$$CH(P) = \left\{ \sum_{i=1}^m \lambda_i x^i \;\middle|\; m \in \mathbb{N}, \sum_{i=1}^m \lambda_i = 1 \text{ and } \forall\, 1 \leq i \leq m : x^i \in P,\ \lambda_i \geq 0 \right\}$$

The relevance of convex hulls for binary classification is due to the following observation (see for example [24, Section 2] for a proof).

Remark 1. Two finite point sets P, Q are linearly separable if and only if $CH(P)$ and $CH(Q)$ do not intersect (see Fig. 1 for an illustration).

The following results in combinatorial convexity will be useful. We use the notation $\mathrm{diam}(P)$ for the diameter of P, namely $\mathrm{diam}(P) \overset{\text{def}}{=} \sup_{x,x' \in P} \|x - x'\|_2$.

Theorem 1 (Carathéodory [7,28]). *Every point in the convex hull of a set $P \subseteq \mathbb{R}^d$ can be expressed as a convex combination of at most $d+1$ points from P.*

Theorem 2 (Approximate Carathéodory [1,28]). *Consider a set $P \subseteq \mathbb{R}^d$. Then, for every point $x \in CH(P)$ and every integer $k \geq 1$, one can find points $x^1, \ldots, x^k \in P$ such that*

$$\left\| x - \frac{1}{k} \sum_{j=1}^k x^j \right\|_2 \leq \frac{\mathrm{diam}(P)}{\sqrt{k}}. \tag{2}$$

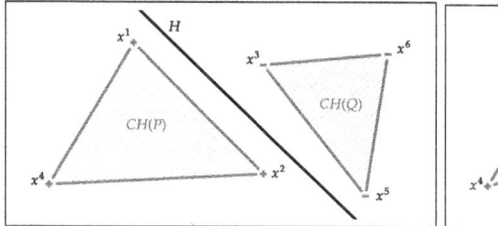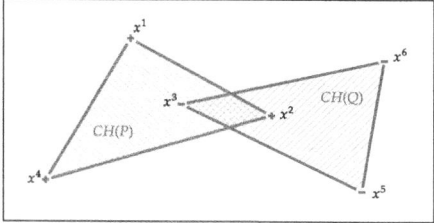

Fig. 1. Examples of linearly separable and inseparable sets. In the left subfigure, the convex hulls do not intersect; thus, it is possible to find a hyperplane H separating the two sets. On the other hand, in the right subfigure, the convex hulls have a non-empty intersection, making the sets linearly inseparable.

Theorem 3 (Kirchberger [13,30]). *Let P and Q be two finite sets of points in $\mathbb{R}^d$. Then, P and Q are linearly separable if and only if, for every set T of $d + 2$ points chosen arbitrarily from $P \cup Q$, there exists a hyperplane separating $P \cap T$ and $Q \cap T$.*

Linear Programming Duality. A *linear program* consists of the problem of minimizing a given linear function over the set of all vectors $v \in \mathbb{R}^d$ that satisfy a set of affine constraints, that is, a system of linear equalities and inequalities. Each linear program can be rewritten as

$$
\begin{aligned}
\text{minimize} \quad & b^\top v \\
\text{subject to} \quad & Av \geq c, \; v \geq 0
\end{aligned}
\tag{3}
$$

where A is a $n \times d$ real matrix, while b and c are vectors in $\mathbb{R}^d$ and $\mathbb{R}^n$, respectively. The problem defined in Eq. (3) will be called the *primal linear program*.

The *dual linear program* relies on maximizing a dual cost function:

$$
\begin{aligned}
\text{maximize} \quad & c^\top z \\
\text{subject to} \quad & A^\top z \leq b, \; z \geq 0.
\end{aligned}
\tag{4}
$$

For further details on deriving the dual from its primal, see [18].

We refer to a vector $v \in \mathbb{R}^d$ satisfying all given constraints as a *feasible solution*. Moreover, each $v^* \in \mathbb{R}^d$ that minimizes the value $b^\top v$ over all feasible solutions is called an *optimal solution* or an *optimum*.

Theorem 4 (Strong Duality [18]). *For the primal and dual programs (3) and (4), exactly one of the following holds:*

1. *Neither the primal nor the dual have a feasible solution.*
2. *The primal is unbounded (from below) and the dual has no feasible solution.*
3. *The primal has no feasible solution and the the dual is unbounded (from above).*

4. *Both the primal and the dual have a feasible solution. Then, both have an optimal solution. Furthermore, if v^* is an optimal solution of the primal and z^* is an optimal solution of the dual, then*

$$b^\top v^* = c^\top z^* \tag{5}$$

Certifying Algorithm (Informal Definition). For the purposes of this work, an informal definition of certifying algorithm will be sufficient. A *(strongly) certifying algorithm* [3,19] for a problem is, loosely speaking, an algorithm that produces, for each input x, either a witness w showing that x does not satisfy the problem's precondition, or an output y and a witness w showing that the pair (x, y) satisfies the problem's postcondition. The witness w is also called a *certificate*. The triple (x, y, w) can be forwarded to a *checker* C, another algorithm that accepts the triple if and only if w indeed proves that either x does not satisfy the precondition, or that (x, y) satisfies the postcondition. A fully formal definition of certifying algorithm can be found in [19, Section 5.1].

Problem Definition. Throughout this work, we assume to be working with n points $x^1, \ldots, x^n$ in $\mathbb{R}^d$ with binary labels $y_i \in \{-1, 1\}$, for $i = 1, \ldots, n$. Furthermore, we assume, without loss of generality, that the given points satisfy $x^i_d = 1$; if not, we can simply add an additional dimension, with the extra coordinate of every point fixed to 1. This assumption is merely for notational convenience, as it allows to rewrite the condition of Eq. (1) as

$$y_i v^\top x^i \geq 1 \text{ for } i = 1, \ldots, n \tag{6}$$

where $v \in \mathbb{R}^d$.

Problem 1 (Certified Linear Classification).
Given: n points $x^1, \ldots, x^n$ in $\mathbb{R}^d$ with binary labels $y_i \in \{-1, 1\}$.
Return: either a vector $v \in \mathbb{R}^d$ satisfying (6), or a subset of $d + 2$ labeled points that no hyperplane can separate.

Example 1. Consider the sets $P = \{x^1, x^2, x^4\}$ and $Q = \{x^3, x^5, x^6\}$ depicted in Fig. 1. In the left subfigure, the convex hulls $CH(P)$ and $CH(Q)$ do not intersect; hence, the sets are linearly separable, and a certifying algorithm may output a (properly scaled) vector normal to the hyperplane H. Conversely, the convex hulls of the sets in the right subfigure have a non-empty intersection, as the sets are linearly inseparable. A certifying algorithm thus may return any $d + 2 = 4$ points that cannot be linearly separated; for example, the subsets $P' = P$ and $Q' = \{x^3\}$.

A Note on the Computational Model. For the purpose of the running time analysis, we assume the unit-cost RAM model [2,8].

3   A Certifying Algorithm

We now discuss a certifying algorithm for linear classification, the pseudo-code of which is given in Algorithm 1. The main intuition behind the algorithm is to exploit linear programming duality to construct, in the inseparable case, a point that acts as a negative witness using Remark 1.

Algorithm 1: Certified Linear Classification

Input: n points $x^1, \ldots, x^n$ in $\mathbb{R}^d$ with binary labels $y_i \in \{-1, 1\}$.
Output: a vector v satisfying Eq. (6), or a set of $d + 2$ labeled points $P' \cup Q'$
that no hyperplane can separate.

1 $Y \leftarrow \mathbf{Diag}(y)$
2 Solve the primal-dual pair (7)–(8)
3 **if** *the primal has a solution* v^* **then**
4 | **return** *(True, v^*)*   // Return a separating hyperplane
5 **else**
6 | $z^* \leftarrow$ a nonzero dual solution   // Lemma 1
7 | $I_{P'} \leftarrow \{i \in \{1, \ldots, n\} \mid y_i = +1\}$ // Separate the indices
8 | $I_{Q'} \leftarrow \{i \in \{1, \ldots, n\} \mid y_i = -1\}$
9 | $x^* \leftarrow \sum_{i \in I_{P'}} z_i^* x^i / \sum_{k \in I_{P'}} z_k^*$ // Lemma 2
10 | **while** $|I_{P'} \cup I_{Q'}| > d + 2$ **do**   // Reduce to $d + 2$ points (Lemma 3)
11 | | **if** *there exists i such that $z_i^* = 0$* **then**
12 | | | $I_{P'} \leftarrow I_{P'} \setminus \{i\}, I_{Q'} \leftarrow I_{Q'} \setminus \{i\}$
13 | | **else**
14 | | | $\alpha \leftarrow$ a nonzero solution of Eqs. (16)-(18)
15 | | | $\rho_{P'} \leftarrow \min\{z_i^*/\alpha_i \mid i \in I_{P'}, \alpha_i > 0\}$
16 | | | $\rho_{Q'} \leftarrow \min\{z_i^*/\alpha_i \mid i \in I_{Q'}, \alpha_i > 0\}$
17 | | | $\rho \leftarrow \min\{\rho_{P'}, \rho_{Q'}\}$
18 | | | $z^* \leftarrow z^* - \rho\alpha$
19 | | | Find i such that $z_i^* = 0$
20 | | | $I_{P'} \leftarrow I_{P'} \setminus \{i\}, I_{Q'} \leftarrow I_{Q'} \setminus \{i\}$
21 | | **end**
22 | **end**
23 | **return** *(False, $I_{P'}, I_{Q'}, x^*, z^*$)*   // Return a negative witness
24 **end**

We begin from a (standard) encoding of the separability problem as a linear optimization problem, which we call the primal problem.

$$
\begin{aligned}
\text{minimize} \quad & 0^\top v \\
\text{subject to} \quad & (YX)v \geq 1, \ v \in \mathbb{R}^d,
\end{aligned}
\tag{7}
$$

Here, $X = (x^1 \cdots x^n)^\top$ is the $n \times d$ matrix of data points, and Y is the $n \times n$ diagonal matrix whose i-th entry is the label y_i of the i-th example x^i (recall that the last column of X consists entirely of ones since we assumed $x_d^i = 1$ for all data points). Any feasible solution v of problem (7) is a vector identifying a separating hyperplane, which can be interpreted as a positive certificate for the linear classification problem. The cost function is identically zero, but writing it as a linear function of v will be convenient for what follows.

To be able to deal with the negative (inseparable) case, we form the dual linear program to (7):

$$
\begin{aligned}
\texttt{maximize} \quad & \mathbf{1}^\top z \\
\texttt{subject to} \quad & (YX)^\top z = 0, \; z \in \mathbb{R}^n_{\geq 0}
\end{aligned}
\tag{8}
$$

Applying strong duality to (7)–(8), we now get the following lemma.

Lemma 1. *Either both (7) and (8) have optimal solutions and the value of both programs is zero, or (7) is infeasible and (8) has a nonzero feasible solution.*

Proof. Observe that the set of feasible solutions for the dual problem in Eq. (8) is non-empty: indeed, the zero vector $0 \in \mathbb{R}^n$ always satisfies the constraints of the dual.

By strong duality (Theorem 4), since the dual has a feasible solution, we can deduce that either both programs have an optimal solution, or the primal has no feasible solution and the dual is unbounded: respectively, cases 3 and 4 in Theorem 4. If both programs have an optimal solution, the programs' value must be zero due to the primal objective and (5). On the other hand, if the dual is unbounded, a nonzero dual feasible solution must exist. $\square$

By problem definition and Lemma 1, if both programs return an optimum, then the points $x^1, \ldots, x^n$ can be linearly separated. Furthermore, the vector v^* output as the primal solution is a normal vector associated with a separating hyperplane, which we can use as a positive certificate for linear separability.

Conversely, if the primal has no feasible solution, the points $x^1, \ldots, x^n$ are not linearly separable. Hence, we would like to extrapolate a negative certificate from the non-zero solution z^* obtained from the dual problem. Using Theorem 3, our goal will be to convert z^* into a subset of $d + 2$ non-linearly-separable points. This part of our algorithm is split into two subsequent steps, discussed in the following subsections.

3.1   Finding a Point in the Convex Hulls' Intersection

Suppose the dual is unbounded and we have computed a non-zero solution $z^* \in \mathbb{R}^n_{\geq 0}$ to the dual problem constraints in Eq. (8), meaning we found a non-zero and non-negative vector in the kernel of $(YX)^\top$. Since z^* is dual feasible,

$$
(YX)^\top z^* = 0.
\tag{9}
$$

A non-zero vector z^* verifying Eq. (9) can also be interpreted as a non-zero weight assignment to the points $x^1, \ldots, x^n$, each multiplied by its associated label y_i, such that the weighted sum is the zero vector:

$$(YX)^\top z^* = 0 \Leftrightarrow \sum_{i=1}^n z_i^* y_i x^i = 0. \tag{10}$$

Let P and Q be the sets containing the points with positive and negative labels, respectively. We can thus rewrite Eq. (10) as

$$\sum_{i \in I_P} z_i^* x^i = \sum_{j \in I_Q} z_j^* x^j, \tag{11}$$

where we used I_P and I_Q to denote the sets of indices of elements in P and Q respectively. Since we are assuming $x_d^i = 1$ for all points, from the vector sums in Eq. (11) we also deduce that the sum of the weights of elements in P equals the sum of weights of elements in Q, that is

$$\sum_{i \in I_P} z_i^* = \sum_{j \in I_Q} z_j^*. \tag{12}$$

Using Eq. (12), we can divide both members of Eq. (11) by their respective weights' sum, obtaining

$$\sum_{i \in I_P} \frac{z_i^*}{\sum_{k \in I_P} z_k^*} x^i = \sum_{j \in I_Q} \frac{z_j^*}{\sum_{k \in I_Q} z_k^*} x^j \tag{13}$$

Since $z^* \in \mathbb{R}_{\geq 0}^n$, the coefficients in both terms of Eq. (13) are non-negative, and their sums are such that

$$\sum_{i \in I_P} \frac{z_i^*}{\sum_{k \in I_P} z_k^*} = \sum_{j \in I_Q} \frac{z_j^*}{\sum_{k \in I_Q} z_k^*} = 1. \tag{14}$$

Therefore, the left and right hand sums in Eq. (13) are indeed two convex combinations in $CH(P)$ and $CH(Q)$, respectively. These convex combinations coincide and so they identify a point $x^* \in CH(P) \cap CH(Q)$, which, by Remark 1, already constitutes a *numerical* certificate of non-separability. We have shown the following.

Lemma 2. *If* (8) *has a non-zero solution* z^*, *then the point*

$$x^* \stackrel{\text{def}}{=} \sum_{i \in I_P} \frac{z_i^*}{\sum_{k \in I_P} z_k^*} x^i \tag{15}$$

satisfies $x^* \in CH(P) \cap CH(Q)$. □

It is this intersection point x^* we are now going to use to compute the $d+2$ inseparable points composing the *combinatorial* part of the certificate.

3.2   Computing the Kirchberger Points

To compute the $d+2$ points guaranteed by Kirchberger's theorem (Theorem 3), we use the following lemma and its proof, which expands on standard proofs of Caratheodory's theorem (Theorem 1); see Haghighatkhah et al. [12] for a similar algorithmic construction in a different context.

Lemma 3. *Let P and Q be two finite sets in $\mathbb{R}^d$ and $x^* \in CH(P) \cap CH(Q)$. Then one can find in time $O(nd^3)$ two subsets $P' \subseteq P$ and $Q' \subseteq Q$ such that $x^* \in CH(P') \cap CH(Q')$ and $|P'| + |Q'| \leq d + 2$, where $n = |P \cup Q|$.*

Proof. Let $p \stackrel{\text{def}}{=} |P|$ and $q \stackrel{\text{def}}{=} |Q|$. If P and Q have a point in common, we can simply return that point, so assume that $P \cup Q = \{x^1, \ldots, x^{p+q}\}$; as before, we use $I_P, I_Q \subseteq \{1, \ldots, p+q\}$ to denote the sets of indices of points in P and Q, respectively. We show that if $p + q \geq d + 3$, then one point from either P or Q can be removed with the new convex hulls still intersecting. Let $x^* \in CH(P) \cap CH(Q)$. By definition of convex hull (Eq. (2)), there are coefficients $\lambda_1, \ldots, \lambda_{p+q} \geq 0$ such that $\sum_{i \in I_P} \lambda_i = 1$, $\sum_{j \in I_Q} \lambda_j = 1$ and

$$\sum_{i \in I_P} \lambda_i x^i = x^* = \sum_{j \in I_Q} \lambda_j x^j.$$

Now, there are two possibilities:

- If any of the λ_i coefficients is zero, then we can eliminate the corresponding point x^i while keeping x^* in the intersection of the convex hulls.
- If none of the λ_i coefficients is zero, we look for coefficients $\alpha_1, \ldots, \alpha_{p+q}$ such that

$$\sum_{i \in I_P} \alpha_i x^i = \sum_{j \in I_Q} \alpha_j x^j \tag{16}$$

$$\sum_{i \in I_P} \alpha_i = 0 \tag{17}$$

$$\sum_{j \in I_Q} \alpha_j = 0. \tag{18}$$

This linear system has $d + 2$ equations and $p + q \geq d + 3$ variables. Hence, nonzero coefficients must exist that satisfy these equations.
Let $\rho_P = \min\{\lambda_i/\alpha_i \mid \alpha_i > 0, i \in I_P\}$, $\rho_Q = \min\{\lambda_j/\alpha_j \mid \alpha_j > 0, j \in I_Q\}$ and $\rho = \min\{\rho_P, \rho_Q\}$. Additionally, define the new coefficients as

$$\lambda'_i = \lambda_i - \rho \alpha_i, \quad \text{for } i = 1, \ldots, p + q. \tag{19}$$

By construction, we have $\lambda'_i \geq 0$ for $1 \leq i \leq p+q$, $\sum_{i \in I_P} \lambda'_i = \sum_{j \in I_Q} \lambda'_j = 1$, and $\sum_{i \in I_P} \lambda'_i x^i = \sum_{j \in I_Q} \lambda'_j x^j = x^*$. But at least one of the λ'_i coefficients is zero, and we can remove the corresponding point.

The above process can be repeated until $p+q = d+2$. Assuming we already have a point $x^* \in CH(P) \cap CH(Q)$ and its decomposition λ in terms of P and Q, finding the $d+2$ points relies on iteratively removing a point from $P \cup Q$ or solving a linear system of equations to eliminate a point and update the coefficients. The computational bottleneck resides in solving the linear system (16)–(18). This linear system can be defined using only $d + 3$ arbitrarily chosen points from $P \cup Q$, and thus can be solved in time $O(d^3)$ using Gaussian elimination (of course, in a practical implementation, any fast linear solver can be used). Hence, computing the desired sets P' and Q' takes at most $O(nd^3)$ time, where $n = |P \cup Q|$. □

In summary, solving the dual problem (8) leads us to a point x^* in $CH(P) \cap CH(Q)$ and its decomposition in terms of elements of P and Q (Line 9 of Algorithm 1, justified by Lemma 2). From here on (Lines 10–22 of the algorithm), we follow the proof of Lemma 3 to eliminate elements from the convex combinations, while maintaining the invariant $x^* \in CH(P) \cap CH(Q)$. By the end of the computation, we are left with two reduced subsets P' and Q' that, by construction, cannot be linearly separated. The indices of points in P' and Q' are returned as a negative certificate, together with the point x^* and its decomposition (Line 23).

Theorem 5. *The Certified Linear Classification problem can be solved in time* $O(n^{3.5}L + nd^3)$, *where L is the input size.*

Proof. The primal and dual problems (problems (7)–(8)) can be efficiently solved using any primal-dual linear programming algorithm; interior-point algorithms applied to (8) run in $O(n^{3.5}L)$ time [14,23]. When P and Q are linearly separable, a separating hyperplane is identified without further computation needed. On the other hand, if P and Q are not linearly separable, we compute $x^* \in CH(P) \cap CH(Q)$ in linear time from the dual solution using (15) and then estimate the $d + 2$ Kirchberger points from x^* following the procedure outlined in Lemma 3, which requires solving at most $n - d$ linear systems in $d + 3$ variables and $d + 2$ constraints each. □

3.3   Accompanying Checker

The last piece to our algorithm is its accompanying checker. A checking algorithm should have a simple definition, and its running time should be lower than the associated certifying algorithm [19]. Here we briefly outline a possible checker for Algorithm 1.

When Algorithm 1 returns `True`, it also returns a vector $v^* \in \mathbb{R}^d$ identifying a candidate separating hyperplane. To check correctness of this answer, the checker can simply test whether all points in P and Q satisfy Eq. (6), by computing n inner products in $\mathbb{R}^d$. The checker thus runs in $O(nd)$ time (i.e. linear in the data size) when Algorithm 1 returns a `True` statement.

On the other hand, when Algorithm 1 returns `False`, it also returns the (at most) $d+2$ points in $P' \cup Q'$, the point x^* and the decomposition vector z^*. The checker can then simply verify that $z^* \geq 0$ (in $O(n)$ time) and that Eqs. from (11) to (15) hold; this proves that $x^* \in CH(P') \cap CH(Q') \subseteq CH(P) \cap CH(Q)$, which by Remark 1 certifies the inseparability of P and Q. Each of the equations can be checked in $O(d^2)$ time, by computing the appropriate linear combination of points. This is again linear in the data size as long as $d = O(n)$.

4   A Dimension-Free Extension

We now discuss an alternative certifying algorithm for linear separation based on the Approximate Carathéodory's theorem (Theorem 2). This approach may reduce the number of points in the negative certificate, as this number becomes independent of the ambient dimension d.

Let us assume once again to be working with n points $x^1, \ldots, x^n$ in $\mathbb{R}^d$ with binary labels $y_i \in \{-1, 1\}$ and $x_d^i = 1$. Furthermore, we keep referring to P and Q as the sets of elements with positive and negative labels, respectively. Our goal is still to determine whether these n points can be separated by some hyperplane $H = \{x \in \mathbb{R}^d \mid v^\top x = 0\}$ where v satisfies Eq. (6). Recall that the *margin* of a hyperplane H is the minimum distance between H and the data points:

$$\min_{x \in P \cup Q, x' \in H} \|x - x'\|_2 .$$

We note that the margin depends on the scale of the data and thus should be related to the diameter of the dataset, $D \overset{\text{def}}{=} \max\{\text{diam}(P), \text{diam}(Q)\}$. If a hyperplane has margin εD for some $\varepsilon > 0$, we will say that it has *normalized margin* ε with respect to the given dataset. To simplify exposition and without loss of generality, we assume that D is known. We remark that an approximation[1] of D within a factor of 2 can be computed in time linear in the data size.

If the points in P and Q can be separated (with *any* margin), our certifying algorithm will identify a separating hyperplane as a positive witness; otherwise, the algorithm will identify $O(1/\varepsilon^2)$ points (possibly with duplicates) that cannot be separated with normalized margin ε as a negative witness. In other words, the algorithm will solve the following problem.

Problem 2 (Approximate Certified Linear Classification).
Parameter: $\varepsilon > 0$ (normalized margin).
Given: n points $x^1, \ldots, x^n$ in $\mathbb{R}^d$ with binary labels $y_i \in \{-1, 1\}$.
Return: a vector $v \in \mathbb{R}^d$ satisfying Eq. (6), or $O(1/\varepsilon^2)$ labeled points that no hyperplane can separate with normalized margin ε.

The main technical ingredient is the following.

Lemma 4. *Let P and Q be two finite sets in $\mathbb{R}^d$ and assume $x^* \in CH(P) \cap CH(Q)$. Then, for any $\varepsilon > 0$, there are two subsets $P' \subseteq P$ and $Q' \subseteq Q$ such that:*

---

[1] One can approximate e.g. $\text{diam}(P)$ as $\tilde{D}_P \overset{\text{def}}{=} 2 \max_{x \in P} \|p - x\|_2$, after selecting an arbitrary $p \in P$, which satisfies $\text{diam}(P) \leq \tilde{D}_P \leq 2\,\text{diam}(P)$ by the triangle inequality. Therefore, $D \leq \max\{\tilde{D}_P, \tilde{D}_Q\} \leq 2D$.

1. $|P'| = |Q'| = O(1/\varepsilon^2)$;
2. P' and Q' cannot be separated with normalized margin ε.

Proof. By Theorem 2 applied to P (respectively, Q) and x^* with $k = \lceil 4/\varepsilon^2 \rceil$, there are k points $p'_1, \ldots p'_k$ in P (resp., $q'_1, \ldots, q'_k$ in Q) such that, if we define $x^+ = (1/k) \sum_i p'_i$ (resp., $x^- = (1/k) \sum_i q'_i$),

$$\begin{aligned}
\left\| x^+ - x^* \right\|_2 &\leq \varepsilon D/2, \\
\left\| x^- - x^* \right\|_2 &\leq \varepsilon D/2.
\end{aligned} \qquad (20)$$

Therefore let us define $P' = \{p'_1, \ldots, p'_k\}$, $Q' = \{q'_1, \ldots, q'_k\}$. We note that by construction, $x^+ \in CH(P')$, $x^- \in CH(Q')$ (in fact, x^+ and x^- are the centroids of P' and Q' respectively). By (20) and the triangle inequality,

$$\left\| x^+ - x^- \right\|_2 \leq \varepsilon D. \qquad (21)$$

But if P' and Q' were separable with normalized margin ε, the distance between any point in $CH(P')$ and any point in $CH(Q')$ would be at least $2\varepsilon D$, contradicting (21). Therefore the claim is proved. □

We give the full algorithm's pseudocode in Algorithm 2. The main steps are the following:

1. Find z^* by solving the primal-dual pair (7)–(8) (lines 1–6 of Algorithm 2);
2. Find $x^* \in CH(P) \cap CH(Q)$ and its convex decompositions in terms of points of P and Q using Eq. (13): $x^* = \sum_i \lambda_i p_i$, $x^* = \sum_i \mu_i q_i$ (lines 7–11);
3. Sample $k \overset{\text{def}}{=} \lceil 16/\varepsilon^2 \rceil$ times the two probability distributions λ and μ, to obtain points $p'_1, \ldots, p'_k$ and $q'_1, \ldots, q'_k$ (lines 12–15).

The first two steps are exactly the same as in Algorithm 1, while the third step is different and is based on random sampling. Indeed, the points $p'_1, \ldots, p'_k$ and $q'_1, \ldots, q'_k$ referred to in the proof of Lemma 4 can be obtained as random samples (with replacement) from the sets P and Q according to the probability distributions that, interpreted as convex combinations, give rise to the point $x^* \in CH(P) \cap CH(Q)$. This idea is implicit in the proof of the Approximate Carathéodory theorem and is sometimes called the "empirical method" of B. Maurey [28, Chapter 0].

Theorem 6. *The Approximate Certified Linear Classification problem can be solved in randomized polynomial time.*

Proof. The only randomized step is the computation of the sets P' and Q'. It is enough to show that, with probability at least (say) $1/2$, the points $x^+ = (1/k) \sum_i p'_i$, $x^- = (1/k) \sum_i q'_i$ satisfy the inequalities (20); this probability can be increased as desired by increasing k. We take $k = \lceil 16/\varepsilon^2 \rceil$ (note that this is slightly larger than the value of k used in the proof of Lemma 4, which was

---

Algorithm 2: Approximate Certified Linear Classification

Input: $\varepsilon > 0$ and n points $x^1, \ldots, x^n$ in $\mathbb{R}^d$ with binary labels $y_i \in \{-1, 1\}$.
Output: a vector v satisfying Equation (6), or $O(1/\varepsilon^2)$ labeled points that no hyperplane can separate with normalized margin ε.

1 $Y \leftarrow \mathbf{Diag}(y)$
2 Solve the primal-dual pair (7)–(8)
3 **if** *the primal has a solution* v^* **then**
4 | **return** *(True, v^*)*   // Return a separating hyperplane
5 **else**
6 | $z^* \leftarrow$ a nonzero dual solution   // Lemma 1
7 | $I_P \leftarrow \{i \in \{1, \ldots, n\} \mid y_i = +1\}$ // Separate the indices
8 | $I_Q \leftarrow \{i \in \{1, \ldots, n\} \mid y_i = -1\}$
9 | $x^* \leftarrow \sum_{i \in I_P} z_i^* x^i / \sum_{k \in I_P} z_k^*$ // Lemma 2
10 | $\lambda \leftarrow \{z_i^* / \sum_{j \in I_P} z_j^* \mid i \in I_P\}$
11 | $\mu \leftarrow \{z_i^* / \sum_{j \in I_Q} z_j^* \mid i \in I_Q\}$
12 | $k \leftarrow \lceil 16/\varepsilon^2 \rceil$
13 | $I_{P'} \leftarrow$ sequence of k indices sampled from distribution λ
14 | $I_{Q'} \leftarrow$ sequence of k indices sampled from distribution μ
15 | **return** *(False, $I_{P'}$, $I_{Q'}$)*   // Return a negative witness
16 **end**

---

only concerned with proving existence). Following the proof of the Approximate Carathéodory theorem as in [28, Chapter 0], we can bound

$$\mathbb{E} \left\| x^+ - x^* \right\|_2^2 \le \frac{D^2}{k},$$
$$\mathbb{E} \left\| x^- - x^* \right\|_2^2 \le \frac{D^2}{k}. \tag{22}$$

Therefore, by (22) and Markov's inequality, if $\gamma \overset{\text{def}}{=} \varepsilon D/2$,

$$\Pr[\|x^+ - x^*\|_2 \ge \gamma] = \Pr[\|x^+ - x^*\|_2^2 \ge \gamma^2] \le \frac{D^2}{k\gamma^2} = \frac{4}{k\varepsilon^2} \le \frac{1}{4},$$

and similarly $\Pr[\|x^- - x^*\|_2 \ge \gamma] \le 1/4$. Thus, by a union bound, (20) is satisfied with probability at least $1/2$. The rest of the argument proceeds as in the proof of Lemma 4 and allows us to conclude that P' and Q' cannot be separated with normalized margin ε. $\square$

Finally, the associated checker to Algorithm 2 retraces the strategy of the checker presented in Subsect. 3.3 in the positive case. In the negative case, on the other hand, one only need compute the centroids of the two sets P' and Q' and check that their distance is at most εD. This requires $O(d/\varepsilon^2)$ time and $O(d)$ time, respectively. Combining the positive and negative cases, the running time of the checker is $O(nd + d/\varepsilon^2)$.

5   Conclusions

In this work we proposed to apply the idea of certifying algorithms to machine learning problems, and we focused on one of the fundamental tasks in this context, binary linear classification. We have shown that both the Certified Linear Classification problem and the Approximate Certified Linear Classification problem can be solved efficiently and that they admit simple checkers. The running time of the checker for Algorithm 1 is at most linear in the data size as long as $d = O(n)$, and the running time of the checker for Algorithm 2 is at most linear in the data size as long as $1/\varepsilon^2 = O(n)$. It would be interesting to study the Certified Linear Classification problem in a high-dimensional regime where $d \gg n$. It would also be interesting to design certifying algorithms for multi-class classification.

More generally, there are clearly other relevant machine learning problems, such as clustering, that one could also investigate from the point of view of certifying algorithms; we hope that this work can stimulate further research in this area.

Acknowledgments. This research was partly supported by project ECS 0000024 of the European Commission, *Rome Technopole*, PNRR grant M4-C2-Inv. 1.5. In particular, manuscript writing and editing were funded by Rome Technopole. The authors would also like to thank the anonymous reviewers for their feedback.

Disclosure of Interests. The authors have no competing interests to declare that are relevant to the content of this article.

References

1. Adiprasito, K., Bárány, I., Mustafa, N.H., Terpai, T.: Theorems of Carathéodory, Helly, and Tverberg without dimension. Discrete Comput. Geom. **64**(2), 233–258 (2020). https://doi.org/10.1007/s00454-020-00172-5
2. Aho, A.V., Hopcroft, J.E., Ullman, J.D.: The Design and Analysis of Computer Algorithms. Addison-Wesley (1974)
3. Alkassar, E., Böhme, S., Mehlhorn, K., Rizkallah, C., Schweitzer, P.: An introduction to certifying algorithms. it - Inf. Technol. **53**(6), 287–293 (2011). https://doi.org/10.1524/itit.2011.0655
4. Barman, S.: Approximating Nash equilibria and dense subgraphs via an approximate version of Carathéodory's theorem. SIAM J. Comput. **47**(3), 960–981 (2018). https://doi.org/10.1137/15M1050574
5. Blum, M., Kannan, S.: Designing programs that check their work. J. ACM **42**(1), 269–291 (1995). https://doi.org/10.1145/200836.200880
6. Boser, B.E., Guyon, I., Vapnik, V.: A training algorithm for optimal margin classifiers. In: Haussler, D. (ed.) Proceedings of the Fifth Annual ACM Conference on Computational Learning Theory, COLT 1992, Pittsburgh, PA, USA, July 27-29, 1992, pp. 144–152. ACM (1992). https://doi.org/10.1145/130385.130401
7. Carathéodory, C.: Über den Variabilitätsbereich der Koeffizienten von Potenzreihen, die gegebene Werte nicht annehmen. Math. Ann. **64**(1), 95–115 (1907). https://doi.org/10.1007/BF01449883

8. Cormen, T.H., Leiserson, C.E., Rivest, R.L., Stein, C.: Introduction to Algorithms, 3rd Edition. MIT Press, Cambridge (2009). http://mitpress.mit.edu/books/introduction-algorithms

9. Cover, T.M.: Geometrical and statistical properties of systems of linear inequalities with applications in pattern recognition. IEEE Trans. Electron. Comput. **14**(3), 326–334 (1965). https://doi.org/10.1109/PGEC.1965.264137

10. Dhiflaoui, M., et al.: Certifying and repairing solutions to large LPs – How good are LP-solvers? In: Proceedings of the Fourteenth Annual ACM-SIAM Symposium on Discrete Algorithms, January 12-14, 2003, Baltimore, Maryland, USA. pp. 255–256. ACM/SIAM (2003). http://dl.acm.org/citation.cfm?id=644108.644152

11. Georgiadis, L., Tarjan, R.E.: Dominator tree certification and divergent spanning trees. ACM Trans. Algorithms **12**(1), 11:1–11:42 (2016). https://doi.org/10.1145/2764913

12. Haghighatkhah, P., Meulemans, W., Speckmann, B., Urhausen, J., Verbeek, K.: Obstructing classification via projection. In: Bonchi, F., Puglisi, S.J. (eds.) 46th International Symposium on Mathematical Foundations of Computer Science (MFCS 2021). Leibniz International Proceedings in Informatics (LIPIcs), vol. 202, pp. 56:1–56:19. Schloss Dagstuhl – Leibniz-Zentrum für Informatik, Dagstuhl, Germany (2021). https://doi.org/10.4230/LIPIcs.MFCS.2021.56

13. Kirchberger, P.: Über Tchebychefsche Annäherungsmethoden. Math. Ann. **57**, 509–540 (1903). https://doi.org/10.1007/BF01445182

14. Kojima, M., Mizuno, S., Yoshise, A.: An $O(\sqrt{n}L)$ iteration potential reduction algorithm for linear complementarity problems. Math. Program. **50**, 331–342 (1991). https://doi.org/10.1007/BF01594942

15. Kratsch, D., McConnell, R.M., Mehlhorn, K., Spinrad, J.P.: Certifying algorithms for recognizing interval graphs and permutation graphs. SIAM J. Comput. **36**(2), 326–353 (2006). https://doi.org/10.1137/S0097539703437855

16. Lipton, Z.C.: The mythos of model interpretability. Commun. ACM **61**(10), 36–43 (2018). https://doi.org/10.1145/3233231

17. Mangasarian, O.L.: Linear and nonlinear separation of patterns by linear programming. Oper. Res. **13**(3), 444–452 (1965). https://www.jstor.org/stable/167808

18. Matoušek, J., Gärtner, B.: Understanding and Using Linear Programming. Springer, Heidelberg (2007)

19. McConnell, R.M., Mehlhorn, K., Näher, S., Schweitzer, P.: Certifying algorithms. Comput. Sci. Rev. **5**(2), 119–161 (2011). https://doi.org/10.1016/j.cosrev.2010.09.009

20. Mehlhorn, K., Näher, S.: LEDA: A Platform for Combinatorial and Geometric Computing. Cambridge University Press, Cambridge (1999). http://www.mpi-sb.mpg.de/%7Emehlhorn/LEDAbook.html

21. Mirrokni, V.S., Leme, R.P., Vladu, A., Wong, S.C.: Tight bounds for approximate Carathéodory and beyond. In: Precup, D., Teh, Y.W. (eds.) Proceedings of the 34th International Conference on Machine Learning, ICML 2017, Sydney, NSW, Australia, 6-11 August 2017. Proceedings of Machine Learning Research, vol. 70, pp. 2440–2448. PMLR (2017). http://proceedings.mlr.press/v70/mirrokni17a.html

22. Mohri, M., Rostamizadeh, A., Talwalkar, A.: Foundations of Machine Learning. Adaptive Computation and Machine Learning, MIT Press, Cambridge (2012)

23. Monteiro, R., Adler, I.: Interior path following primal-dual algorithms. Part I: Linear Program. Math. Program. **44**(1–3), 27–41 (1989). https://doi.org/10.1007/BF01587075

24. Rosen, J.B.: Pattern separation by convex programming. J. Math. Anal. Appl. **10**(1), 123–134 (1965). https://doi.org/10.1016/0022-247X(65)90150-2

25. Schmidt, J.M.: Contractions, removals, and certifying 3-connectivity in linear time. SIAM J. Comput. **42**(2), 494–535 (2013). https://doi.org/10.1137/110848311
26. Schrijver, A.: Combinatorial Optimization. Springer (2004)
27. Seshia, S.A., Sadigh, D., Sastry, S.S.: Toward verified artificial intelligence. Commun. ACM **65**(7), 46–55 (2022). https://doi.org/10.1145/3503914
28. Vershynin, R.: High-Dimensional Probability: An Introduction with Applications in Data Science. Cambridge Series in Statistical and Probabilistic Mathematics, Cambridge University Press, Cambridge (2018). https://doi.org/10.1017/9781108231596
29. Watson, D.: A refinement of theorems of Kirchberger and Carathéodory. J. Aust. Math. Soc. **15**(2), 190–192 (1973). https://doi.org/10.1017/S1446788700012957
30. Webster, R.J.: Another simple proof of Kirchberger's theorem. J. Math. Anal. Appl. **92**(1), 299–300 (1983). https://doi.org/10.1016/0022-247X(83)90286-X

Improved Sublinear-Time Moment Estimation Using Weighted Sampling

Anup Bhattacharya$^{(\boxtimes)}$ and Pinki Pradhan

National Institute of Science Education and Research, An OCC of Homi Bhabha
National Institute, Bhubaneswar, India
{anup,pinki.pradhan}@niser.ac.in

Abstract. In this work we study the *moment estimation* problem using weighted sampling. Given sample access to a set A with n weighted elements, and a parameter $t > 0$, we estimate the t-th moment of A given as $S_t = \sum_{a \in A} w(a)^t$. For $t = 1$, this is the *sum estimation* problem for which sublinear time algorithms are known. The moment estimation problem along with a number of its variants have been extensively studied in streaming, sublinear and distributed communication models. Despite being well studied, we don't yet have a complete understanding of the sample complexity of the moment estimation problem in the sublinear model and in this work, we make progress on this front. On the algorithmic side, our upper bounds match the known upper bounds for the problem for $t > 1$. To the best of our knowledge, no sublinear algorithms were known for this problem for $0 < t < 1$. We design a sublinear algorithm for this problem for $t > 1/2$ and show that no sublinear algorithms exist for $t \leq 1/2$. We prove a $\Omega(\frac{n^{1-1/t} \ln 1/\delta}{\epsilon^2})$ lower bound for moment estimation for $t > 1$, and show optimal sample complexity bound $\Theta(\frac{n^{1-1/t} \ln 1/\delta}{\epsilon^2})$ for moment estimation for $t \geq 2$. Hence, we obtain a complete understanding of the sample complexity for moment estimation using proportional sampling for $t \geq 2$. We also study the moment estimation problem in the beyond worst-case analysis paradigm and identify a new *moment-density* parameter of the input that characterizes the sample complexity of the problem using proportional sampling and derive tight sample complexity bounds with respect to that parameter. We also study the moment estimation problem in the *hybrid sampling* framework in which one is given additional access to a uniform sampling oracle. We show access to a hybrid sampling framework does not provide any additional gain for this problem over a proportional sampling oracle in the worst case.

Keywords: Sublinear algorithms · moment estimation · sample complexity · lower bounds · beyond-worst case analysis

1 Introduction

Let A be any weighted set of n elements. Each element in A is associated with a weight using the function $w : A \rightarrow [0, \infty)$. Given an input parameter $t > 0$,

I. Finocchi and L. Georgiadis (Eds.): CIAC 2025, LNCS 15679, pp. 170–186, 2025.
https://doi.org/10.1007/978-3-031-92932-8_12

the problem of estimating the t-th moment of A, expressed as $S_t = \sum_{a \in A} w(a)^t$, is called the *moment estimation* problem[1]. In this work we study the moment estimation problem in a model in which we don't have direct access to the weights of the elements in A, instead we get indirect access to the weights using samples from an oracle. Our objective is to obtain a good approximation of the tth moment S_t of A while making a small number of queries to the oracle.

Estimation tasks on very large datasets are often performed using samples from the dataset. One common objective in these settings is to compute an approximate answer using only a small number of samples. Sublinear algorithms are algorithms that access only a tiny portion of the data (sublinear in the input size) and compute an approximate answer. For many problems of interest, uniform sampling-based algorithms require a lot of samples, and hence might not be very useful in designing sublinear algorithms where as weighted sampling-based algorithms often give better performance guarantees. This is the case, for example, for the sum estimation problem. Uniform sampling-based approaches require $\Omega(n)$ samples for the sum estimation problem in the worst-case where as weighted sampling-based algorithms require only $O(\sqrt{n}/\epsilon)$ samples [5, 14]. This, however, uses the assumption of access to a stronger query oracle that returns samples according to a weighted distribution. In applications where access to such a stronger query oracle is available, one might use the above algorithms with better performance guarantees. One such use case is the parameter estimation problems on graphs where one is given access to a random edge sampling oracle. Details about using weighted sampling for graph parameter estimation problems and other application areas can be found in [2, 5, 14].

In this work we study the sample complexity of estimators for the moment estimation problem assuming access to a proportional sampling oracle on the set A. Proportional sampling on set A returns an element $a \in A$ with probability proportional to $w(a)$. Let $W = \sum_{a \in A} w(a)$. Using proportional sampling an element $a \in A$ is chosen with probability $w(a)/W$. Next, we formally define the moment estimation problem that we study in this work.

Definition 1. $((\epsilon, \delta) - Moment\ Estimation)$ *Given sample access to a set A of n weighted elements and input parameters $\epsilon, \delta \in (0, 1)$, $t > 0$, design an algorithm ALG that returns $ALG(A, t, \epsilon, \delta)$ such that*

$$\Pr[(1 - \epsilon)S_t \leq ALG(A, t, \epsilon, \delta) \leq (1 + \epsilon)S_t] \geq 1 - \delta$$

The moment estimation problem is one of the fundamental problems with a number of variants that are studied in the literature of sublinear algorithms. For $t = 1$, this is the *sum estimation* problem. Motwani *et al.* [14] initiated the study of designing sublinear algorithms for the sum estimation problem assuming access to a proportional sampling oracle. Recently, Beretta and Tětek [5] have improved the sample complexity of the sum estimation problem to $O(\sqrt{n}/\epsilon)$ using proportional sampling. Estimating frequency moments in streams is one

---

[1] The t-th moment of A is also defined as $S_t = 1/n \cdot \sum_{a \in A} w(a)^t$. We follow $S_t = \sum_{a \in A} w(a)^t$ for simplicity of calculations [7].

of the most well-studied problems in the streaming literature [3]. Moment estimation is also well studied in the distributed communication models [12]. This problem can also be thought of as the vector norm estimation problem where the weights correspond to the entries of a vector. When the set of elements corresponds to the set of vertices in a graph and the weights correspond to the degrees of the vertices in the graph, this problem is known as the *degree distribution moment estimation* problem. For $t = 1$, this is the edge estimation problem in graphs for which Feige [8] and Goldreich and Ron [10] designed sublinear algorithms. For the related degree distribution moment estimation problem in graphs, the authors in [6,7,11] designed sublinear time algorithms in the sparse graph model in which one is allowed to make uniform vertex, degree and neighbour queries. Aliakbarpour *et al.* [2] designed a sublinear algorithm for estimating $\sum_{a \in A} \binom{w(a)}{t}$ using $O(\frac{n^{1-1/t} \ln 1/\delta}{\epsilon^2})$ queries in the sparse graph model with additional access to a random edge oracle, where $w(a)$ denotes the degree of vertex a. Various applications of the moment estimation problem including its variants are discussed in [2,5,6].

A number of recent works [4,9,15] highlighted the importance of designing sublinear algorithms with optimal dependence not only on n but also on ϵ and δ. Beretta and Tětek have improved the sample complexity of the sum estimation problem from $\tilde{O}(\sqrt{n}/\epsilon^{7/2})$ proportional samples in Motwani *et al.* [14] to $\Theta(\sqrt{n}/\epsilon)$ proportional samples [5]. Assadi and Nguyen designed a sublinear algorithm to estimate h-index with optimal sample complexity dependence on all input parameters [4]. Despite a lot of work on sublinear algorithms for the moment estimation problem, we don't have a complete understanding of the sample complexity of the moment estimation problem using weighted sampling. One of the primary motivations of this work is to make progress in our understanding of the moment estimation problem in sublinear models. Next, we discuss the main results of this work.

1.1   Our Contributions

Moment Estimation Using Proportional Sampling. Our results for the (ϵ, δ)-moment estimation problem using proportional sampling for $t > 1$ are as follows.

Theorem 1. *There exists an algorithm ALG that given proportional sampling access to the weights of the elements of a set A and parameters $t > 1$, $\epsilon, \delta \in (0,1)$, provides an (ϵ, δ)-estimate of S_t using $O((\frac{\sqrt{n}}{\epsilon} + \frac{n^{1-1/t}}{\epsilon^2}) \ln \frac{1}{\delta})$ samples.*

For $t \geq 2$, the sample complexity of our algorithm is $O(\frac{n^{1-1/t} \ln \frac{1}{\delta}}{\epsilon^2})$. Our next result shows that this bound is tight.

Theorem 2. *For any $\epsilon, \delta \in (0,1)$ and $t > 1$, any randomized algorithm that computes an (ϵ, δ)-estimate of S_t requires $\Omega(\frac{n^{1-1/t} \ln \frac{1}{\delta}}{\epsilon^2})$ proportional samples.*

For $0 < t < 1$, we show the following. We design an (ϵ, δ)-estimator for S_t when $t > 1/2$.

Theorem 3. *There exists an algorithm ALG that given proportional sampling access to the weights of the elements of a set A and parameters $1/2 < t < 1$, $\epsilon, \delta \in (0,1)$, provides an (ϵ, δ)-estimate of S_t using $O((\frac{\sqrt{n}}{\epsilon} + \frac{n^{\frac{1}{t}-1}}{\epsilon^2}) \ln \frac{1}{\delta})$ samples.*

We show that when $t \leq 1/2$, no sublinear algorithms exist for this problem. More specifically, we show the following.

Theorem 4. *For any $\epsilon > 0$ and $t \leq 1/2$, any randomized algorithm that computes an $(\epsilon, 1/3)$-estimate of S_t requires $\Omega(n)$ proportional samples.*

Significance of Our Results and Comparisons with Known Results.

- **Case $t > 1$:** Eden *et al.* [6,7] designed sublinear algorithms for the degree distribution moment estimation problem in the sparse graph model assuming access to uniform vertex, degree and neighbour queries. Aliakbarpour *et al.* [2] improved the sample complexity bound for this problem assuming additional access to a random edge oracle. Aliakbarpour *et al.* observed that estimating the number of t-stars in a graph can be seen as a variant of the moment estimation problem. Let $w(a)$ denote the degree of vertex $a \in A$. Then, $\sum_{a \in A} \binom{w(a)}{t}$ counts the number of t-stars in the graph. The authors designed sublinear algorithms for this problem assuming access to a random edge sampling oracle in the sparse graph model. To the best of our knowledge, this problem of estimating $\sum_{a \in A} \binom{w(a)}{t}$ of Aliakbarpour *et al.* seems to be the most closely related problem to ours. In our setting, the result of Aliakbarpour *et al.* gives a sublinear-time algorithm for moment estimation using $O(\frac{n^{1-1/t} \ln 1/\delta}{\epsilon^2})$ proportional samples.[2] They also proved a $\Omega(n^{1-1/t})$ sample complexity lower bound for this problem. In the following we discuss the key differences of our results with theirs.

  1. Upper bound for moment estimation: We note that for $t > 1$ our upper bound matches with that of Aliakbarpour *et al.* [2] and the algorithmic ideas and the analysis of these works are similar. We do not claim much technical novelty for our upper bound. However, we point out that our setup is strictly more general than that of Aliakbarpour *et al.* and their result can be recovered in our setup. We also note that our algorithm uses only proportional samples where as the algorithm of Aliakbarpour *et al.* uses queries in the sparse graph model with additional access to random edge samples. Since it is known that using uniform edge samples one can sample vertices with probabilities proportional to their degrees, it appears that access to proportional samples enables one to design sublinear algorithms for the moment estimation problem.

  2. Lower bound for moment estimation: Aliakbarpour *et al.* gave a $\Omega(n^{1-1/t})$ lower bound for this problem. We show an improved lower bound of $\Omega(\frac{n^{1-1/t} \ln \frac{1}{\epsilon}}{\epsilon^2})$ for $t > 1$. For $t \geq 2$, this settles the sample complexity

---

[2] The authors stated their bound as $O(n^{1-1/t}/\epsilon^3)$ for constant success probability, but we believe using [5], it can be improved to $O(n^{1-1/t}/\epsilon^2)$.

of $\Theta(\frac{n^{1-1/t}\ln\frac{1}{\delta}}{\epsilon^2})$ for the moment estimation problem with optimal dependence on all input parameters n, ϵ, δ.

– **Case $t < 1$:** Moment estimation for $0 < t < 1$ is a well motivated problem with a number of applications [12]. However, to the best of our knowledge, no sublinear algorithms were known for the moment estimation problem when $t < 1$. The moment estimation algorithms of Eden *et al.* [6,7] and Aliakbarpour *et al.* [2] work only for $t \geq 1$. We design the first sublinear algorithm for the moment estimation problem for $t > 1/2$. We also show that for $t \leq 1/2$, no sublinear algorithms exist for this problem.

Characterization of Sample Complexity. The moment estimation problem using proportional sampling requires $\Omega(\frac{n^{1-1/t}\ln\frac{1}{\delta}}{\epsilon^2})$ samples in the worst case. We study the moment estimation problem in a beyond worst-case analysis paradigm and identify a parameter of the input that characterizes the sample complexity of the problem using proportional sampling. For the degree distribution moment estimation problem in graphs, Eden *et al.* [7] identified the *arboricity* of a graph as the relevant parameter and obtained sample complexity bounds in terms of the arboricity of the graph.

We introduce a new *moment-density* parameter ρ of the input that characterizes the sample complexity for the moment estimation problem. For $W = \sum_{a \in A} w(a)$ and $S_t = \sum_{a \in A} w(a)^t$, we define the parameter ρ as

$$\rho = \max_{L \subseteq A} \frac{\frac{\sum_{a \in L} w(a)^t}{\sum_{a \in L} w(a)}}{\frac{\sum_{a \in A} w(a)^t}{\sum_{a \in A} w(a)}} = \max_{L \subseteq A} \frac{\sum_{a \in L} w(a)^t}{\sum_{a \in L} w(a)} \cdot \frac{W}{S_t}$$

The motivation for writing the moment-density parameter ρ in the above form is as follows. The key idea behind the lower bound instances for the moment estimation problem using proportional sampling (described in Sect. 4) is to assign large weights on only a few input elements such that these elements are not easily detected using proportional sampling but the tth power of their weights dominate the moment value of the instance. This is why proportional sampling requires a lot of samples on these kind of instances. Now, if we were allowed to sample elements of A with probabilities proportional to their tth power of the weights, then the complexity of the moment estimation problem becomes the same as the complexity of the sum estimation problem using proportional sampling. But, when we sample using proportional sampling, there might be some elements with slightly larger weights having outsized influence in the moment value of the instance. The moment-density parameter ρ is defined such that its value will be large on those instances. For any subset $L \subseteq A$, let ρ_L denote the ratio of the fractional contribution of the elements in L to the moment value and the probability that an element in L is going to be sampled using proportional sampling. We have $\rho = \max_{L \subseteq A} \rho_L$. Alternatively, ρ captures the maximum contribution of any subset L to the t-th moment S_t relative to the sum of the weights of elements in L. For scaling we divide it by S_t/W.

We show an upper bound for (ϵ, δ)-estimate of the moment using proportional sampling where we write the sample complexity in terms of ϵ, δ and ρ.

Theorem 5. *There exists an algorithm ALG that given proportional sampling access to the weights of the elements in a set A with moment-density parameter ρ and parameters $\epsilon, \delta \in (0, 1), t > 1$, provides an (ϵ, δ)-estimate of S_t using $O((\sqrt{n}/\epsilon + \frac{\rho}{\epsilon^2}) \ln 1/\delta)$ samples.*

Next, we show an almost tight lower bound on the sample complexity for the moment estimation problem on instances with moment-density parameter ρ in terms of ρ, ϵ, δ.

Theorem 6. *For any $\epsilon, \delta \in (0, 1)$ and $t > 1$, any randomized algorithm for (ϵ, δ)-estimate of S_t on an instance with moment-density parameter ρ requires $\Omega(\frac{\rho \ln 1/\delta}{\epsilon})$ proportional samples.*

Moment Estimation Using Hybrid Sampling. Proportional sampling-based algorithms for the moment estimation problem require $\Omega(\frac{n^{1-1/t} \ln 1/\delta}{\epsilon^2})$ samples for $t > 1$ in the worst case. One natural idea to design algorithms with improved sample complexity bounds is to give more power to the algorithm designer in the form of access to a stronger query oracle. In this section we explore whether additional access to a uniform sampling oracle allows one to design an algorithm with improved sample complexity for this problem. The hybrid sampling framework allows one to use both proportional and uniform samples. This study is motivated by the fact that, for the sum estimation problem, Motwani *et al.* [14] and Beretta and Tětek [5] exploited access to a hybrid sampling oracle to design algorithms that make $\tilde{O}(n^{1/3}/\epsilon^{9/2})$ and $O(n^{1/3}/\epsilon^{4/3})$ samples, respectively. This is in contrast to the $\Theta(\sqrt{n}/\epsilon)$ sample complexity bound for the sum estimation problem using only proportional samples. In this work we explore whether hybrid sampling might give us better sample complexity bounds for the moment estimation problem. We prove a lower bound result showing that no improved algorithm with better sample complexity bounds exists for the moment estimation problem using hybrid sampling. We state the result next and prove it in Sect. 5.

Theorem 7. *(Lower bound using hybrid sampling) For any $\epsilon, \delta \in (0, 1)$ and $t > 1$, any algorithm having access to a hybrid sampling oracle requires at least $\Omega(\frac{n^{1-1/t} \ln 1/\delta}{\epsilon^2})$ samples to compute an (ϵ, δ)-estimate for S_t.*

1.2 Techniques

The main idea behind the upper bound is fairly standard and we describe it as follows. Suppose we obtain a sample $a \in A$ of weight $w(a)$ using proportional sampling. Let us set $X = w(a)^t$. Then, $\mathbb{E}[X] = \sum_{a \in A} w(a)^t p_a$, where $p_a = \frac{w(a)}{W}$ denotes the probability of sampling the element a and $W = \sum_{a \in A} w(a)$. Suppose we know p_a for each $a \in A$, then $X = w(a)^t/p_a$ would give us an unbiased estimator of S_t. However, we don't know W and hence don't know about these probabilities p_a. Our crucial observation here is to use the sum estimation algorithm in Beretta and Tětek [5] to obtain an $(\epsilon_1, \delta/2)$ estimate $\widetilde{W}$ of W, and use this in turn to obtain estimates $\tilde{p}_a$ for p_a. Our estimator built in this manner would not be an unbiased estimator of S_t but we can reduce the bias of the estimator considerably by choosing an appropriate ϵ_1. For our algorithmic results, we use $\epsilon_1 = \epsilon/2$. We use similar techniques for designing algorithms in related settings. For the lower bound we use Yao's minimax lemma to construct families of instances that cannot be distinguished using a small number of proportional samples. A number of lower bound results were known for variants of the moment estimation problem [2,6,11]. Our lower bound constructions are motivated from these lower bound constructions. Unlike the earlier lower bounds though, our lower bounds have optimal dependence in n, ϵ, δ.

1.3   Related Works

Motwani *et al.* [14] initiated the study of the sum estimation problem using access to a proportional sampling oracle and designed the first sublinear algorithm for this problem that uses $\tilde{O}(\sqrt{n})$ queries. They also designed sublinear algorithms in the hybrid sampling framework using $\tilde{O}(n^{1/3})$ queries. Beretta and Tětek [5] have recently improved these results; they prove $\Theta(\frac{\sqrt{n}}{\epsilon})$ sample complexity bound using proportional sampling and in the hybrid sampling setting gives almost tight sample complexity bound of $O(n^{1/3}/\epsilon^{4/3})$. Variants of these problems are also studied in the graph parameter estimation literature, where the elements of the set correspond to the vertices of a graph and the weights correspond to the degrees of the vertices and we are given query access to the graph. The sum estimation problem in this context becomes the *edge estimation* problem and moment estimation is known as the degree distribution moment estimation problem. Eden *et al.* [6,7] studied the *degree distribution moment estimation* problem in the graph query model and designed sublinear algorithms with improved sample complexity bounds. Aliakbarpour *et al.* [2] studied the estimation of the number of t-stars in a graph and showed this problem to be closely related to the moment estimation problem. Moment estimation problem is also studied in streaming and distributed communication models [3,12].

2   Preliminaries

We use the following well known facts in the analysis.

Fact 1. *For any vector $x \in \mathcal{C}^n$, and for any $0 < r < p$, we have $||x||_p \leq ||x||_r \leq n^{(1/r-1/p)}||x||_p$.*

Lemma 1. *(Chernoff bound [13]) Let $Z_1, Z_2, \ldots, Z_v$ be independent and identically distributed Bernoulli random variables. Let $Z = \sum_{i=1}^{v} Z_i$. Then, $\Pr[Z \leq (1 - \gamma)\mathbb{E}[Z]] \leq e^{-1/2 \cdot \gamma^2 \cdot \mathbb{E}[Z]}$.*

3 Estimation of Moments Using Proportional Sampling

We describe our algorithm for the moment estimation problem using proportional sampling. We mentioned earlier that for $t > 1$, our upper bounds match those of Aliakbarpour *et al.* [2]. However, since our algorithm works in strictly more general settings and the algorithm for $1/2 < t < 1$ uses the same ideas, we describe it in detail. Let A be a set of n weighted elements. We assume access to a proportional sampling oracle on the weights of the elements in A. For a proportional sample, the oracle returns an element $a_j \in A$ with probability $w(a_j)/W$, where $W = \sum_{a_j \in A} w(a_j)$. Given parameters $t > 1, \epsilon, \delta \in (0, 1)$, we design an (ϵ, δ)-estimate of $S_t = \sum_{a_j \in A} w(a_j)^t$.

Theorem 8. *There exists an algorithm ALG that given proportional sampling access to the weights of the elements in a set A and parameters $t > 1, \epsilon, \delta \in (0, 1)$, provides an (ϵ, δ)-estimate of S_t using $O(\frac{\sqrt{n} \log 1/\delta}{\epsilon} + \frac{n^{1-1/t} \log 1/\delta}{\epsilon^2})$ samples.*

Algorithm 1. Moment Estimation using Proportional Sampling

1: **procedure** MOMENTESTIMATOR(A, t, ϵ, δ)
2:     Let $\widetilde{W}$ denote an $(\epsilon_1 = \epsilon/2, \delta/2)$-estimate of W using the sum estimation algorithm of [5]. This step requires $480 \cdot \frac{\sqrt{n} \log(2/\delta)}{\epsilon}$ proportional samples.
3:     **for do** $r = 1$ to $v = 48 \cdot \log 2/\delta$
4:         **for do** $j = 1$ to $l = 48 \cdot n^{1-1/t}/\epsilon^2$
5:             Let a_j denote a proportional sample of weight $w(a_j)$.
6:             Compute $\tilde{p}_j = \frac{w(a_j)}{\widetilde{W}}$.
7:             Set $X_j = \frac{w(a_j)^t}{\tilde{p}_j}$
8:         **end for**
9:         $Y_r = \frac{\sum_{j=1}^{l} X_j}{l}$
10:    **end for**
11:    **return** median$(Y_1, \dots, Y_v)$
12: **end procedure**

Algorithm 1 first computes an $(\epsilon_1, \delta/2)$-estimate $\widetilde{W}$ of the sum W of the weights of elements in A using the sum estimation algorithm in [5][3]. Let the probability of sampling element a_j using proportional sampling be given as $p_j = \frac{w(a_j)}{W}$. Note that we don't know p_j, however we can obtain a good approximation of it as follows. For a proportional sample a_j, we know its weight $w(a_j)$, and $\widetilde{W}$ gives us an approximation of W. Using this we get a approximation for p_j as $\tilde{p}_j = \frac{w(a_j)}{\widetilde{W}}$. Let $\mathcal{E}$ denote the event that $\widetilde{W} \in [(1 - \epsilon_1)W, (1 + \epsilon_1)W]$. In what follows we condition on event $\mathcal{E}$.

---

[3] [5] states the sample complexity for probability of success at least $2/3$. Here, we are stating the bounds for an $(\epsilon_1, \delta/2)$-estimate. This is obtained using an application of the medians of means technique.

Claim. Conditioned on $\mathcal{E}$, for any $j \in [n]$, we have $\frac{p_j}{1+\epsilon_1} \leq \tilde{p}_j \leq \frac{p_j}{1-\epsilon_1}$.

Proof. Conditioned on $\mathcal{E}$, $(1-\epsilon_1)W \leq \widetilde{W} \leq (1+\epsilon_1)W$. The above inequality follows.

Conditioned on $\mathcal{E}$, for any j, we have $\frac{p_j}{1+\epsilon_1} \leq \tilde{p}_j \leq \frac{p_j}{1-\epsilon_1}$. Given a proportional sample a_j, we define a random variable X_j with value $\frac{w(a_j)^t}{\tilde{p}_j}$. Then, we have $(1-\epsilon_1)S_t \leq \mathbb{E}[X_j] \leq (1+\epsilon_1)S_t$. Here, X_j is not an unbiased estimator of S_t. Next, we bound the variance of this estimator given as $\text{Var}[X_j] = \mathbb{E}[X_j^2] - \mathbb{E}^2[X_j] \leq \mathbb{E}[X_j^2]$.

$$
\begin{aligned}
\mathbb{E}[X_j^2] &= \sum_{a_j \in A} \frac{w(a_j)^{2t}}{\tilde{p}_j^2} p_j \\
&\leq (1+\epsilon_1)^2 \sum_{a_j \in A} \frac{w(a_j)^{2t}}{p_j} \\
&= (1+\epsilon_1)^2 \cdot W \cdot \sum_{a_j \in A} \frac{w(a_j)^{2t}}{w(a_j)} \qquad \text{(Substituting } p_j \text{ with } \frac{w(a_j)}{W}) \\
&= (1+\epsilon_1)^2 \cdot W \cdot \sum_{a_j \in A} w(a_j)^{2t-1} \qquad\qquad\qquad\qquad (1)
\end{aligned}
$$

Let us obtain l independent samples using proportional sampling and let these random variables be $X_1, \ldots, X_l$. Let $X = \frac{1}{l} \sum_{j=1}^{l} X_j$. We have $(1-\epsilon_1)S_t \leq \mathbb{E}[X] = \mathbb{E}[X_j] \leq (1+\epsilon_1)S_t$, and $\text{Var}[X] = \frac{\text{Var}[X_j]}{l}$. Using Chebyshev's inequality, we have $\Pr[|X - S_t| > \epsilon S_t] \leq \Pr[|X - \mathbb{E}[X]| > (\epsilon - \epsilon_1)S_t]] \leq \frac{\text{Var}[X]}{(\epsilon-\epsilon_1)^2 S_t^2}$. For appropriate choice of parameter l, we show this probability to be at most a small constant using the following claim.

Claim. $\frac{\text{Var}[X]}{(\epsilon-\epsilon_1)^2 S_t^2} \leq \frac{(1+\epsilon_1)^2}{l(\epsilon-\epsilon_1)^2} \cdot n^{1-1/t}$.

Proof.

$$
\begin{aligned}
&\frac{\text{Var}[X]}{(\epsilon - \epsilon_1)^2 S_t^2} \\
&\leq \frac{\mathbb{E}[X_j^2]}{l(\epsilon - \epsilon_1)^2 \cdot S_t^2} \\
&\leq \frac{(1+\epsilon_1)^2}{l(\epsilon - \epsilon_1)^2} \cdot \frac{W \cdot \sum_{a_j \in A} w(a_j)^{2t-1}}{S_t^2} \qquad \text{(Using Eq. (1))} \\
&= \frac{(1+\epsilon_1)^2}{l(\epsilon - \epsilon_1)^2} \cdot \frac{W \cdot \|w(A)\|_{2t-1}^{2t-1}}{S_t^2} \qquad \text{(vector } w(A) \text{ has length } n) \\
&\leq \frac{(1+\epsilon_1)^2}{l(\epsilon - \epsilon_1)^2} \cdot \frac{W \cdot (\|w(A)\|_t^t)^{2-1/t}}{S_t^2} \qquad \text{(Using Fact 1, } \|w(A)\|_{2t-1} \leq \|w(A)\|_t) \\
&= \frac{(1+\epsilon_1)^2}{l(\epsilon - \epsilon_1)^2} \cdot \frac{\|w(A)\|_1 \cdot (\|w(A)\|_t^t)^{2-1/t}}{(\|w(A)\|_t^t)^2}
\end{aligned}
$$

$$= \frac{(1 + \epsilon_1)^2}{l(\epsilon - \epsilon_1)^2} \cdot \frac{||w(A)||_1}{||w(A)||_t}$$

$$\leq \frac{(1 + \epsilon_1)^2}{l(\epsilon - \epsilon_1)^2} \cdot n^{1 - 1/t} \qquad\qquad \text{(Using Fact 1, } ||w(A)||_1 \leq n^{1 - 1/t}||w(A)||_t)$$

Let $\epsilon_1 = \epsilon/2$. For $l = 48n^{1-1/t}/\epsilon^2$, this failure probability is at most $1/3$. Using the standard median trick, we show that for $v = 48 \log 2/\delta$, this failure probability can be reduced to be at most $\delta/2$. Let us define independent Bernoulli random variables $Z_1, \ldots, Z_v$ such that $\Pr[Z_i = 1] = 2/3$ for all i. Let $Z = \sum_{r=1}^{v} Z_r$. Now, conditioned on $\mathcal{E}$, the probability that the output of Algorithm 1 does not lie in the interval $[(1 - \epsilon)S_t, (1 + \epsilon)S_t]$ is the same as the probability that the median of $Y_1, \ldots, Y_v$ lies outside the interval $[(1 - \epsilon)S_t, (1 + \epsilon)S_t]$. This probability is at most $\Pr[Z < v/2]$. Using a standard application of a Chernoff bound, given in Lemma 1, we have $\Pr[Z < v/2] \leq \delta/2$.

Correctness and Sample Complexity Bounds We need to show that Algorithm 1 returns an estimate $ALG(A, t, \epsilon, \delta)$ for which with probability at least $1 - \delta$, we have $(1 - \epsilon)S_t \leq ALG(A, t, \epsilon, \delta) \leq (1 + \epsilon)S_t$. Step 2 of Algorithm 1 uses $\Theta(\frac{\sqrt{n} \log(2/\delta)}{\epsilon_1})$ proportional samples to obtain an $(\epsilon_1, \delta/2)$ estimate $\widetilde{W}$ of W. Algorithm 1 fails if either $\mathcal{E}$ does not hold or the estimate in Step 10 is incorrect. Since both of these failure probabilities are at most $\delta/2$, the algorithm succeeds with probability at least $1 - \delta$.

Algorithm 1 uses $\Theta(\frac{\sqrt{n} \log(2/\delta)}{\epsilon_1})$ proportional samples in Step 2 and uses $O(\frac{n^{1-1/t} \log 1/\delta}{\epsilon^2})$ proportional samples in Steps 3 and 4. Therefore, the required number of proportional samples is $O(\frac{\sqrt{n} \log 1/\delta}{\epsilon_1} + \frac{n^{1-1/t} \log 1/\delta}{\epsilon^2})$. For $\epsilon_1 = \epsilon/2$, this gives $O(\frac{\sqrt{n} \log 1/\delta}{\epsilon} + \frac{n^{1-1/t} \log 1/\delta}{\epsilon^2})$.

4 Lower Bound for Moment Estimation Using Proportional Sampling

We use Yao's minimax lemma to prove the sample complexity lower bound for obtaining an (ϵ, δ) estimate of S_t using any randomized algorithm. We construct two families of instances on which S_t differs by at least a $(1 \pm \epsilon)$-factor and show that it is hard to distinguish these two instances using a small number of proportional samples.

Our lower bound constructions are as follows. There are n_1 elements of weight d_1 and n_2 elements of weight d_2 in both the instances, where the values of the parameters n_1, n_2 and d_1 are the same in both instances, and the instances differ in the value of parameter d_2. The exact values of these parameters will be set below. In one instance we set $d_2 = n$, where as for the other instance $d_2 = 0$. These choices for parameter values creates a gap of a multiplicative $(1 \pm \epsilon)$ factor between the S_t values of the two instances. One can differentiate these two instances only when an element of weight $d_2 = n$ is sampled using proportional sampling, and we show that this requires a lot of samples.

Theorem 9. *For any $\epsilon, \delta \in (0,1)$ and $t > 1$, any randomized algorithm that computes an (ϵ, δ)-estimate of S_t requires $\Omega(\frac{n^{1-1/t} \ln 1/\delta}{\epsilon^2})$ proportional samples.*

Proof. We construct two families of instances which we show are hard to be distinguished using a few proportional samples by Yao's lemma. In the first instance we have n_1 elements with weight d_1 and n_2 elements of weight 0, and in the second instance, there are n_1 elements of weight d_1 and n_2 elements of weight d_2, where the following values are used. The parameter values used in the lower bound constructions of the two instances are given as follows.

$$
n_1 = \frac{n^2}{n + \epsilon^{\frac{2t-1}{t-1}}}
\qquad\qquad
n_1 = \frac{n^2}{n + \epsilon^{\frac{2t-1}{t-1}}}
$$

$$
d_1 = n^{1-1/t}\epsilon^{1/(t-1)}
\qquad\qquad
d_1 = n^{1-1/t}\epsilon^{1/(t-1)}
$$

$$
n_2 = \frac{n\epsilon^{\frac{2t-1}{t-1}}}{n + \epsilon^{\frac{2t-1}{t-1}}}
\qquad\qquad
n_2 = \frac{n\epsilon^{\frac{2t-1}{t-1}}}{n + \epsilon^{\frac{2t-1}{t-1}}}
$$

$$
d_2 = 0
\qquad\qquad\qquad
d_2 = n
$$

The S_t value for the first instance is given as follows.

$$
n_1 \cdot d_1^t + n_2 \cdot 0 = \frac{n^2}{n + \epsilon^{\frac{2t-1}{t-1}}} \cdot n^{t-1}\epsilon^{t/(t-1)}
$$

$$
= \frac{n^{t+1}\epsilon^{t/(t-1)}}{n + \epsilon^{\frac{2t-1}{t-1}}}
$$

The S_t value for the second instance is

$$
n_1 \cdot d_1^t + n_2 \cdot d_2^t = \frac{n^{t+1}\epsilon^{t/(t-1)}}{n + \epsilon^{\frac{2t-1}{t-1}}} + \frac{n\epsilon^{\frac{2t-1}{t-1}}}{n + \epsilon^{\frac{2t-1}{t-1}}} \cdot n^t
$$

$$
= \frac{n^{t+1}\epsilon^{t/(t-1)}}{n + \epsilon^{\frac{2t-1}{t-1}}} + \epsilon\frac{n^{t+1}\epsilon^{t/(t-1)}}{n + \epsilon^{\frac{2t-1}{t-1}}}
$$

$$
= (1 + \epsilon)\frac{n^{t+1}\epsilon^{t/(t-1)}}{n + \epsilon^{\frac{2t-1}{t-1}}}
$$

The above two instances differ in their S_t values by a multiplicative factor of $(1 + \epsilon)$. In order to distinguish these two instances, one is required to sample an element of weight $d_2 = n$. Using proportional sampling, the probability of sampling an element of weight n in the above instance is given to be at least

$$\frac{n_2 d_2}{n_2 d_2 + n_1 d_1} = \frac{\frac{n\epsilon^{\frac{2t-1}{t-1}}}{n+\epsilon^{\frac{2t-1}{t-1}}} \cdot n}{\frac{n\epsilon^{\frac{2t-1}{t-1}}}{n+\epsilon^{\frac{2t-1}{t-1}}} \cdot n + \frac{n^2}{n+\epsilon^{\frac{2t-1}{t-1}}} \cdot n^{1-1/t}\epsilon^{1/(t-1)}}$$

$$= \frac{n^2 \epsilon^{\frac{2t-1}{t-1}}}{n^2 \epsilon^{\frac{2t-1}{t-1}} + n^2 \cdot n^{1-1/t} \cdot \epsilon^{\frac{1}{t-1}}}$$

$$= \frac{1}{1 + \frac{n^{1-1/t}}{\epsilon^2}}$$

Let $p = \frac{1}{1+\frac{n^{1-1/t}}{\epsilon^2}}$. The lower bound on the sample complexity for this instance distinguishing problem is given as the number of samples required to observe a *success* with probability at least $(1-\delta)$ while drawing independent samples from $Geom(p)$. Here, $Geom(p)$ denotes a geometric distribution with success probability p. The number of samples required to observe one *success* from $Geom(p)$ with probability at least $(1-\delta)$ is at least $\Omega(\frac{\ln 1/\delta}{p})$. Therefore, $\Omega(\frac{n^{1-1/t}\ln 1/\delta}{\epsilon^2})$ samples are required to distinguish these two instances with probability at least $1-\delta$.

5   Estimation of Moments Using Hybrid Sampling

In this section we prove a lower bound showing that for the moment estimation problem, the hybrid sampling framework does not provide any significant advantage over access to just the proportional sampling oracle. In contrast, note that for the sum estimation problem, hybrid sampling-based algorithms in fact give much better sample complexity bounds over proportional sampling [5,14]. We prove the following result.

Theorem 10. *For any $\epsilon, \delta > 0$ and $t > 1$, any algorithm having access to a hybrid sampling oracle requires to make at least $\Omega(\frac{n^{1-1/t}\ln 1/\delta}{\epsilon^2})$ queries to compute an (ϵ, δ)-estimate for S_t.*

We show that the lower bound instance described in Sect. 4 yields a lower bound for the hybrid sampling as well. In order to distinguish these instances, one is required to sample an element of weight n. We have seen that using just proportional sampling $\Omega(\frac{n^{1-1/t}\log 1/\delta}{\epsilon^2})$ samples are required. The probability of sampling an element of weight d_2 using uniform sampling is given as $\frac{n_2}{n_1+n_2}$. This probability using the values of the parameters from Sect. 4 equals $\frac{n_2}{n_1+n_2} = \frac{1}{1+\frac{n}{\epsilon^{\frac{2t-1}{t-1}}}}$. The instances are distinguished if an element of weight n is sampled using either proportional sampling or uniform sampling. These two probabilities are given as $\frac{1}{1+\frac{n^{1-1/t}}{\epsilon^2}}$ and $\frac{1}{1+\frac{n}{\epsilon^{\frac{2t-1}{t-1}}}}$, respectively. Overall, we get a lower bound of

$\Omega(\min\{\frac{n^{1-1/t}}{\epsilon^2}, \frac{n}{\epsilon^{\frac{2t-1}{t-1}}}\}\ln 1/\delta) = \Omega(\frac{n^{1-1/t}\ln 1/\delta}{\epsilon^2})$ for the (ϵ, δ) moment estimation problem using hybrid sampling.

6   Characterization of Sample Complexity

We define a *moment-density* parameter of the input that governs the sample complexity of the moment estimation problem using proportional sampling. For $S_t = \sum_{a \in A} w(a)^t$ and $W = \sum_{a \in A} w(a)$, we define the moment-density parameter as

$$\rho = \max_{L \subseteq A} \frac{\frac{\sum_{a \in L} w(a)^t}{S_t}}{\frac{\sum_{a \in L} w(a)}{W}} = \max_{L \subseteq A} \frac{\sum_{a \in L} w(a)^t}{\sum_{a \in L} w(a)} \cdot \frac{W}{S_t}$$

We give an upper bound for an (ϵ, δ)-estimator of S_t using $O((\sqrt{n}/\epsilon + \rho/\epsilon^2) \ln 1/\delta)$ proportional samples.

Theorem 11. *There exists an algorithm ALG that given proportional sampling access to the weights of elements of A having moment-density parameter ρ and parameters $t > 1, \epsilon, \delta \in (0, 1)$, provides an (ϵ, δ)-estimate of S_t using $O((\sqrt{n}/\epsilon + \rho/\epsilon^2) \ln 1/\delta)$ proportional samples.*

Proof. Algorithm 1 gives the above sample complexity bound. We adapt the calculations from Sect. 3. Following Eq. 1, we can write the sample complexity of our $(\epsilon, 1/3)$ estimator to be at most $O(\sqrt{n}/\epsilon + \frac{1}{\epsilon^2} \frac{W \cdot \sum_{a \in A} w(a)^{2t-1}}{S_t^2})$. We will show that $\frac{W \cdot \sum_{a \in A} w(a)^{2t-1}}{S_t^2} \leq \rho$.

$$\frac{W \cdot \sum_{a \in A} w(a)^{2t-1}}{S_t^2} = \frac{W}{S_t} \cdot \frac{\sum_{a \in A} w(a)^{2t-1}}{\sum_{a \in A} w(a)^t}$$

$$\leq \frac{W}{S_t} \cdot \max_{a \in A} \frac{w(a)^t}{w(a)}$$

$$= \rho$$

Therefore, the sample complexity of our $(\epsilon, 1/3)$-estimate of S_t is $O(\sqrt{n}/\epsilon + \rho/\epsilon^2)$. The $(\epsilon, 1/3)$-estimate of the moment can be improved to an (ϵ, δ)-estimate with a multiplicative $\ln 1/\delta$-factor sample complexity overhead using the standard median trick.

Next, we show a $\Omega(\frac{\rho \ln 1/\delta}{\epsilon})$ lower bound on the sample complexity of any algorithm for the (ϵ, δ)-moment estimation problem on any instance with the moment-density parameter ρ. We construct two families of input distributions that are hard to be distinguished using a small number of proportional samples. The lower bound instances for the moment estimation using proportional sampling described in Sect. 4 have the moment-density parameter values as $\rho = 1$ and $\rho = O(n^{1-1/t}/\epsilon)$. Next, we show that similar lower bounds hold even when we restrict the input instances to have their moment-density parameter values to be within a constant factor.

Theorem 12. *For any ρ, $\epsilon, \delta > 0$ and $t > 1$, any randomized algorithm for an (ϵ, δ)-estimate of S_t on an instance with the moment-density parameter ρ requires $\Omega(\frac{\rho \ln 1/\delta}{\epsilon})$ proportional samples.*

Proof. Given any ρ, ϵ, δ as input, we construct two families of instances for which their moment-density parameter values $\rho_1, \rho_2 = O(\rho)$ and their moment values S_t differ by a $(1 \pm \epsilon)$ factor. There are n elements in both the instances. In the first instance n_1 elements will have weight d_1 and n_2 elements will have value d_2. In the second instance there are n_1 elements of weight d_1, $\frac{n_2}{3}$ elements of weight d_2, and $\frac{2n_2}{3}$ elements of weight 0. We set the values of the parameters n_1, n_2, d_1, d_2 as follows:

$$n_1 = \frac{n^2}{n + (3\epsilon)^{\frac{2t-1}{t-1}}} \qquad d_1 = n^{1-1/t}(3\epsilon)^{1/(t-1)}$$

$$n_2 = \frac{n(3\epsilon)^{\frac{2t-1}{t-1}}}{n + (3\epsilon)^{\frac{2t-1}{t-1}}} \qquad d_2 = n$$

The moment value for the first instance is:

$$S_t = n_1 d_1^t + n_2 d_2^t = \frac{n^2}{n + (3\epsilon)^{\frac{2t-1}{t-1}}} \cdot n^{t-1}(3\epsilon)^{\frac{t}{t-1}} + \frac{n^{t+1}(3\epsilon)^{\frac{2t-1}{t-1}}}{n + (3\epsilon)^{\frac{2t-1}{t-1}}}$$

$$= \frac{n^{t+1}(3\epsilon)^{\frac{t}{t-1}} + n^{t+1}(3\epsilon)^{\frac{2t-1}{t-1}}}{n + (3\epsilon)^{\frac{2t-1}{t-1}}}$$

$$= \frac{n^{t+1}(3\epsilon)^{\frac{t}{t-1}}}{n + (3\epsilon)^{\frac{2t-1}{t-1}}}(1 + 3\epsilon)$$

The moment value for the second instance is:

$$S_t = n_1 d_1^t + \frac{n_2}{3} d_2^t + \frac{2}{3} n_2 \cdot 0 = \frac{n^2}{n + (3\epsilon)^{\frac{2t-1}{t-1}}} \cdot n^{t-1}(3\epsilon)^{\frac{t}{t-1}} + \frac{1}{3} \frac{n^{t+1}(3\epsilon)^{\frac{2t-1}{t-1}}}{n + (3\epsilon)^{\frac{2t-1}{t-1}}}$$

$$= \frac{n^{t+1}(3\epsilon)^{\frac{t}{t-1}}}{n + (3\epsilon)^{\frac{2t-1}{t-1}}}(1 + \frac{3\epsilon}{3})$$

$$= \frac{n^{t+1}(3\epsilon)^{\frac{t}{t-1}}}{n + (3\epsilon)^{\frac{2t-1}{t-1}}}(1 + \epsilon)$$

The moment values of the above two instances differ by a factor of $\frac{1+3\epsilon}{1+\epsilon} > 1 + \epsilon$. Let ρ_1 and ρ_2 be the moment-density parameters for these two instances, respectively. We compute the ρ values of the instances as follows.

$$\rho_1 = \frac{n^t}{n} \cdot \frac{n_1 d_1 + n_2 d_2}{n_1 d_1^t + n_2 d_2^t}$$

$$= n^{t-1} \cdot \frac{\frac{n^2}{n+(3\epsilon)^{\frac{2t-1}{t-1}}} \cdot n^{1-1/t}(3\epsilon)^{1/(t-1)} + \frac{n(3\epsilon)^{\frac{2t-1}{t-1}}}{n+(3\epsilon)^{\frac{2t-1}{t-1}}} \cdot n}{\frac{n^2}{n+(3\epsilon)^{\frac{2t-1}{t-1}}} \cdot n^{t-1}(3\epsilon)^{t/(t-1)} + \frac{n(3\epsilon)^{\frac{2t-1}{t-1}}}{n+(3\epsilon)^{\frac{2t-1}{t-1}}} \cdot n^t}$$

$$= \frac{n^{1-1/t} + (3\epsilon)^2}{3\epsilon + (3\epsilon)^2} = \frac{n^{1-1/t} + 9\epsilon^2}{3\epsilon + 9\epsilon^2}$$

Similarly, we compute ρ_2 as

$$\rho_2 = \frac{n^t}{n} \cdot \frac{n_1 \cdot d_1 + n_2/3 \cdot d_2 + 2n_2/3 \cdot 0}{n_1 \cdot d_1^t + n_2/3 \cdot d_2^t + 2n_2/3 \cdot 0}$$

$$= \frac{n^{1-1/t} + \frac{1}{3}(3\epsilon)^2}{3\epsilon(1+\epsilon)} = \frac{n^{1-1/t} + 3\epsilon^2}{3\epsilon + 3\epsilon^2}$$

We have two instances as given above where their ρ values are within a constant factor of each other and the moment values differ by at least a factor of $(1 + \epsilon)$. These instances using proportional sampling remain indistinguishable until an element with weight d_2 is sampled. The probability of sampling an element of weight d_2 using proportional sampling is at most:

$$\frac{n_2 d_2}{n_1 d_1 + n_2 d_2} = \frac{1}{1 + \frac{n_1 d_1}{n_2 d_2}} = \frac{1}{1 + \frac{n^2 n^{1-1/t}(3\epsilon)^{1/(t-1)}}{n^2(3\epsilon)^{\frac{2t-1}{t-1}}}} = \frac{1}{1 + \frac{n^{1-1/t}}{9\epsilon^2}} \le \frac{1}{1 + O(\frac{\rho}{\epsilon})}$$

Therefore, we require at least $\Omega(\rho/\epsilon)$ samples to distinguish the two instances with constant probability of success. Any algorithm giving an (ϵ, δ)-estimate would need to make $\Omega(\frac{\rho \ln 1/\delta}{\epsilon})$ proportional samples.

7   Moment Estimation for $0 < t < 1$

7.1   Moment Estimation Using Proportional Sampling for $1/2 < t < 1$

We use Algorithm 1 to estimate moment S_t for $1/2 < t < 1$. We adapt the analysis in Sect. 3 to prove the following result.

Theorem 13. *There exists an algorithm ALG that given proportional sampling access to the weights of the elements of a set A and parameters $1/2 < t < 1$, $\epsilon \in (0, 1)$, provides an $(\epsilon, 1/3)$-estimate of S_t using $O(\frac{\sqrt{n}}{\epsilon} + \frac{n^{\frac{1}{t}-1}}{\epsilon^2})$ samples.*

The sample complexity of the algorithm for $1/2 < t < 1$ is given as follows. Note that the sample complexity of our algorithm is given as $\frac{W(||w(A)||_{2t-1})^{2t-1}}{\epsilon^2(||w(A)||_t)^{2t}}$. We can rewrite it as $\frac{||w(A)||_1}{\epsilon^2 ||w(A)||_t} \cdot (\frac{||w(A)||_{2t-1}}{||w(A)||_t})^{2t-1}$. Using Fact 1, this is upper bounded as $n^{\frac{1}{t}-1}/\epsilon^2$. Overall sample complexity of the $(\epsilon, 1/3)$-estimator of the moment is $O(\sqrt{n}/\epsilon + n^{\frac{1}{t}-1}/\epsilon^2)$. Using the median trick we transform the $(\epsilon, 1/3)$-estimator to an (ϵ, δ)-estimator with sample complexity $O((\sqrt{n}/\epsilon + n^{\frac{1}{t}-1}/\epsilon^2)\ln 1/\delta)$.

7.2   Lower Bound for $t \le 1/2$

We show that there are no sublinear algorithms for the problem when $t \le 1/2$.

Theorem 14. *For any $\epsilon > 0$ and $t \leq 1/2$, any randomized algorithm that computes an $(\epsilon, 1/3)$-estimate of S_t requires $\Omega(n)$ proportional samples.*

Proof. We construct two families of instances which we show are hard to be distinguished using proportional sampling. In one instance we have one element with weight $(n-1)$, and the rest $(n-1)$ elements with weight 0. In another instance, we have one element of weight $(n-1)$, and $(n-1)$ elements have weight $\frac{\epsilon^{1/t}}{n-1}$. The moment values of the two instances are given as $(n-1)^t$ and $(n-1)^t + \epsilon(n-1)^{1-t} \geq (1+\epsilon)(n-1)^t$ as $t \leq 1/2$. Hence, there exists a $(1+\epsilon)$-multiplicative gap between the moment values of these two instances.

These two instances are distinguished using proportional sampling once we sample an element of weight $\frac{\epsilon^{1/t}}{n-1}$. The probability of sampling such an element is at most $\frac{\epsilon^{1/t}}{\epsilon^{1/t}+(n-1)}$. Hence, one requires $\Omega(n)$ samples to distinguish the two instances with constant probability.

8   Conclusion

In this paper we improve the results for moment estimation using proportional sampling and proved optimal sample complexity bounds for moment estimation using proportional sampling for $t \geq 2$. Since hybrid sampling does not provide better guarantees for the moment estimation problem, one possible approach to obtain better bounds might be to explore whether conditional sampling-based approaches give better results. Conditional sampling has been used to design algorithms for the sum estimation problem [1] but to the best of our knowledge no such algorithm is known for the moment estimation problem.

Acknowledgments. Anup Bhattacharya is supported by the Science and Engineering Research Board (SERB) via the project (CRG/2023/002119). The authors thank the reviewers for suggestions to improve the manuscript.

References

1. Acharya, J., Canonne, C.L., Kamath, G.: Adaptive estimation in weighted group testing. In: 2015 IEEE International Symposium on Information Theory (ISIT), pp. 2116–2120. IEEE (2015)
2. Aliakbarpour, M., Biswas, A.S., Gouleakis, T., Peebles, J., Rubinfeld, R., Yod-pinyanee, A.: Sublinear-time algorithms for counting star subgraphs via edge sampling. Algorithmica **80**, 668–697 (2018)
3. Alon, N., Matias, Y., Szegedy, M.: The space complexity of approximating the frequency moments. J. Comput. Syst. Sci. **58**(1), 137–147 (1999)
4. Assadi, S., Nguyen, H.A.: Asymptotically optimal bounds for estimating h-index in sublinear time with applications to subgraph counting. arXiv preprint arXiv:2209.08114 (2022)

5. Beretta, L., Tětek, J.: Better sum estimation via weighted sampling. In: Proceedings of the 2022 Annual ACM-SIAM Symposium on Discrete Algorithms (SODA), pp. 2303–2338. SIAM (2022)
6. Eden, T., Ron, D., Seshadhri, C.: Sublinear time estimation of degree distribution moments: The degeneracy connection. In: 44th International Colloquium on Automata, Languages, and Programming (ICALP 2017). Schloss Dagstuhl-Leibniz-Zentrum fuer Informatik (2017)
7. Eden, T., Ron, D., Seshadhri, C.: Sublinear time estimation of degree distribution moments: the arboricity connection. SIAM J. Discret. Math. **33**(4), 2267–2285 (2019)
8. Feige, U.: On sums of independent random variables with unbounded variance and estimating the average degree in a graph. SIAM J. Comput. **35**(4), 964–984 (2006)
9. Goldreich, O.: Introduction to Property Testing. Cambridge University Press, Cambridge (2017)
10. Goldreich, O., Ron, D.: Approximating average parameters of graphs. Random Struct. Algorithms **32**(4), 473–493 (2008)
11. Gonen, M., Ron, D., Shavitt, Y.: Counting stars and other small subgraphs in sublinear-time. SIAM J. Discret. Math. **25**(3), 1365–1411 (2011)
12. Jayaram, R., Woodruff, D.: Towards optimal moment estimation in streaming and distributed models. ACM Trans. Algorithms **19** (2023). https://doi.org/10.1145/3596494
13. Mitzenmacher, M., Upfal, E.: Probability and Computing: Randomization and Probabilistic Techniques in Algorithms and Data Analysis. Cambridge University Press, Cambridge (2017)
14. Motwani, R., Panigrahy, R., Xu, Y.: Estimating sum by weighted sampling. In: International Colloquium on Automata, Languages, and Programming, pp. 53–64. Springer (2007)
15. Tětek, J., Thorup, M.: Edge sampling and graph parameter estimation via vertex neighborhood accesses. In: Proceedings of the 54th Annual ACM SIGACT Symposium on Theory of Computing, pp. 1116–1129 (2022)

Structural Parameterization of Locating-Dominating Set and Test Cover

Dipayan Chakraborty[1,2,3]([✉]) [iD], Florent Foucaud[1] [iD], Diptapriyo Majumdar[4] [iD], and Prafullkumar Tale[5] [iD]

[1] Université Clermont Auvergne, CNRS, Mines Saint-Étienne, Clermont Auvergne INP, LIMOS, 63000 Clermont-Ferrand, France
{dipayan.chakraborty,florent.foucaud}@uca.fr
[2] Department of Mathematics and Applied Mathematics,
University of Johannesburg, Auckland Park, Johannesburg 2006, South Africa
[3] Department of Computer Science and Mathematics, Lebanese American University,
Beirut, Lebanon
[4] Indraprastha Institute of Information Technology Delhi, New Delhi, India
diptapriyo@iiitd.ac.in
[5] Indian Institute of Science Education and Research Pune, Pune, India
prafullkumar@iiserb.ac.in

Abstract. We investigate structural parameterizations for two identification problems in graphs and set systems: LOCATING-DOMINATING SET and TEST COVER. In the first problem, an input is a graph G and an integer k, and one asks whether there is a subset S of k vertices such that any two distinct vertices not in S are dominated by distinct subsets of S. In the second problem, an input is a set of items U, a collection of subsets $\mathcal{F}$ of U called *tests*, and an integer k, and one asks whether there is a solution set S of at most k tests such that each pair of items belongs to a distinct subset of tests of S.

In a related work [ISAAC 2024], we proved that both problems admit a conditional double-exponential lower bound and a matching algorithm when parameterized by the treewidth of the input graph. We continue this line of investigation and consider parameters larger than treewidth, like vertex cover number and feedback edge set number. We design a nontrivial dynamic programming scheme for TEST COVER in "slightly super-exponential" time $2^{\mathcal{O}(|U|\log|U|)}(|U|+|\mathcal{F}|)^{\mathcal{O}(1)}$ in the number $|U|$ of items, and also LOCATING-DOMINATING SET in time $2^{\mathcal{O}(\mathsf{vc}\log\mathsf{vc})}\cdot n^{\mathcal{O}(1)}$, where vc is the vertex cover number and n the order of the graph. Thus, the known lower bounds with respect to treewidth cannot be extended to the vertex cover number. We also show that when parameterized by the feedback edge set number, LOCATING DOMINATING SET admits a linear kernel, answering an open question from [Cappelle et al., LAGOS 2021]. Finally, we show that neither LOCATING-DOMINATING SET nor TEST COVER is likely to admit a compression algorithm returning an input with a subquadratic number of bits.

Keywords: Locating-Dominating Set · Test Cover · Parameterized Complexity · Structural Parameterization

I. Finocchi and L. Georgiadis (Eds.): CIAC 2025, LNCS 15679, pp. 187–204, 2025.
https://doi.org/10.1007/978-3-031-92932-8_13

1   Introduction

We study two discrete identification problems, LOCATING-DOMINATING SET and TEST COVER, that are extensively studied since the 1970s. This type of problems, in which one aims at distinguishing all the elements of a discrete structure by a solution set, is popular in both combinatorics [6, 26] and algorithms [5, 7], and has many natural applications e.g. to network monitoring [25], or machine learning [13]. In a recent paper, Chakraborty et al. [11] revisited the parameterized complexity of LOCATING-DOMINATING SET and TEST COVER when parameterized by the solution size or the treewidth (tw) of the input graph. In the case of TEST COVER, the authors considered the treewidth of the natural auxiliary bipartite graph (which we specify later). They proved that LOCATING-DOMINATING SET and TEST COVER admit an algorithm running in time $2^{2^{O(tw)}} \cdot \text{poly}(n)$, where n is the number of elements/vertices in the system/graph. More interestingly, they proved that these algorithms are tight, i.e., neither problem admits an algorithm running in time $2^{2^{o(tw)}} \cdot \text{poly}(n)$, unless the ETH fails. These results add LOCATING-DOMINATING SET and TEST COVER to the small list of NP-complete problems, recently initiated by Foucaud et al. [19], that admit a conditional double-exponential lower bound when parameterized by the treewidth. We continue this line of research and study the parameterized complexity of these problems for parameters that are larger than the treewidth. Before moving forward, we formally define the two studied problems.

LOCATING-DOMINATING SET
Input: A graph G on n vertices and an integer k.
Question: Does there exist a locating-dominating set of size k in G, that is, a dominating set S of G such that for any two different vertices $u, v \in V(G) \backslash S$, their neighbourhoods in S are different, i.e., $N(u) \cap S \neq N(v) \cap S$?

TEST COVER
Input: A set of items U, a collection of subsets of U called *tests* and denoted by $\mathcal{F}$, and an integer k.
Question: Does there exist a collection of at most k tests such that for each pair of items, there is a test that contains exactly one of the two items?

As TEST COVER is defined over set systems, to consider structural graph parameters, we define the *auxiliary graph* of a set system $(U, \mathcal{F})$ in the natural way: The bipartite graph G on n vertices with bipartition $\langle R, B \rangle$ of $V(G)$ such that sets R and B contain a vertex for every set in $\mathcal{F}$ and for every item in U, respectively, and where $r \in R$ and $b \in B$ are adjacent if and only if the set corresponding to r contains the element corresponding to b.

We refer readers to [11] for a comprehensive overview about the motivations, applicability and known literature regarding the two problems, as we only mention the most relevant results here. LOCATING-DOMINATING SET [14] and TEST COVER [20, Problem SP6] are both NP-complete. They are also trivially FPT for the parameter solution size (see [8, 9, 16] or the discussion before Theorem 2 and Theorem 3 in [11]). Hence, it becomes interesting to study these problems under more refined angles, such as kernelization, fine-grained complexity, and alternative parameterizations. Naturally, these lines of research have been pursued in the literature. TEST COVER was studied within the framework of "above/below guarantee" parameterizations in [4,15,16,21] and kernelization in [4,15,21]. These results have shown an intriguing behaviour for TEST COVER, with some nontrivial techniques being developed to solve the problem [4,16]. In the case of LOCATING-DOMINATING SET, fine-grained complexity results regarding the number of vertices and solution size were obtained in [3,8,9], and structural parameterizations of the input graph have been studied in [8,9]. In particular, it was shown there that the problem admits a linear kernel for the parameter max leaf number, however (under standard complexity assumptions) no polynomial kernel exists for the solution size, combined with either the vertex cover number or the distance to clique. They also provide a double-exponential kernel for the parameter distance to cluster.

Our Contributions. We extend the above results by systematically studying the fine-grained complexity and kernelization of LOCATING-DOMINATING SET and TEST COVER for the parameters number of vertices/elements/tests as well as structural parameters. Our results contribute to the complexity landscape for structural parameterizations of LOCATING-DOMINATING SET started in [8,9]. See Fig. 1 for an overview of the known results for (structural) parameterizations of LOCATING-DOMINATING SET, using a Hasse diagram of standard graph parameters. Our first result is an FPT algorithm for LOCATING-DOMINATING SET parameterized by the vertex cover number (vc) of the input graph, and for TEST COVER when parameterized by the number $|U|$ of items.

Theorem 1. *LOCATING-DOMINATING SET admits an algorithm running in time $2^{\mathcal{O}(vc \log vc)} \cdot n^{\mathcal{O}(1)}$, where vc is the vertex cover number of the input graph. Also, TEST COVER admits an algorithm running in time $2^{\mathcal{O}(|U| \log |U|)} \cdot (|U| + |\mathcal{F}|)^{\mathcal{O}(1)}$.*

The above result shows that unlike treewidth, for which we cannot expect a running time of the form $2^{2^{o(tw)}}$ [11], if we choose vc, a parameter larger than treewidth, then we can significantly improve the running time. Moreover, we show that the above result for LOCATING-DOMINATING SET also extends to the parameters distance to clique and twin-cover number. For TEST COVER, the result is reminiscent of the known $\mathcal{O}(2^n mn)$ dynamic programming scheme for SET COVER on a universe of size n and m sets (see [18, Theorem 3.10]), and improves upon the naive $2^{\mathcal{O}(|U|^2)} \cdot (|U| + |\mathcal{F}|)^{\mathcal{O}(1)}$ brute-force algorithm.

Our next result is a linear vertex kernel when feedback edge set number (fes) is considered as the parameter. It solves an open problem raised in [8,9].

Theorem 2. *LOCATING-DOMINATING SET admits a kernel with $\mathcal{O}(fes)$ vertices and edges, where fes is the feedback edge set number of the input graph.*

Our final result is about the incompressibility of both problems.

Theorem 3. *Neither* Locating-Dominating Set *nor* Test Cover *admits a polynomial compression of size* $\mathcal{O}(n^{2-\epsilon})$ *for any* $\epsilon > 0$, *unless* $\mathsf{NP} \subseteq \mathsf{coNP}/poly$, *where* n *denotes the number of vertices and the number of items and tests of the input, respectively.*

The reduction used to prove Theorem 3 also provides an alternative proof that Locating-Dominating Set cannot be solved in time $2^{o(n)}$ (see [3] for an earlier proof), and thus, in time $2^{o(\mathsf{vc})}$ (under the ETH). Hence, the bound of Theorem 1 is optimal up to a logarithmic factor. The reduction used to prove Theorem 3 also yields the following results. First, it proves that Locating-Dominating Set, parameterized by the vertex cover number of the input graph and the solution size, does not admit a polynomial kernel, unless $\mathsf{NP} \subseteq \mathsf{coNP}/poly$. Our reduction is arguably a simpler argument than the one from [8,9] to obtain the latter result. Second, it implies that Test Cover, parameterized by the number of items $|U|$ and the solution size k, does not admit a polynomial kernel, unless $\mathsf{NP} \subseteq \mathsf{coNP}/poly$. Additionally, it also implies that Test Cover, parameterized by the number $|\mathcal{F}|$ of tests, does not admit a polynomial kernel, unless $\mathsf{NP} \subseteq \mathsf{coNP}/poly$. Since we can assume $k \leq |\mathcal{F}|$ in any non-trivial instance, this improves upon the result in [21], that states that there is no polynomial kernel for the problem when parameterized by k alone.

Organization. We use the Locating-Dominating Set problem to demonstrate key technical concepts, and explain how to adapt the arguments for Test Cover. Theorem 1 is addressed in Sect. 2 and Theorem 2 in Sect. 3. We present the reduction used to prove Theorem 3 in Sect. 4. Finally, we conclude the paper with some open problems in Sect. 5. All details and proofs missing from this article can be found in its full version [10].

2   Slightly Super-Exponential Algorithm Parameterized by Vertex Cover

We describe an algorithm to prove Theorem 1, for Locating-Dominating Set, i.e., to prove that it admits an algorithm running in time $2^{\mathcal{O}(\mathsf{vc} \log \mathsf{vc})} \cdot n^{\mathcal{O}(1)}$, where vc is the vertex cover number of the input graph. Our algorithm is based on a reduction to a dynamic programming scheme for a generalized partition refinement problem, that can also be used to Test Cover. We start with the following reduction rule which, using a result from [22], can be shown to be applicable in polynomial time.

Reduction Rule 4. *Let* (G, k) *be an instance of* Locating-Dominating Set. *If there exist three vertices* u, v, x *of* G *such that any two of* u, v, x *are twins, then delete* x *from* G *and decrease* k *by one.*

Our algorithm starts by finding a minimum vertex cover, say U, of G by an algorithm in time $1.2528^{\mathsf{vc}} \cdot n^{\mathcal{O}(1)}$ [23]. Then, $\mathsf{vc} = |U|$. Let $R = V \setminus U$ be the corresponding independent set in G. The algorithm first applies Reduction

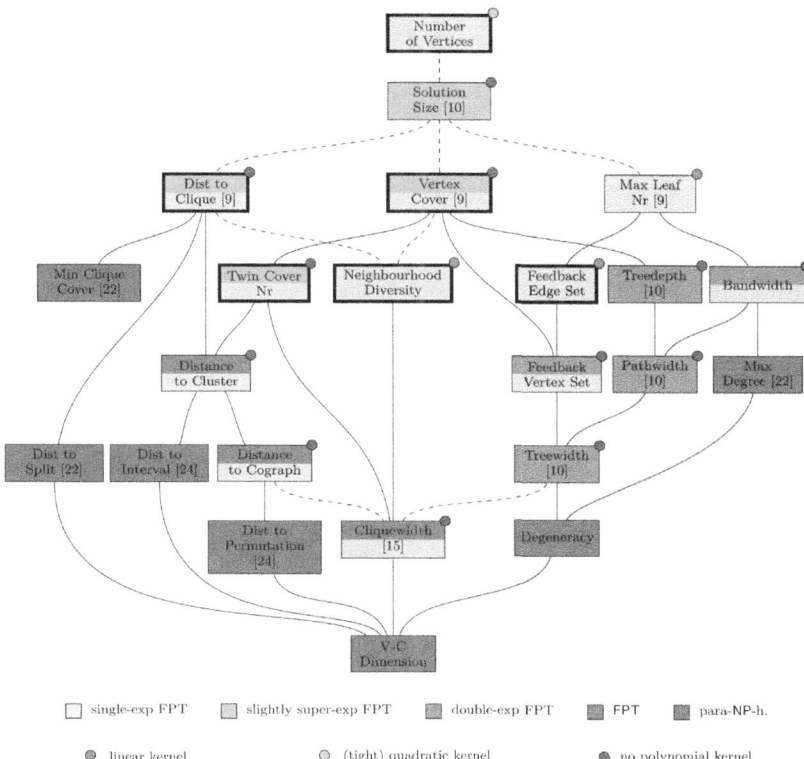

Fig. 1. Hasse diagram of graph parameters and associated results for LOCATING-DOMINATING SET. An edge from a lower parameter to a higher parameter indicates that the lower one is upper bounded by a function of the higher one. If the line is dashed, then the bound is exponential; otherwise, it is polynomial. Colors correspond to the known FPT complexity status with respect to the highlighted parameter: the upper half of the box represents the upper bound, and the lower half of the box represents the lower bound. By "single-exponential", "slightly super-exponential" and "double-exponential", we mean functions of the form $2^{\mathcal{O}(p)}$, $2^{\mathcal{O}(p \log p)}$ or $2^{\mathcal{O}(p^2)}$, and $2^{2^{\mathcal{O}(p)}}$, respectively. The parameters for which the running time is not known to be tight are thus striped. The red circle in the upper-right corner means that LOCATING-DOMINATING SET does not admit a polynomial kernel when parameterized by the marked parameter unless $\mathsf{NP} \subseteq \mathsf{coNP}/poly$; the yellow one means that a (tight) quadratic kernel exists, and the green one, that a linear kernel exists. The bold borders highlight parameters that are covered in this paper. (Color figure online)

Rule 4 exhaustively to reduce every twin-class to size at most 2. For simplicity, we continue to call the reduced instance of LOCATING-DOMINATING SET as (G, k).

Consider an optimal but hypothetical locating-dominating set L of G. The algorithm constructs a partial solution Y_L and, in subsequent steps, expands it

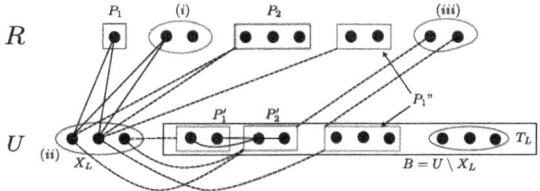

Fig. 2. An instance (G, k) of LOCATING-DOMINATING SET where U is a vertex cover of G. The dotted edge denotes that a vertex is adjacent with all the vertices in the set. For the sake of brevity, we do not show all the edges. The vertices in green ellipses are part of solution because of (i) being a part of false-twins, (ii) guessed intersection with U, and (iii) the requirement of solution to be a dominating set. Vertices T_L are not dominated by the partial solution Y_L. Parts P_1, P_2, P'_1, P'_2, and P''_1 are parts of partition of $R \cup B$ induced by Y_L. (Color figure online)

to obtain L. The algorithm initializes Y_L as follows: for all pairs u, v of twins in G, it adds one of them in Y_L. Slater proved that for any set S of vertices of a graph G such that any two vertices in S are twins, any locating-dominating set contains at least $|S| - 1$ vertices of S [28]. Hence, it is safe to assume that all the vertices in Y_L are present in any locating-dominating set. Next, the algorithm guesses the intersection of L with U. Formally, it iterates over all the subsets of U, and for each such set, say X_L, it computes a locating-dominating set of appropriate size that contains all the vertices in X_L and no vertex in $U \backslash X_L$. Consider a subset X_L of U and define $B = U \backslash X_L$ and $R' = R \backslash (N(X_L) \cup Y_L)$. As L is also a dominating set, it is safe to assume that $R' \subseteq L$. The algorithm updates Y_L to include $X_L \cup R'$.

At this stage, Y_L dominates all the vertices in R. However, it may not dominate all the vertices in B. The remaining (to be chosen) vertices in L are part of R and are responsible for dominating the remaining vertices in B and to locate all the vertices in $R \cup B$. See Fig. 2 for an illustration. As the remaining solution, i.e., $L \backslash Y_L$, does not intersect B, it is safe to ignore the edges both of whose endpoints are in B. The vertices in Y_L induce a partition of the vertices of the remaining graph, according to their neighborhood in Y_L. We can redefine the objective of selecting the remaining vertices in a locating-dominating set as to *refine* this partition such that each part contains exactly one vertex. Partition refinement is a classical concept in algorithms, see [22]. However in our case, the partition is not standalone, as it is induced by a solution set. To formalize this intuition, we introduce the following notation.

Definition 1 (Partition Induced by C and Refinement). *For a subset $C \subseteq V(G)$, define an equivalence relation $\sim_C$ on $V' \subseteq V(G) \backslash C$ as follows: for any pair $u, v \in V'$, $u \sim_C v$ if and only if $N(u) \cap C = N(v) \cap C$. Then, for any set $V \subseteq V(G)$, the partition of V induced by C, denoted by $\mathcal{P}(C)$, is defined as:*

$$\mathcal{P}(C) = \{\{c\} \mid c \in V \cap C\} \cup \{S \mid S \text{ is an equivalence class of } \sim_C \text{ defined on } V \backslash C\}.$$

Moreover, for two partitions $\mathcal{P}$ and $\mathcal{Q}$ of V, a refinement *of $\mathcal{P}$ by $\mathcal{Q}$, denoted by $\mathcal{P} \pitchfork \mathcal{Q}$, is the partition defined as* $\{P \cap Q : P \in \mathcal{P}, Q \in \mathcal{Q}\}$.

Suppose $V = V(G)$ and $\mathcal{P}$ and $\mathcal{Q}$ are two partitions of $V(G)$, then $\mathcal{P} \pitchfork \mathcal{Q}$ is also a partition of $V(G)$. In addition, it can be checked that, if $C_1, C_2 \subseteq V(G)$, then $\mathcal{P}(C_1 \cup C_2) = \mathcal{P}(C_1) \pitchfork \mathcal{P}(C_2)$ on any subset $V \subseteq V(G)$. We say a partition $\mathcal{P}$ is the *identity partition* if every part of $\mathcal{P}$ is a singleton set. This implies that a set $C \subseteq V(G)$ for any graph G has the property $N(u) \cap C \neq N(v) \cap C$ for all distinct $u, v \in V(G) \backslash C$ if and only if the partition $\mathcal{P}(C)$ of $V(G)$ is the identity partition. With these definitions, we define the following auxiliary problem.

ANNOTATED RED-BLUE PARTITION REFINEMENT

Input: A bipartite graph G with bipartition $\langle R, B \rangle$ of $V(G)$; a partition $\mathcal{Q}$ of $R \cup B$; a collection of forced solution vertices $C_0 \subseteq R$; a collection of vertices $T_L \subseteq B$ that needs to be dominated; and an integer λ.
Question: Does there exist a set C of size at most λ such that $C_0 \subseteq C \subseteq R$, C dominates T_L, and $\mathcal{Q} \pitchfork \mathcal{P}(C)$ is the identity partition of $R \cup B$?

Suppose there is an algorithm $\mathcal{A}$ that solves ANNOTATED RED-BLUE PARTITION REFINEMENT in time $f(|B|) \cdot (|R| + |B|)^{\mathcal{O}(1)}$. Then, there is an algorithm that solves LOCATING-DOMINATING SET in time $2^{\mathcal{O}(\mathsf{vc})} \cdot f(\mathsf{vc}) \cdot n^{\mathcal{O}(1)}$. Consider the algorithm described so far in this subsection. The algorithm then calls $\mathcal{A}$ as a subroutine with the bipartite graph $\langle R, B \rangle$ obtained from G, where $B = U \backslash X_L$ and $R = V \backslash U$ (notice that, since the hypothetical solution L has no vertices from B, the edges of G with both endpoints in B are irrelevant to the locating property of L). It sets $\mathcal{Q}$ as the partition of $R \cup B$ induced by Y_L, $C_0 = Y_L \cap R$, $T_L = B \backslash N[Y_L]$, and $\lambda = k - |X_L|$. Note that T_L is the collection of vertices that are *not* dominated by the partial solution Y_L. The correctness of the algorithm follows from the correctness of $\mathcal{A}$ and the fact that for two sets $C_1, C_2 \subseteq V(G)$, $C_1 \cup C_2$ is a locating-dominating set of G if and only if $C_1 \cup C_2$ is a dominating set and $\mathcal{P}(C_1 \cup C_2) = \mathcal{P}(C_1) \pitchfork \mathcal{P}(C_2)$ is the identity partition of $V(G)$. Hence, it suffices to prove the following lemma.

Lemma 1. *There is an algorithm that solves* ANNOTATED RED-BLUE PARTITION REFINEMENT *in time* $2^{\mathcal{O}(|B| \log |B|)} \cdot (|R| + |B|)^{\mathcal{O}(1)}$.

The remainder of the section is devoted to prove Lemma 1. We present the algorithm that can roughly be divided into three parts. In the first part, the algorithm processes the partition $\mathcal{Q}$ of $R \cup B$ with the goal of reaching to a refinement of partition $\mathcal{Q}$ such that every part is completely contained either in R or in B. Then, we introduce some terms to obtain a 'set-cover' type dynamic programming. We also introduce three conditions that restrict the number of dynamic programming states we need to consider to $2^{\mathcal{O}(|B| \log(|B|))} \cdot |R|$. Finally, we state the dynamic programming procedure.

Pre-processing the Partition. Consider the partition $\mathcal{P}(C_0)$ of $R \cup B$ induced by C_0, and define $\mathcal{P}_0 = \mathcal{Q} \pitchfork \mathcal{P}(C_0)$. We classify the parts in $\mathcal{P}_0$ into three classes

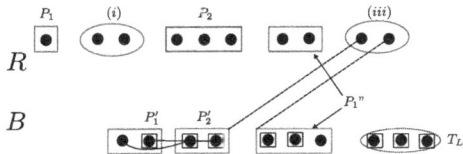

Fig. 3. Instance of the ANNOTATED RED-BLUE PARTITION REFINEMENT . Vertices in green ellipse denote set C_0 whereas dotted ellipse denotes vertices in T_L, and the partition $\mathcal{Q} = \{P_1, P_2, P_1', P_2', P_1''\}$. The vertices of T_L° are in rectangles. (Color figure online)

depending on whether they intersect with R, B, or both. Let $P_1, P_2, \ldots, P_t$ be an arbitrary but fixed order on the parts of $\mathcal{P}_0$ that are completely in R. Similarly, let $P_1', P_2', \ldots, P_{t'}'$ be an arbitrary but fixed order on the parts that are completely in B. Also, let $P_1'', P_2'', \ldots, P_{t''}''$ be the collection of parts that intersect R as well as B. Formally, we have $P_j \subseteq R$ for any $j \in [t]$, $P_{j'}' \subseteq B$ for any $j' \in [t']$, and $P_{j''}'' \cap R \neq \emptyset \neq P_{j''}'' \cap B$ for any $j'' \in [t'']$. See Fig. 3 for an illustration.

Recall that T_L is the collection of vertices in B that are required to be dominated. The algorithm first expands this set so that it is precisely the collection of vertices that need to be dominated by the solution C. In other words, at present, the required condition is $T_L \subseteq N(C)$, whereas, after the expansion, the condition is $T_L = N(C)$. Towards this, it first expands T_L to include $N(C_0)$. It then uses the property that the final partition needs to be the identity partition to add some more vertices in T_L.

Consider parts $P_{j'}'$ or $P_{j''}''$ of $\mathcal{P}_0$. Suppose that $P_{j'}'$ (respectively, $P_{j''}'' \cap B$) contains two vertices that are *not* dominated by the final solution C, then these two vertices would not be separated by it and hence would remain in one part, a contradiction. Therefore, any feasible solution C needs to dominate all vertices but one in $P_{j'}'$ (respectively in $P_{j''}'' \cap B$).

– The algorithm modifies T_L to include $N(C_0)$. Then, it iterates over all the subsets T_L° of $B \backslash T_L$ such that for any $j' \in [t']$ and $j'' \in [t'']$, $|P_{j'}' \backslash (T_L^\circ \cup T_L)| \leq 1$ and $|(P_{j''}'' \cap B) \backslash (T_L^\circ \cup T_L)| \leq 1$.

Consider a part P_j of $\mathcal{P}_0$, which is completely contained in R. As R is an independent set and the final solution C is not allowed to include any vertex in B, for any part P_j, there is at most one vertex outside C, i.e., $|P_j \backslash C| \leq 1$. However, unlike the previous step, the algorithm cannot enumerate all the required subsets of R in the desired time. Nonetheless, it uses this property to safely perform the following sanity checks and modifications.

– Consider parts P_j or $P_{j''}''$ of $\mathcal{P}_0$. Suppose there are two vertices in P_j (respectively $P_{j''}'' \cap R$) that are adjacent with some vertices in $B \backslash (T_L^\circ \cup T_L)$. Then, discard this guess of T_L° and move to the next subset.
– Consider parts P_j or $P_{j''}''$ of $\mathcal{P}_0$. Suppose there is a unique vertex, say r, in P_j (respectively, in $P_{j''}'' \cap R$) that is adjacent with some vertex in $B \backslash (T_L^\circ \cup T_L)$. Then, modify C_0 to include all the vertices in $P_j \backslash \{r\}$ or in $(P_{j''}'' \cup R) \backslash \{r\}$.

Consider a part $P''_{j''}$ of $\mathcal{P}_0$ and suppose $|(P''_{j''} \cap B) \backslash (T^\circ_L \cup T_L)| = 0$. Alternately, every vertex in $P''_{j''} \cap B$ is adjacent to some vertex in the final solution C. As R is an independent set and $C \subseteq R$, this domination condition already refines the partition $P''_{j''}$, i.e. every vertex in $(P''_{j''} \cap B)$ needs to be adjacent with some vertex in C whereas no vertex in $(P''_{j''} \cap R)$ can be adjacent with C. This justifies the following modification.

– For a part $P''_{j''}$ of $\mathcal{P}_0$ such that $|(P''_{j''} \cap B) \backslash (T^\circ_L \cup T_L)| = 0$, the algorithm modifies the input partition $\mathcal{P}_0$ by removing $P''_{j''}$ and adding two new parts $P_j := P''_{j''} \cap R$ and $P'_{j'} := P''_{j''} \cap B$.

Consider a part $P''_{j''}$ of $\mathcal{P}_0$ and suppose $|(P''_{j''} \cap B) \backslash (T^\circ_L \cup T_L)| = 1$. Let b be the unique vertex in $(P''_{j''} \cap B) \backslash (T^\circ_L \cup T_L)$. Suppose there is a (unique) vertex, say r, in $(P''_{j''} \cap R)$ that is not part of the final solution C. However, this implies that both b and r are in the same part, a contradiction.

– For part $P''_{j''}$ of $\mathcal{P}_0$ such that $|(P''_{j''} \cap B) \backslash (T^\circ_L \cup T_L)| = 1$, the algorithm includes all the vertices $(P''_{j''} \cap R)$ into C_0 (which, by definition, modifies $\mathcal{P}_0$). Also, it removes $P''_{j''}$ from $\mathcal{P}_0$ and adds $P'_{j'} := P''_{j''} \cap B$.

These modifications ensure that any part of $\mathcal{P}_0$ is either completely contained in R or completely contained in B.

Setting up the Dynamic Programming. Define $\ell = |R \backslash C_0|$ and suppose $\{r_1, r_2, \ldots, r_\ell\}$ be the reordering of vertices in $R \backslash C_0$ such that any part in P_j contains the consecutive elements in the order. Formally, for any $i_1 < i_2 < i_3 \in \{1, 2, \ldots, |R \backslash C_0|\}$ and $j \in \{1, 2, \ldots, t\}$, if $r_{i_1} \in P_j$ and $r_{i_3} \in P_j$, then $r_{i_2} \in P_j$. Define a function $\pi : \{1, 2, \ldots, |R \backslash C_0|\} \mapsto \{1, 2, \ldots, t\}$ such that $\pi(r_i) = j$ if and only if vertex r_i is in part P_j.

For the remaining part, the algorithm relies on a 'set cover' type dynamic programming.[1] For every integer $i \in [0, \ell]$, define $R_i = \{r_1, r_2, \ldots, r_i\}$. Then, a 'set cover'-type dynamic programming tries to find the optimum size of a subset of R_i that dominates all the vertices in S, which is a subset of T_L. However, two different optimum solutions in R_i will induce different partitions of $R \cup B$. To accommodate this, we define the notion of *valid tuples*. For an integer $i \in [0, \ell]$ and a subset $S \subseteq T_L$, we say a tuple $(i, \mathcal{P}, S)$ is a *valid tuple* if $\mathcal{P}$ is a refinement of $\mathcal{P}_0$ and it satisfies the three properties mentioned below. Note that these properties are the consequences of the fact that we require $\mathcal{P}$ to be such that, for any (partial) solution C with vertices upto r_i in the ordering of $R \backslash C_0$, we can have $\mathcal{P}(C) = \mathcal{P}$ and $N(C) = S$.

1. For any $i \in [0, \ell]$, the partition $\mathcal{P}$ restricted to any P_j, where $j \in [\pi(i) - 1]$, is the identity partition of P_j. This is because, if on the contrary, $\mathcal{P}$ restricted to P_j is not the identity partition, it would imply that there exists a part P, say, of $\mathcal{P}$ with $|P| \geq 2$ which cannot be refined by picking vertices of R in the final solution C (upto r_ℓ), as R is an independent set.

---

[1] See [18, Theorem 3.10] for a simple $\mathcal{O}(mn2^n)$ dynamic programming scheme for SET COVER on a universe of size n and m sets.

2. There may be at most one vertex r. say, in $P_{\pi(i)} \cap R_i$ which is not yet picked in C. This is because if two vertices r', r'' of $P_{\pi(i)} \cap R_i$ are not picked in C, the final solution C cannot induce the identity partition, since $\{r'\}, \{r''\} \notin \mathcal{P}(C)$. Thus, since for any $r' \in P_{\pi(i)} \cap R_i \backslash \{r\}$, we have $\{r'\} \in \mathcal{P}(C)$, we therefore require the partition P of $\mathcal{P}$ to be a singleton set if $P \cap P_{\pi(i)} \cap R_i \neq \emptyset$ and $r \notin P$. Moreover, since $(\{r\} \cup P_{\pi(i)} \backslash R_i) \in \mathcal{P}(C)$, we require that for $r \in P$, the part P must be $(\{r\} \cup P_{\pi(i)} \backslash R_i) \in \mathcal{P}(C)$.

3. Since any (partial) solution upto r_i does not refine any of the parts $P_{\pi(i)+1}, \ldots, P_t$, for every part $P \in \mathcal{P}$ that does not intersect $P_1 \cup P_2 \cup \ldots \cup P_{\pi(i)}$, P must be of the form P_x, where $x \in [\pi(i) + 1, t]$.

From the above three properties of how the refinements $\mathcal{P}$ of $\mathcal{P}_0$ can look like, we find that $\mathcal{P}$ restricted to B can be any partition on B (which can be at most $\mathcal{O}(2^{\mathcal{O}(|B| \log |B|)})$ many) and the sets $P \in \mathcal{P}$ containing the vertex $r \in P_{\pi(i)}$ can be as many as the number of vertices in $P_{\pi(i)} \cap R_i$, that is, at most $|R|$ many. Let $\mathcal{X}$ be the collection of all the valid tuples. Then, for any $i \in [\ell]$ and any $S \subseteq T$, the number of partitions $\mathcal{P}$ for which $(i, \mathcal{P}, S) \in \mathcal{X}$ is $\mathcal{O}(2^{\mathcal{O}(|B| \log |B|)} \cdot |R|)$.

Dynamic Programming. For every valid tuple $(i, \mathcal{P}, S) \in \mathcal{X}$, we define $\mathsf{opt}[i, \mathcal{P}, S]$ equal to the minimum cardinality of a set C that is *compatible* with the valid triple $(i, \mathcal{P}, S)$, i.e., (i) $C \subseteq R_i$ (ii) $\mathcal{P}_0 \cap \mathcal{P}(C) = \mathcal{P}$ & (iii) $N(C) = S$.

If no such $(i, \mathcal{P}, S)$-compatible set exists, then we let the value of $\mathsf{opt}[i, \mathcal{P}, S]$ to be ∞. Moreover, any $(i, \mathcal{P}, S)$-compatible set C of minimum cardinality is called a *minimum $(i, \mathcal{P}, S)$-compatible set*. The quantities $\mathsf{opt}[i, \mathcal{P}, S]$ for all $(i, \mathcal{P}, S) \in \mathcal{X}$ are updated inductively. Then finally, the quantity $\mathsf{opt}[\ell, \mathcal{I}(R \cup B), T]$ gives us the required output of the problem. To start with, we define the quantity $\mathsf{opt}[0, \mathcal{P}_0, N(C_0)] = |C_0|$ and set $\mathsf{opt}[i, \mathcal{P}, S] = \infty$ for each triple $(i, \mathcal{P}, S) \in \mathcal{X}$ such that $(i, \mathcal{P}, S) \neq (0, \mathcal{P}_0, N(C_0))$. Then, the value of each $\mathsf{opt}[i, \mathcal{P}, S]$ updates by the following dynamic programming formula.

$$\mathsf{opt}[i, \mathcal{P}, S] = \min \begin{cases} \mathsf{opt}[i-1, \mathcal{P}, S], \\ 1 + \min_{\substack{\mathcal{P}' \cap \mathcal{P}(r_i) = \mathcal{P}, \\ S' \cup N(r_i) = S}} \mathsf{opt}[i-1, \mathcal{P}', S']. \end{cases} \tag{1}$$

The proof of correctness and running time of formula (1) can be found in the full version [10] of this paper.

3   Linear Kernel for LOCATING-DOMINATING SET parameterized by Feedback Edge Set Number

We prove Theorem 2. A set of edges X of a graph is a *feedback edge set* if $G - X$ is a forest. The minimum size of a feedback edge set is known as the *feedback edge set number* fes. If a graph G has n vertices, m edges, and r components, then $\mathsf{fes}(G) = m - n + r$ [17]. Our proof is nonconstructive, that is, it shows the *existence* of a large (but constant) number of gadget types with which the kernel can be constructed. For missing parts of the proof, we refer the reader to [10]. We first describe graphs of given feedback edge set number.

Proposition 1 ([24, Observation 8]). *Any graph G with feedback edge set number* fes *is obtained from a multigraph $\widetilde{G}$ with at most $2\textsf{fes} - 2$ vertices and $3\textsf{fes} - 3$ edges by first subdividing edges of $\widetilde{G}$ an arbitrary number of times, and then, repeatedly attaching degree 1 vertices to the graph, and $\widetilde{G}$ can be computed from G in $\mathcal{O}(n + \textsf{fes})$ time.*

By Proposition 1, we compute the multigraph $\widetilde{G}$ in $\mathcal{O}(n + \textsf{fes})$ time, and we let $\widetilde{V}$ be the set of vertices of $\widetilde{G}$. For every edge $v_1 v_2$ of $\widetilde{G}$, we have either:

1. $v_1 v_2$ corresponds to an edge in the original graph G, if $v_1 v_2$ has not been subdivided when obtaining G from $\widetilde{G}$ (then $v_1 \neq v_2$ since G is loop-free), or
2. $v_1 v_2$ corresponds to a component C of the original graph G with $\{v_1, v_2\}$ removed. Moreover, C induces a tree, with two (not necessarily distinct) vertices c_1, c_2 in C, where c_1 and c_2 are the only vertices of C adjacent to v_1 and v_2, respectively. (Possibly, $v_1 = v_2$ and the edge $v_1 v_2$ is a loop of $\widetilde{G}$.)

To avoid dealing with the case of loops, for every component C of $G \backslash \{v_1\}$ with a loop at v_1 in $\widetilde{G}$ and with c_1, c_2 defined as above, we select an arbitrary vertex x of the unique path between c_1 and c_2 in $G[C]$ and add x to $\widetilde{V}$. As there are at most fes loops in $\widetilde{G}$, we can assume that $\widetilde{G}$ is a loopless multigraph with at most $3\textsf{fes} - 2$ vertices and $4\textsf{fes} - 3$ edges.

The kernelization has two steps. First, we need to handle *hanging trees*, that is, parts of the graph that induce trees and are connected to the rest of the graph via a single edge. Those correspond to the iterated addition of degree 1 vertices to the graph from $\widetilde{G}$ described in Proposition 1. Step 2 will deal with the subgraphs that correspond to the edges of $\widetilde{G}$.

Step 1 – Handling the Hanging Trees. By [27] (also [2]), there is a linear-time dynamic programming algorithm to solve LOCATING-DOMINATING SET on trees. To this end, one can define the following five types of (partial) solutions. For a tree T rooted at a vertex v, we define five types for a subset L of $V(T)$ as follows:

- Type A: L dominates all vertices of $V(T) \backslash \{v\}$ and any two vertices of $V(T) \backslash (L \cup \{v\})$ are located.
- Type B: L is a dominating set of T and any two vertices of $V(T) \backslash (L \cup \{v\})$ are located;
- Type C: L is a locating-dominating set of T;
- Type D: L is a locating-dominating set of T with $v \in L$;
- Type E: L is a locating-dominating set of T such that $v \in L$ and there is no vertex w of T with $N(w) \cap L = \{v\}$;

Consider the lexicographic ordering $A < B < C < D < E$. Then for $X < X'$, if L is a solution of type X', it is also one of type X. For a tree T rooted at v and $X \in \{A, B, C, D, E\}$, let $opt_X(T, v)$ be the minimum size of a set $L \subseteq V(T)$ of type X. Note that $opt_C(T, v)$ is the location-domination number of T.

Proposition 2. *There is a linear-time dynamic programming algorithm that, for any tree T and root v, computes $opt_X(T, v)$ for all $X \in \{A, B, C, D, E\}$.*

Lemma 2. *Let T be a tree with vertex v, and let $X, X' \in \{A, B, C, D, E\}$ with $X < X'$. We have $opt_{X'}(T, v) \leq opt_X(T, v) \leq opt_{X'}(T, v) + 1$.*

By Lemma 2, for any rooted tree (T, v), the five values $opt_X(T, v)$ (for $X \in \{A, B, C, D, E\}$) differ by at most one, and we know that they increase along with X: $opt_A(T, v) \leq opt_B(T, v) \leq \ldots \leq opt_E(T, v)$. Hence, for any rooted tree (T, v), there is one value $X \in \{A, B, C, D, E\}$ so that, whenever $X' \leq X$, we have $opt_{X'}(T, v) = opt_X(T, v)$, and whenever $X' > X$, we have $opt_{X'}(T, v) = opt_X(T, v) + 1$. Thus, we can introduce the following definition.

Definition 2 (Rooted tree classes). *For $X \in \{A, B, C, D, E\}$, we define the class $\mathcal{T}_X$ as the set of pairs (T, v) such that T is a tree with root v, and there is an integer k with $opt_{X'}(T, v) = k$ whenever $X' \leq X$, and $opt_{X'}(T, v) = k + 1$ whenever $X' > X$ (note that k may differ across the trees in $\mathcal{T}_X$).*

Note that for trees in $\mathcal{T}_E$, all five types of optimal solutions have the same size. Lemma 2 implies that the set of all possible (T, v), where T is a tree and v a vertex of T, can be partitioned into the five classes $\mathcal{T}_X$, where $X \in \{A, B, C, D, E\}$. Besides, it is not difficult to construct small rooted trees for all the five classes.

Definition 3 (Rooted tree gadgets). *For $X \in \{A, B, C, D, E\}$, let (T_X, v_X) be the rooted tree of class $\mathcal{T}_X$ from Table 1, and let $k_X = opt_X(T_X, v_X)$.*

Table 1. The five rooted tree gadgets with their values of the five opt_X functions.

	v_A	v_B	v_C	v_D	v_E
	T_A	T_B	T_C	T_D	T_E
opt_A	0	1	2	1	3
opt_B	1	1	2	1	3
opt_C	1	2	2	1	3
opt_D	1	2	3	1	3
opt_E	1	2	3	2	3

We are now ready to present our first reduction rule.

Reduction Rule 5. *Let (G, k) be an instance of* LOCATING-DOMINATING SET *such that G can be obtained from a graph G' and a tree T by identifying a vertex v of T with a vertex w of G' with (T, v) being in the class $\mathcal{T}_X$, $X \in \{A, B, C, D, E\}$, and $opt_X(T, v) = t$. Then, we remove all vertices of $T \backslash \{v\}$ from G (this results in G'), we consider G and a copy of the rooted tree gadget (T_X, v_X), and we identify v_X with v, to obtain G''. The reduced instance is $(G'', k - t + k_X)$.*

Lemma 3. *For an instance* (G, k) *of* LOCATING-DOMINATING SET, *Reduction Rule 5 can be applied exhaustively in time* $\mathcal{O}(n^2)$, *and for each application,* (G, k) *is a YES-instance of* LOCATING-DOMINATING SET *if and only if* $(G'', k - t + k_X)$ *is a YES-instance of* LOCATING-DOMINATING SET.

Step 2 – Handling the Subgraphs Corresponding to Edges of $\widetilde{G}$. In the second step of our algorithm, we construct tree gadgets similar to those defined in Step 1, but with *two* distinguished vertices. We thus extend the above terminology to this setting. Let (T, v_1, v_2), be a tree T with two distinguished vertices v_1, v_2. We define types of trees as in the previous subsection. To do so, we first define the types for each of v_1 and v_2, and then we combine them. For $X \in \{A, B, C, D, E\}$, a subset L of $V(T)$ is of *type* $(X, -)$ *with respect to* v_1 if:

- Type $(A, -)$: L dominates all vertices of $V(T) \backslash \{v_1, v_2\}$ and any two vertices of $V(T) \backslash (L \cup \{v_1, v_2\})$ are located.
- Type $(B, -)$: L dominates all vertices of $V(T) \backslash \{v_2\}$ and any two vertices of $V(T) \backslash (L \cup \{v_1, v_2\})$ are located;
- Type $(C, -)$: L dominates all vertices of $V(T) \backslash \{v_2\}$ and any two vertices of $V(T) \backslash (L \cup \{v_2\})$ are located;
- Type $(D, -)$: L dominates all vertices of $V(T) \backslash \{v_2\}$, $v_1 \in L$, and any two vertices of $V(T) \backslash (L \cup \{v_2\})$ are located;
- Type $(E, -)$: L dominates all vertices of $V(T) \backslash \{v_2\}$, $v_1 \in L$, any two vertices of $V(T) \backslash (L \cup \{v_2\})$ are located, and there is no vertex w of $V(T) \backslash \{v_2\}$ with $N(w) \cap L = \{v_1\}$;

Thus, the idea for a set L of type $(X, -)$ with respect to v_1 is that there are specific constraints for v_1, but no constraint for v_2 (and the usual domination and location constraints for the other vertices). We say that L if of type (X, Y) with respect to (T, v_1, v_2) if L is of type $(X, -)$ with respect to v_1 and of type $(Y, -)$ with respect to v_2. Thus, we have a total of 25 such possible types.

Let $opt_{X,Y}(T, v_1, v_2)$ be the smallest size of a subset L of T of type (X, Y). Again, there is a linear-time dynamic programming algorithm that, for any tree T and vertices v_1, v_2, computes $opt_{X,Y}(T, v_1, v_2)$ for all $X, Y \in \{A, B, C, D, E\}$.

Lemma 4. *Let* T *be a tree with vertices* v_1, v_2 *of* T, *and let* $X, X', Y, Y' \in \{A, B, C, D, E\}$ *with* $X \leq X'$ *and* $Y \leq Y'$. *If* $X = X'$ *and* $Y < Y'$ *or* $X < X'$ *and* $Y = Y'$, *we have* $opt_{X,Y}(T, v_1, v_2) \leq opt_{X',Y'}(T, v_1, v_2) \leq opt_{X,Y}(T, v_1, v_2) + 1$. *If* $X < X'$ *and* $Y < Y'$, *we have* $opt_{X,Y}(T, v_1, v_2) \leq opt_{X',Y'}(T, v_1, v_2) \leq opt_{X,Y}(T, v_1, v_2) + 2$.

Definition 4 (Doubly rooted tree classes). *For a function* $g : \{A, B, C, D, E\}^2 \rightarrow \{0, 1, 2\}$, *we define the class* $\mathcal{T}_g$ *as the set of triples* (T, v_1, v_2) *such that* T *is a tree,* v_1, v_2 *are two of its vertices, and for every* $(X, Y) \in \{A, B, C, D, E\}^2$, *we have* $opt_{X,Y}(T, v_1, v_2) - opt_{A,A}(T, v_1, v_2) = g(X, Y)$.

There are at most 3^{25} possible classes of triples, since that is the number of possible functions g. By Lemma 4, these classes define a partition of all sets of triples (T, v_1, v_2). We next define suitable tree gadgets for the above classes of triples. As there is a very large number of possible types of triples, we cannot give a concrete definition of such a gadget, but rather, an existential one.

Definition 5 (Doubly-rooted tree gadgets). *For a function* g : $\{A, B, C, D, E\}^2 \to \{0, 1, 2\}$, *let* (T_g, v_1, v_2) *be the smallest tree in* $\mathcal{T}_g$ *(if such a tree exists), and let* $k_g = opt_{A,A}(T_g, v_1, v_2)$.

The next reduction replaces some triples (T, v_1, v_2) by gadgets in Definition 5.

Reduction Rule 6. *Let* (G, k) *be an instance of* LOCATING-DOMINATING SET *with*

- *an induced subgraph* G' *of* G *with two vertices* w_1, w_2;
- *a tree* T *with two vertices* v_1, v_2 *with* (T, v_1, v_2) *in class* $\mathcal{T}_g$ *with* g : $\{A, B, C, D, E\}^2 \to \{0, 1, 2\}$, *and with* $opt_{A,A}(T, v_1, v_2) = t$;
- w_1, w_2 *do not belong to a hanging tree of* G;
- G *can be obtained by taking a copy of* T *and a copy of* G' *and identifying, respectively,* v_1 *with* w_1, *and* v_2 *with* w_2.

Then, we remove all vertices of $T \setminus \{v_1, v_2\}$ *from* G *(that is, we compute* G'*), we consider* G' *and a copy of doubly-rooted tree gadget* (T_g, v_X, v_Y) *and identify* v_X *with* v_1 *and* v_Y *with* v_2, *to obtain* G''. *The reduced instance is* $(G'', k-t+k_g)$.

Lemma 5. *For an instance* (G, k) *of* LOCATING-DOMINATING SET, *Reduction Rule 6 can be applied to every tree component corresponding to an edge of* $\widetilde{G}$ *in time* $\mathcal{O}(n\mathsf{fes})$, *and* (G, k) *is a YES-instance of* LOCATING-DOMINATING SET *if and only if* $(G'', k - t + k_g)$ *is a YES-instance of* LOCATING-DOMINATING SET.

Completion of the Proof. By Lemmas 3 and 5, we can apply Reduction Rule 5 and Reduction Rule 6 in time $\mathcal{O}(n^2 + n\mathsf{fes}) = \mathcal{O}(mn)$. Since $\widetilde{G}$ has at most $3\mathsf{fes} - 2$ vertices and $4\mathsf{fes} - 3$ edges, we replaced each vertex of $\widetilde{G}$ by a constant-size tree-gadget from Definition 3 and each edge of $\widetilde{G}$ by a constant-size doubly-rooted tree gadget from Definition 5, the resulting graph has $\mathcal{O}(\mathsf{fes})$ vertices and edges.

4   Incompressiblity of LOCATING-DOMINATING SET

In this section, we present the reduction used to prove Thereom 3 which says that LOCATING-DOMINATING SET does not admit a polynomial compression of size $\mathcal{O}(n^{2-\epsilon})$ for any $\epsilon > 0$, unless NP $\subseteq$ coNP/*poly*, where n is the number of vertices of the input graph. The reduction takes as input an instance $(G', \langle R', B' \rangle, k')$ of RED-BLUE DOMINATING SET (a restricted version of DOMINATING SET, which also does not admit a compression with $\mathcal{C}(n^{2-\epsilon})$ bits unless NP $\subseteq$ coNP/*poly* [1]) and constructs an instance (G, k) of LOCATING-DOMINATING SET. Suppose we have $R' = \{r_1', r_2', \dots, r_{|R|}'\}$ and $B' = \{b_1', b_2', \dots, b_{|B|}'\}$. The reduction constructs graph G in the following steps. See Fig. 4 for an illustration.

- It adds sets of vertices R and B to G, where R contains a vertex corresponding to each vertex in R' and B contains two vertices corresponding to each vertex in B'. Formally, $R = \{r_i| \ i \in [|R'|]\}$ and $B = \{b_i^\circ, b_i^\star| \ i \in [|B'|]\}$.
- The reduction adds a *bit representation gadget* to locate set R. Informally, it adds some supplementary vertices such that it is safe to assume these vertices are present in a locating-dominating set, and they locate every vertex in R.
  - Set $q := \lceil \log(|R|)\rceil + 1$. Then q allows to uniquely represent each integer in $[|R|]$ by its bit-representation in binary starting with 1 and not 0.
  - For every $i \in [q]$, it adds two vertices $y_{i,1}$ and $y_{i,2}$ and edge $(y_{i,1}, y_{i,2})$.
  - For every integer $\ell \in [|R|]$, let $\texttt{bit}(\ell)$ denote the binary representation of ℓ using q bits. It connects $r_\ell \in R$ with $y_{i,1}$ if the ith bit in $\texttt{bit}(\ell)$ is 1.
  - It adds two vertices $y_{0,1}$ and $y_{0,2}$, and edge $(y_{0,1}, y_{0,2})$, and makes every vertex in R adjacent with $y_{0,1}$.
  Let $\texttt{bit-rep}(R)$ be the collection of the vertices $y_{i,1}$ for all $i \in \{0\} \cup [q]$.
- Similarly, the reduction adds a bit representation gadget to locate set B. However, it adds vertices such that for any pair $b_j^\circ, b_j^\star$, the supplementary vertices adjacent to them are identical.
  - The reduction sets $p := \lceil \log(|B|/2)\rceil + 1$ and for every $i \in [p]$, it adds two vertices $z_{i,1}$ and $z_{i,2}$ and edge $(z_{i,1}, z_{i,2})$.
  - For every integer $j \in [|B|/2]$, let $\texttt{bit}(j)$ be the binary representation of j using p bits. Connect $b_j^\circ, b_j^\star \in B$ with $z_{i,1}$ if the ith bit in $\texttt{bit}(j)$ is 1.
  - It add two vertices $z_{0,1}$ and $z_{0,2}$, and edge $(z_{0,1}, z_{0,2})$. It also makes every vertex in B adjacent with $z_{0,1}$.
  Let $\texttt{bit-rep}(B)$ be the collection of the vertices $z_{i,1}$ for all $i \in \{0\} \cup [p]$.
- Finally, for every pair of vertices $r_i' \in R'$ and $b_j' \in B'$,
  - if r_i' is *not* adjacent with b_j', it adds both (r_i, b_j°) and $(r_i, b_j^\star)$ to $E(G)$,
  - if r_i' is adjacent with b_j' then it adds (r_i, b_j°) only, i.e., it does not add an edge with endpoints r_i and $b_j^\star$.

This completes the construction of G. The reduction sets $k = k' + |\texttt{bit-rep}(R)| + |\texttt{bit-rep}(B)| = k' + \lceil \log(|R|)\rceil + 1 + 1 + \lceil \log(|B|/2)\rceil + 1 + 1$, and returns (G, k) as an instance of LOCATING-DOMINATING SET.

We present an overview of the proof of correctness. Let $(G', \langle R', B'\rangle, k')$ be a YES-instance of RED-BLUE DOMINATING SET and let $S' \subseteq R'$ be a solution. We claim that $S = S^\star \cup \texttt{bit-rep}(R) \cup \texttt{bit-rep}(B)$ is a locating-dominating set of G with $|S| \leq k$, where $S^\star = \{r_i : r_i' \in S'\}$. Note that vertices $y_{0,1}$ and $z_{0,1}$ force each pendant vertex adjacent with a vertex in $\texttt{bit-rep}(R) \cup \texttt{bit-rep}(B)$ to have a unique neighbour in S. Thus, apart from those of the form $(b_j^\circ, b_j^\star)$, every other vertex pair is resolved by vertices in $\texttt{bit-rep}(R) \cup \texttt{bit-rep}(B)$. By the construction of G, a vertex $r_i \in R$ resolves a pair $b_j^\circ, b_j^\star$ if and only if r_i' and b_j' are adjacent. This concludes the forward direction of the proof. In the reverse direction, by [12, Lemma 5], consider a locating dominating set S containing $\texttt{bit-rep}(R) \cup \texttt{bit-rep}(B)$, as each vertex in it is adjacent to a pendant vertex. We can modify S to another locating-dominating set S_1 such that $S_1 \backslash (\texttt{bit-rep}(R) \cup \texttt{bit-rep}(B)) \subseteq R$. It can also be verified that $S_1 \backslash (\texttt{bit-rep}(R) \cup \texttt{bit-rep}(B))$ is a dominating set of G'. This concludes the overview.

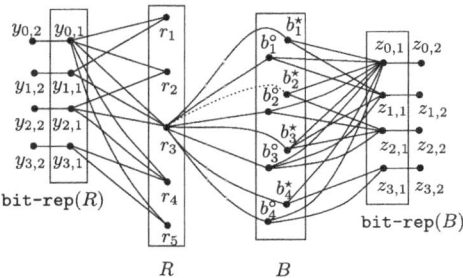

Fig. 4. An illustrative example of the graph constructed by the reduction. Adjacency to $y_{1,1}$, $y_{2,1}$, $y_{3,1}$ correspond to 1 in the bit representation from left to right (most significant to least significant) bits. In this example, the bit representation of r_1 is $\langle 1, 0, 0 \rangle$, r_2 is $\langle 0, 1, 0 \rangle$, r_3 is $\langle 1, 1, 0 \rangle$, etc. For brevity, we only show the edges incident on r_3. The dotted line represents a non-edge. In this example, r_3' is only adjacent to b_2' in G'. It is easy to see that r_3 can only resolve pair $(b_2^\circ, b_2^\star)$.

5　Conclusion

We investigated the structural parameterization of Locating-Dominating Set and Test Cover. We presented several results about the algorithmic complexity of these two problems. However, our main technical contribution is an FPT algorithm when parameterized by the vertex cover number, a linear kernel when parameterized by the feedback edge set number, and a non-compressibility result. Our first two results imply that the double-exponential lower bound when parameterized by treewidth in [11] cannot be extended to larger parameters like vertex cover number and feedback edge set number. The reduction used to prove the third result also provides simplified proofs of some known results.

We do not know whether our $2^{\mathcal{O}(\mathrm{vc}\,\log\,\mathrm{vc})}n^{\mathcal{O}(1)}$-time FPT algorithm for Locating-Dominating Set parameterized by vertex cover (and similarly, the algorithm for Test Cover parameterized by $|U|$) is optimal, or whether there exist single-exponential algorithms. Another open question in this direction, is whether this algorithm can be generalized to the parameter distance to cluster. As our linear kernel for parameter feedback edge set number is not explicit, it would be nice to obtain a concrete kernel.

Acknowledgement. This research was supported by the French government IDEX-ISITE initiative 16-IDEX-0001 (CAP 20-25), International Research Center "Innovation Transportation and Production Systems" of the I-SITE CAP 20-25, by the ANR project GRALMECO (ANR-21-CE48-0004). The research of Diptapriyo Majumdar was supported by the Science and Engineering Research Board (SERB) grant SRG/2023/001592 and the research of Prafullkumar Tale by the INSPIRE Faculty Fellowship of DST, India. The work of Dipayan Chakraborty was supported in part by the Lebanese American University under the Presidents Intramural Research Fund PIRF0056.

References

1. Agrawal, A., Kanesh, L., Saurabh, S., Tale, P.: Paths to trees and cacti. Theor. Comput. Sci. **860**, 98–116 (2021)
2. Argiroffo, G.R., Bianchi, S.M., Lucarini, Y., Wagler, A.K.: Linear-time algorithms for three domination-based separation problems in block graphs. Discret. Appl. Math. **281**, 6–41 (2020)
3. Barbero, F., Isenmann, L., Thiebaut, J.: On the distance identifying set meta-problem and applications to the complexity of identifying problems on graphs. Algorithmica **82**(8), 2243–2266 (2020)
4. Basavaraju, M., Francis, M.C., Ramanujan, M.S., Saurabh, S.: Partially polynomial kernels for set cover and test cover. SIAM J. Discret. Math. **30**(3), 1401–1423 (2016)
5. Berman, P., DasGupta, B., Kao, M.: Tight approximability results for test set problems in bioinformatics. J. Comput. Syst. Sci. **71**(2), 145–162 (2005)
6. Bondy, J.A.: Induced subsets. J. Comb. Theory, Ser. B **12**(2), 201–202 (1972)
7. Bontridder, K., et al.: Approximation algorithms for the test cover problem. Math. Program. **98**(1–3), 477–491 (2003)
8. Cappelle, M.R., de C. M. Gomes, G., dos Santos, V.F.: Parameterized algorithms for locating-dominating sets. CoRR **abs/2011.14849** (2020)
9. Cappelle, M.R., Gomes, G.C.M., dos Santos, V.F.: Parameterized algorithms for locating-dominating sets. In: Proceedings of the XI Latin and American Algorithms, Graphs and Optimization Symposium, (LAGOS). Procedia Computer Science, vol. 195, pp. 68–76 (2021)
10. Chakraborty, D., Foucaud, F., Majumdar, D., Tale, P.: Structural parameterization of locating-dominating set and test cover. CoRR **abs/2411.17948** (2024)
11. Chakraborty, D., Foucaud, F., Majumdar, D., Tale, P.: Tight (double) exponential bounds for identification problems: Locating-dominating set and test cover. In: 35th International Symposium on Algorithms and Computation, (ISAAC). LIPIcs, vol. 322, pp. 19:1–19:18 (2024)
12. Chakraborty, D., Hakanen, A., Lehtilä, T.: The $n/2$-bound for locating-dominating sets in subcubic graphs. CoRR **abs/2406.19278** (2024)
13. Chlebus, B.S., Nguyen, S.H.: On finding optimal discretizations for two attributes. In: Proceedings of the First International Conference on Rough Sets and Current Trends in Computing, vol. 1424, pp. 537–544 (1998)
14. Colbourn, C., Slater, P.J., Stewart, L.K.: Locating-dominating sets in series-parallel networks. Congr. Numer. **56**, 135–162 (1987)
15. Crowston, R., Gutin, G.Z., Jones, M., Muciaccia, G., Yeo, A.: Parameterizations of test cover with bounded test sizes. Algorithmica **74**(1), 367–384 (2016)
16. Crowston, R., Gutin, G., Jones, M., Saurabh, S., Yeo, A.: Parameterized study of the test cover problem. In: Rovan, B., Sassone, V., Widmayer, P. (eds.) MFCS 2012. LNCS, vol. 7464, pp. 283–295. Springer, Heidelberg (2012). https://doi.org/10.1007/978-3-642-32589-2_27
17. Diestel, R.: Graph Theory, 4th Edition, Graduate Texts in Mathematics, vol. 173. Springer, Cham (2012)
18. Fomin, F.V., Kratsch, D.: Exact Exponential Algorithms. Texts in Theoretical Computer Science. An EATCS Series, Springer (2010)
19. Foucaud, F., et al.: Problems in NP can admit double-exponential lower bounds when parameterized by treewidth or vertex cover. In: 51st International Colloquium on Automata, Languages, and Programming, (ICALP). LIPIcs, vol. 297, pp. 66:1–66:19 (2024)

20. Garey, M.R., Johnson, D.S.: Computers and Intractability - A Guide to NP-Completeness. W.H, Freeman and Company (1979)
21. Gutin, G.Z., Muciaccia, G., Yeo, A.: (non-)existence of polynomial kernels for the test cover problem. Inf. Process. Lett. **113**(4), 123–126 (2013)
22. Habib, M., Paul, C., Viennoti, L.: A synthesis on partition refinement: A useful routine for strings, graphs, Boolean matrices and automata. In: Morvan, M., Meinel, C., Krob, D. (eds.) STACS 1998. LNCS, vol. 1373, pp. 25–38. Springer, Heidelberg (1998). https://doi.org/10.1007/BFb0028546
23. Harris, D.G., Narayanaswamy, N.S.: A faster algorithm for vertex cover parameterized by solution size. In: 41st International Symposium on Theoretical Aspects of Computer Science, (STACS). LIPIcs, vol. 289, pp. 40:1–40:18 (2024)
24. Kellerhals, L., Koana, T.: Parameterized complexity of geodetic set. J. Graph Algorithms Appl. **26**(4), 401–419 (2022)
25. Rao, N.: Computational complexity issues in operative diagnosis of graph-based systems. IEEE Trans. Comput. **42**(4), 447–457 (1993)
26. Rényi, A.: On random generating elements of a finite Boolean algebra. Acta Scientiarum Mathematicarum Szeged **22**, 75–81 (1961)
27. Slater, P.J.: Domination and location in acyclic graphs. Networks **17**(1), 55–64 (1987)
28. Slater, P.J.: Dominating and reference sets in a graph. J. Math. Phys. Sci. **22**(4), 445–455 (1988)

Improved Bounds for Group Testing in Arbitrary Hypergraphs

Annalisa De Bonis[✉]

Dipartimento di Informatica, Università di Salerno, 84084 Fisciano, SA, Italy
adebonis@unisa.it

Abstract. In *group testing*, the problem is to determine the defective members of a given set of elements O by performing tests on properly selected subsets of O. A response to a test is "yes" if the tested group contains one or more defective elements, and is "no" otherwise. The classical model of group testing has been recently generalized in a way such that the potentially contaminated sets are the hyperedges of a given hypergraph $\mathcal{F} = (V, E)$. This model takes into account how social and geographical clusterings affect the spreading of contaminations. The paper studies group testing algorithms with little adaptiveness, i.e., algorithms where tests are performed in stages and all tests performed in the same stage are decided at the beginning of the stage. In particular, the paper presents the first two-stage algorithm that uses $o(d \log |E|)$ tests for arbitrary hypergraphs with hyperedges of size at most d, and a three-stage algorithm that improves by a $d^{1/6}$ factor on the number of tests of the best so far known three-stage algorithm. These algorithms are special cases of an s-stage algorithm designed for an arbitrary positive integer $s \leq d$. The design of this algorithm resorts to a new non-adaptive algorithm (one-stage algorithm), i.e., an algorithm where all tests must be decided beforehand. A major contribution of the paper consists in a non-existential result for non-adaptive group testing. Remarkably, this result implies the first lower bound for non-adaptive group testing that improves on the information theoretic lower bound $\Omega(\log |E|)$ and gets very close to the number of tests used by the best non-adaptive group testing algorithm.

Keywords: Group Testing · Superimposed Codes · Separable Codes · Hypergraphs · Lower Bounds · Algorithms

1 Introduction

In *group testing*, the problem is to determine the defective members of a given set of elements O by performing tests on appropriately selected subsets of O. A "yes" response to a test indicates that the tested group contains one or more defective elements, whereas a "no" response indicates that the group contains no defective element. The parameter to optimize is the number of tests used to find all defective elements. This search paradigm was introduced by Dorfman

I. Finocchi and L. Georgiadis (Eds.): CIAC 2025, LNCS 15679, pp. 205–221, 2025.
https://doi.org/10.1007/978-3-031-92932-8_14

[11] during World War II as a mass blood testing methodology, and since then it has found applications in many fields ranging from quality control in product testing [25], conflict resolution algorithms for multiple-access systems [27], fault diagnosis in optical networks [18], failure detection in wireless sensor networks [22], data compression [19], molecular biology [12], and many others. In classical group testing, the set of defectives is any of the possible subsets of size less than or equal to a certain parameter d. In the present paper, we consider a more general version of group testing, parameterized by a hypergraph $\mathcal{F} = (V, E)$, with the contaminated set being one of the hyperedges of E. More precisely, in this generalized version of group testing, the goal is to detect the hyperedge that consists of all the defective elements. This model of group testing takes into account how social and geographical clusterings affect the spreading of contaminations. [2,23]. In this contexts the hyperedges of E represent potentially infected sets that might correspond to groups of friends, families, neighbours, and so on.

1.1 Related Work

We first review the main results for classical group testing with respect to algorithms with different levels of adaptiveness [12]. It is well known that the information theoretic lower bound $\Omega(d \log(n/d))$ is achieved by the best *adaptive strategies*, i.e. algorithms that at each step decide which group to test by looking at the responses of previous tests, whereas, *non-adaptive* strategies, i.e., strategies in which all tests are decided beforehand, are much more costly than their adaptive counterparts. The minimum number of tests used by the non-adaptive procedures is estimated by the minimum length of certain combinatorial structures known under the name of *separable codes* [12], or by that of *d-superimposed codes*, also referred to as *d-cover free families* and *strongly selective families* [5,13,14,21]. The known bounds for these combinatorial structures [1,3,13,24] imply that the number of tests performed by any non-adaptive group testing algorithm is lower bounded by $\Omega((d^2/\log d) \log n)$, and that there exist non-adaptive group testing algorithms that use $O(d^2 \log n)$ tests. Interestingly, it has been proved [10] that the information theoretic lower bound can already be attained by two-stage group testing algorithms, i.e., algorithms in which the tests are performed in two stages each consisting in a non-adaptive algorithm.

The study of group testing in arbitrary hypergraphs has been initiated only recently in [17] and continued in [9,26]. Similar search models were previously considered by the authors of [23] who assumed a known community structure in virtue of which the population is partitioned into separate families and the defective hyperedges are those that contain elements from a certain number of families. While in that paper the information on the structure of potentially infected groups is used to improve on the efficiency of the group testing algorithms, other papers, [20,28], exploit this knowledge to improve on the efficiency of decoding the tests' responses.

The definition of group testing in arbitrary hypergraphs, as studied in the present paper, has been given by the authors of [17] who investigated both adaptive and non-adaptive group testing. For the adaptive setting, when hyperedges in E are of size exactly d, they give an $O(\log |E| + d \log^2 d)$ algorithm that is close to the $\Omega(\log |E|)$ information theoretic lower bound. In the non-adaptive setting, they exploit a random coding technique to prove an $O(\frac{d}{p} \log |E|)$ upper bound on the number of tests, where d is the maximum size of a hyperedge $e \in E$ and p is a lower bound on the size of the difference $e' \backslash e$ between any two hyperedges $e, e' \in E$. This upper bound implies an $O(d \log |E|)$ bound, if no assumption is made on the size of the difference between any pair of hyperedges. In [26], the author presents a new adaptive algorithm for generalized group testing, which is asymptotically optimal if $d = o(\log |E|)$ and, for $d = 2$, gives an asymptotically optimal algorithm that works in three stages. The author of [9] formally defines binary codes that are equivalent to non-adaptive algorithms for group testing in arbitrary hypergraphs. These combinatorial structures are a generalized version of *classical* separable codes [12] and are parameterized by the set of hyperedges that correspond to the potentially contaminated sets. Paper [9] introduces also a notion of selectors, parameterized by a set of hyperedges E, that generalizes the (k, m, n)-selectors of [10] by enforcing the desired properties only on subsets of codewords associated with the hyperedges in E, similarly to what happens with the selectors of [7] and [15], and with the ad-hoc selective families of [4]. The combinatorial constructions in [9] allow to achieve, in the non-adaptive setting, the same $O(\frac{d}{p} \log |E|)$ upper bound of [17], and to design an $O(\sqrt{d} \log |E|)$ three-stage algorithm and a two-stage algorithm that achieves the $\Omega(\log |E|)$ information theoretic lower bound, provided that, for some constant q and for any $q + 1$ distinct hyperedges $e, e'_1 \ldots, e'_q$, it holds that $|\bigcup_{i=1}^{q} e'_i \backslash e| = \Omega(d)$.

1.2   Contribution of the Paper

The paper studies slightly adaptive group testing algorithms, i.e., algorithms where tests are performed in few completely non-adaptive stages. In particular, the paper presents the first two-stage algorithm that uses $o(d \log |E|)$ tests for arbitrary hypergraphs with hyperedges of size at most d. We remark that, differently from the two-stage algorithm in [9], the two-stage algorithm given in the present paper does not resort on any assumption on the size of the differences among hyperedges. The present paper gives also a three-stage algorithm that improves by a $d^{1/6}$ factor on the number of tests of the three-stage algorithm in [9]. These algorithms are special cases of an s-stage algorithm, designed for an arbitrary positive integer $s \leq d$, whose performance in terms of the number of tests decreases with s, thus providing a trade-off between the number of tests and adaptiveness. The design of this algorithm resorts to a new non-adaptive algorithm that uses $O(\frac{b}{p} \log |E|)$ to discard all hyperedges e that contain at least p non-defective vertices provided that the size of the difference $e' \backslash e$ between any two hyperedges $e, e' \in E$ is at most b. As far as it concerns the problem of identifying the defective hyperedge non-adaptively, the present paper mainly focuses on non-existential results, for which very little is known in the literature. The paper presents a lower bound that, for E being sufficiently large and $p \leq \frac{1}{3}d$, is the first lower bound that improves on the information theoretic lower bound

$\Omega(\log|E|)$ and gets close to the best upper bound for non-adaptive group testing, by exhibiting an $O(\log\frac{d-p}{p})$ gap with the $O(\frac{d}{p}\log|E|)$ upper bound. This gap resembles the gap existing between the best upper and lower bounds for the non-adaptive case in classical group testing. In classical group testing, E consists of all possible subsets of up to d elements in $\{1,\ldots,n\}$, and consequently, it holds that $p = 1$. We remark that by setting $p = 1$ in the above gap, one recovers the $O(\log d)$ gap holding for classical group testing.

2   Notations and Terminology

For a finite set V and a family of subsets E of V, the pair $\mathcal{F} = (V, E)$ denotes the hypergraph having V as set of vertices and E as set of hyperedges. If all hyperedges of E have the same size d then the hypergraph is said to be d-uniform. For any positive integer m, we denote by $[m]$ the set $\{1,\ldots,m\}$. Unless specified differently, the hypergraph specifying the set of potentially contaminated sets is assumed to have $V = [n]$. We remark that in our group testing problem, given an input hypergraph $\mathcal{F} = ([n], E)$, every vertex of $[n]$ is contained in at least one hyperedge of E. If otherwise, one could remove the vertex from the hypergraph without changing the collection of potentially defective hyperedges. As a consequence, for a given hypergraph $\mathcal{F} = ([n], E)$ we need only to specify its set of hyperedges E to characterize both the input of the problems and the related combinatorial tools.

Throughout the paper, we will denote by e the base of the natural logarithm $e = 2.7182...$, and unless specified differently, all logarithms are in base 2.

3   The Group Testing Model

In group testing for arbitrary hypergraphs, one is given a hypergraph $\mathcal{F} = (V, E)$ whose hyperedges contain at most a certain number d of vertices. The set V corresponds to a set of elements among which there are at most d defective elements. The *unknown* defective subset, i.e., the subset of all defective elements, is one of the hyperedges in E. The goal is to find the defective hyperedge by testing groups of elements of V. The response to a test is "yes" if the tested group contains one or more defective elements, and a "no" otherwise. This version of the group testing problem corresponds to *classical group testing* when E is the set of all possible subsets of V of size at most d. The following section illustrates the correspondence between non-adaptive group testing algorithms and binary codes of size $|V|$, or equivalently families of size $|V|$.

3.1   Non-adaptive Group Testing for Arbitrary Hypergraphs

Let $\mathcal{F} = (V, E)$ be a hypergraph with $V = [n]$ and hyperedges of size at most d. We first illustrate the correspondence between families of n subsets and non-adaptive group testing algorithms. Given a family $\mathrm{F} = \{F_1,\ldots,F_n\}$ with $F_i \subseteq$

$[t]$, we design a non-adaptive group testing strategy as follows. We denote the elements in the input set by the integers in $[n] = \{1, \ldots, n\}$, and for $i = 1, \ldots, t$, we define the group $T_i = \{j : i \in F_j\}$. Obviously, $T_1, \ldots, T_t$ can be tested in parallel and therefore the resulting algorithm is non-adaptive. Conversely, given a non-adaptive group testing strategy for an input set of size n that tests $T_1, \ldots, T_t$, we define a family $F = \{F_1, \ldots, F_n\}$ by setting $F_j = \{i \in [t] : j \in T_i\}$, for $j = 1, \ldots, n$. Alternatively, any non-adaptive group testing algorithm for an input set of size n that performs t tests can be represented by a binary code of size n with each codeword being a binary vector of length t. This is due to the fact that any family of size n on the ground set $[t]$ is associated with the binary code of length t whose codewords are the characteristic vectors of the members of the family. Given such a binary code $C = \{\mathbf{c}_1, \ldots, \mathbf{c}_n\}$, one has that j belongs to pool T_i if and only if the i-th entry $\mathbf{c}_j(i)$ of $\mathbf{c}_j$ is equal to 1. Such a code can be represented by a $t \times n$ binary matrix M such that $M[i, j] = 1$ if and only if element j belongs to T_i. We represent the responses to tests on $T_1, \ldots, T_t$ by a binary vector whose i-th entry is equal to 1 if and only if T_i tests positive. We call this vector the *response vector*. For any input set E of hyperedges with vertices in $[n]$, the response vector is the bitwise OR of the columns associated with the vertices of the defective hyperedge $e \in E$. It follows that a non-adaptive group testing strategy successfully detects the defective hyperedge in E if and only if for any two distinct hyperedges $e, e' \in E$ we obtain two different response vectors. In terms of the associated binary matrix M, this means that the bitwise OR of the columns with indices in e and the OR of the columns with indices in e' are distinct.

4 Combinatorial Structures for Group Testing in Arbitrary Hypergraphs

The following definition [9] provides a combinatorial tool which is essentially equivalent to a non-adaptive group testing algorithm in our search model.

Definition 1. *[9] Given a hypergraph $\mathcal{F} = (V, E)$ with $V = [n]$ with hyperedges of size at most d, we say that a $t \times n$ matrix M with entries in $\{0, 1\}$ is an E-separable code if for any two distinct hyperedges $e, e' \in E$, it holds that $\bigvee_{j \in e} c_j \neq \bigvee_{j \in e'} c_j$, where c_j denotes the column of M with index j. The integer t is the length of the E-separable code.*

Having in mind the correspondence between binary codes and non-adaptive algorithms illustrated in Sect. 3.1, one can see that a non-adaptive algorithm successfully determines the contaminated hyperedge in E if and only if the binary code associated to the algorithm is E-separable. Therefore, the minimum number of tests of such an algorithm coincides with the minimum length of an E-separable code. In Sect. 5, we present a lower bound on the minimum length of E-separable codes for the case when all hyperedges of E contain exactly d vertices. This lower bound translates into a lower bound on the minimum number of tests needed to find the defective hyperedge in a non-adaptive fashion.

5   Lower Bound on the Number of Rows of E-Separable Codes

Let E be a set of hyperedges of size d on the set of vertices $[n]$ and let M be a $t \times n$ E-separable code with set of columns $\{c_1, \ldots, c_n\}$. We define $G = \{G_1, \ldots, G_n\}$ as the family whose members are the subsets of $[t]$ having the columns of M as characteristic vectors, i.e., $G_j = \{i : c_j(i) = 1\}$. Notice that since M is an E-separable code, for any two edges e, e', it holds that $\bigcup_{j \in e'} G_j \neq \bigcup_{j \in e} G_j$. We say that G is an E-separable family.

The proof of the lower bound on the length of E-separable codes uses a reduction technique that consists in reducing the size of the members of the family by subtracting up to a certain number of "large" members of the family from the remaining members. While in analogous proofs of non-existential results for classical cover free families [1, 24], one is free to choose the members to be subtracted among all "large" members of the family, here we need the "large" members to be associated to vertices of large degree so as to ensure that the compressed members form a separable family for a hypergraph of suitable size. We need the following lemma.

Lemma 1. *Let $\mathcal{F}(V, |E|)$ be a hypergraph with $|V| = n$, $|e| = d$ for any $e \in E$. Moreover, let f be a positive integer with $f \leq |E|d/n$, and n_s be the number of vertices of $\mathcal{F}$ with degree smaller that f. It holds that $n_s \leq n - 1$, and that the number of hyperedges of $\mathcal{F}$ consisting only of vertices with degree larger than or equal to f is at least $|E| - (f - 1)n_s$.*

Proof. Let E_g denote the subset of hyperedges of E consisting only of vertices with degree larger than or equal to f and let E_{mix} be the subset of hyperedges of E that contain at least one vertex with degree smaller than f. Since hyperedges in E_{mix} contain at least one vertex with degree smaller than f, the number of the hyperedges in E_{mix} is smaller than or equal to the sum of the degrees of these vertices. Then, having denoted with n_s the number of vertices of E with degree smaller than f, it holds that $|E_{mix}| \leq (f - 1)n_s$. Since $|E| = |E_g| + |E_{mix}|$, we have that $|E_g| \geq |E| - (f - 1)n_s$. Notice that among the vertices in E, there is at least a vertex of maximum degree Δ. Since the sum of the degrees of all vertices in V is equal to $|E|d$ and it is smaller than or equal to $n\Delta$, the maximum degree Δ is at least $|E|d/n$ and consequently, for $f \leq |E|d/n$, it holds that $n_s < n$. □

Theorem 1. *Let $\mathcal{F} = ([n], E)$ be a d-uniform hypergraph, and let v be a positive integer, with $v < d \leq \frac{3v}{2c}$, and c being a constant in $[1, 3/2)$. It holds that the minimum length of an E-separable code is at least*

$$\frac{v}{c(d - v/c) \log\left(\frac{ev}{c(d - v/c)}\right)} \log\left(\frac{|E|}{n^{v/c}}\right).$$

Proof. Let G be a maximum size E-separable family on the ground set $[t]$.

We will show how to reduce the size of the members of G while preserving separability with respect to a hypergraph with suitable parameters. Let $z = v/c$, for some constant in $[1, 3/2)$. For $q = 0, \ldots, z$, we define the set of hyperedges E^q as follows

$$E^q = \begin{cases} E & \text{if } q = 0 \\ \{e \backslash \{i_{q-1}\} : i_{q-1} \in e, e \in E^{q-1}\}, & \text{if } 1 \leq q \leq z, \end{cases} \tag{1}$$

where, for each $q = 0, \ldots, z$, i_q is a vertex that occurs in at least $\frac{|E^q|}{n-q}$ hyperedges of E^q. Notice that, for $q \geq 0$, each hyperedge of E^q contains $d - q$ vertices in $[n] \backslash \{i_0, \ldots, i_{q-1}\}$. Notice also that, in virtue of Lemma 1, one has that the number of vertices of degree smaller than $\frac{|E^q|}{n-q}$ is at most $n - q - 1$, thus implying that vertex i_q in (1) exists. Moreover, for $q \geq 1$, it holds that $|E^q| \geq \frac{|E^{q-1}|}{(n-q+1)}$, from which we get that

$$|E^q| \geq \frac{|E|}{n(n-1)\cdots(n-q+1)}. \tag{2}$$

Let us denote with E_g^q the subset of hyperedges of E^q that contain only vertices with degree at least $\frac{|E^q|}{n-q}$. By Lemma 1, the size of E_g^q is

$$|E_g^q| \geq |E^q| - \frac{|E^q|}{n-q}(n-q-1) + (n-q-1) > \frac{|E^q|}{n-q}. \tag{3}$$

In addition to sets $E^1, \ldots, E^z$, we define families $G^0, \ldots, G^z$ by setting $G^0 = G$ and $G^q = \{G \backslash G_{i_{q-1}} : G \in G^{q-1} \backslash \{G_{i_{q-1}}\}\}$, for $q = 1, \ldots, z$, where $G_{i_{q-1}}$ is the member of G^{q-1} associated with the vertex i_{q-1} in (1). For each q, $0 \leq q \leq z$, one can see that G^q is E^q-separable by using induction. For $q = 0$, G^q is obviously E^q-separable since $E^0 = E$ and $G^0 = G$. Suppose by induction hypothesis that G^q is E^q-separable up to a certain $q \geq 0$. In order to see that G^{q+1} is E^{q+1}-separable, assume by contradiction that G^{q+1} is not E^{q+1}-separable. By this assumption, there exists two hyperedges $e_1', e_2' \in E^q$ such that $\bigcup_{k \in e_1'} G_k' = \bigcup_{k \in e_2'} G_k'$, where G_k' is the member of G^{q+1} associated to vertex k. It follows that $\bigcup_{k \in e_1'} G_k' \cup G_{i_q} = \bigcup_{k \in e_2'} G_k' \cup G_{i_q}$. By construction of E^{q+1} and G^{q+1}, both the hyperedges $e_1 = e_1' \cup \{i_q\}$ and $e_2 = e_2' \cup \{i_q\}$ belong to E^q and it holds that $\bigcup_{k \in e_1'} G_k' \cup G_{i_q} = \bigcup_{k \in e_1} G_k$ and $\bigcup_{k \in e_2'} G_k' \cup G_{i_q} = \bigcup_{k \in e_2} G_k$, where G_k is the member of G^q associated to k. Therefore, it holds that $\bigcup_{k \in e_1} G_k = \bigcup_{k \in e_2} G_k$, and this contradicts the induction hypothesis that G^q be E^q-separable.

Suppose that G contains at least one member of size larger than t/z. If this is the case, we start a process aimed at bounding from above the cardinality of G by that of a family with members of size smaller than or equal to t/z. In the following discussion, we assume that z divides t and that c divides v. One can easily convince herself that the proof works also in the case when t is not a multiple of z and v is not a multiple of c. We start by setting $G^0 = G$ and define families with members of progressively smaller size as follows. For $q \geq 0$, we replace E^q by E^{q+1}, if there exists a vertex i_q with degree at least $\frac{|E^q|}{n-q}$ and $|G_{i_q}| > t/z$, where G_{i_q} is the member of G^q associated with the vertex i_q. Notice that i_q is as required by (1). We have seen that for each q, $0 \le q \le z$, $\mathbf{G}^q$ is E^q-separable. If for some $q \ge 0$, $\mathbf{G}^q$ contains no member of size larger than t/z associated with a vertex of degree at least $\frac{|E^q|}{n-q}$ in E^q, then the compression process stops and, if there exist members of E^q of size larger than t/z, then the set E^q is replaced by E_g^q i.e., by the subset of the hyperedges of E^q consisting only of vertices with degree at least $\frac{|E^q|}{n-q}$. In this case, we replace $\mathbf{G}^q$ by the family $\mathbf{G}_g^q$ obtained from $\mathbf{G}^q$ by removing the members associated to vertices with degree smaller than $\frac{|E^q|}{n-q}$. The family $\mathbf{G}_g^q$ is obviously E_g^q-separable. By (3), we have that $|E_g^q| > \frac{|E^q|}{n-q}$.

Now, we need to show that, at the end of the compression process, we are left with a set of hyperedges that is not "too small". Notice that after at most $q = z$ steps, the compression process terminates since, at each step, at least t/z elements are removed from the ground set of G, and therefore, after at most z steps, one is left with a family G_q whose members have size at most t/z. However, it might happen that the process terminates at some step $q < z$ because $\mathbf{G}^q$ contains no member of size larger than t/z that is associated with a vertex occurring in at least $\frac{|E^q|}{n-q}$ hyperedges of E^q. We have seen that, in this case, if $\mathbf{G}^q$ contains some member of size larger than t/z, then E^q is replaced by E_g^q and the family $\mathbf{G}^q$ is replaced by $\mathbf{G}_g^q$, i.e., by the family obtained by taking only the members of $\mathbf{G}^q$ associated with vertices of degree at least $\frac{|E^q|}{n-q}$ and all of them have size at most t/z. Let us denote by $\tilde{E}$ the set of hyperedges resulting from the above process, and let $\tilde{G}$ denote a maximum size $\tilde{E}$-separable family.

Suppose that the compression process stops after $j < z$ steps. We further reduce the number of vertices in the hyperedges by defining

$$\tilde{E}^q = \begin{cases} \tilde{E} & \text{if } q = j \\ \{e \backslash \{i_{q-1}^*\} : i_{q-1}^* \in e, e \in \tilde{E}^{q-1}\}, & \text{if } j+1 \le q \le z, \end{cases} \tag{4}$$

where, for each $q = j, \ldots, z-1$, i_q^* is a vertex with largest degree in E^q. Notice that in the above definition we do not need i_q^* to be associated to a large member of $\mathbf{G}^q$. Moreover, for $q = j, \ldots, z-1$, we set $\tilde{\mathcal{G}}^{q+1} = \{G \backslash G_{i_q^*} : G \in \tilde{\mathcal{G}}^q \backslash \{G_{i_q^*}\}\}$, where $G_{i_q^*}$ is the member of $\tilde{\mathcal{G}}^q$ associated with the vertex i_q^*. By the same argument we have used to show that G_q is E^q-separable, one can see that $\tilde{\mathcal{G}}^q$ is an $\tilde{E}^q$ separable family,

In order to limit the size of $\tilde{E}^z$ from below, we assume that the compression process stops with j as large as possible, i.e., $j = z$. In this case, at the end of compression process, the algorithm is left with $\tilde{E} = E^z$ and it holds that

$$|E^z| \ge \frac{|E|}{n(n-1)\ldots(n-z+1)}. \tag{5}$$

Since, all members of $\mathbf{G}^z$ have size at most $t/z = ct/v$, the union $\bigcup_{k \in e} G_k$, for any $e \in E^z$, has size at most $(d-z)t/z = (d-v/c)ct/v$, where G_k denotes the member of $\mathbf{G}^z$ associated with vertex k. Further, since $\mathbf{G}^z$ is E^z-separable, we have that the following inequality holds:

$$|E^z| \leq \sum_{i=0}^{\lfloor (d-v/c)ct/v \rfloor} \binom{t}{i}.$$

By hypothesis, it is $d \leq \frac{3v}{2c}$, and consequently, it holds that $(d - v/c)ct/v \leq t/2$, and we can exploit the following well known inequality, holding for $a \leq t/2$,

$$\sum_{i=0}^{a} \binom{t}{i} \leq 2^{tH(a/t)},$$

where $H(\frac{a}{t})$ denotes the binary entropy $H(\frac{a}{t}) = -\frac{a}{t} \log \frac{a}{t} - (1 - \frac{a}{t}) \log(1 - \frac{a}{t})$. Therefore, we obtain the following upper bound:

$$|E^z| \leq \sum_{i=0}^{\lfloor (d-v/c)ct/v \rfloor} \binom{t}{i} \leq 2^{tH\left(\frac{c(d-v/c)}{v}\right)}. \tag{6}$$

For $a/t \leq 1/2$, it holds the following inequality (see (4) in [8] for a proof of it): $H\left(\frac{a}{t}\right) \leq \frac{a}{t} \log \frac{et}{a}$. This inequality implies that

$$H\left(\frac{c(d-v/c)}{v}\right) \leq \left(\frac{c(d-v/c)}{v}\right) \log\left(\frac{ev}{c(d-v/c)}\right). \tag{7}$$

From (6) and (7) it follows that $|E^z| \leq 2^{t\left(\frac{c(d-v/c)}{v}\right) \log\left(\frac{ev}{c(d-v/c)}\right)}$, and consequently, one has that

$$t \geq \left(\frac{v}{c(d-v/c)}\right) \frac{\log|E^z|}{\log\left(\frac{ev}{c(d-v/c)}\right)}. \tag{8}$$

Finally, by (5) and (8), it holds that

$$t \geq \left(\frac{v}{c(d-v/c)}\right) \frac{\log|E^{v/c}|}{\log\left(\frac{ev}{c(d-v/c)}\right)}$$

$$\geq \frac{v}{c(d-v/c) \log\left(\frac{ev}{c(d-v/c)}\right)} \log\left(\frac{|E|}{n(n-1)\cdots(n-v/c+1)}\right)$$

$$\geq \frac{v}{c(d-v/c) \log\left(\frac{ev}{c(d-v/c)}\right)} \log\left(\frac{|E|}{n^{v/c}}\right). \tag{9}$$

$\square$

Notice that if in the statement of Theorem 1, E is such that, for any two distinct hyperedges $e, e' \in E$, it holds that $|e \cap e'| \leq \overline{\lambda} = v$, then we have that the following corollary of Theorem 1 holds.

Corollary 1. *Let $\mathcal{F} = ([n], E)$ be a d-uniform hypergraph, and let $\overline{\lambda}$ be a positive integer, with $\overline{\lambda} < d \leq \frac{3\overline{\lambda}}{2c}$ and c being a constant in $[1, 3/2)$. If for any two distinct hyperedges $e, e' \in E$, it holds that $|e \cap e'| \leq \overline{\lambda}$, then the minimum length of an E-separable code is at least*

$$\frac{\overline{\lambda}}{c(d - \overline{\lambda}/c) \log\left(\frac{e\overline{\lambda}}{c(d-\overline{\lambda}/c)}\right)} \log\left(\frac{|E|}{n^{\overline{\lambda}/c}}\right).$$

Notice that, if a d-uniform hypergraph has n vertices and its hyperedges pairwise intersect in at most $\overline{\lambda}$ vertices, then it contains at most $\frac{\binom{n}{\overline{\lambda}+1}}{\binom{d}{\overline{\lambda}+1}}$ hyperedges (see Lemma 2.4.1 in [12]). Therefore, the set of hyperedges E in the statement of Corollary 1 has size at most $\frac{\binom{n}{\overline{\lambda}+1}}{\binom{d}{\overline{\lambda}+1}}$, and the term $\log\left(\frac{|E|}{n^{\overline{\lambda}/c}}\right)$ in the lower bound of that corollary might be very small. However, if the set of hyperedges E has size $|E| > n^{\frac{\overline{\lambda}c'}{c}}$, for some positive constant $c' \in (1, c)$, then it holds that $\frac{|E|}{n^{\overline{\lambda}/c}} \geq |E|^{\frac{c'-1}{c'}}$, and by setting $\tilde{c} = \frac{c'}{c}$, we have that the upper bound of Corollary 1, along with the information theoretic lower bound, implies the following corollary:

Corollary 2. *Let $\mathcal{F} = ([n], E)$ be a d-uniform hypergraph, and let $\overline{\lambda}$ be a positive integer, with $\overline{\lambda} < d \leq \frac{3\overline{\lambda}}{2c}$ and c being a constant in $[1, 3/2)$. If for any two distinct hyperedges $e, e' \in E$, it holds that $|e \cap e'| \leq \overline{\lambda}$ and $|E| \geq n^{\overline{\lambda}\tilde{c}}$, for some constant $\tilde{c} \in (\frac{1}{c}, 1)$, then the minimum length of an E-separable code is at least*

$$\Omega\left(\max\left\{\frac{\overline{\lambda}}{(d - \overline{\lambda}) \log\left(\frac{e\overline{\lambda}}{d-\overline{\lambda}}\right)} \log|E|, \log|E|\right\}\right).$$

The lower bound stated by Corollary 1 and Corollary 2, translate into lower bounds on the minimum number of tests needed by any non-adaptive algorithm that finds the defective hyperedge in a set of hyperedges E such that $\max\{|e' \cap e| : e, e' \in E\} \leq \overline{\lambda}$, and $|e| = d$ for any $e \in E$. Papers [9,17] give non-adaptive algorithms that determine the defective hyperedge by $O(\frac{d}{p} \log|E|)$ tests, when the set E is such that $\min\{|e' \setminus e| : e, e' \in E\} \geq p$. If the hyperedges in such a set E have size equal to d, then the hyperedges pairwise intersect in at most $\overline{\lambda} = d - p$ vertices, i.e., the quantity $d - \overline{\lambda}$ in our lower bounds is equal to the parameter p in the above mentioned upper bound of [9,17]. We remark that when $p = d - \overline{\lambda} = \Theta(d)$, the algorithms in [9,17] achieve the $\Omega(\log|E|)$ information theoretic lower bound, and consequently, they also achieve the lower bound of Corollary 2 that, for $d - \overline{\lambda} = \Theta(d)$, is equal to the information theoretic lower bound. On the other hand, for $p = d - \overline{\lambda} = o(d)$, we observe that the ratio between the $O(\frac{d}{p} \log|E|)$ upper bound in [9,17] and the lower bound stated by Corollary 2 is $\frac{d}{\overline{\lambda}} \log\left(\frac{e\overline{\lambda}}{d-\overline{\lambda}}\right) \leq \frac{3}{2} \log\left(\frac{e\overline{\lambda}}{d-\overline{\lambda}}\right) = \frac{3}{2} \log\left(\frac{e(d-p)}{p}\right)$.

Notice that if E consists of all possible subsets of d vertices in $[n]$, then an E-separable code is indeed a classical separable code. In this case, it holds that $\overline{\lambda} = d - 1$. For $d \geq 3$ and $n \geq d^d$, it holds that $\overline{\lambda} = d - 1 \geq \frac{2}{3}d$ and $|E| = \binom{n}{d} \geq n^{d-1} = n^{\overline{\lambda}}$. Therefore, we have that E satisfies the hypothesis of Corollary 2 thus obtaining an $\Omega((d^2/\log d)\log n)$ lower bound on the length of classical separable codes. Remarkably, this is the best lower bound known in the literature on the minimum length of classical separable codes [1,3,13,24], and consequently, on the minimum number of tests performed by any non adaptive algorithm for classical group testing.

We point out that the authors of [16,17] give an example of a hypergraph for which the minimum number of tests needed to find the defective hyperedge is $\Omega\left(\frac{d}{\log d}\log|E|\right)$. However that result does not imply a lower bound for arbitrary hypergraphs.

5.1   An Improved Non-adaptive Group Testing Algorithm for Arbitrary Hypergraphs

In this section, we present a new non-adaptive algorithm that discards all hyperedges e that contain at least p non-defective vertices provided that the size of the difference $e'\backslash e$ between any two hyperedges e and e' is at most b. In order to design this algorithm, we need the following result of [9].

Theorem 2. *[9] Let $\mathcal{F} = (V, E)$ be a hypergraph with $V = [n]$ with all hyperedges in E of size at most d. For any positive integer $p \leq d$, there exists a non-adaptive algorithm that allows to discard all but those hyperedges e such that $|e\backslash e^*| < p$, where e^* is the defective hyperedge, and uses $t = O\left(\frac{d}{p}\log|E|\right)$ tests.*

Notice that if it holds $|e\backslash e'| \geq p$, for any two distinct hyperedges in E, then the algorithm of Theorem 2 allows to detect the defective hyperedge. Since, for any set E, it holds that $|e\backslash e'| \geq 1$, for any two distinct hyperedges in E such that $e \not\subset e'$, one has that by setting $p = 1$ in the algorithm of Theorem 2, one obtains the $O(d\log|E|)$ upper bound in [9,17] on the minimum number of tests needed to determine the defective hyperedge in a non-adaptive fashion. Notice that if E contains hyperedges that are properly contained in other hyperedges, then the algorithm might end up with one or more hyperedges in addition to the defective one. This is possible only if the additional hyperedges are properly contained in the defective hyperedge e^*. Since we assume that the defective hyperedge is the one that contains all defective vertices, the algorithm returns the largest of the above said hyperedges.

In the following theorem, we present a new non-adaptive algorithm that can be used to reduce the number of hyperedges that are candidate to be the defective hyperedge.

Theorem 3. *Let d and n be integers with $1 \leq d \leq n$, and let E be a set of hyperedges of size at most d on $[n]$. Moreover, let b be an integer such that $1 \leq \max\{|e'\backslash e| : e, e' \in E\} < b$. For any positive integer $p \leq b - 1$, there exists a non-adaptive group testing algorithm that allows to discard all but those hyperedges e such that $|e\backslash e^*| < p$, where e^* is the defective hyperedge, and uses*

$t = O\left(\frac{b}{p}\log|E| + d\right)$ *tests. Moreover, the algorithm returns a set of hyperedges* $\hat{E}$ *such that any hyperedge of* $\hat{E}$ *has size at most* $b - 1$ *and is a subset of a hyperedge of* E *that has not been discarded by the algorithm. For any hyperedge* $\hat{e} \in \hat{E}$, *it holds that* $|\hat{e} \backslash \hat{e}^*| < p$, *where* $\hat{e}^*$ *is the unique hyperedge of* $\hat{E}$ *such that* $\hat{e}^* \subseteq e^*$.

Proof. The algorithm chooses a hyperedge $\tilde{e}$ of maximum size. Since it holds that $|\tilde{e} \backslash e^*| < b$, at least $|\tilde{e}| - b + 1$ of the vertices of $\tilde{e}$ are defective. The algorithm replaces every hyperedge e by $e \backslash \tilde{e}$. Let $\hat{E}$ denote the set of these hyperedges after removing the duplicates. From the hypothesis that $\max\{|e' \backslash e| : e, e' \in E\} < b$, it follows that any hyperedge in $\hat{E}$ has size at most $b - 1$. For each hyperedge $e \in E$, let us denote with $\hat{e}$ the corresponding hyperedge of $\hat{E}$, i.e., $\hat{e} = e \backslash \tilde{e}$. Notice that a hyperedge $\hat{e} \in \hat{E}$ might correspond to more than one hyperedge in E. Theorem 2 implies that there exists an $O(\frac{b}{p}\log|\hat{E}|)$ non-adaptive algorithm $\hat{A}$ that discards all but the hyperedges $\hat{e} \in \hat{E}$ such that $|\hat{e} \backslash \hat{e}^*| < p$, where $\hat{e}^*$ is the hyperedge in $\hat{E}$ corresponding to the original defective hyperedge e^*, i.e., $\hat{e}^* = e^* \backslash \tilde{e}$. Notice that there might be non-defective hyperedges e such that $e \backslash \tilde{e} = e^* \backslash \tilde{e} = \hat{e}^*$. Let $\hat{M}$ be the binary matrix associated with $\hat{A}$ and let M' be the matrix obtained by adding $|\tilde{e}|$ distinct rows at the bottom of $\hat{M}$, with each of these rows having an entry equal to 1 in correspondence of a distinct vertex of $\tilde{e}$ and all the other entries equal to 0. Let us consider the non-adaptive algorithm associated to M'. The tests associated to the rows of $\hat{M}$ allow to discard all non-defective hyperedges of $\hat{E}$ but those for which $|\hat{e} \backslash \hat{e}^*| < p$, i.e., the non-defective hyperedges are not discarded if they contain less than p non-defective vertices. Notice that for any hyperedge e that originally contained at least p non-defective vertices, one has that either the corresponding hyperedge $\hat{e} \in \hat{E}$ contains at least p non-defective vertices, or it holds that $e \cap \tilde{e}$ contains one or more non-defective vertices. In the former case $\hat{e}$ is discarded by $\hat{A}$, whereas in the latter case, $\hat{e}$ is discarded after inspecting the responses to the tests associated with the last $|\tilde{e}|$ rows of M'. Indeed, these responses uncover the non-defective vertices in $\tilde{e}$ and the algorithm discards $\hat{e}$ if it contains one or more of these non-defective vertices. Notice that if the problem admits that E contains hyperedges that are properly contained in e^*, then it holds that $\hat{e} \neq \hat{e}^*$, i.e., $\hat{e}$ is properly contained in $\hat{e}^*$. Otherwise if it were $\hat{e} = \hat{e}^*$, there should be a defective vertex in $\tilde{e}$ that does not belong to e. The algorithm gets rid of such a hyperedge by discarding all hyperedges in $\hat{E}$ that do not contain all the defective vertices in $\tilde{e}$. Those vertices, if any, are uncovered once again by the responses to the tests associated with the last $|\tilde{e}|$ rows of M'. Notice that if all responses to these tests are positive, then it holds that $e^* = \tilde{e}$, since all elements in $\tilde{e}$ are defective and $\tilde{e}$ is the largest such hyperedge. $\square$

The above theorem asymptotically improves on the upper bound of Theorem 2 when it holds that $\max\{|e' \backslash e| : e, e' \in E\} < b$ for $b = o(d)$. In the next section, we resort to Theorem 2 and Theorem 3 to design an s-stage algorithm, for an arbitrary $s \leq d$.

6   An s-Stage Group Testing Algorithms for Arbitrary Hypergraphs

In this section we present an s-stage algorithm, for an arbitrary $s \geq 1$, i.e., an algorithm consisting in s non-adaptive stages. For $s \geq 3$, the algorithm improves on the three-stage algorithm given in [9] and for $s = 2$ it achieves the same asymptotic performance of the three-stage algorithm in [9].

Theorem 4. *Let $\mathcal{F} = (V, E)$ be a hypergraph with $V = [n]$ with hyperedges of size at most d. Let s be any positive integer smaller that or equal to d, and let $b_1, \ldots, b_s$ be s integers with $d > b_1 > \ldots > b_s = 1$. There exists an s-stage algorithm that finds the defective hyperedge in E and uses $O(\frac{d}{b_1} \log |E| + \sum_{i=2}^{s} (\frac{b_{i-1}}{b_i}) \log |E| + b_{i-2})$ tests.*

Proof. Let us describe the s stage algorithm achieving the stated upper bound. Let $b_0 = d$ and let $b_1, \ldots, b_s$ be s integers with $b_0 > b_1 > \ldots > b_s = 1$. In the following discussion e^* denotes the defective hyperedge in E. The first stage aims at restricting the search to hyperedges of a set $E_1 \subseteq E$ such that $\max\{|e'\backslash e| : e, e' \in E_1\} < b_1$. From Theorem 2, one has that there is an $O(\frac{d}{b_1} \log |E|)$ non-adaptive algorithm $\mathcal{A}_1$ that discards all but the hyperedges e such that $|e\backslash e^*| < b_1$, where e^* is the defective hyperedge. Let stage 1 execute algorithm $\mathcal{A}_1$. After running algorithm $\mathcal{A}_1$, the algorithm gets rid of any not yet discarded hyperedge e for which there exists another non-discarded hyperedge e' such that $|e'\backslash e| \geq b_1$. Indeed, e cannot be the defective hyperedge. To see this, observe that if e were the defective hyperedge then e' would have been discarded by algorithm $\mathcal{A}_1$. After getting rid of these hyperedges, the algorithm is left only with a set of hyperedges $E_1 \subset E$ such that for any two distinct hyperedges e and e' it holds that $|e'\backslash e| < b_1$ and $|e\backslash e'| < b_1$. Stage 2 aims at restricting the search to a set E_2 of hyperedges each of which contains at most $b_1 - 1$ vertices and such that $\max\{|e'\backslash e| : e, e' \in E_2\} < b_2$. Notice that E_1 satisfies the hypothesis of Theorem 3 with $b = b_1$. By that theorem, there exists an $O(\frac{b_1}{b_2} \log |E_1| + b_0)$ non-adaptive algorithm $\mathcal{A}_2$ that discards all but the hyperedges $e \in E_1$ such that $|e\backslash e^*| < b_2$. Let stage 2 run algorithm $\mathcal{A}_2$. By Theorem 3, after running algorithm $\mathcal{A}_2$, stage 2 is left with a set $\tilde{E}_2$ of hyperedges of size at most $b_1 - 1$ such that each hyperedge $\hat{e}$ of $\tilde{E}_2$ is subset of a hyperedge of E_1 and such that $|\hat{e}\backslash\hat{e}^*| < b_2$, where $\hat{e}^*$ is the unique hyperedge of $\tilde{E}_2$ for which it holds that $\hat{e}^*$ consists only of defective vertices. Stage 2, discards any hyperedge $\hat{e}$ of $\tilde{E}_2$ for which there exists another non-discarded hyperedge $\hat{e}'$ such that $|\hat{e}'\backslash\hat{e}| \geq b_1$. Indeed, by the same argument used above, one can see that $\hat{e}$ cannot be $\hat{e}^*$. After discarding these hyperedges from $\tilde{E}_2$, stage 2 is left with a set E_2 of hyperedges of size at most $b_1 - 1$ that contains a unique hyperedge $\hat{e}^*$ that is subset of e^*. Moreover, for any two hyperedges $\hat{e}, \hat{e}' \in E_2$, it holds that $|\hat{e}\backslash\hat{e}'| < b_2$. By continuing in this way, we design an algorithm that in stage i, for any $i \geq 2$, performs $O(\frac{b_{i-1}}{b_i} \log |E_{i-1}| + b_{i-2})$ tests, and returns a set E_i of hyperedges each of which has size at most $b_{i-1} - 1$ and is a subset of a hyperedge of E_{i-1}. Moreover, for any two distinct hyperedges $\hat{e}, \hat{e}' \in E_i$, it holds that $|\hat{e}\backslash\hat{e}'| < b_i$, and there exist a unique hyperedge $\hat{e}^*$ of E_i such that $\hat{e}^* \subseteq e^*$. We prove by induction on i that this invariant holds at any stage $i \geq 2$. We have seen that this invariant is true for $i = 2$. Let us prove that it is true for any $i > 2$. Suppose that the invariant is true up to a certain stage $i \geq 2$ and let us prove that it holds also for stage $i + 1$. Since by induction hypothesis stage i satisfies the invariant, one can see that stage i returns a set E_i of hyperedges each of which has size at most $b_{i-1} - 1$ and such that, for any hyperedges $\hat{e}, \hat{e}' \in E_i$, it holds that $|\hat{e} \backslash \hat{e}'| < b_i$. Moreover, E_i contains a unique hyperedge $\hat{e}^*$ such that $\hat{e}^* \subseteq e^*$. The set E_i satisfies the hypothesis of Theorem 3 with $d = b_{i-1} - 1$ and $b = b_i$. That theorem implies that there exists an $O(\frac{b_i}{b_{i+1}} \log |E_i| + b_{i-1})$ non-adaptive algorithm $\mathcal{A}_{i+1}$ that returns a set of hyperedges $\tilde{E}_{i+1}$ such that any hyperedge of E_{i+1} has size at most $b_i - 1$, and for any hyperedge it holds that $|\hat{e} \backslash \hat{e}^*| \leq b_{i+1}$, where $\hat{e}^*$ is the unique hyperedge in $\tilde{E}_{i+1}$ such that $\hat{e}^* \subseteq e^*$. After running algorithm $\mathcal{A}_{i+1}$, stage $i + 1$ discards any hyperedge $\hat{e}$ of $\tilde{E}_{i+1}$ for which there exists another hyperedge $\hat{e}'$ of $\tilde{E}_{i+1}$ such that $|\hat{e}' \backslash \hat{e}| \geq b_{i+1}$. Indeed, by the same argument used for stage 1, one can see that $\hat{e}$ cannot be $\hat{e}^*$. After discarding these hyperedges from $\tilde{E}_{i+1}$, stage $i + 1$ is left with a set E_{i+1} of hyperedges of size at most $b_i - 1$ that contains a unique hyperedge $\hat{e}^*$ that is entirely contained in e^*. Moreover, for any two hyperedges $\hat{e}, \hat{e}' \in E_2$, it holds that $|\hat{e} \backslash \hat{e}'| < b_{i+1}$. Notice also that stage $i + 1$ performs $O(\frac{b_i}{b_{i+1}} \log |E_i| + b_{i-1})$ tests, Therefore, we have proved that the invariant holds also for $i + 1$. At stage s, the algorithm surely determines the defective hyperedge since we have set $b_s = 1$. Indeed, this means that at stage s, the algorithm of Theorem 3 discards all hyperedges of E_s but those such that $|\hat{e} \backslash \hat{e}^*| < 1$ and returns a set of hyperedges E_s such that any hyperedge of E_s has size at most $b_{s-1} - 1$ and is a subset of a hyperedge of E_{s-1}. Moreover, for any hyperedge $\hat{e} \in E_s$, it holds that $|\hat{e} \backslash \hat{e}^*| < b_s = 1$, where $\hat{e}^*$ is the unique hyperedge of E_s such that $\hat{e}^* \subset e^*$. In other words, E_s contains only the hyperedge $\hat{e}^*$. In order to recover e^* the algorithm needs only to add back to $\hat{e}^*$ the vertices that have been removed from e^* through the s stages.

By summing up the number of tests performed by the s stages, we get that the total number of tests performed by the algorithm is $O(\frac{d}{b_1} \log |E| + \sum_{i=2}^{s}(\frac{b_{i-1}}{b_i} \log |E_{i-1}| + b_{i-2}))$. $\qquad\square$

By setting $b_i = d^{\frac{s-i}{s}}$, for $i = 1, \ldots, s$, we have that $b_1, \ldots, b_s$ satisfies the hypothesis of Theorem 4, and we get the following corollary.

Corollary 3. *Let $\mathcal{F} = (V, E)$ be a hypergraph with $V = [n]$ with hyperedges of size at most d. Let s be any positive integer smaller than or equal to d. There exists an s-stage algorithm that finds the defective hyperedge in E and uses $O(sd^{\frac{1}{s}} \log |E| + sd)$ tests.*

Notice that by setting $s = \lceil \log d \rceil$, the above corollary implies that there exists a $\lceil \log d \rceil$-stage algorithm that finds the defective hyperedge in E and uses $O((\log d)(\log |E|) + d)$ tests.

By setting $s = 2$ in Corollary 3, we obtain a two-stage algorithm that achieves the same asymptotic number of tests of the three-stage algorithm of [9]. Interestingly, this existential result is independent from the size of the differences between hyperedges. This result is stated in the following corollary.

Corollary 4. *Let $\mathcal{F} = (V, E)$ be a hypergraph with $V = [n]$ with hyperedges of size at most d. There exists a two-stage algorithm that finds the defective hyperedge in E and uses $O(\sqrt{d} \log |E| + d)$ tests.*

By setting $s = 3$ in Corollary 3, we obtain a three stage algorithm that improves by a $d^{1/6}$ factor on the $O(\sqrt{d} \log |E|)$ upper bound on the number of tests of the three-stage algorithm in [9].

Corollary 5. *Let $\mathcal{F} = (V, E)$ be a hypergraph with $V = [n]$ with hyperedges of size at most d. There exists a three-stage algorithm that finds the defective hyperedge in E and uses $O(d^{\frac{1}{3}} \log |E| + d)$ tests.*

We end this section by comparing the two-stage algorithm given in [9] with the one in Corollary 4.

Theorem 5. *[9] Let $\mathcal{F} = (V, E)$ be a hypergraph with $V = [n]$ and all hyperedges in E of size at most d. Moreover, let q and χ be positive integers such that $1 \leq q \leq |E| - 1$, $\chi = \min\{|\bigcup_{i=1}^{q} e'_i \backslash e|,$ for any $q + 1$ distinct $e, e'_1, \ldots, e'_q \in E\}$. There exists a (trivial) two-stage algorithm that uses*

$$t < \frac{2e(d+\chi)}{\chi} \left(1 + \ln \left(\binom{d+\chi-1}{d+\chi-d-1} \beta \right) \right) + dq \text{ tests, where } \beta = \min \left\{ e^q |E| \left(\frac{|E|-1}{q} \right)^q, \right.$$
$$\left. e^{d+\chi-1} \left(\frac{n+d-1}{d+\chi-1} \right)^{d+\chi} \right\} \text{ and } e \text{ denotes the base of the natural logarithm..}$$

We observe that, in order for the algorithm in Theorem 5 to outperform asymptotically the two-stage algorithm of Corollary 4, there should exist a constant q such that

$$\min\{| \bigcup_{i=1}^{q} e'_i \backslash e|, \text{ for any } q + 1 \text{ distinct } e, e'_1, \ldots, e'_q \in E\} = \omega(\sqrt{d}).$$

We remark that the upper bounds proved for our s-stage algorithms do not rely on any particular feature of the hypergraph.

7   Conclusion

In this paper, we have provided the first two-stage algorithm that uses $o(d \log |E|)$ tests for arbitrary hypergraphs with hyperedges of size at most d. An interesting feature of this algorithm is that, differently from the two-stage algorithm in [9], it does not resort on any assumption on the size of the differences among hyperedges. We have also given a three-stage algorithm that improves by a $d^{1/6}$ factor on the number of tests of the three-stage algorithm in [9]. These algorithms are special cases of an s-stage algorithm, designed for an arbitrary positive integer $s \leq d$, whose performance in terms of the number of tests decreases with s, thus providing a trade-off between the number of tests and adaptiveness. As far as it concerns one-stage group testing, i.e., the problem of identifying the defective hyperedge non-adaptively, the paper provides a lower bound for E being sufficiently large and $|e' \backslash e| \leq \frac{1}{3}d$, for any two distinct $e, e' \in E$. This lower bound is the first lower bound that improves on the $\Omega(\log |E|)$ lower bound and exhibits an $O(\log \frac{d-p}{p})$ gap with the $O(\frac{d}{p} \log |E|)$ upper bound. We notice that, in the classical setting, $|E|$ consists of all possible subsets of $\{1, \ldots, n\}$, and by setting $p = 1$ in the above gap, one recovers the $O(\log d)$ gap holding between the best upper and lower bounds for classical group testing.

References

1. Alon, N., Asodi, V.: Learning a hidden subgraph. SIAM J. Discrete Math. **18**(4), 697–712 (2005)
2. Arasli, B. and Ulukus, S.: Graph and cluster formation based group testing. In: 2021 IEEE ISIT, pp. 1236–1241 (2021)
3. Chen, H.B., Hwang, F.K.: Exploring the missing link among d-separable, $\bar{d}$-separable, and d-disjunct matrices. Discrete Appl. Math. **155**, 662–664 (2007)
4. Clementi, A.E.F., Crescenzi, P., Monti, A., Penna, P., Silvestri, R.: On Computing Ad-hoc selective families. In: RANDOM-APPROX 2001, pp. 211–222 (2001)
5. Clementi, A.E.F., Monti, A., Silvestri, R.: Selective families, superimposed codes, and broadcasting on unknown radio networks. In: Twelfth Annual ACM-SIAM Symposium on Discrete Algorithms, pp. 709–718 (2001)
6. Coja-Oghlan, A., et al.: Optimal group testing. In: Thirty Third Conference on Learning Theory, pp. 1374–1388 (2020)
7. De Bonis, A.: Conflict resolution in arbitrary hypergraphs. In: 19th International Symposium, ALGOWIN 2023, Lecture Notes in Computer Science, vol. 14061, pp. 13–36. Springer (2023)
8. De Bonis, A.: Constraining the number of positive responses in adaptive, non-adaptive, and two-stage group testing. J. Comb. Optim. **32**(4), 1254–1287 (2016)
9. De Bonis, A.: Group testing in arbitrary hypergraphs and related combinatorial structures. In: Fernau, H., Gaspers, S., Klasing, R. (eds.) SOFSEM 2024: Theory and Practice of Computer Science. SOFSEM 2024. Lecture Notes in Computer Science, vol. 14519, pp 154–168. Springer, Cham (2024)
10. De Bonis, A, Gąsieniec, L, Vaccaro, U.: Optimal two-stage algorithms for group testing problems. SIAM J. Comput. 34(5), 1253–1270 (2005)
11. Dorfman, R.: The detection of defective members of large populations. Ann. Math. Statist. **14**, 436–440 (1943)
12. Du, D.Z., Hwang, F. K.: Pooling design and nonadaptive group testing. Ser. Appl. Math. **18** (2006)
13. D'yachkov, A.G., Rykov, V.V.: A survey of superimposed code theory. Probl. Control Inform. Theory, **12**, 229–242 (1983)
14. Erdös, P., Frankl, P., Füredi, Z.: Families of finite sets in which no set is covered by the union of r others. Israel J. Math. **51**, 79–89 (1985)
15. Gargano, L., Rescigno, A.A., Vaccaro, U.: On k-strong conflict–free multicoloring. In: Gao, X., Du, H., Han, M. (eds.) Combinatorial Optimization and Applications. COCOA 2017. Lecture Notes in Computer Science, vol. 10628, pp. 276–290. Springer, Cham (2017)
16. Gonen, M., Langberg, M., Sprintson A.: Group Testing on General Set-Systems. Manuscript, Available at https://arxiv.org/abs/2202.04988

17. Gonen, M., Langberg, M., Sprintson A.: Group testing on general set-systems. In: 2022 IEEE International Symposium on Information, pp. 874–879 (2022)
18. Harvey, N.J.A., Patrascu, M., Wen, Y., Yekhanin, S., Chan, V.W.S.: Non-adaptive fault diagnosis for all-optical networks via combinatorial group testing on graphs. In: 26th IEEE International Conference on on Computer Communications, pp. 697–705 (2007)
19. Hong, E.S., Ladner, R.E.: Group testing for image compression. IEEE Trans. Image Process. **11**(8), 901–911 (2002)
20. Goenka R., Cao S.J., Wong C.W., Rajwade A., Baron D.: Contact tracing enhances the efficiency of covid-19 group testing. In: ICASSP 2021, pp. 8168–8172 (2021)
21. Kautz, W.H., Singleton, R.C.: Nonrandom binary superimposed codes. IEEE Trans. Inf. Theory **10**, 363–377 (1964)
22. Lo, C., Liu, M., Lynch, J.P., Gilbert, A.C.: Efficient sensor fault detection using combinatorial group testing. In: 2013 IEEE International Conference on Distributed Computing in Sensor Systems, pp. 199–206 (2013)
23. Nikolopoulos, P., Srinivasavaradhan, S.R., Guo, T., Fragouli, C., Diggavi S.: Group testing for connected communities. In: The 24th International Conference on Artificial Intelligence and Statistics, vol. 130, pp. 2341–2349. PMLR (2021)
24. Ruszinkó, M.: On the upper bound of the size of the r-cover-free families. J. Combin. Theory Ser. A **66**, 302–310 (1994)
25. Sobel, M., Groll, P.A.: Group testing to eliminate efficiently all defectives in a binomial sample. Bell Syst. Tech. J. **38**, 1179–1252 (1959)
26. Vorobyev, I.: Note on generalized group testing (2022). https://doi.org/10.48550/arXiv.2211.04264
27. Wolf, J.: Born again group testing: multiaccess communications. IEEE Trans. Inf. Theory **31**, 185–191 (1985)
28. Zhu, J., Rivera, K., and Baron, D.: Noisy pooled PCR for virus testing (2020). https://doi.org/10.48550/arXiv.2004.02689

A Parameterized Perspective
of All-Colors

Václav Blažej, Satyabrata Jana[✉], and Peter Strulo

University of Warwick, Coventry, UK
{vaclav.blazej,peter.strulo}@warwick.ac.uk, satyamtma@gmail.com

Abstract. In this paper, we study the All-Colors problem: given a graph G each of whose vertices is equipped with a button and assigned a color value from the set $\{0, 1, \ldots, m-1\}$ and an integer k, can we reach color value 1 (mod m) on every vertex of G by pressing the button at most k times. The rule we follow is the following: if a button of a corresponding vertex is pressed one time, then the color values of the vertex and its neighbors are incremented by 1. This problem is known to be NP-hard on bipartite graphs even when $m = 2$ [Theor. Comput. Sci., 2007], although linear time solvable on trees [SIAM J. Comput., 2004]. In this work, we study this problem in the realm of parameterized complexity with respect to several parameters. In particular, we show the following for All-Colors.
- W[1]-hard when parameterized by solution size (k).
- FPT algorithm parameterized by solution size + maximum degree.
- FPT algorithm parameterized by tree width, clique-width.
- NP-hard on sub-cubic planar graphs.

Keywords: FPT · Clique-width · Treewidth · All-colors problem · All-ones problem · Planar graph

1 Introduction

In the All-Colors problem we have an undirected graph G and an integer k where each vertex has a button and some initial color from $\mathbb{Z}_m$. Pressing the button on a vertex v increments the color value on $N[v]$ (v and its neighbors) by 1 (mod m). The goal is to determine whether it is possible to press the buttons up to k times so that the color at each vertex becomes 1.

> All-Colors (AC)
>
> *Input:*   A graph $G = (V, E)$, a color function $c \colon V \to \mathbb{Z}_m$ where $m > 1$, and an integer k.
>
> *Question:* Is there a press function $p \colon V \to \mathbb{Z}_m$ such that $\sum_{v \in V} p(v) \leq k$ and $c(v) + \sum_{w \in N[v]} p(w) \equiv 1 \mod m$ for every vertex $v \in V$?

© The Author(s), under exclusive license to Springer Nature Switzerland AG 2025
I. Finocchi and L. Georgiadis (Eds.): CIAC 2025, LNCS 15679, pp. 222–239, 2025.
https://doi.org/10.1007/978-3-031-92932-8_15

Zhang and Wang [32] defined ALL-COLORS as an extension of the special case where $m = 2$ and all vertices start with color 0, known as the ALL-ONES problem. In this case we can extend the button metaphor by giving each vertex a lamp and call the two states of each vertex *lit* and *unlit*. This case was studied in various fields under different names: σ^+ *games* in cellular automata [2,28,30], *odd parity cover* [29], *odd dominating set* [4,6] in graph theory, and also has connections to coding theory [14,17]. The term ALL-ONES was introduced by Sutner (see [26]).

Although ALL-COLORS inherits NP-hardness from ALL-ONES [27] the reduction requires k to be "big" (proportional to the input size) and little is known when k is "small". To the best of our knowledge this is the first study of the ALL-COLORS and ALL-ONES problems from the perspective of parameterized complexity. In Sect. 4, we establish that one cannot solve ALL-ONES in $f(k) \cdot n^{\mathcal{O}(1)}$ time assuming FPT $\neq W[1]$.

Theorem 1. ALL-ONES *is* W[1]-*hard with respect to the solution size* k.

Moreover, assuming the Exponential Time Hypothesis (ETH), we obtain a time complexity lower bound which is close to the brute-force $\mathcal{O}(n^{k+1})$ time algorithm.

Corollary 1. *There is no algorithm solving* ALL-ONES *in* $f(k) \cdot n^{o(k/\log k)}$ *time for some function* f, *where* k *is the size of the solution and* n *is the input length, unless the Exponential Time Hypothesis fails.*

To get some tractability one may attempt to combine k with other parameters. In Sect. 5 we show that ALL-COLORS is FPT with respect to k and the maximum degree. We complement this tractability later in Sect. 8 by showing NP-hardness on planar graphs with maximum degree 3.

Theorem 2. ALL-COLORS *admits an* FPT *algorithm with respect to the combined parameter of size of the solution* k *and maximum degree* Δ *and that runs in* $(k \cdot \Delta^2)^k \cdot n^{\mathcal{O}(1)}$ *time.*

Aside from k being a parameter we also investigate the problem of minimizing the possible k when the input instance is parameterized in other ways. In Sect. 6, we show that ALL-COLORS can be solved in $\mathcal{O}(m^{6tw} \cdot n)$ time for graphs of bounded treewidth. This generalizes the ALL-ONES results that led to an algorithm for treewidth at most 4 [8,22] and provides an explicit (linear in n and single exponential in tw for any fixed m) algorithm in place of the treewidth algorithm implied by Courcelle's theorem [9].

Theorem 3. *Given an* n-*vertex graph* G *with treewidth* tw, ALL-COLORS *with* m *colors can be solved in* $m^{\mathcal{O}(tw)} \cdot n$ *time.*

In order to generalize the treewidth result we push into parameters that can be bounded even for dense graphs. In Sect. 7 we show a dynamic programming algorithm for clique-width. Intuitively, the clique-width of a graph serves as a measure of its structural complexity. Although it may appear that this result subsumes the algorithm for treewidth note that $\mathtt{cw} \leq 2^{\mathtt{tw}+1} + 1$ [10], hence, the dependence on $\mathtt{tw}$ from this inference would be doubly exponential.

Theorem 4. *Given a graph G and a term t that witnesses that clique-width of G is at most $\mathtt{cw}$, ALL-COLORS with m colors can be solved in $(m^4)^{\mathtt{cw}} \cdot n^{\mathcal{O}(1)}$ time.*

We conclude with showing an NP-hardness reduction for ALL-ONES. This NP-hardness is known for bipartite graphs [4]. While the reduction of Theorem 1 runs in polynomial time, which implies NP-hardness, we aim to show hardness for as restricted graph class as possible. The reduction produces planar graphs that have maximum degree 3 which serves to complement our results two-fold. First, this complements Theorem 2 as dropping k from the parameter leads to instances that are NP-hard even for constant Δ. Second, as further generalization of Theorem 4 one could attempt an algorithm parameterized by twin-width, a parameter that gained popularity in recent years. Planar graphs, however, are known to have constant twin-width [3,18] so unless P=NP there is no XP algorithm for twin-width.

Theorem 5. ALL-ONES *is* NP-*hard in sub-cubic planar graphs.*

Corollary 2. *Assuming P≠NP, no algorithm can solve ALL-COLORS with m colors in an n-vertex graph G of twin-width at most $\mathtt{tww}$ in $n^{f(\mathtt{tww},m)}$ time, for any computable function f.*

2   Related Work

An equivalent version of the ALL-ONES problem was introduced by Peled in [25]. Numerous studies have explored the ALL-ONES problem, including works by Sutner [30,31], and Barua and Ramakrishnan [2], along with other references therein. While the definition does not immediately imply this, the ALL-ONES problem invariably has a solution. This assertion was established by Sutner [29], who employed linear algebra to demonstrate that it is always possible to light every lamp in any graphs. Another insightful proof, also utilizing linear algebra, was provided by Lossers [21]. Furthermore, Erikisson et al. [13] provide a short and elegant graph-theoretic proof.

In graph theory terms, the ALL-ONES problem can be described as follows: given a graph $G = (V, E)$, is there a subset X of vertices in V such that, for each vertex v in $V(G)$, the number of vertices in X that intersects the closed neighborhood of v is odd? This subset X is termed an *odd parity cover* as per [29], or an *odd dominating set* according to [4]. Sutner [26] questioned whether graph-theoretic techniques could solve the ALL-ONES problem specifically for trees. Galvin [15] provided a positive answer to this.

In [27], Sutner demonstrated that determining whether an instance of ALL-ONES has a solution size of at most k is a NP-complete problem. Later, Broersma and Li showed that this holds even for bipartite graphs [4]. This naturally leads to the investigation of specific graph classes for which polynomial time algorithms are possible. For certain graph structures like unicyclic and bicyclic graphs [7], there exists a linear time algorithm to address this issue, as well as for series-parallel graphs [1], binomial trees, butterfly networks, and Beneš networks [23]. Caro et al. investigated other graph classes such as grid graphs, exclusive graphs [5,6]. Lu and Li [22] tackled the minimization version for graphs with a maximum treewidth of 4 (a superclass of trees and series-parallel graphs) and developed a linear time solution. Broersma and Li [4] established that All-Ones can be defined in MSOL and hence is FPT when parameterized by treewidth by Courcelle's theorem. Li et al. [20] explored a more general version of All-Ones where the requirement for all vertices to start unlit is removed and derived a linear time algorithm specifically for trees. Fleischer and Yu conducted an extensive survey on this version [14].

Zhang and Wang [32] instead drop the requirement that $m = 2$ and consider the problem on graph classes such as caterpillars, paths, radioactive trees, and graphs with a path that is 3-connected. They are interested in determining if a given graph has some solution *of any size* when the initial state is all zeros so their definition of All-Colors is slightly different to ours in that it only has a graph as input and does not include a budget k or an initial color function c. They use a straightforward algebraic approach to solve this problem: "*A graph G has a solution for the* All-Colors *problem if and only if the linear equation system* $(A + I)X = (\mathbf{1}) \pmod{m}$ *is solvable over* $\mathbb{Z}_m$". Here, A represents the adjacency matrix of G, I is the identity matrix, X is an $n \times 1$ vector of variables, and $\mathbf{1} \pmod{m}$ is an $n \times 1$ vector each element of which is $(1) \pmod{m}$.

3 Preliminaries

Let us use shorthand $f(X) = \sum_{u \in X} f(u)$. For a tuple x let us denote by $x.i$ the i-th element, e.g., for $p = (5, 8)$ we have $p.1 = 5$ and $p.2 = 8$.

The few equivalent definitions of All-Colors partially stem from the fact that a process of gradually pressing buttons is equivalent to simply choosing how many times each button is pressed in the end. We use the inverse principle in Sect. 5.

Consider an assignment $f : V(G) \to \mathbb{Z}_m$. We can check whether f is a solution to All-Colors by trying whether for each $u \in V(G)$ we have $c(v) + f(N[v]) \equiv 1$ mod m, clearly a polynomial procedure. This implies that to solve AC we may employ a simple brute-force algorithm that iterates through all assignments and for each checks whether it constitutes a solution. There are m^n assignments and we check each in $n^{O(1)}$ time. Note that this also implies a trivial $2^n \cdot n^{O(1)}$ single exponential algorithm for All-Ones. Moreover, if we seek solution of size no bigger than k then this gives an algorithm with $\binom{n}{k} \cdot n^{O(1)}$ time complexity; we will see in Corollary 3 that one cannot do much better.

Before more complex results we make a simple observation and a simple helpful lemma.

Observation 1. *A solution X to* All-Ones *is a dominating set.*

Lemma 1. *Let us have an* ALL-ONES *instance* $\mathcal{I} = (G, k)$. *If there exists a set of vertices* $R \subseteq V(G)$ *such that* $|R| = k$ *and for every* $u, v \in R, u \neq v$ *we have* $N[u] \cap N[v] = \emptyset$, *then every solution* X *to* $\mathcal{I}$ *has* $|X| = k$ *and for each* $v \in R, |X \cap N[v]| = 1$.

Proof. The solution X forms a dominating set in G, hence, neighborhood of every vertex in R must contain at least one vertex of X. As $|X| \leq k = |R|$ we conclude that $|X| = k$ and so every vertex of R must contain in its neighborhood exactly one vertex of X. □

Fixed Parameter Tractable. A parameterized problem Π is a subset of $\Gamma^* \times \mathbb{N}$ for some finite alphabet Γ. An instance of a parameterized problem consists of (X, k), where k is called the parameter. A parameterized problem L is considered to have a fixed parameter tractable (FPT) algorithm if there is an algorithm $\mathcal{A}$ that can determine whether $(X, k) \in L$ in time $f(k) \cdot n^{\mathcal{O}(1)}$ for some computable function f, where n is the size of the input.

4   ALL-ONES Parameterized by Solution Size

Definition 1. *For a graph* G *and its subset of vertices* $C \subseteq V(G)$ *let building a* clique gadget $\mathcal{C}$ *over* C *mean that we create two new vertices* u *and* v *called* pendant, *and create edges* $\{vw, uw \mid w \in C\}$ *and* $\{ww' \mid w, w' \in C\}$. *We call* C *the* body *of the clique gadget* $\mathcal{C}$.

Assuming we build a set of clique gadgets in G all at once (so that we cannot have pendant vertices of one gadget as body of another), then in each clique gadget its pendant vertices form an independent set while its body forms a clique in G.

Theorem 1. ALL-ONES *is* W[1]-*hard with respect to the solution size* k.

Proof. We show a reduction from PARTITIONED SUBGRAPH ISOMORPHISM (PSI) $\mathcal{I}' = (G', H', \psi)$ [24]. In this problem, we are given graphs G', H' and a "color" map $\psi \colon V(G') \to V(H')$, our task is to find a map $\phi \colon V(H') \to V(G')$ such that for all $uv \in E(H')$ we have $\phi(u)\phi(v) \in E(G')$ and $\psi \circ \phi$ is the identity on $V(H')$. Equivalently, map ψ can be thought of as coloring of $V(G')$ by colors $V(H')$ and the goal is to find a copy of H' in G' which preserves vertex colors. PSI can easily be seen to be W[1]-hard because it is a generalization of MULTICOLORED CLIQUE which is a standard W[1]-hard problem [12].

Let C_i be vertices of color i and let $C_{i,j}$ be edges that have incident vertices of colors i and j, i.e., $C_i = \{u \mid u \in V(G), \psi(u) = i\}$ and $C_{i,j} = \{uv \mid uv \in E(G), \psi(u) = i, \psi(v) = j\}$. We construct an instance $\mathcal{I}$ of ALL-ONES as follows, see Fig. 1.

Let $k = |V(H')| + |E(H')| + 1$. For convenience, we abuse the notation by using vertices and edges of G' to denote vertices of G. To distinguish edges of G we denote them by sets explicitly, e.g. $\{u, v\}$. The vertex set of G is initially made of four sets: $V(G')$, $E(G')$, $L = \{(u,c) \mid u \in V(G'), c \in [k']\backslash\psi(u)\}$, and a vertex q. For each vertex $(u,c) \in L$ we add edges $\{u,(u,c)\}$ and $\{q,(u,c)\}$. For every $uv \in E(G')$ we add edges $\{uv,(u,\psi(v))\}$ and $\{uv,(v,\psi(u))\}$. In the last step we build clique gadgets, let $\mathcal{C}_i$ be a clique gadget over C_i for each $i \in |V(H')|$, let $\mathcal{C}_{i,j}$ be a clique gadget over $C_{i,j}$ for each $ij \in E(H')$, and let $\mathcal{Q}$ be a clique gadget over q.

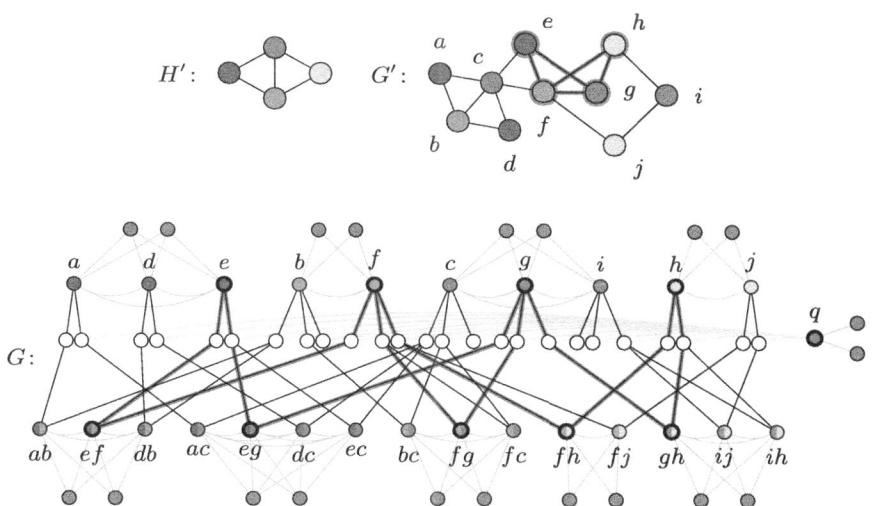

Fig. 1. Reduction from PSI to AO. Vertex colors represent $V(H')$. Gray vertices are the pendant vertices of the clique gadgets (and q). The thick edges and vertices of G show a solution which represents a copy of H' in G'. (Color figure online)

Intuition: The clique gadgets force the solution to contain q and a vertex from C_i for each i and a vertex from $C_{i,j}$ for each i,j. This choice directly represents the copy of H' in G' and no other solution is feasible due to a tight budget.

To prove the correctness we show the equivalence one direction at a time.

$\Rightarrow$ Assume that there is a PSI solution $S' \subseteq G'$, i.e., S' is isomorphic to H' (including colors). We claim that $X = S' \cup \{q\} \cup \{uv \mid uv \in E(S')\}$ is an AO solution in G. First observe that $|X| = |V(H')|+1+|E(H')| = k$. For each clique gadget $\mathcal{C}$ we have for each $u \in V(\mathcal{C})$ exactly one vertex of its closed neighborhood in X. The vertices outside of the clique gadgets form the set L. Each vertex $(u,c) \in L$ is neighboring q which is in X, moreover, if (u,c) neighbors u which is in X then it also neighbors uv such that $\gamma(v) = c$ which is in X. Hence, for each $u \in V(G)$ we have $|N[u] \cap X| \equiv 1 \mod 2$ and X is a solution to AO.

$\Leftarrow$ Assume that there exists a solution X to AO on G. Note that in total we constructed k clique gadgets – $\mathcal{Q}$, $\mathcal{C}_i$, and $\mathcal{C}_{i,j}$. Let us create a set R that contains a single pendant vertex of every clique gadget in G. For each $u, v \in R$, $u \neq v$ we have $N[u] \cap N[v] = \emptyset$ because these sets are entirely contained in the respective clique gadgets and the clique gadgets in our construction are over disjoint sets. By $|R| = k$ and the above disjointness argument we can apply Lemma 1 to conclude that the solution X has exactly one vertex in each clique gadget. For a gadget $\mathcal{C}$ we note that its vertex in X cannot be the pendant vertex as the other pendant vertex would have no vertex of X in its neighborhood, hence, $V(\mathcal{C}) \cap X$ is in the body of $\mathcal{C}$. All of the budget is already used to have X in the clique gadgets so $L \cap X = \emptyset$. With such an assignment, all vertices in the clique gadgets have exactly one neighbor in X so it remains to check whether vertices of L are satisfied. Let us fix one $(u, c) \in L$ and argue about $|N[(u, c)] \cap X|$. Vertex (u, c) neighbors q ($q \in X$), then it neighbors u, and last it neighbors uv if and only if $\gamma(v) = c$, i.e., $\{uv \mid uv \in E(G'), \gamma(v) = c\}$ – this is a subset of vertices in the body of the clique gadget $\mathcal{C}_{\gamma(u),c}$ so at most one of them is in X. Assuming $|N[(u, c)] \cap X| \equiv 1 \mod 2$ we conclude that if a neighbor of (u, c) in $\mathcal{C}_{\gamma(u),c}$ is in X, then u must be in X as well. Let us pull X back to S', a set of edges and vertices in G' of the PSI instance. The above argument translates to saying that whenever S' contains an edge $uv \in G'$ then it also must contain vertices u and v. This condition is sufficient to conclude that S' indeed forms a copy of H' within G'.   □

The above reduction also gives us a conditional lower bound based on Exponential Time Hypothesis (ETH) which says that n-variable 3SAT has no algorithm that runs in $2^{o(n)}$ time. Marx [24, Corollary 6.3] showed that if PSI can be solved in time $f(G) \cdot n^{o(k'/\log k')}$, where f is an arbitrary function and $k' = |E(H')|$, then Exponential Time Hypothesis (ETH) fails. As we reduced from PSI to AO and as we can assume $|V(H')| \leq 2|E(H')|$ (by a simple pre-processing of isolated vertices), we get $k = |E(H')| + |V(H')| + 1 \leq 3|E(H')| + 1$ and consequently the following corollary.

Corollary 3. *If* ALL-ONES *can be solved in time* $f(G) \cdot n^{o(k/\log k)}$, *where* f *is an arbitrary function and* k *is the solution size, then* ETH *fails.*

5   ALL-COLORS w.r.t. Solution Size and Maximum Degree

Definition 2. *For a solution* p *to* ALL-COLORS *with budget* k *let a* solution sequence $S = v_1, \ldots, v_k$ *be a list of* k *vertices such that the number of times* v *appears in the sequence is* $p(v)$ *for every vertex* v. *Given a solution sequence* S, *let* $f_0^S, f_1^S, \ldots, f_k^S$ *be the sequence of functions where* $f_0^S = \{c(v) \mid v \in V(G)\}$, *and for each* $i \in \{1, \ldots, k\}$ *we have*

$$f_i^S(u) = \begin{cases} (f_{i-1}^S(u) + 1) \bmod m & \text{if } u \in N[v_i] \text{ and} \\ f_{i-1}^S(u) & \text{otherwise.} \end{cases}$$

We write simply f_i *when the solution sequence is clear from the context.*

In other words, a solution sequence represents a list of vertices such that by going through the list and for each vertex we increase values in its closed neighborhood (modulo m) we end up with the color at each vertex being 1, while the functions f_i represent partial functions after increasing values for only vertices v_j with $j \leq i$. The final function f_k will be equal to 1 on every vertex.

Lemma 2. *Let p be a solution to ALL-COLORS such that the used budget $k = \sum_{v \in V(G)} p(v)$ is minimized, then there exists a solution sequence $w_1, \ldots, w_k$ of p such that for each w_i there exists $u_i \in N[w_i]$ where $f_{i-1}(u_i) \not\equiv 1 \mod m$.*

Proof. Towards a contradiction, assume that p has no such solution sequence. Let the *light index* for a fixed solution sequence $S = v_1, \ldots, v_k$ be its minimum index a where the vertex v_a has $f_{a-1}^S(u_a) \equiv 1 \mod m$ for all $u_a \in N[v_a]$. Let us take a solution sequence $S = v_1, \ldots, v_k$ which maximizes the light index, a. The values of f_{a-1}^S depend only on vertices $v_1, \ldots, v_{a-1}$ so by permuting the remaining $v_a, \ldots, v_k$ we obtain a family $\mathcal{F}$ of solution sequences that have $f_{a-1}^T = f_{a-1}^S$ for all $T \in \mathcal{F}$. As the sequence S maximized its light index and as we permuted only $v_a, \ldots, v_k$ we conclude that each sequence in $\mathcal{F}$ has light index equal to a. In particular, for every $u \in \{v_a, \ldots, v_k\}$ there is a sequence $S_u \in \mathcal{F}$ such that u is its a-th element – it occupies the light index – and implies that $f_{a-1}^{S_u}(v) \equiv 1 \mod m$ for each $v \in N[u]$. As f_k^S (which is equal to 1 on every vertex) can differ from f_{a-1}^S only in the vertices that are in the neighborhoods of $\{v_a, \ldots, v_k\}$, $f_{a-1}^{S_u} = f_{a-1}^S$, and $f_{a-1}^{S_u}(v) \equiv 1$ for all such vertices, we conclude that $v_1, \ldots, v_{a-1}$ is a solution sequence as well but with only $a - 1 < k$ elements, which is in a contradiction with p being a solution that uses the minimum budget. $\square$

Theorem 2. ALL-COLORS *admits an* FPT *algorithm with respect to the combined parameter of size of the solution k and maximum degree Δ and that runs in $(k \cdot \Delta^2)^k \cdot n^{\mathcal{O}(1)}$ time.*

Proof. We devise a branching algortihm. We aim to find a solution sequence such that at each step $i \in [k]$ the vertex v_i has in its neighborhood a vertex u with $f_{i-1}(u) \not\equiv 1 \mod m$. As pressing button at any vertex can change at most $\Delta+1$ vertex values we know that a solution of size k can change values in at most $k \cdot (\Delta+1)$ vertices. Let R be the set of vertices $u \in V(G)$ where $c(u) \not\equiv 1 \mod m$. If R contains more than $k \cdot (\Delta + 1)$ then we immediately answer no (return no from the branch). Otherwise, we do as follows. Let $N[R] = \{N[u] \mid u \in R\}$, note $|N[R]| \leq k \cdot (\Delta + 1)^2$. From Lemma 2 we know that there exists a solution sequence with $v_1 \in N[R]$. Hence, we branch on the choice of $v_1 \in N[R]$ and apply recursion with the value of k decreased by 1. If a branch encounters $R = \emptyset$ then it returns an empty sequence as a success. If any branch returns a sequence $v_2, \ldots, v_k$ (possibly empty) then we return $v_1, v_2, \ldots, v_k$. Every step branches into at most $k \cdot (\Delta + 1)^2$ vertices and decreases k by one, so the procedure runs in $(k \cdot \Delta^2)^k \cdot n^{\mathcal{O}(1)}$ which is FPT. $\square$

As a side note; an FPT algorithm for ALL-COLORS opens a question of having a (polynomial) sized kernel (see Cygan et al. [12] for details), we can resolve this question by a simple observation for ALL-ONES.

Recall that in ALL-ONES we have $c(u) = 0$ for all $u \in V(G)$ and $m = 2$. In the above proof we observed that $k \cdot (\Delta + 1)$ is the maximum number of vertices that can have changed values by the solution. This immediately gives a trivial polynomial kernel for ALL-ONES which preserves G if $|V(G)| \leq k \cdot (\Delta + 1)$ or gives a trivial no instance which is a disjoint union of $k + 1$ vertices.

Corollary 4. ALL-ONES *has a polynomial kernel with respect to the combined parameter size of the solution and maximum degree.*

6   ALL-COLORS by Treewidth

To solve ALL-COLORS we employ a standard dynamic programming approach on a nice tree decomposition. We now explain the basic intuition while assuming basic knowledge of tree decompositions (see full version for the definition). The DP table state keeps for each vertex two values in $\mathbb{Z}_m$, the first representing how many times the vertex is pressed in the final solution, the second representing its current color in the graph that has been "introduced" so far by the decomposition. By employing a nice tree decomposition it is then sufficient to show how to update a state when 1) a vertex is introduced, 2) a vertex is removed, 3) two states are joined. For example, when a vertex is introduced the color of its neighbors change by the how many times the vertex is pressed. We compute the states bottom up while trying all possibilities, i.e., for an introduced vertex all possible number of its presses. The formal description with technical details are in the full version of the paper due to space restrictions and gives a proof of the following lemma.

Lemma 3. *Let G be an n-vertex graph with its tree decomposition of width* tw. *Then the* ALL-COLORS *with m colors in G is solvable in $\mathcal{O}(m^{3\mathtt{tw}} \cdot n)$ time.*

The following result is known due to Korhonen [19].

Proposition 1. [19, Theorem I.1.] *There is an algorithm, that given an n-vertex graph G and an integer* tw, *in time $2^{\mathcal{O}(\mathtt{tw})} \cdot n$ either outputs a tree decomposition of G of width at most $2\mathtt{tw} + 1$ or determines that the treewidth of G is larger than* tw.   □

Lemma 3 and Proposition 1 together imply the following result.

Theorem 3. *Given an n-vertex graph G with treewidth* tw, ALL-COLORS *with m colors can be solved in $m^{\mathcal{O}(tw)} \cdot n$ time.*

7   ALL-COLORS by Clique-Width

In this section, we show that if the input graph has bounded clique-width cw and we are given its clique decomposition, then we can solve AO in FPT time with respect to a combined parameter cw and m. Let us first recall the definition of clique-width and its decomposition.

Definition 3 (Clique-width, [11]). *Let $\mathscr{L}$ be a countable set of labels. A labeled graph is a pair (G, γ) where $\gamma \colon V(G) \to \mathscr{L}$. Graph is C-labeled if $C \subseteq \mathscr{L}$ is finite and $\gamma(V(G)) \subseteq C$. We introduce the following symbols: a nullary symbol a for every $a \in \mathscr{L}$; a unary symbol $\rho_{a \to b}$ for $a, b \in \mathscr{L}$ with $a \neq b$; a unary symbol $\eta_{a,b}$ for $a, b \in \mathscr{L}$ with $a \neq b$; a binary symbol $\oplus$. For $C \subseteq \mathscr{L}$ we denote by $T(C)$ the set of finite well-formed terms written with the symbols $\oplus, a, \rho_{a \to b}, \eta_{a,b}$ for $a, b \in C, a \neq b$. Let $\mathrm{val}(t)$ for $t \in T(C)$ be the abstract labeled graph that corresponds to the set of all graphs that can be created as follows (this set contains the same graphs up to vertex relabeling).*

1. *$t = a \in C$: $\mathrm{val}(t) = \big((\{u\}, \emptyset), (u, a)\big)$ for a new vertex u, i.e., graph with a single vertex labeled a;*
2. *$t = t_1 \oplus t_2$: $\mathrm{val}(t) = (G_1 \cup G_2, \gamma_1 \cup \gamma_2)$ where $(G_1, \gamma_1) = \mathrm{val}(t_1)$, $(G_2, \gamma_2) = \mathrm{val}(t_2)$, G_1 is disjoint from G_2, and $\gamma_1 \cup \gamma_2$ for u is $\gamma_1(u)$ if $u \in V(G_1)$ and $\gamma_2(u)$ if $u \in V(G_2)$, i.e., a disjoint union of graphs retrieved from t_1 and t_2;*
3. *$t = \rho_{a \to b}(t')$: $\mathrm{val}(t) = (G, \gamma')$ where $(G, \gamma) = \mathrm{val}(t')$ and $\gamma' = \{(u, \delta_{a \to b}(\ell)) \mid (u, \ell) \in \gamma\}$ where $\delta_{a \to b}(a) = b$ and $\delta_{a \to b}(c) = c$ for $c \in C \setminus \{a\}$, i.e., the graph is obtained by replacing all vertex labels a with b;*
4. *$t = \eta_{a,b}(t')$: $\mathrm{val}(t) = (G + E, \gamma)$ where $(G, \gamma) = \mathrm{val}(t')$ and $E = \{uv \mid u \neq v, \gamma(u) = a, \gamma(v) = b\}$, i.e., the graph with an edge added between every pair of vertices where one is labeled a and the other is labeled b.*

Clique-width of G is $cwd(G) = \min\{|C| \mid (G, \gamma) \in \mathrm{val}(t), t \in T(C)\}$, i.e., it is the minimum number of labels needed to construct G.

Intuitively, terms in the above definition can be viewed as nodes of a binary tree. Each node is labeled with an operation but an operation can be given to only nodes with prescribed number of children. Our graph then gets gradually constructed by a bottom-up evaluation of the tree where each node creates a subgraph by applying operations on graphs returned by its children, eventually giving the desired graph in the root node. Each graph during the construction has labelled vertices. To design an FPT algorithms for clique-width, we aim to associate a small state with labels instead of vertices and use these to compute the optimum answer. The following theorem shows such an algorithm.

In short, we associate m^2 states with each label. Each state (i, j) for $i, j \in \mathbb{Z}_m$ represents the case where sum of f over all vertices that have the label is $\equiv i$ mod m and for each vertex with the label the sum f over its closed neighborhood is $\equiv j \mod m$.

Theorem 4. *Given a graph G and a term t that witnesses that clique-width of G is at most cw, ALL-COLORS with m colors can be solved in $(m^4)^{cw} \cdot n^{\mathcal{O}(1)}$ time.*

Proof. We have a set of labels $C \subseteq \mathcal{L}$ of size $\mathtt{cw}$ and the witnessing term $r \in T(C)$. Let $(G_t, \gamma_t) = \mathrm{val}(t)$ be the labelled subgraph of G that is constructed by t, where t is a subterm of r. For each $u \in V(G_t)$ let $N_t[u]$ be the closed neighborhood of u in G_t. Let $G_{t,a}$ be the induced subgraph of G_t on vertices labeled with a, i.e., $G_{t,a} = G_t[A]$ where $A = \{u \in V(G_t) \mid \gamma_t(u) = a\}$.

Observe that once two vertices have the same label then any further operations performed on them (connecting them to other vertices, relabeling) are the same. We claim that for a fixed solution configuration f to AO there is a value $v_{t,a} \in \mathbb{Z}_m$ such that all vertices of $u \in V(G_{t,a})$ have $f(N_t[u]) = v_{t,a}$. It suffices to show that the value is the same for any pair of vertices $u, v \in V(G_{t,a})$ due to transitivity. We prove the claim by a contradiction, assume that for some pair of vertices $u, v \in V(G_{t,a})$ we have $f(N_t[u]) \neq f(N_t[v])$. Then as all operations are based on labels of u and v and as $\gamma(u) = \gamma(v)$ we have that any neighbors they attain via operations performed by terms that contain t are neighbors of both, therefore, $f(N_r[u]) - f(N_t[u]) \equiv f(N_r[v]) - f(N_t[v]) \mod m$. We know that their values are not equal in t, it follows that these values are not equal in r, i.e., $f(N_r[u]) \not\equiv f(N_r[v]) \mod m$, but both ought to have $f(N_r[u]) \equiv 1 \mod m$ and $f(N_r[v]) \equiv 1 \mod m$, a contradiction.

We showed that there is $v_{t,a} \in Z_m$ where for each $u \in V(G_{t,a})$ we have $f(N_t[u]) \equiv v_{t,a} \mod m$, let us also have $n_{t,a} \equiv f(V(G_{t,a})) \mod m$. Consider how these two values are set and changed when we perform the operation defined by a term symbol t to obtain $(G_t, \gamma_t) \in \mathrm{val}(t)$.

1. $t = a \in C$ which sets G_t to be a new vertex u with label a, then $n_{t,a} \equiv v_{t,a} \equiv f(u) + c(u) \mod m$.
2. $t = t_1 \oplus t_2$ is disjoint union so for each $a \in C$ we have $n_{t,a} \equiv n_{t_1,a} + n_{t_2,a} \mod m$ and by the above claim we have that $v_{t,a} \equiv v_{t_1,a} \equiv v_{t_2,a} \mod m$.
3. $t = \rho_{a \to b}(t')$ relabels a to b so $n_{t,b} \equiv n_{t',a} + n_{t',b} \mod m$ while again, by the above claim we have $v_{t',a} \equiv v_{t',b} \mod m$, and for all other labels the values do not change.
4. $t = \eta_{a,b}(t')$ pair-wise connects all vertices in $G_{t',a}$ with the vertices in $G_{t',b}$ so $v_{t,a} \equiv v_{t',a} + n_{t',b} \mod m$ and $v_{t,b} \equiv v_{t',b} + n_{t',a} \mod m$ while other values do not change.

We see that the above equalities are not satisfied only by the solution configuration f but also by any configuration f' that on G_t results in values that satisfy our claim. This will not introduce a problem due to an exchange argument that we show later.

We use the equations shown above to compute the solution in the following way. For a term t let $\mathcal{D}_t \colon (C \to Z_m^2) \to \mathbb{N} \cup \{\infty\}$. Our aim is to compute $\mathcal{D}_t$ for each t subterm of r so that $\mathcal{D}_t(a \mapsto (n_{t,a}, v_{t,a})) = f(V(G_t))$. To do so, we follow the construction of the term r and compute the values of $\mathcal{D}_t$ as follows.

1. $t = a \in C$ for $d \in \mathbb{Z}_m$ let $\mathcal{D}_t(\{(a, (d, c(u)+d))\} \cup \{(b, (0,0)) \mid b \in C \backslash \{a\}\}) = d$ and let $\mathcal{D}_t(g) = \infty$ for all other functions g.

2. $t = t_1 \oplus t_2$ let

$$\mathcal{D}_t(g) = \min_{g_1 \in C \to \mathbb{Z}_m^2} \{\mathcal{D}_{t_1}(g_1) + \mathcal{D}_{t_2}(g_2) \bmod m \mid \forall a \in C,$$

$$g_2(a) = g(a).1 - g_1(a).1 \bmod m,$$

$$g_2(a).2 = g_1(a).2 = g(a).2\}.$$

3. $t = \rho_{a \to b}(t')$ let $\mathcal{D}_t(g) = \min_{i \in \mathbb{Z}_m} \mathcal{D}_{t'}(g_i)$ where

$$g_i(c) = \begin{cases} \infty & \text{if } c = a, \\ \infty & \text{if } c \neq a \text{ and } g_i(c).2 \neq g(c).2, \\ (g(b).1 + g(a).1, g(b).2), & \text{if } c = b \text{ and } g_i(c).2 = g(c).2, \text{ and} \\ g(c) & \text{otherwise.} \end{cases}$$

4. $t = \eta_{a,b}(t')$ let $\mathcal{D}_t(g) = \mathcal{D}_{t'}(g')$ where

$$g'(c) = \begin{cases} (g(a).1, g(a).2 - g(b).1 \bmod m) & \text{if } c = a, \\ (g(b).1, g(b).2 - g(a).1 \bmod m) & \text{if } c = b, \\ g(c) & \text{otherwise.} \end{cases}$$

The above computations 1. and 4. require $\mathcal{O}((m^2)^{\mathtt{cw}})$ time while 2. requires $\mathcal{O}((m^2)^{2 \cdot \mathtt{cw}})$ and 3. requires $\mathcal{O}(m \cdot (m^2)^{\mathtt{cw}})$ time. Altogether, the term r contains $\mathcal{O}(n)$ subterms so we can compute the values $\mathcal{D}_t$ for each t in $(m^2)^{2 \cdot \mathtt{cw}} n^{\mathcal{O}(1)}$ time. Last, we iterate through every function $g \in \mathcal{F}$ such that $g \colon C \to \mathbb{Z}_m^2$ and $g(a).1 = 1$ for all $a \in C$ in $m^{\mathtt{cw}}$ time and we return $s = \min_{g \in \mathcal{F}} \mathcal{D}_r(g)$.

We claim that the returned value s is equal to $f(V(G))$ for an optimum configuration f. Note that the above computation contains computation of f as one of the possible computation branches. Towards contradiction, consider that s differs from $f(V(G))$. As f is assumed to have an optimum value, and the computation of f is a computation contained in the computation of $\mathcal{D}_r$ we must have that $s < f(V(G))$. Let us trace how s was created throughout the equations within the subterms of r. In particular, note that a non-infinite value of any $\mathcal{D}_t(g)$ can be traced back to values assigned via nullary symbols because other symbols merely sum and take minimum out of the assigned values. We find a minimal term t where the computed value $s_t = \mathcal{D}_t(a \mapsto (n_{t,a}, v_{t,a}))$ is smaller than $f(G_t)$. Let f' be f where assignment on G_t is altered to the assignment on nullary symbols in t implied by s_t. The evaluation of both assignments resulted in the same g function in $\mathcal{D}_t(g)$ and so this replacement in f doesn't change its evaluation aside from the resulting value. We have $f'(V(G)) < f(V(G))$, a contradiction that finishes the correctness proof. □

8   NP-Hardness of ALL-ONES on Subcubic Planar Graphs

Theorem 5. ALL-ONES *is* NP-*hard in sub-cubic planar graphs.*

Proof. We show a reduction from Planar 3-SAT. Given an instance of Planar 3-SAT $\mathcal{I}' = (V, C)$ we construct an instance $\mathcal{I} = (G, k)$ of AO as follows. We choose $k = |V| + 3|C| + 3 \sum_{\ell \in L} (r_\ell - 1)$ where L is the set of literals and r_ℓ is the number of times a literal ℓ appears across all clauses. The graph G is made up of 3 types of gadgets: a variable gadget V_x for each variable $x \in V$, a clause gadget W_y for each clause $y \in C$, and split gadgets $S_{\ell,t}$ for each $\ell \in L$ and $t \in [r_\ell - 1]$.

We show a lower bound on the number of solution vertices that are chosen from each gadget that, when summed, exactly matches this choice of budget and therefore forces each gadget to have exactly that many solution vertices.

Each gadget contains in- and out-vertices. By connecting gadget A to B we mean adding an edge between an out-vertex of A to an in-vertex of B. Note that an out-vertex need not be lit by the solution on its gadget since it can be lit by the adjacent in-vertex in the connected gadget. In our reduction the out-vertices are never chosen as the solution but may be lit by the solution on the gadget. If not then this forces the corresponding in-vertex in the connected gadget to be chosen which allows us to propagate a choice. We use this by giving the variable gadgets two possible choices of solution both of which have one out-vertex lit and leave the other out-vertex unlit. This choice is then propagated through the split gadgets to the in-vertices of the clause gadgets which restrict these choices. For overview of the resulting structure, see Fig. 3.

Gadgets. A variable gadget V_x has 4 vertices a, b, c, d and edges ca, ab, bd with out-vertices c and d and no in-vertices, see Fig. 2a.

This gadget is used to model a choice between two mutually exclusive options: setting a variable x to true or to false (and therefore $\neg x$ to true). When x is true the solution contains a which makes c lit while d remains unlit by this gadget, and when x is false solution contains b which makes d lit while c remains unlit.

A split gadget $S_{\ell,q}$ has 10 vertices a, b, c, d, e, f, g, h, i, j and edges ab, bc, cd, de, ef, bg, gh, hi, ij with in-vertex a and out vertices f and j, see Fig. 2b.

We connect a split gadget to an out vertex v of a variable gadget A whenever we need the literal represented by v being lit by A to appear in more than one clause. Whenever the out-vertex that is connected to a split gadget B is unlit by its gadget, the in-vertex of the split gadget has to be chosen. This stops B's out-vertices being lit by the solution on B, and therefore replicates the situation at its in-vertex. We can chain these split gadgets to allow literals to appear arbitrarily many times.

A clause gadget W_y has 10 vertices a, b, c, d, e, f, g, h, i, j and edges ab, bc, ac, de, ef, df, gh, hi, gi, bj, ej, hj with in-vertices a, d, and g, and no out-vertices, see Fig. 2c.

Each triangle represents one of the literals in the clause and vertex j being lit represents the clause being satisfied.

Reduction. We refer to a vertex a of a gadget V_x as $V_{x,a}$. We create in G a variable gadget V_x for each variable $x \in V$, a clause gadget W_y for each clause $y \in C$, and split gadgets $S_{\ell,t}$ for each $\ell \in L$ and $t \in [r_\ell - 1]$. We construct a set of out-vertices that represent a literal being true, T_ℓ for each literal $\ell \in L$ in the following way.

Case $r_\ell = 1$: In this case, we simply use the out-vertex of the variable gadget so if ℓ represents a positive literal x for some $x \in V$, then $T_x = \{V_{x,c}\}$, otherwise ℓ represents $\neg x$ and we set $T_{\neg x} = \{V_{x,d}\}$.

Case $r_\ell > 1$: Here we need to use split gadgets $S_{\ell,1}, \ldots, S_{\ell,r_\ell-1}$ to make T_ℓ larger. We connect $S_{\ell,t,j}$ to $S_{\ell,t+1,a}$ for each $t \in [r_\ell - 2]$. Then if ℓ represents x for some $x \in V$ we connect $V_{x,c}$ to $S_{\ell,1,a}$, otherwise we connect $V_{x,d}$ to $S_{\ell,1,a}$. Finally we set $T_\ell = \{S_{\ell,1,f}, \ldots, S_{\ell,r_\ell-1,f}\} \cup \{S_{\ell,r_\ell-1,j}\}$.

Note that $|T_\ell| = r_\ell$ in both cases.

Now we consider each clause $y \in C$. Let y_j for $j \in [3]$ be the three literals that make up the clause y. We connect an unused vertex of T_{y_1} to $W_{y,a}$, an unused vertex of T_{y_2} to $W_{y,d}$, and an unused vertex of T_{y_3} to $W_{y,g}$. This completes the reduction.

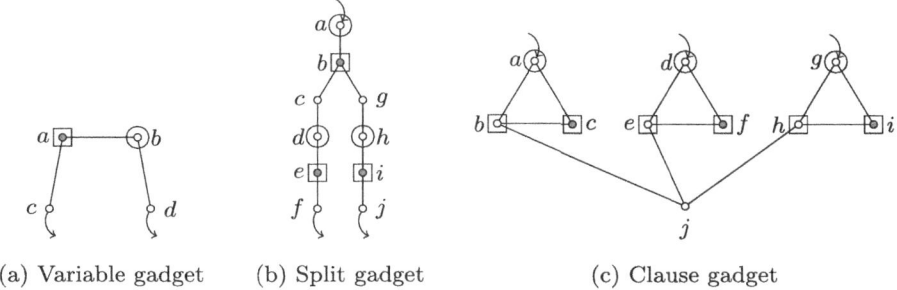

(a) Variable gadget   (b) Split gadget   (c) Clause gadget

Fig. 2. The three gadgets. Out and in vertices are represented with outgoing and incoming curved arrows respectively. Shaded vertices are in R. The boxed and circled vertices show the possible solution vertices: in the variable gadget these choices represent true and false for the variable respectively, in the split gadget these represent the literal it is attached to being true or false, and in the clause gadget each triangle has three possible solutions two of which represents its literal being true and the other false.

Correctness. Suppose we have a satisfying assignment for an instance of Planar 3-SAT. We construct a solution to our instance of AO as follows. For each variable x assigned true we choose vertex $V_{x,a}$ and for each variable assigned false we choose $V_{x,b}$. For each split gadget we are either forced to choose vertex $S_{\ell,t,a}$ and hence also vertices $S_{\ell,t,d}$ and $S_{\ell,t,h}$ or we are not in which case we choose $S_{\ell,t,b}$, $S_{\ell,t,e}$, and $S_{\ell,t,i}$. For each triangle in each clause gadget we choose the in-vertex ($W_{y,a}$, $W_{y,d}$, or $W_{y,g}$) if we are forced to by the connected split or variable gadget, otherwise, if $W_{y,j}$ is not yet lit we choose the vertex adjacent to $W_{y,j}$ ($W_{y,b}$, $W_{y,e}$, or $W_{y,h}$) else $W_{y,j}$ is already lit so we choose the other

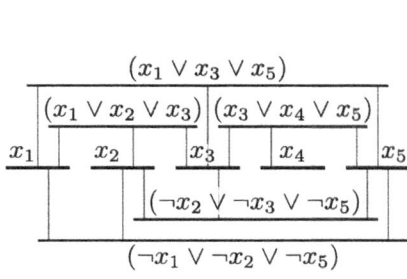

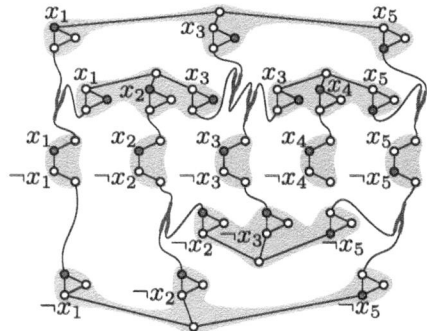

(a) Planar 3-SAT instance   (b) Resulting ALL-ONES instance

Fig. 3. Example of an instance created by the NP-hardness reduction. In (b) note how connections from variable gadgets (middle, green background) to clause gadgets (above and below, purple background) can follow planar embedding of (a) while splitting from a single connection to multiple using the split gadgets (simplified, shown as small red forks). Dark vertices show a solution devised from SAT solution that assigns $x_1, x_2, x_3, x_4, \neg x_5$ to be true. (Color figure online)

vertex ($W_{y,c}$, $W_{y,f}$, or $W_{y,i}$). Since the assignment satisfies every clause, at least one literal in every clause is true. Hence, the top vertex of at least one triangle is not required to be chosen so vertex $W_{y,j}$ in every clause gadget is lit.

Conversely suppose we have a solution X to our instance of AO. We construct a set R from the vertices $V_{x,a}$ for each $x \in V$, $S_{\ell,t,b}$, $S_{\ell,t,e}$, $S_{\ell,t,i}$ for each $\ell \in L$ and $t \in [r_\ell - 1]$, and $W_{y,c}$, $W_{y,f}$, and $W_{y,i}$ for each $y \in C$ (see Fig. 2). As our budget $k = |R|$ and as the closed neighborhoods of vertices in R are disjoint, by Lemma 1 we have that X contains exactly one vertex from the neighborhood of each vertex of R. For each variable gadget, $V_{x,c} \notin X$ since this would require another vertex from the gadget to light $V_{x,b}$ which would exceed the budget. So we have two possibilities: $V_{x,a} \in X$ in which case we set x to true or $V_{x,b} \in X$ in which case we set it to false. For each clause gadget exactly one vertex is chosen from each triangle. One or three of these must be adjacent to $W_{y,j}$ and hence at least one of $W_{y,a}$, $W_{y,d}$, and $W_{y,g}$ is not in X. We now have two cases:

Case 1: The triangle represents a literal ℓ that only appears once across all clauses. In this case, this in-vertex is connected to an out-vertex of the variable gadget that represents ℓ: $V_{x,c}$ if $\ell = x$ and $V_{x,d}$ if $\ell = \neg x$. Since the in-vertex is not in X its connected out-vertex must be a neighbor of a solution vertex in the variable gadget, i.e., (for $\ell = x$) $V_{x,a} \in X$ and hence x is set to true in our assignment for a positive literal and vice versa for $\ell = \neg x$.

Case 2: The triangle represents a literal ℓ that appears more than once. Here, this in-vertex is connected to an out-vertex of a split gadget. This means that the $S_{\ell,t,e}$ (resp. $S_{\ell,t,i}$) vertex of the split gadget is chosen and therefore $S_{\ell,t,b}$ must also be chosen in order to light $S_{\ell,t,c}$ (resp. $S_{\ell,t,g}$). We get that the in-vertex of $S_{\ell,t}$ is not chosen so we can repeat this argument with any further split gadgets until we reach a variable gadget and can apply Case 1.

Since we have at least one literal in every clause that is set to true in our assignment, it satisfies the instance of the Planar 3-SAT.

The reduction is from Planar 3-SAT and the gadgets are planar subgraphs. If we can place the gadgets according to the planar embedding of the Planar 3-SAT instance we get a planar AO instance. Finally, all three gadgets have maximum degree three but there are also vertices of degree 2 so the resulting graph is subcubic.   □

This reduction also works for the Dominating Set problem (and Exact Dominating Set problem) which was previously shown to be NP-hard with these restrictions in an unpublished work by Garey and Johnson, see [16, p. 190].

9   Future Work

We have shown that All-Colors is W[1]-hard with respect to solution size, yet it is fixed-parameter tractable in relation to various structural parameters like treewidth, clique-width, and the combination of solution size with maximum degree. The natural consideration in exploring the parameterized complexity of All-Colors would be to investigate the following questions:

- Does All-Colors have a polynomial kernel with respect to any structural parameter, e.g., vertex cover, feedback vertex set?
- Are there more graph classes where All-Colors admits an FPT algorithm when parameterized by solution size?

Acknowledgements. We would like to thank anonymous referees for their helpful comments. All author acknowledges Engineering and Physical Sciences Research Council for supporting this research via grant EP/V044621/1.

References

1. Amin, A.T., Slater, P.J.: Neighborhood domination with parity restrictions in graphs. Congressus Numerantium, 19–19 (1992)
2. Barua, R., Ramakrishnan, S.: Sigma-game, sigma+-game, and two-dimensional additive cellular automata. Theor. Comput. Sci., 154(2), 349–366 (1996). https://doi.org/10.1016/0304-3975(95)00091-7
3. Bekos, M.A., Lozzo, G.D., Hliněný, P., Kaufmann, M.: Graph product structure for h-framed graphs. In ISAAC. LIPIcs, vol. 248, pp. 23:1–23:15. Schloss Dagstuhl - Leibniz-Zentrum für Informatik (2022)
4. Broersma, H., Li, X.: On the complexity of dominating set problems related to the minimum all-ones problem. Theor. Comput. Sci. **385**(1–3), 60–70 (2007). https://doi.org/10.1016/J.TCS.2007.05.027

5. Caro, Y., Klostermeyer, W., Goldwasser, J.: Odd and residue domination numbers of a graph. Discussiones Mathematicae Graph Theory **21**(1), 119–136 (2001)

6. Caro, Y., Klostermeyer, W.F.: The odd domination number of a graph. J. Comb. Math. Comb. Comput. **44**, 65–84 (2003)

7. Chen, W., Li, X., Wang, C., Zhang, X.: Linear time algorithms to the minimum all-ones problem for unicyclic and bicyclic graphs. Electron. Notes Discret. Math. **17**, 93–98 (2004). https://doi.org/10.1016/J.ENDM.2004.03.018

8. Chen, W., Li, X., Wang, C., Zhang, X.: The minimum all-ones problem for trees. SIAM J. Comput. **33**(2), 379–392 (2004). https://doi.org/10.1137/S0097539703421620

9. Courcelle, B.: The monadic second-order logic of graphs. i. recognizable sets of finite graphs. Inf. Comput. **85**(1), 12–75 (1990). https://doi.org/10.1016/0890-5401(90)90043-H

10. Courcelle, B., Olariu, S.: Upper bounds to the clique width of graphs. **101**(1), 77–114 (2000). https://doi.org/10.1016/S0166-218X(99)00184-5

11. Courcelle, B., Olariu, S.: Upper bounds to the clique width of graphs. Discret. Appl. Math. **101**(1), 77–114 (2000). https://doi.org/10.1016/S0166-218X(99)00184-5

12. Cygan, M., et al.: Parameterized Algorithms. Springer, Cham (2015)

13. Eriksson, H., Eriksson, K., Sjöstrand, J.: Note on the lamp lighting problem. Adv. Appl. Math. **27**(2–3), 357–366 (2001)

14. Fleischer, R., Yu, J.: A survey of the game "lights out!". Space-Efficient Data Structures, Streams, and Algorithms: Papers in Honor of J. Ian Munro on the Occasion of His 66th Birthday, pp. 176–198 (2013)

15. Galvin, F.: Solution to problem 88-8. Math. Intelligencer **11**(2), 31–32 (1989)

16. Garey, M.R., Johnson, D.S.: Computers and Intractability: A Guide to the Theory of NP-Completeness. W. H. Freeman (1979)

17. Goldwasser, J., Klostermeyer, W.: Maximization versions of "lights out" games in grids and graphs. Congressus Numerantium, 99–112 (1997)

18. Hliněný, P., Jedelský, J.: Twin-width of planar graphs is at most 8. *CoRR*, abs/2210.08620 (2022)

19. Korhonen, T.: A single-exponential time 2-approximation algorithm for treewidth. In: 62nd IEEE Annual Symposium on Foundations of Computer Science, FOCS 2021, Denver, CO, USA, February 7-10, 2022, pp. 184–192. IEEE (2021). https://doi.org/10.1109/FOCS52979.2021.00026

20. Li, X., Wang, C., Zhang, X.: The general sigma all-ones problem for trees. Discret. Appl. Math. **156**(10), 1790–1801 (2008). https://doi.org/10.1016/J.DAM.2007.08.042

21. Lossers, O.P.: Solution to problem 10197 [1992, 162]-an all-ones problem. Amer. Math. Monthly **100**(8), 806–807 (1993)

22. Lu, Y., Li, Y.: The minimum all-ones problem for graphs with small treewidth. In: Dress, A., Xu, Y., Zhu, B. (eds.) COCOA 2007. LNCS, vol. 4616, pp. 335–342. Springer, Heidelberg (2007). https://doi.org/10.1007/978-3-540-73556-4_35

23. Manuel, P., Rajasingh, I., Rajan, B., Prabha, R.: The all-ones problem for binomial trees, butterfly and benes networks. Int. J. Math. Soft Comput. **2**(2), 1–6 (2012)

24. Marx, D.: Can you beat treewidth? **6**(5), 85–112. https://doi.org/10.4086/toc.2010.v006a005

25. Peled, U.: Problem 10197. Amer. Math. Monthly **99**(2), 162 (1992)

26. Sutner, K.: Problem 88-8. Math. Intelligencer **10** (1988)

27. Sutner, K.: Additive automata on graphs. Complex Syst. **2**(6), 649–661 (1988)

28. Sutner, K.: On σ-automata. Complex Syst. **2**(1), 1–28 (1988)

29. Sutner, K.: Linear cellular automata and the garden-of-eden. Math. Intelligencer **11**(2), 49–53 (1989). https://doi.org/10.1007/BF03023823
30. Sutner, K.: The σ-game and cellular automata. Am. Math. Mon. **97**(1), 24–34 (1990)
31. Sutner, K.: sigma-automata and chebyshev-polynomials. Theor. Comput. Sci. **230**(1–2), 49–73 (2000). https://doi.org/10.1016/S0304-3975(97)00242-9
32. Zhang, X., Wang, C.: Solutions to all-colors problem on graph cellular automata. Complex, **2019**, 3164692:1–3164692:11 (2019). https://doi.org/10.1155/2019/3164692

On the Discrete and Semi-continuous Versions of the Two-Watchtower Problem in the Plane

Leonidas Palios$^{(\boxtimes)}$ [ID]

Department of Computer Science and Engineering, University of Ioannina, Ioannina, Greece
palios@cse.uoi.gr

Abstract. We consider the two-watchtower problem in the plane in which, given a 1.5-dimensional terrain T, we want to compute the minimum common height of two watchtowers that are erected at points of T and together guard the entire T. The problem is encountered in three versions, the *discrete*, the *semi-continuous*, and the *continuous* [1,3,5].

In this paper, we focus on the discrete and semi-continuous versions of the problem. We prove important geometric properties of the problem which enable us to describe a simpler and faster decision process for each of these two versions; this leads to a faster $O(n^2 \log^2 n)$-time algorithm for the discrete two-watchtower problem and an algorithm matching the complexity of the algorithm in [1] for the semi-continuous version. Additionally, for the discrete version, we present a simple algorithm that does not use parametric search; it runs in $O(n^3 \log n)$ time, which improves upon the currently best algorithm for this problem.

Keywords: Visibility · Plane · 1.5-Dimensional terrain · Watchtower · Guard · Discrete version · Semi-continuous version

1 Introduction

Guarding a terrain by watchtowers is a special case of the well-known Art Gallery Problems, which have been extensively studied in the literature and find many applications, e.g., in surveillance, navigation, computer vision, GIS, graphics, and modeling [18].

A (polyhedral) *terrain* in $\mathbb{R}^d$ is the graph of a continuous, piecewise-linear $(d-1)$-variate function. In particular, in $\mathbb{R}^2$, a *terrain* T (called a 1.5-dimensional terrain) is a polygonal chain that is x-monotone (i.e., it is intersected by any vertical line in at most one point). A *watchtower* is a vertical line segment whose bottom endpoint (base) lies on the terrain T. A watchtower *sees* a point t of T (or the point is visible, seen, or guarded by the watchtower) if the top endpoint of the watchtower sees t, that is, the line segment connecting the top endpoint of the watchtower to t does not contain a point below T. Two watchtowers on the

I. Finocchi and L. Georgiadis (Eds.): CIAC 2025, LNCS 15679, pp. 240–257, 2025.
https://doi.org/10.1007/978-3-031-92932-8_16

terrain T are said to *guard* T if *each* point of T is visible from the top endpoint of at least one of the towers.

In this paper, we study the two-watchtower problem for terrains in $\mathbb{R}^2$, which is defined as follows: Given a 1.5-dimensional terrain T, find the smallest height $h \geq 0$ for which there exist two points $t_1, t_2 \in T$ such that the watchtowers of height h built at t_1 and t_2 guard T. The problem is encountered in three versions (see Figure 1 in [1]):

- in the *discrete* version, the bases t_1, t_2 are vertices of T;
- in the *semi-continuous* version, one of t_1, t_2 is a vertex of T and the other can be located anywhere on T;
- in the *continuous* version, t_1, t_2 can be located anywhere on T.

The problem admits many interesting variants, such as asking for the minimum height of just one or any number of watchtowers to guard T, or requiring that the guards are on vertices of T or points on T (or points in any given subset of T) as well as extending it to higher dimensional space.

Related Results. The problem of guarding an n-edge terrain T in $\mathbb{R}^2$ by two watchtowers has received considerable attention. Bespamyatnikh et al. [5] showed that for the discrete case of deciding whether there exist two watchtowers of a given height h that guard the entire terrain can be done in $O(n^3)$ time, which led to an $O(n^3 \log^2 n)$-time algorithm for the optimization problem using parametric search; they also presented an $O(n^4)$-time algorithm that avoids parametric search. Ben-Moshe et al. [3] also addressed the discrete version of the problem in $\mathbb{R}^2$ and gave an $O(n^{3/2}M(n)^{1/2} \log^2 n) \simeq O(n^{2.6858} \log^2 n)$-time algorithm based on parametric search, where $M(n)$ is the time to multiply two $n \times n$ matrices. The continuous version for $\mathbb{R}^2$ is solved in [5] by an algorithm that requires $O(n^4 \log^2 n)$ time, using parametric search as well. Agarwal et al. [1] presented improved parametric search algorithms running in $O(n^2 \log^4 n)$ time for the discrete and semi-continuous two-watchtower problem and in $O(n^3\alpha(n) \log^3 n)$ time for the continuous version. They also showed that the problem of guarding a set of m points on the terrain by two watchtowers can be solved in $O(mn \log^4 n)$ time.

Given a terrain T and a positive integer k, Bahoo et al. [2] proposed an algorithm to place a *single* watchtower w with minimum height such that the line segment from w to any point on T crosses T at most k times; their algorithm runs in $O((n^2 + \nu) \log n)$ time where ν denotes the number of vertices on the boundary of the k-kernel of T (for $k = 2$, it holds that $\nu = O(n^2)$, whereas for arbitrary k, $\nu = O(n^4)$); they also presented an $O(n^3)$-time algorithm for the discrete version of the problem, i.e., the watchtower is placed at a vertex of T.

Placing watchtowers has been considered for terrains in $\mathbb{R}^3$ as well. The case of a single watchtower guarding such a terrain has been shown by Sharir [16] to be solvable in $O(n \log^2 n)$ time; an $O(n \log n)$-time algorithm for this problem was later given by Zhu [19]. Tripathi et al. [17] developed an algorithm to decide whether, for a given height h and an integer $k > 2$, it is feasible to guard the entire terrain using k watchtowers on terrain vertices; the algorithm runs in $O(n^{k+3}k^2\alpha^2(n)\log n)$ time, and is used to yield an $O(n^{k+3}k^2\alpha^2(n)\log n + n^7\alpha^3(n)\log n)$-time algorithm for the discrete version of the k-watchtower problem on a 2.5-dimensional terrain. A review on known algorithms for the placement of towers for guarding terrains in $\mathbb{R}^2$ and $\mathbb{R}^3$ was given by Gewali and Dahal [10], who also presented an $O(n^2)$-time algorithm for computing the location of a single watchtower of a given height to maximize the visible part of the given terrain in $\mathbb{R}^2$. Agarwal et al. [1] studied the problem of guarding a terrain by two watchtowers in $\mathbb{R}^3$, and presented an $O(n^{11/3}polylog(n))$-time algorithm for the discrete two-watchtower problem using parametric search. Seth et al. [15] introduced the acrophobic guard watchtower problem for a polyhedral terrain, in which a square axis-aligned platform is placed on the top of a minimum-height tower such that every point of the terrain is visible from the platform and they showed that the problem can be solved in $O(n)$ time in $\mathbb{R}^2$ and in $O(n\log n)$ time in $\mathbb{R}^3$.

Attention has also been given to the problem of guarding a two-dimensional terrain with the *minimum* number of guards placed on the terrain. The problem is known to be NP-hard both in $\mathbb{R}^2$ [14] and in $\mathbb{R}^3$. In $\mathbb{R}^2$, Ben-Moshe, Katz, and Mitchell [4], and, independently, Clarkson and Varadarajan [6], proposed constant-factor approximation algorithms for the problem. Another 4-approximation algorithm has been given by King [13] and Friedrichs et al. [9] described a polynomial time approximation scheme and formulated the problem as an Integer Linear Program. Khodakarami et al. [12] introduced the "terrain guard range", demonstrated that the problem is fixed-parameter tractable with respect to this parameter, and presented an exact $O(4^k k^2 n)$-time algorithm for the problem where k is the guard range of the given terrain.

Our Contribution. In this paper we focus on the discrete and semi-continuous versions of the two-watchtower problem. We prove important geometric properties of the problem which enable us to describe simple and faster decision procedures for these two versions which, for an n-edge terrain, run in $O(n^2)$ time and require $O(n)$ space. The decision procedure for the discrete version can be used together with the algorithm of [1] to yield an improved $O(n^2 \log^2 n)$-time algorithm using parametric search, making progress towards the resolution of the open problem posed in [1] asking for an $O(n^2 \log n)$-time algorithm. For the semi-continuous version, the resulting algorithm matches the complexity of the one in [1]. Additionally, for the discrete version, we present a simple algorithm that avoids parametric search; it requires $O(n^3 \log n)$ time and $O(n)$ space, which improves upon the currently best algorithm of Bespamyatnikh et al. [5] for this problem. Our approach is of independent interest and is useful in obtaining efficient algorithms not only for the semi-continuous version without using parametric search, but also for extensions of the problem, e.g., for the problem for $k > 2$ watchtowers, and for guarding a given set of points on the terrain.

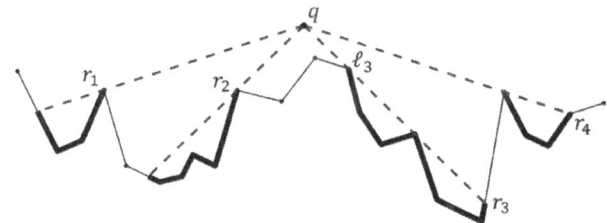

Fig. 1. The q-non-visible chains of the terrain T are shown in thick line.

2 Theoretical Framework

For any 2-dimensional point p, we use $p.x$ and $p.y$ to denote its x- and y-coordinate respectively.

Terrain. In the following, we will be considering 1.5-dimensional terrains, to which we will refer as terrains for simplicity. The terrain consists of edges joined together at vertices. We assume that each edge or portion of an edge of the terrain is *open*, i.e., it does not contain its endpoints.

Let T be a given terrain. The definition of the terrain implies that none of its edges is vertical; therefore the notion of a point being above/below the line supporting any edge of the terrain is well defined. Moreover, the x-monotonicity of the terrain T allows us to easily define the points above/below T: a point p is *above* (*below*, resp.) the terrain T if p does not lie on T and the downward-pointing vertical ray emanating from p intersects (does not intersect, resp.) the terrain T.

A *chain* of a terrain T is a *connected* subset of T; we note that the endpoints of a chain need not be vertices of T. A chain is *closed* if it contains its endpoints; a closed chain of T from point a to the left to point b to the right will be denoted by $T[a..b]$.

As in [1], for a point t of a terrain T and a real number $h \geq 0$, we denote by $t(h)$ the point on the vertical line through t at height h with respect to T.

Visibility. For any two points p, q on or above the terrain T, point q *sees* point p (or point p is *visible* from q) if the line segment pq connecting them contains no point below the terrain T. This definition allows for "grazing" visibility, that is, for any set A of points of T collinear with the visibility point q, the point in A closest to q does not occlude the remaining points in A (in Fig. 1, point r_3 is visible from q).

Crucial in our decision procedures is the notion of an "upper tangent": the *upper tangent* from a point q on or above T to the (closed) chain $T[a, b]$ of T where $q.x = a.x < b.x$ is a ray emanating from q, pointing towards b, that goes

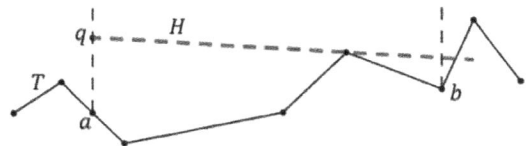

Fig. 2. The upper tangent H from point q to the chain $T[a, b]$.

through at least one point of $T[a, b]$ other than a, and has no point of $T[a, b]$ above it (Fig. 2); note that q may coincide with point a of T. In a symmetric fashion, we define the upper tangent from point q to the chain $T[a, b]$ where $a.x < b.x = q.x$.

Let us consider a point q on or above T. The fact that a point of T is visible from point q relies on the following observation.

Observation 1. *Let T be a terrain, q a point on or above T, and q_T the vertical projection of q on T. Then a point $s \neq q_T$ of T is visible from q if and only if q is on or above the upper tangent from s to the closed chain of T with endpoints s and q_T.*

In general, not the entire T is visible from point q. Clearly, q entirely sees the terrain edge that contains its vertical projection q_T on T or has q_T as an endpoint. For the remaining edges, the definition of the terrain implies the following:

Observation 2. *For each (open) terrain edge e to the left (right resp.) of the vertical projection q_T of a point q on or above the terrain, if q sees a point $t \in e$, it sees all the points of e to the left (right resp.) of t.*

Therefore, each edge e of T may be entirely visible from q, entirely non-visible, or may contain a point t such that t and e's part to one side of t is visible whereas the remaining part is non-visible (the third case occurs only for edges of T between the two watchtowers; see Lemma 3 in [5]). The non-visible portions of the edges of T can be grouped to form non-visible chains. In fact, each *maximal* chain of T such that each of its open segments is non-visible from q and the union of their closure is connected is called a *q-non-visible chain*[1] of T (in Fig. 1 there are four q-non-visible chains, one of which is the chain $T[\ell_3..r_3]$); although no point of any segment of a q-non-visible chain is visible from q, both its left and right endpoints are visible, as may be some of its intermediate edge endpoints (see chain $T[\ell_3..r_3]$ in Fig. 1).

It is not difficult to see that the lines of sight from q that define q-non-visible chains are angularly ordered around q and that the right endpoint of each of the q-non-visible chains to the left of q and the left endpoint of each of the q-non-visible chains to the right of q is a vertex of the terrain T (Fig. 1). Additionally, the following observation holds:

---

[1] A non-visible chain is referred to as an invisibility region in [1].

Observation 3. *Let $T[\ell_{k+1}..r_{k+1}], \ldots$ be the q-non-visible chains to the right of q ordered by increasing x-coordinate of their right endpoints along T. Then, for each $i \geq k+1$ (see Fig. 1):*

(a) the right endpoint r_i is below the line through q, ℓ_j, r_j for each $j > i$;
(b) the right endpoint r_{i+1} is below the upper tangent from r_i to the chain $T[r_i..r_{i+1}]$ of T.

Upper Envelope of a Set of Non-vertical Lines. The *upper envelope* of a set L of non-vertical lines is the 0th level of the arrangement of the lines in L, i.e., the polygonal line of points on the lines of L that have no line strictly above them ([7], Chapter 8); clearly, if L contains a single line, then the upper envelope coincides with the line in L. For completeness, we mention that for the upper envelope of a set of n non-vertical lines, it holds that:

○ it covers the entire x-axis (i.e., for every real number α, the envelope contains a point with x-coordinate equal to α);
○ it is convex and hence is x-monotone and the slopes of its edges from left to right form an increasing sequence;
○ it can be computed in $O(n \log n)$ time whereas if the lines are sorted by slope, it can be computed in $O(n)$ time.

The linear-time computation of the upper envelope for lines sorted by slope results from the fact that the standard point-line duality transform [7] maps the upper envelope of lines in increasing slope to the lower hull of points in increasing x-coordinate ([11], Chapter 25).

2.1   Building the Second Tower

Suppose that a tower of height h has been placed at vertex u of the terrain T. If the top point $u(h)$ of the tower does not see the entire T, a second tower needs to be built at an appropriate location p and to be of the appropriate *minimum* height h' so that each $u(h)$-non-visible chain of T is seen by the top point $p(h')$ of the second tower. As mentioned in [1] (Section 2.2, second paragraph), this can be ensured as follows:

Observation 4. *[1] Let T be a terrain. Consider a tower at $u \in T$ of height $h \geq 0$ and let P denote the set of all the endpoints of the portions of the edges of T that are not visible from $u(h)$. Then, $u(h)$ and another tower $v(h')$ (with a possibly different height h') guard T if and only if $v(h')$ sees all the points in P.*

In this section, for a watchtower of height h at a terrain vertex u, we establish properties for the height h' of the second watchtower erected at a point p of the terrain T so that the two towers guard the entire T. We note that, in the optimal solution, $p \neq u$ since the maximum of the tower heights h, h' can be reduced by placing the lower tower at a terrain vertex on the visibility line that determines the height of the higher tower.

Let us now consider the $u(h)$-non-visible chains of T (Fig. 1). Since u is visible from $u(h)$, u does not belong to any such chain. Next, we prove conditions for the top point $p(h')$ of the second tower to watch each of the $u(h)$-non-visible chains of T to the left of p, containing p, and to the right of p.

Lemma 1. *Let T be the given terrain and consider two towers at $u, p \in T$ (with u to the left of p along T) of heights $h, h' \geq 0$ respectively. If σ is any (open) $u(h)$-non-visible chain to the* left *of p along T, then σ is entirely visible from $p(h')$ if and only if*

(a) for each edge e of σ, $p(h')$ lies on or above the line supporting e and
(b) $p(h')$ lies on or above the upper tangent from the right *endpoint r of σ to the chain $T[r..p]$ of T.*

Proof. By Observation 4, the entire chain σ is visible from the top point $p(h')$ of the second tower if and only if each endpoint a of each edge of σ is visible from $p(h')$ which, by Observation 1, is equivalent to the fact that for each such a, $p(h')$ is on or above the upper tangent from a to the (non-empty) closed chain $T[a..p]$. Therefore, it suffices to prove that for each endpoint a of each edge of σ, $p(h')$ is on or above the upper tangent from a to the chain $T[v..p]$ if and only if conditions (a) and (b) of the statement of the lemma hold.

($\Longrightarrow$) Suppose that for each endpoint a of each edge of the chain σ, $p(h')$ is on or above the upper tangent from a to the chain $T[a..p]$. Then, condition (b) of the statement of the lemma clearly holds. Consider now any edge e of the chain σ and let b be the left endpoint of e. The fact that e is an edge of the chain $T[b..p]$ implies that the slope of the line L supporting e is no greater than the slope of the upper tangent H_b from b to the chain $T[b..p]$; hence, to the right of b, H_b is no lower than L. Then, since p is to the right of b along T and since $p(h')$ is on or above the tangent H_b from b to the chain $T[b..p]$, $p(h')$ is on or above the line L supporting e. Therefore, both conditions (a) and (b) of the statement of the lemma hold.

($\Longleftarrow$) Suppose that conditions (a) and (b) of the statement of the lemma hold. We will show that for each edge endpoint a of the chain σ, $p(h')$ is on or above the upper tangent from a to the chain $T[a..p]$. This is clearly true for the right endpoint r of σ due to condition (b).

Now suppose, for contradiction, that there is an edge endpoint z of σ other than r such that $p(h')$ is below the upper tangent H_z from z to the chain $T[z..p]$. The edge with left endpoint z belongs to σ and thus is not visible from $u(h)$. Let c be the *leftmost* point of tangency of the tangent H_z from z to the chain $T[z..p]$ other than z. If c belongs to the chain σ, then c is an edge endpoint of σ; let e be the edge of σ with right endpoint c. The definition of c implies that e lies below H_z; this and the fact that c belongs to H_z in turn imply that the slope of H_z is smaller than the slope of the line L supporting e, and thus, to the right of c, the line L is above H_z. Since $p(h')$ is on or above the lines supporting the edges of σ, it is on or above the line L supporting e, which, because $p(h')$ is to the right of c, implies that it is above the tangent H_z, a contradiction to $p(h')$ being below H_z.

Next, suppose that c does not belong to σ; then, since $z \in \sigma$, c is to the right of r. Moreover, r lies below the upper tangent H_z and c is a terrain vertex other than r belonging of the chain $T[r..p]$. Then, the slope of the line through r and c is greater than the slope of the tangent H_z and is at most equal to the slope of the upper tangent from r to the chain $T[r..p]$. But then since p is not to the left of c, and because of condition (b) in the statement of the lemma, $p(h')$ is above the upper tangent H_z, a contradiction. Therefore, each endpoint of each edge of σ is visible from $p(h')$.   □

On the other hand, for the $u(h)$-non-visible chains of T to the right of p or containing p, only the first condition is needed, as we show in the next lemma.

Lemma 2. *Let T be the given terrain and consider two towers at $u, p \in T$ (with u to the left of p along T) of heights $h, h' \geq 0$ respectively. If σ is any $u(h)$-non-visible chain that contains p or is to the right of p along T, then σ is entirely visible from $p(h')$ if and only if for each edge e of σ, $p(h')$ lies on or above the line supporting e.*

Now we are ready to prove the main result on which our algorithms rely.

Theorem 1. *Let T be the given terrain watched by two towers at $u, p \in T$ (with u to the left of p along T) of heights $h, h' \geq 0$ respectively, and let $r_1, \ldots, r_k, \ldots, r_\ell$, where $0 \leq k \leq \ell$, be the ordering (from left to right along T) of the right endpoints of the $u(h)$-non-visible chains to the left of p among which $r_1, \ldots, r_k$ are to the left of u. Then, the terrain T is entirely visible from the two towers with top points $u(h)$ and $p(h')$ if and only if $p(h')$ is on or above the (non-vertical) lines in $L_e \cup L_r$ where*

$$L_e = \{ \text{ line supporting } e \mid \forall \text{ (partially/entirely) } u(h)\text{-non-visible edge } e \text{ of } T \}$$

and

$$L_r = \{ \text{ upper tangent from } r_i \text{ to the chain } T[r_i..r_{i+1}] \mid i = k, \ldots, \ell - 1 \}$$
$$\cup \{ \text{ upper tangent from } r_\ell \text{ to the chain } T[r_\ell..p] \}.$$

Proof. From Lemmas 1 and 2, we conclude that the terrain T is entirely visible from the two towers with top points $u(h)$ and $p(h')$ if and only if $p(h')$ is on or above the (non-vertical) lines in $L_e \cup L'$ where

$$L' = \{ \text{upper tangent from } r_i \text{ to the chain } T[r_i..p] \mid i = 1, \ldots, \ell \}.$$

We recall that the points $r_1, r_2, \ldots, r_\ell, p$ are encountered in that order from left to right along the terrain T and thus each of the chains $T[r_i..r_{i+1}]$ (and a fortiori the chains $T[r_i..p]$) is non-empty and thus the upper tangents from r_i to each of them is well defined.

First, we prove that for each $j = 1, 2, \ldots, \ell$, the terrain T is entirely visible from the two towers with top points $u(h)$ and $p(h')$ if and only if $p(h')$ is on or above the lines in $L_e \cup L_j$ where

$$L_j = \{\text{upper tangent from } r_i \text{ to the chain } T[r_i..r_{i+1}] \mid i = 1, \ldots, j-1 \}$$
$$\cup \ \{\text{upper tangent from } r_i \text{ to the chain } T[r_i..p] \mid i = j, \ldots, \ell \}.$$

We use induction on j. For the base case, $j = 1$ and

$$L_1 = \{\text{upper tangent from } r_i \text{ to the chain } T[r_i..p] \mid i = 1, \ldots, \ell \};$$

the observation holds as the result of Lemmas 1 and 2. For the inductive hypothesis, suppose that the terrain T is entirely visible from the two towers if and only if $p(h')$ is on or above the lines in $L_e \cup L_{j_0}$ for some $1 \leq j_0 < \ell$, and for the inductive step, we will show that T is entirely visible from the two towers if and only if $p(h')$ is on or above the lines in $L_e \cup L_{j_0+1}$; note that

$$L_{j_0+1} = \left(L_{j_0} \cup \{\text{upper tangent from } r_{j_0} \text{ to the chain } T[r_{j_0}..r_{j_0+1}] \} \right)$$
$$\setminus \ \{\text{upper tangent from } r_{j_0} \text{ to the chain } T[r_{j_0}..p] \}.$$

Consider the upper tangent H_{j_0} from r_{j_0} to the chain $T[r_{j_0}..p]$ and let b be the *leftmost* point of tangency of H_{j_0} with the chain $T[r_{j_0}..p]$ other than r_{j_0}. We distinguish two cases.

- *Point b belongs to the chain $T[r_{j_0}..r_{j_0+1}]$:* Then, the upper tangents from r_{j_0} to the chain $T[r_{j_0}..r_{j_0+1}]$ and to the chain $T[r_{j_0}..p]$ coincide; thus, $L_e \cup L_{j_0} = L_e \cup L_{j_0+1}$.
- *Point b does not belong to the chain $T[r_{j_0}..r_{j_0+1}]$:* Then, the slope of the upper tangent from r_{j_0} to the chain $T[r_{j_0}..r_{j_0+1}]$ is smaller than the slope of the upper tangent H_{j_0} from r_{j_0} to the chain $T[r_{j_0}..p]$, and b is a vertex of the terrain T to the right of r_{j_0+1} and up to p (inclusive). Moreover, r_{j_0+1} is below H_{j_0}; this implies that *(i)* the slope of H_{j_0} is smaller than the slope of the line through r_{j_0+1} and b, which in turn is no greater than the slope of the upper tangent H_{j_0+1} from r_{j_0+1} to the chain $T[r_{j_0+1}..p]$ and *(ii)* the points of intersection of the upper tangent from r_{j_0} to the chain $T[r_{j_0}..r_{j_0+1}]$ with H_{j_0+1} and of H_{j_0} with H_{j_0+1} have x-coordinates no greater than the x-coordinate of b. The latter fact, the ordering of slopes of the upper tangents, and the fact that p is not to the left of b along T imply that the point of intersection of the vertical line V through p with H_{j_0+1} is no lower than both the points of intersection of V with the upper tangent from r_{j_0} to the chain $T[r_{j_0}..r_{j_0+1}]$ and with H_{j_0}.

Then, in either case, the point $p(h')$ is on or above the lines in $L_e \cup L_{j_0}$ if and only if $p(h')$ is on or above the lines in $L_e \cup L_{j_0+1}$. This together with the inductive hypothesis implies that T is entirely visible from the two towers if and only if $p(h')$ is on or above the lines in $L_e \cup L_{j_0+1}$, as desired.

The inductive proof is complete and we have proved that the terrain T is entirely visible from the two towers if and only if $p(h')$ is on or above the lines in $L_e \cup L'_r$ where

$L'_r =\{$ upper tangent from r_i to the chain $T[r_i..r_{i+1}] \mid i = 1, \ldots, \ell - 1\}$
$\cup$ $\{$ upper tangent from r_ℓ to the chain $T[r_\ell..p]\}$.

Next, for $i = 1, \ldots, k - 1$, we note that the point r_i corresponds to a $u(h)$-non-visible chain to the left of u and that the chain $T[r_i..r_{i+1}]$ is a subchain of the chain $T[r_i..u]$. Thus, the slope of the upper tangent from r_i to the chain $T[r_i..r_{i+1}]$ is no greater that the slope of the line through r_i and $u(h)$, which in turn is smaller than the slope of the (non-visible) edge e of T having r_i as its right endpoint. Then, if the point $p(h')$ is on or above the lines in L_e supporting the (partially/entirely) non-visible edges from $u(h)$, this implies that $p(h')$ is above the upper tangent from r_i to the chain $T[r_i..r_{i+1}]$. The theorem follows.   □

3   Decision Procedures

Since the top point $p(h')$ in Theorem 1 needs to be on or above the lines in $L_e \cup L_r$, it needs to be on or above the upper envelope of these lines; in fact, the minimum height h' for a watchtower at point $p \in T$ is equal to the vertical distance of p to the upper envelope of the lines in $L_e \cup L_r$.

3.1   Discrete Two-Watchtower Problem

We will apply a more general version of Theorem 1.

Corollary 1. *Suppose that the conditions of the statement of Theorem 1 hold and the points r_i ($i = 1, \ldots, k, \ldots, \ell$) and the sets L_e, L_r are as defined in it. For each $i = k, \ldots, \ell - 1$, let B_i be the set of upper tangents from r_i to each of the chains $T[r_i..z]$ where z is a vertex of the terrain T belonging to the chain $T[r_i..r_{i+1}]$ and let C_i be any subset of B_i. Then, the terrain T is entirely visible from the two towers with top points $u(h)$ and $p(h')$ if and only if $p(h')$ is on or above the (non-vertical) lines in $L_e \cup L_r \cup L_z$ where $L_z = \bigcup_{i=k}^{\ell-1} C_i$.*

Our decision procedure for the discrete version exploits Corollary 1 and, for any vertex u of the given terrain T and a height h, it gives a method to compute the shortest second tower erected at a vertex of T, which, together with a tower of height h at u, guard the entire terrain. The procedure starts by sorting the terrain edges by slope and stores them in an array from which we can compute the ordering of any subset of edges in time linear in the size of the terrain. Then, we compute the upper envelope U of the lines supporting the terrain edges that are (partially/entirely) non-visible from $u(h)$, and of the line supporting the upper tangent from r_k to the chain $T[r_k..u]$ where r_k is the right endpoint of the rightmost $u(h)$-non-visible chains (if any) to the left of u. Next, for each terrain vertex v to the right of u, we perform the following four actions:

▷ we compute the upper tangent H from the right endpoint $prev\_r$ of the rightmost $u(h)$-non-visible chain to the left of v (if any) to the chain $T[prev\_r..t]$ where t is either v or the right endpoint r of a $u(h)$-non-visible chain such that r coincides with v or belongs to the terrain edge having v as its right endpoint;

▷ if the slope of H is greater than the slope computed so far, then we update the slope and check if the envelope U needs updating as well, and if yes, we do so;

▷ we compute the vertical distance from vertex v to the upper envelope U and if it is smaller than the distance computed so far, we update it and set the location of the second tower to v;

▷ if there exists a right endpoint coinciding with v or belonging to the edge having v as its right endpoint, we update the right endpoint $prev\_r$.

The correctness of the decision procedure follows from Corollary 1 and the fact that for each pair of consecutive right endpoints r_i, r_{i+1} between the tower bases u and v, r_{i+1} is below the upper tangent H_i from r_i to the chain $T[r_i..r_{i+1}]$ of T (Observation 3(b)) and thus the upper tangent from r_{i+1} to $T[r_{i+1}..r_{i+2}]$ or $T[r_{i+1}..v]$ needs to have slope greater than that of H_i in order to participate in the upper envelope U.

For the time complexity of the algorithm, we note that updating the upper tangent H takes $O(1)$ time per processed vertex v. Moreover, by taking into account that the clipping is done about tangents of increasing slope and by maintaining a pointer to the vertex of the upper envelope U at which a line of the slope computed so far is tangent to U, we update U in $O(1)$ time plus time proportional to the deleted part of U due to the clipping; since if a line participating to U is deleted, it cannot re-appear in U, the total time for updating U over all vertices v is $O(n)$. Since the vertices to the right of u are processed in increasing x-coordinate, it suffices to use another pointer on a vertex of U which walks along the boundary of U from left to right following vertex v in order to compute the vertical distance from v to U in $O(1)$ time (this pointer is updated in $O(1)$ time if the vertex at which it is located is clipped off from U).

The following theorem summarizes the results of our decision procedure for the discrete version.

Theorem 2. *(a) Given a terrain T in the plane with n edges, an array of T's edges sorted by increasing slope, a vertex u of T, and a height $h \geq 0$, we can compute, in $O(n)$ time and space, the shortest second tower erected at a vertex of T, which, together with $u(h)$, guard the entire terrain.*

(b) Given a terrain T in the plane with n edges and a height $h \geq 0$, we can determine, in $O(n^2)$ time and $O(n)$ space, whether h is smaller than, equal to, or greater than the optimum height h^* for the discrete two-watchtower problem for T.

Then, the use of the parametric search part of the algorithm in [1] for the discrete two-watchtower problem with our decision procedure leads to an $O(n^2 \log^2 n)$-time and $O(n)$-space algorithm improving over the current best algorithm for the discrete two-watchtower problem.

Corollary 2. *The discrete two-watchtower problem for a terrain in the plane with n edges can be solved in $O(n^2 \log^2 n)$ time and $O(n)$ space.*

3.2   Semi-continuous Two-Watchtower Problem

We work as in the discrete version. However, in addition to the computations in that algorithm, we need do the following:

▷ when processing vertex v, we also process the (open) terrain edge e with v as its right endpoint in which case we consider the vertices of the upper envelope U with x-coordinate between the x-coordinates of the endpoints of e and compute the shortest among their vertical distances to e [1,8];
▷ we repeat the above processing from right to left in order to consider solutions in which the tower located at a terrain vertex is to the right;
▷ we return the overall minimum height and the locations of the two watchtowers.

Then, we have:

Theorem 3. *(a) Given a terrain T in the plane with n edges, an array of the edges of T sorted by increasing slope, a vertex u of T, and a height $h \geq 0$, we can compute, in $O(n)$ time and space, the shortest second tower erected at any point of T, that, together with $u(h)$, see the entire terrain.*
(b) Given a terrain T in the plane with n edges and a height $h \geq 0$, we can determine, in $O(n^2)$ time and $O(n)$ space, whether h is smaller than, equal to, or greater than the optimum height h^* for the semi-continuous two-watchtower problem for T.

The use of the parametric search part of the algorithm in [1] for the semi-continuous two-watchtower problem with our decision procedure leads to an $O(n^2 \log^4 n)$-time and $O(n)$-space algorithm matching the complexity of the algorithm in [1].

4   Avoiding the Parametric Search in the Discrete Version

In this section, we present a simple algorithm for the discrete two-watchtower problem which avoids the use of parametric search. The algorithm computes the minimum common height of two watchtowers that are erected at vertices of the given terrain T and guard the entire T; let u, v be the vertices at the tower bases with v being to the right of u.

First, for each vertex w of the terrain T, we compute a set of "critical" heights at which the sets of entirely visible terrain edges change (as in [1]); these critical heights, to which we will refer as "type 1 critical heights", are precisely the heights of the points of intersection of the vertical upward-pointing ray emanating from w and the lines supporting the edges of the shortest path tree from u to the vertices of T (Fig. 3). As a result, we have:

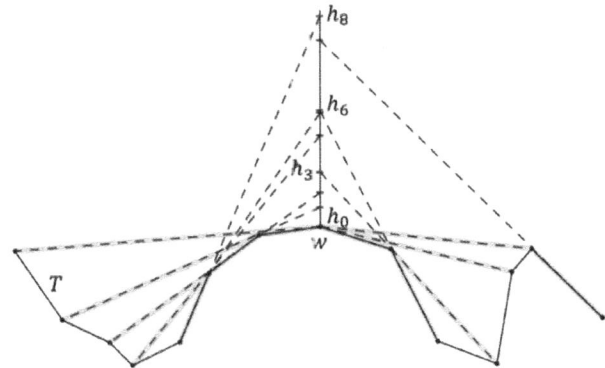

Fig. 3. The shortest path tree of w and w's critical heights $h_0 = 0, h_1, \ldots, h_8$.

Observation 5. *Let u be a vertex of the terrain T and let $h_0(u) = 0, h_1(u), \ldots$ be the ordering of the distinct critical heights of u by increasing value. Then, for any two consecutive type 1 critical heights $h_i(u), h_{i+1}(u)$ of u, the set of entirely visible edges of T from each point $u(h)$, where $h_i(u) \le h < h_{i+1}(u)$, is the same.*

The type 1 critical heights are useful to ensure that the $u(h)$-non-visible edges of T are *entirely* seen by the top point $v(h')$ of the second tower. However, it may be the case that an edge is visible in part by both towers (see Fig. 4); these edges are found between the two towers (Lemma 3 in [5]) and we call them *jointly visible* edges.

Let us study the edges between the two towers at u, v in more detail. For any point r of the closure of a terrain edge e between u, v, let $h(r)$ ($h'(r)$ resp.) be the height of the shortest tower at u (v, resp.) whose top point sees r; then, the height $h(r)$ ($h'(r)$ resp.) is equal to the height of the point of intersection of the upper tangent from r to the chain $T[u..r]$ ($T[r..v]$ resp.) with the vertical line through u (v, resp.). It is crucial to observe that *as r moves from e's left endpoint towards e's right endpoint, the height $h(r)$ decreases whereas the height $h'(r)$ increases.* In fact, for such an edge e with left endpoint a and right endpoint b, one of the following three cases holds:

1. $h(a) \le h'(a)$: Then, $h(b) < h(a) \le h'(a) < h'(b)$ and for the *minimum* height $h_{u,e,v}$ so that the entire edge e is guarded, it holds that $h_{u,e,v} = h(a)$; in this case, the edge e is entirely visible from the top point of the tower at u.
2. $h(b) \ge h'(b)$: This case is symmetric to the previous one; then, $h_{u,e,v} = h'(b)$ and the edge e is entirely visible from the top point of the tower at v.
3. $h(a) > h'(a)$ and $h(b) < h'(b)$: Then, the edge is jointly visible and there is a *unique minimum* height $h_{u,e,v} \in (h(b), h(a)) \cap (h'(a), h'(b))$ such that the top points $u(h_{u,e,v}), v(h_{u,e,v})$ of two towers at u, v of height $h_{u,e,v}$ guard the entire e. For the computation of $h_{u,e,v}$ where vertex v is to the right of u along

T, we exploit the geometry in Fig. 4: the collinearity of $r, s, u(h) = u(h_{u,e,v})$ and the collinearity of $r, t, v(h') = v(h_{u,e,v})$ yield two linear equations on $h_{u,e,v}, r.x, r.y$, whereas a third equation is obtained from the fact that r lies on the edge e. The three equations yield a quadratic equation on $h_{u,e,v}$ whose largest root is the sought value.

Since $h_{u,e,v}$ is the minimum such height, it holds that:

Observation 6. *For any two vertices u, v and edge e of the terrain T as in Fig. 4, let $h_{u,e,v}$ be as defined above. Then, the edge e is entirely visible from the top points $u(h), v(h)$ of the two towers if and only if $h \geq h_{u,e,v}$.*

The observation implies that for the minimum common height $h^*_{u,v}$ for towers at u and v such that their top points $u(h^*_{u,v}), v(h^*_{u,v})$ guard the entire terrain T, it must hold that $h^*_{u,v} \geq h_{u,e,v}$ for each edge e between u, v. The heights $h_{u,e,v}$ introduce an additional set of critical heights to which we will refer as "type 2 critical heights." Then, for the sought height $h^*_{u,v}$, it holds:

Lemma 3. *Let u, v be two vertices of the given terrain T as in Fig. 4. Let $H(u)$ be the set of type 1 and type 2 critical heights of u, and $h_{j_u}(v)$ be the minimum element of $H(u)$ such that $h' \leq h_{j_u}(v)$ where h' is the minimum height of a tower at v for which the top points $u(h_{j_u}(v))$ and $v(h')$ guard the entire T; similarly, let $H(v)$ be the set of type 1 and type 2 critical heights of v, and $h_{j_v}(u)$ be the corresponding minimum element of $H(v)$. Then, the minimum common height $h^*_{u,v}$ for towers at u and v to guard the entire terrain T is the minimum of $h_{j_u}(v), h_{j_v}(u)$.*

Proof. First, we show that the minimum common height $h^*_{u,v}$ is equal to a (type 1 or type 2) critical height of u or v. Suppose for contradiction that $h^*_{u,v}$ differs from all critical heights in $H(u) \cup H(v)$ and let $h_j, h'_{j'}$ be the *greatest* critical height in $H(u), H(v)$ respectively that is less than $h^*_{u,v}$. Then, in light of Observations 5 and 6, the set of (jointly or not) entirely visible edges of T from the top points $u(h^*_{u,v})$ and $v(h^*_{u,v})$ is the same as the set of (jointly or not) entirely visible edges of T from the top points $u(h_{u,v})$ and $v(h_{u,v})$ where $h_{u,v} = \max\{h_j, h'_{j'}\}$; since towers at u and v of height $h^*_{u,v}$ guard the entire terrain T, then so do towers at u and v of height $h_{u,v} < h^*_{u,v}$, in contradiction to the minimality of $h^*_{u,v}$. Therefore, the minimum common height $h^*_{u,v}$ is equal to a (type 1 or type 2) critical height of u or v.

The lemma follows from the definition of $h_{j_u}(v)$ and $h_{j_v}(u)$: it suffices to take into account Observations 5 and 6 in light of the fact that $h' \leq h_{j_u}(v)$ and $h' \leq h_{j_v}(u)$, and to note that the minimality of $h_{j_u}(v)$ and $h_{j_v}(u)$ implies that towers of height equal to critical heights less than $h_{j_u}(v)$ and $h_{j_v}(u)$ do not guard the entire T. □

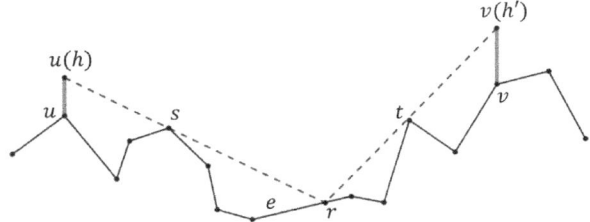

Fig. 4. Jointly guarding the edge e.

4.1   The Algorithm

Based on the preceding discussion, the algorithm for the discrete version of the two-watchtower problem proceeds as follows: We start by sorting the edges of the given terrain T by non-decreasing slope and by storing them in an array so that the ordering of any subset of edges is computed in time linear in the size of T. Next, for each vertex u of T:

1. we compute the shortest path tree of u and from this, the type 1 critical heights of u;
2. for each vertex v of T to the right of u
   a. we compute the shortest path tree and the type 1 critical heights of v;
   b. we insert the type 1 critical heights of u and of v in an empty array B;
   c. for each edge e between u and v along T, we maintain the shortest paths from u and v to the endpoints of e, from which we compute the minimum height $h_{u,e,v}$ of towers at u, v to guard e (type 2 critical height for u, v, e) and we insert it in the array B;
   d. we sort the elements of the array B by non-decreasing height;
   e. we apply binary search on the array B and for the current height h:
      ▷ we apply the decision procedure of the discrete version (Sect. 3.1) for vertex u and height h up to vertex v, and compute the minimum height h' such that the top points $u(h), v(h')$ guard the entire T;
      ▷ the binary search is driven by the relation of the heights h, h' of the two towers (Theorem 2(b)) until we find the *smallest* critical height $h^*_{u,v}$ in the array B such that the corresponding minimum height of the second tower is no greater than $h^*_{u,v}$ (Lemma 3);

The minimum among the heights $h^*_{u,v}$ over all pairs of vertices u, v gives the solution to the discrete version of the two-watchtower problem for the given terrain T.

The correctness of the algorithm follows from Lemma 3 and the preceding discussion. For a terrain edge $e = ab$ between u, v, we can determine in $O(1)$ time whether Cases 1 and 2 apply by using the lines supporting the edges of the shortest paths of u and v to both endpoints a and b of e. If Case 3 applies (i.e.,

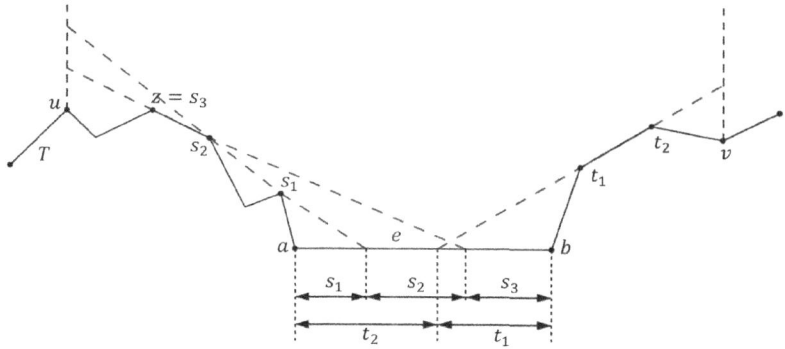

Fig. 5. Computing the points of tangency s and t for the height $h_{u,e,v}$.

e is jointly visible), the computation of $h_{u,e,v}$ takes $O(1)$ time provided that the points of tangency s, t are known (Fig. 4). For the computation of s, t, we observe that the point s (t resp.) is a vertex of the upper convex hull of the chain $T[u..a]$ ($T[b..v]$ resp.) or equivalently is a vertex of the shortest path from u to a (from v to b resp.). Let us denote by $\rho_{u,a}$ the shortest path from u to a, and let z be the *rightmost* common vertex of $\rho_{u,a}$ and the shortest path from u to b. Then, because a vertex w of $\rho_{u,a}$ is a point of tangency s for all the points of the edge e between the points of intersection of e with the lines supporting the edges of $\rho_{u,a}$ incident on w, it holds that *(i)* the point s is one of the vertices of $\rho_{u,a}$ from z (inclusive) to a (not inclusive), and *(ii)* the points of intersection of e with the lines supporting the edges of the shortest path from z to a define a partition of e into segments each of which is associated with a shortest path vertex as the corresponding point of tangency s (Fig. 5). Symmetric properties hold for the point of tangency t. Based on the above, we compute the points of tangency s, t for the height $h_{u,e,v}$ by overlaying the two partitions of e, and by finding the segment cd of the overlay (with c to the left of d) for which $h(c) \geq h'(c)$ and $h(d) \leq h'(d)$; the sought vertices s, t are the points of tangency associated with the segment cd.

    In this way, for a pair of vertices u, v, the computation of $h_{u,e,v}$ for all terrain edges e between u, v takes time linear in the sum of the sizes of the shortest path trees of u and v, which is $O(n)$. Moreover, because the total number of (type 1 and type 2) critical heights for u, v is $O(n)$, we have:

Theorem 4. *Given a terrain T in the plane with n edges, we can compute the solution to the discrete two-watchtower problem for T in $O(n^3 \log n)$ time and $O(n)$ space.*

5   Open Problems

The quest for more efficient algorithms is a staple of computer science research. So, it is immediate to ask whether we can construct a parametric search

$O(n^2 \log n)$-time algorithm for the discrete version of the two watchtowers problem, as it was asked in [1], or even a faster one. Can we improve the complexity of the parametric-search algorithm for the semi-continuous version and the non-parametric-search algorithm for the discrete version? Can we get simple and fast non-parametric-search algorithms for the semi-continuous and continuous versions? We hope that our results will contribute towards these directions.

References

1. Agarwal, P.K., et al.: Guarding a terrain by two watchtowers. Algorithmica **58**(2), 352–390 (2010). https://doi.org/10.1007/s00453-008-9270-3
2. Bahoo, Y., Bose, P., Durocher, S.: Watchtower for k-crossing visibility. In Friggstad, Z., De Carufel, J.-L. (eds.) 31st Canadian Conference on Computational Geometry (CCCG), pp. 203–209 (2019)
3. Ben-Moshe, B., Carmi, P., Katz., M.J.: Computing all large sums-of-pairs in $\mathbb{R}^n$ and the discrete planar two-watchtower problem. Inf. Process. Lett. **89**(3), 137–139 (2004). https://doi.org/10.1016/j.ipl.2003.07.008
4. Ben-Moshe, B., Katz, M.J., Mitchell, J.S.B.: A constant-factor approximation algorithm for optimal 1.5d terrain guarding. SIAM J. Comput. **36**(6), 1631–1647 (2007). https://doi.org/10.1137/S0097539704446384
5. Bespamyatnikh, S., Chen, Z., Wang, K., Zhu, B.: On the planar two-watchtower problem. In: Wang, J. (ed.) COCOON 2001. LNCS, vol. 2108, pp. 121–130. Springer, Heidelberg (2001). https://doi.org/10.1007/3-540-44679-6_14
6. Clarkson, K.L., Varadarajan, K.: Improved approximation algorithms for geometric set cover. Discret. Comput. Geom. **37**(1), 43–58 (2006). https://doi.org/10.1007/s00454-006-1273-8
7. de Berg, M., Cheong, O., van Kreveld, M.J., Overmars, M.H.: Computational Geometry: Algorithms and Applications, 3rd edn. Springer, Heidelberg (2008). https://doi.org/10.1007/978-3-540-77974-2. ISBN 978-3-540-77973-5
8. Edelsbrunner, H.: Computing the extreme distances between two convex polygons. J. Algorithms **6**(2), 213–224 (1985). https://doi.org/10.1016/0196-6774(85)90039-2
9. Friedrichs, S., Hemmer, M., King, J., Schmidt, C.: The continuous 1.5d terrain guarding problem: discretization, optimal solutions, and PTAS. J. Comput. Geom. **7**(1), 256–284 (2016). https://doi.org/10.20382/jocg.v7i1a13
10. Gewali, L., Dahal, B.: Algorithms for tower placement on terrain. In: Latifi, S. (ed.) 16th International Conference on Information Technology-New Generations (ITNG 2019). AISC, vol. 800, pp. 551–556. Springer, Cham (2019). https://doi.org/10.1007/978-3-030-14070-0_77
11. Har-Peled, S.: Geometric Approximation Algorithms. Mathematical Surveys and Monographs, vol. 173, pp. 362. American Mathematical Society (2011)
12. Khodakarami, F., Didehvar, F., Mohades, A.: 1.5d terrain guarding problem parameterized by guard range. Theor. Comput. Sci. **661**, 65–69 (2017). https://doi.org/10.1016/j.tcs.2016.11.015
13. King, J.: A 4-approximation algorithm for guarding 1.5-dimensional terrains. In: Correa, J.R., Hevia, A., Kiwi, M. (eds.) LATIN 2006. LNCS, vol. 3887, pp. 629–640. Springer, Heidelberg (2006). https://doi.org/10.1007/11682462_58
14. King, J., Krohn, E.: Terrain guarding is NP-hard. SIAM J. Comput. **40**(5), 1316–1339 (2011). https://doi.org/10.1137/100791506

15. Seth, R., Maheshwari, A., Nandy, S.C.: Acrophobic guard watchtower problem. Comput. Geom. **109**, 101918 (2023). https://doi.org/10.1016/j.comgeo.2022.101918
16. Sharir, M.: The shortest watchtower and related problems for polyhedral terrains. Inf. Process. Lett. **29**(5), 265–270 (1988). https://doi.org/10.1016/0020-0190(88)90120-2
17. Tripathi, N., Pal, M., De, M., Das, G., Nandy, S.C.: Guarding polyhedral terrain by k-watchtowers. In: Chen, J., Lu, P. (eds.) FAW 2018. LNCS, vol. 10823, pp. 112–125. Springer, Cham (2018). https://doi.org/10.1007/978-3-319-78455-7_9
18. Urrutia, J.: Art gallery and illumination problems. In: Sack, J.R., Urrutia, J. (eds.) Handbook of Computational Geometry, pp. 973–1027. North Holland (2000). https://doi.org/10.1016/B978-044482537-7/50023-1
19. Zhu, B.: Computing the shortest watchtower of a polyhedral terrain in $O(n \log n)$ time. Comput. Geom. **8**, 181–193 (1997). https://doi.org/10.1016/S0925-7721(96)00009-0

Degree Realization by Bipartite Cactus Graphs

Amotz Bar-Noy[1], Toni Böhnlein[2], David Peleg[3], Yingli Ran[4],
and Dror Rawitz[5(✉)]

[1] City University of New York (CUNY), New York, USA
amotz@sci.brooklyn.cuny.edu
[2] Huawei, Zurich, Switzerland
toniboehnlein@web.de
[3] Weizmann Institute of Science, Rehovot, Israel
david.peleg@weizmann.ac.il
[4] Zhejiang Normal University, Jinhua, China
ranyingli@zjnu.edu.cn
[5] Bar Ilan University, Ramat-Gan, Israel
dror.rawitz@biu.ac.il

Abstract. The DEGREE REALIZATION problem with respect to a graph
family $\mathcal{F}$ is defined as follows. The input is a sequence d of n positive
integers, and the goal is to decide whether there exists a graph $G \in \mathcal{F}$
whose degrees correspond to d. The main challenges are to provide a
precise characterization of all the sequences that admit a realization in
$\mathcal{F}$ and to design efficient algorithms that construct one of the possible
realizations, if one exists.

This paper studies the problem of realizing degree sequences by bipartite cactus graphs (where the input is given as a single sequence, without
the bi-partition). A characterization of the sequences that have a cactus
realization is already known [30]. In this paper, we provide a systematic way to obtain such a characterization, accompanied by a realization
algorithm. This allows us to derive a characterization for bipartite cactus graphs, and as a byproduct, also for several other interesting sub-
families of cactus graphs, including bridge-less cactus graphs and core
cactus graphs, as well as for the bipartite sub-families of these families.

1 Introduction

1.1 Background and Motivation

We study graph realization problems in which for some specified graph family
$\mathcal{F}$, a sequence of integers d is given, and the requirement is to construct a graph
from $\mathcal{F}$ whose degrees abide by d. More formally, the DEGREE REALIZATION
problem with respect to a graph family $\mathcal{F}$ is defined as follows. The input is a
sequence $d = (d_1, \ldots, d_n)$ of positive integers, and the goal is to decide whether

This work was supported by US-Israel BSF grant 2022205.

I. Finocchi and L. Georgiadis (Eds.): CIAC 2025, LNCS 15679, pp. 258–275, 2025.
https://doi.org/10.1007/978-3-031-92932-8_17

there exists a graph $G \in \mathcal{F}$ whose degrees correspond to d, i.e., where the vertex set is $V = \{1, \ldots, n\}$ and $\deg_G(i) = d_i$, for every $i \in V$. If such a graph exists, then d is called $\mathcal{F}$-*graphic*. Observe that while every graph $G \in \mathcal{F}$ corresponds to a unique degree sequence d, a degree sequence may be realized by more than one graph in $\mathcal{F}$. For example, in the family of bipartite graphs, the sequence $(2, 2, 2, 2, 2, 2, 2, 2)$ can be realized by a 8-vertex cycle or by two 4-vertex cycles.

There are two fundamental challenges that arise in this context. The first is to provide an algorithm that *decides* whether a given sequence can be realized by a graph from $\mathcal{F}$, and furthermore to provide a characterization of all the realizable sequences. The second is to design an efficient algorithm that *constructs* one of the realizations if one exists.

We consider sub-families of the family of *cactus graphs* (*cacti*). A cactus graph is a connected graph in which any edge may be a member of at most one cycle, which means that different cycles do not share edges, but may share one vertex. Cacti are an important and interesting graph family with many applications, for instance in modeling electric circuits [28,38], communication networks [4], and genome comparisons [29]. We provide a characterization for the DEGREE REALIZATION problem with respect to bipartite cactus graphs (bi-cacti), which implies a linear time algorithm for the decision problem. Furthermore, we provide a linear time realization algorithm for degree realization by bipartite cactus graphs. We introduce a systematic way to obtain such a characterization, which allows us to obtain the known characterization for cactus graphs [30] and in addition also some new (previously unknown) characterizations for several other interesting sub-families of cactus graphs, including bridge-less cactus graphs and core cactus graphs, as well as the bipartite sub-families of these families.

The characterization of families of sparse graphs, such as cactus graphs, may assist in finding a characterization for the family of planar graphs and the family of outer-planar graphs, both of which are open problems for about half a century.

1.2   Related Work

General graphs. DEGREE REALIZATION with respect to the family of all graphs was studied extensively in the past. Erdős and Gallai [18] gave a necessary and sufficient condition (which implies an $O(n)$ decision algorithm) for a sequence to be realizable, or *graphic*. (Alternative proofs were given in [3,16,17,35–37].) Havel [24] and Hakimi [23] (independently) gave another characterization for graphic sequences, which also implies an efficient $O(m)$ time algorithm for constructing a realizing graph for a graphic sequence, where m is the number of edges in the graph. A variant of this realization algorithm is given in [39].

Bipartite Graphs. The history of DEGREE REALIZATION with respect to the family of bipartite graphs is as long as the one for general graphs. In this problem, a sequence is given as input and the goal is to find a realizing bipartite graph. This problem has a variant in which the input consists of two sequences representing the degree sequences of the two sides of a bipartite realization. This variant was solved by Gale and Ryser [21,32] even before Erdős and Gallai's characterization of graphic sequences. However, the general bipartite realization problem remains open despite being mentioned as open over 40 years ago [31]. Recent attempts solve special cases and emphasize *approximate* realizations [8,9]. The sequence d is called *forcibly $\mathcal{F}$-graphic* if every realization of d is in $\mathcal{F}$. A characterization of sequences that are forcibly bipartite-graphic was given in [10].

Sparse Graphs. The last and most relevant category is of families which contain graphs with a linear number of edges. The problem is straightforward with respect to trees [22], forests, and unicyclic graphs [13]. A characterization of Halin graphs was given in [12]. Rao [30] provided a characterization for cactus graphs. He also gave a characterization for cactus graphs whose cycles are triangles and for connected graphs whose blocks are cycles of k vertices. Beineke and Schmeichel [11] characterized cacti degree sequences with up to four cycles and also provided a sufficient condition for cactus realization.

Rao [31] mentioned DEGREE REALIZATION with respect to planar graphs and related families as open. A characterization is known for regular sequences [25] and for sequences with $d_1 - d_n = 1$, where d is assumed to be in non-increasing order [33]. Partial results are known if $d_1 - d_n = 2$ [19,20,33]. A characterization of bi-regular sequences with respect to the family of bipartite planar graphs is given in [1]. A sufficient condition for planarity was given in [6]. As for outerplanar graphs, only partial results are known. Several necessary conditions were given in [26,34]. Choudum [15] gave a characterization for forcibly outerplanar sequences. In [7] it was shown that any sequence that satisfies a certain necessary condition for outerplanarity is either non-outerplanaric or has a 2-page book embedding. A sufficient condition was given in [5]. Sufficient conditions for the realization of 2-trees were given in [27]. Bose et at. [14] gave a characterization for 2-trees with a linear time realization algorithm.

1.3   Our Results and Techniques

As opposed to the approach taken in [30], the characterizations and realization algorithms of this paper were developed by starting with simple graph families and gradually coping with families that are more involved. Specifically, Sect. 3 contains characterizations and realization algorithms for *unicyclic* graphs and *bi-unicyclic* graphs. Coping with the simplest non-trivial cacti provides the basic techniques needed for the more general cases, but this also serves as a light introduction to degree realization. *Bridge-less cacti* are studied in Sect. 5, which contains characterizations and realization algorithms for *bridge-less cactus* graphs and *bridge-less bi-cactus* graphs. The next family we consider is that of *core cactus* graphs (see definition in Sect. 2). Section 6 contains characterizations and realization algorithms for *core cactus* graphs and *core bi-cactus* graphs. Finally, in Sect. 7 we provide a characterization for degree realization by *cactus* graphs and *bi-cactus* graphs.

The crux in developing a necessary and sufficient condition for cactus and bi-cactus realizability of a given sequence is to bound the number of possible edges in the realizing graph as shown in Sect. 4. This is obtained when the number of *bridges* in the graph (see definition in Sect. 2) is minimized. The above condition depends on a *bridge parameter*, which is defined as

$$\beta \triangleq \max \left\{ \omega_1, \tfrac{1}{2}(\omega_1 + \omega_{odd}) \right\},$$

where ω_1 is the multiplicity of 1 in d and ω_{odd} is the number of odd integers greater than 1 in d. We note that this parameter is implicit in [30]. The decision about a given sequence depends only on the volume $\sum_i d_i$, n, and the bridge parameter (see Theorem 11). We believe that this parameter may be of independent interest.

The decision and realization algorithms of all the above mentioned families work in linear time. Our algorithms are reminiscent of the minimum pivot version of the Havel-Hakimi algorithm [23,24] for realizing sequences by general graphs. However, in our algorithms pivots are not connected to the vertices with the *maximum* residual degrees in the sequence d. Hence, our analysis is not based on swapping arguments. In particular, degree-1 vertices should be connected to odd degree vertices, rather than to even degree vertices, even if the latter degrees are larger. When the sequence does not contain degree-1 vertices, pairs of degree-2 vertices are used to construct a triangle that lowers the degree of another vertex by 2. Again, smaller odd degree vertices are preferred over larger even degree vertices. Throughout the paper, when dealing with bi-cactus graphs, we adapt the techniques used in the cactus case to avoid constructing odd-length cycles. This task turned out to be more involved, since in this case a realization may require one extra bridge edge.

Given a graph family, realizability of sequences may depend on certain parameters. There are two extremes. One extreme is the elaborate test of Erdős-Gallai that examines the relationship among all the degrees before determining if a sequence is graphic. The other extreme is for forests in which the length of the sequence n and the sequence sum are the only two interesting parameters, i.e., a sequence d is realizable by a forest if $\sum_i d_i \leq 2n - 2$. The results for cacti and bi-cacti are not that simple, but still depend only on four parameters: the multiplicities of 1's and of odd numbers, the sequence length, and the sequence sum. Our structured proof demonstrates the roles of these two additional parameters, through the bridge parameter. Hopefully, the next step will be to add additional parameters, e.g., the multiplicity of 2's, to obtain characterizations of sequences that can be realized by other families of sparse graphs, such as planar graphs and outerplanar graphs. Both of which are long standing open problems.

2   Preliminaries

Definitions and Notation. We consider simple graphs $G = (V, E)$ where $V = \{1, \ldots, n\}$. The degree of a vertex $i \in V$, denoted by $\deg_G(i)$, is its number of neighbors. The degree sequence of graph G is $\deg(G) = (\deg_G(1), \ldots, \deg_G(n))$. Let $d = (d_1, \ldots, d_n)$ be a sequence of positive integers. If there exists a graph G such that $\deg(G) = d$, then it is said that G *realizes* d. A sequence d that has a realization G is called *graphic*. We refer to $\sum_i d_i$ as the *volume* of d. Define

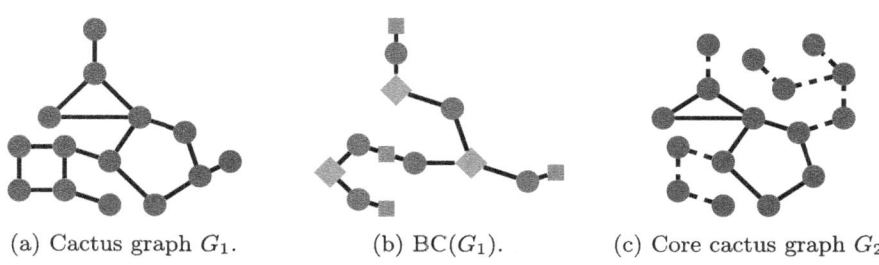

(a) Cactus graph G_1.   (b) BC(G_1).   (c) Core cactus graph G_2.

Fig. 1. G_1 is a cactus graph; BC(G_1) is the block point-cut graph of G_1; G_2 is a core-cactus graph. In BC(G_1) circles are cut-vertices, squares are bridge blocks, and diamonds are cycle blocks. In G_2 dashed lines represent bridges, and solid lines are the edges of the bridge-less core.

$m \triangleq \frac{1}{2} \sum_i d_i$. Notice that if d is graphic, then m is the number of edges in any realization of d. A sequence d is called a *degree sequence* if $d_i \in \{1, \ldots, n-1\}$, for every i, and the volume $\sum_i d_i$ is even. Throughout the paper, we assume that $d_i \geq d_{i+1}$, for every $1 \leq i \leq n-1$. For brevity, we use a^k as a shorthand for a subsequence of k consecutive a's (e.g., 2^3 represents $2, 2, 2$). Given a degree sequence d, let ω_i be the number of times the integer i appears in d. Finally, ω_{odd} is the number of odd integers that are larger than 1 in d, namely $\omega_{odd} = \sum_{k \geq 1} \omega_{2k+1}$.

Graph Families. A graph G is *connected* if there is a path between every two vertices in the graph. A *cut-vertex* of a connected graph is a vertex whose removal disconnects the graph. A *bridge* in a connected graph is an edge whose removal disconnects the graph. A *block* of G is a maximal connected subgraph of G that does not have cut-vertices. That is, it is a maximal subgraph which is either an isolated vertex, a bridge edge, or a 2-connected subgraph.

A graph G is called a *pseudo-tree* if it is connected and it contains at most one cycle. It is called *unicyclic* if it contains exactly one cycle. A graph G is called a *pseudo-forest* if each of its connected component is a pseudo-tree. A *cactus* graph is a connected graph in which any edge may be a member of at most one cycle, which means that different cycles do not share edges, but may share one vertex. An alternative definition is that a graph G is a (non-trivial) cactus if and only if every block of G is either a simple cycle or a bridge. See an example in Fig. 1a. A cactus G is called *bridge-less* if it has no bridges. In this case every edge belongs to exactly one cycle in G. A cactus G is called a *triangulated cactus* if all the cycles are of length three and each edge belongs to a cycle. A cactus graph G is called a *core cactus* if there are no bridges that split the graph such that each of the two components contain a cycle. In other words, when all the bridges of a core cactus are removed, what remains is a bridge-less cactus. See an example in Fig. 1c. A graph G is a *bipartite cactus* or a *bi-cactus* if G is a cactus graph and also a bipartite graph. Bi-pseudo-trees, bridge-less bi-cactus, and core bi-cactus are defined in a similar manner.

Given a connected graph G, the *block-cutpoint graph* $BC(G) = (V', E')$ of a graph G is defined as follows [2]. Let $C(G) \subseteq V$ be the set of cut vertices, and let $\mathcal{B}(G)$ be the set of blocks in G. Then, $V' = C(G) \cup \mathcal{B}(G)$ and $E' = \{(v, B) : v \in C(G), B \in \mathcal{B}(G), v \in V(B)\}$. Observe that $BC(G)$ is a tree. See an example in Fig. 1b.

3   Realization by Pseudo-trees and Bi-pseudo-Trees

In this section we give a characterization for degree realization by pseudo-trees and bi-pseudo-trees. These results are used later on, but also serve as a warm-up.

Observation 1. *If G is a pseudo-forest, then $\sum_i d_i \in \{2(n - c), \ldots, 2n\}$, where c is the number of connected components in G. If G is a pseudo-tree, then $\sum_i d_i \in \{2(n - 1), 2n\}$.*

The realization problem is straightforward on trees and forests.

Theorem 1 ([22])**.** *Let d be a degree sequence such that $\sum_i d_i \leq 2n - 2$. Then d has a forest realization with $(2n - \sum_i d_i)/2$ components. If $\sum_i d_i = 2n - 2$, then it has a tree realization.*

The following observation considers the case, where $\sum_i d_i = 2n$ and $n \geq 3$.

Observation 2. *Let G be a pseudo-tree such that $n \geq 3$ and $\sum_i d_i = 2n$. Then, $d_3 \geq 2$.*
Proof. If $d_3 = 1$, then $d_1 + d_2 = 2n - (n - 2) = n + 2$. Hence, $d = (k, n + 2 - k, 1^{n-2})$, where $k \in \{n - 1, \ldots, \lceil \frac{1}{2}(n + 2) \rceil\}$. For any k, the sequence d cannot be realized because ω_1 must be at least n to satisfy the degree requirements d_1 and d_2 even if the vertices whose degrees are d_1 and d_2 are connected.   □

3.1   Unicyclic Realization

We show that there is a realization by a unicyclic graph, if $\sum_i d_i = 2n$ and $d_3 \geq 2$. This was proven before in [13]. In this paper, we give a constructive proof that illustrates our approach for subsequent results. More specifically, we use the minimum pivot version of the Havel-Hakimi algorithm [23,24] as long as the sequence contains a degree of 1.

Theorem 2 ([13])**.** *Let d be a degree sequence such that $\sum_i d_i = 2n$ and $d_3 \geq 2$. Then, the sequence d has a unicyclic realization.*
Proof. By induction on $n - \omega_2$. The base case is a sequence (2^n), for $n \geq 3$, for which there is a realization of d consisting of one cycle that contains all the vertices. For the inductive step, assume that there is a unicyclic realization for sequences d' such that $n' - \omega_2' < n - \omega_2$. Since $\sum_i d_i = 2n$ and $d \neq (2^n)$, it must be that $d_1 \geq 3$ and $d_n = 1$. Moreover, $n > 3$, since $d_3 \geq 2$. Let d' be the sequence which is obtained by removing vertex n and subtracting 1 from d_1. Notice that $\sum_i d_i' = \sum_i d_i - 2 = 2n - 2 = 2n'$ and that $d_3' \geq 2$. Also, $n' - \omega_2' \leq n - \omega_2 - 1$. Hence, by the inductive hypothesis there is a unicyclic realization G' of d'. Obtain a realization G of d by adding the edge $(1, n)$.   □

The above proof describes an algorithm that creates a cycle containing all vertices whose degree is larger than 1. Then, it adds degree-1 vertices as leaves to any vertex whose degree is greater than 2. Hence, Theorem 1 and the proof of Theorem 2 imply the following.

Corollary 1. *Let d be a degree sequence such that $\sum_i d_i = 2n$ and $d_3 \geq 2$. There is a linear time algorithm for computing a unicyclic realization of d that contains a cycle of all vertices whose degree is greater than 1 (a.k.a. closed caterpillar).*

3.2   Bi-unicyclic Realization

In the case of bi-unicyclic graph one needs to observe that there cannot be a realization if $d = (2^n)$, where n is odd. Hence, a realization algorithm should avoid such sequences.

Theorem 3. *Let d be a degree sequence such that $\sum_i d_i = 2n$ and $d_3 \geq 2$. The sequence d has a bi-unicyclic realization if and only if $d \neq (2^n)$ for an odd n.*

Proof. If $d = (2^n)$, where n is odd, then the only connected realization is a cycle. There is no bi-unicyclic realization for d, since bipartite graphs cannot have odd cycles as sub-graphs. The rest of the proof is similar to the proof of Theorem 2, and proceeds by induction on $n - \omega_2$. The base case is a sequence $d = (2^n)$, where $n \geq 3$ and n is even. In this case, there is a realization of d consisting of one cycle that contains all the vertices.

For the inductive step, assume that there is a bi-unicyclic realization for sequences d' such that $n' - \omega_2' < n - \omega_2$. Since $d \neq (2^n)$, it must be that $d_1 \geq 3$ and $d_n = 1$. Moreover, $n > 3$, since $d_3 \geq 2$. There are two cases.

First, suppose that $d = (3, 2^{n-2}, 1)$ and $n \geq 3$ is even. In this case, d can be realized by a cycle of length $n-2$, which is connected to a P_2, i.e., by the following edges set $E = \{(i, i+1) : i = 1, \ldots, n-3\} \cup \{(1, n-2), (1, n-1), (n-1, n)\}$.

Otherwise, let d' be the sequence which is obtained by removing vertex n and subtracting 1 from d_1. Notice that $\sum_i d_i' = \sum_i d_i - 2 = 2n - 2 = 2n'$ and that $d_3' \geq 2$. Also notice that $d' \neq (2^n)$, where n is odd. In addition, $n' - \omega_2' \leq n - \omega_2 - 1$. Hence, by the inductive hypothesis there is a bi-unicyclic realization G' of d'. We obtain a realization G of d by adding the edge $(1, n)$. $\square$

The following is implied by the proof of Theorem 3.

Corollary 2. *Let d be a degree sequence such that $\sum_i d_i = 2n$, $d_3 \geq 2$, and $d \neq (2^n)$, where n is odd. There is a linear time algorithm that computes bi-unicyclic realization of d that contains a cycle of all vertices whose degree is larger than 1, maybe with the exception of one such vertex.*

4 Bounds on the Number of Edges

4.1 Bound on the Number of Edges in a Cactus

Given a cactus graph G, let c be the number of cycles in G (not counting the outside face), let t be the number of edges in G that belong to a cycle, and let b be the number of bridges in G. Notice that $m = b + t$.

The next observation is implied by the fact that a cactus graph is connected and planar. More specifically, it is a direct implication of Euler's Formula.

Observation 3. *Let G be a cactus graph. Then $m = n + c - 1$.*

Proof. Since G is planar, given an embedding of G in the plane, Euler's formula implies that $m = n + f - 2$, where f is the number of faces. As G is a cactus graph, all faces in the embedding but the outside face are cycles, thus $f = c + 1$. □

Next, we give an upper bound on the number of edges in a cactus graph.

Lemma 4. *Let G be a cactus graph. Then, $m \leq \left\lfloor \frac{3(n-1)-b}{2} \right\rfloor$.*

Proof. Each edge is part of at most one cycle, so $c \leq (m - b)/3$. Observation 3 implies that $m = n + c - 1 \leq n + (m - b)/3 - 1$, or $2m \leq 3n - b - 3$. □

We now consider bridge-less cactus graphs and triangulated cactus graphs.

Lemma 5. *Let G be a bridge-less cactus graph. Then, $m \leq \left\lfloor \frac{3(n-1)}{2} \right\rfloor$. In particular, if G is a triangulated cactus graph, then n is odd and $m = 1.5(n - 1)$.*

Proof. The first bound is a direct implication of Lemma 4. Assume that G is a triangulated cactus. Then n is odd and $m = t = 3c$, and thus by Lemma 4 we have that $m = 1.5(n - 1)$. □

Let G be a cactus graph and let $d = \deg(G)$. Recall that ω_1 is the number of 1's in d, and that ω_{odd} is the number of odd integers that are larger than 1 in d. Define the *bridge parameter* of a sequence d as follows: $\beta \triangleq \max\left\{\omega_1, \frac{1}{2}(\omega_1 + \omega_{odd})\right\}$. Note that β is an integer since $\omega_1 + \omega_{odd}$ is even. For example, the cactus graph in Fig. 1a has $\omega_1 = 3$, $\omega_{odd} = \omega_3 = 5$, and $\beta = \frac{1}{2}(\omega_1 + \omega_{odd}) = 4$.

We show that β serves as a lower bound on the number of bridges in a cactus.

Lemma 6. *Let G be a cactus graph, where $n > 2$. Then, $b \geq \beta$.*

Proof. Any odd degree vertex must be connected to at least one bridge. Hence, $b \geq \frac{\omega_1 + \omega_{odd}}{2}$. In particular, the edge which is attached to a degree-1 vertex (a leaf) must be bridge, and due to connectivity it must be connected to a vertex whose degree is greater than 1. Thus, $b \geq \omega_1$. The lemma follows. □

Lemmas 4 and 6 imply the following bound the number of edges in a cactus graph. We note that this bound is implicit in [30].

Theorem 4. *Let G be a cactus graph and $d = \deg(G)$. Then $m \leq \left\lfloor \frac{3(n-1)-\beta}{2} \right\rfloor$.*

4.2   Bound on the Number of Edges in a Bi-cactus

An obvious requirement from a bipartite graph is that all cycles have even length.

Observation 7. *A cactus graph G is bipartite if and only if all its cycles are of even length. In particular, each cycle contains at least 4 edges. Moreover, if G is bridge-less, then $m = \frac{1}{2} \sum_i d_i$ must be even.*

The following lemma is the version of Lemma 4 for bi-cacti. Its proof is somewhat more complicated.

Lemma 8. *Let G be a bi-cactus, where $n \geq 4$. Then, $m \leq 2 \left\lfloor \frac{2(n-1-b)}{3} \right\rfloor + b$.*

Proof. Since G is a bi-cactus each edge is part of at most one cycle, and by Observation 7 each cycle contains at least 4 edges. It follows that $c \leq m/4$. However, one may obtain a tighter bound. Consider the block cut-point tree $BC(G)$, where the root is a cycle. We remove blocks from G according to $BC(G)$ in a bottom up manner. When one removes a bridge, both the number of edges and the number of vertices in G are reduced by 1. When one removes a cycle of size k from G, k edges and $k-1$ vertices are removed, where k is even and thus $k \geq 4$. The highest ratio between the number of removed edges and the number of removed vertices is obtained when $k = 4$, i.e., a ratio of $4/3$. Assume that one is able to obtain this ratio of $4/3$ for all cycle edges. Then, the last cycle may be of size 4, 6 or 8, depending on the remainder of dividing $n - b$ by 3. Hence we get this ratio of 4 edges per 3 vertices from $n - b - k'$ vertices, where k' is the size of the last cycle. Hence, the highest number of cycles is $(n - b - k')/3 + 1 = (n - b - k' + 3)/3$.

By Observation 3, $m = n + c - 1 \leq n + (n - b - k')/3$. Let $n - 1 - b = 3q - r$, where $q = \left\lceil \frac{n-1-b}{3} \right\rceil$ and $r = 3q - (n - 1 - b)$. Observe that $k' - 4 = 2r$. Hence,

$$3m \leq 4(n-1) - 2r - b = 4(n-1) - 2\left(3\left\lceil \tfrac{n-1-b}{3} \right\rceil - (n-1-b)\right) - b$$

$$= 6(n-1-b) - 6\left\lceil \tfrac{n-1-b}{3} \right\rceil + 3b = 6\left\lfloor \tfrac{2(n-1-b)}{3} \right\rfloor + 3b,$$

where the last equality is due to $x = \left\lfloor \frac{2x}{3} \right\rfloor + \left\lceil \frac{x}{3} \right\rceil$. The lemma follows. □

In a bridge-less bi-cactus $b = 0$, and thus we obtain the following lemma.

Lemma 9. *Let G be a bridge-less bi-cactus, where $n \geq 4$. Then, $m \leq 2\left\lfloor \frac{2(n-1)}{3} \right\rfloor$.*

The next example shows that one cannot replace b with β in the bound of Lemma 8 as was done in the cactus case (see Theorem 4). Consider the sequence $d = (4, 3, 2^6, 1)$. If we replace b with β in the bound of Lemma 8, we get an upper bound of $m \leq 2\left\lfloor \frac{2(n-1-\beta)}{3} \right\rfloor + \beta = 2\left\lfloor \frac{2(9-1-1)}{3} \right\rfloor + 1 = 9$. However, d can be realized using 10 edges as depicted in Fig. 2. Notice that there is an even degree vertex which is adjacent to two bridges. In the sequel we show that one such "correction" for changing b to β in the bound of Lemma 8 is enough.

The following two technical lemmas are required for obtaining a bound on the number of edges in bi-cactus graphs. Their proofs are omitted for lack of space.

Fig. 2. A bi-cactus realization of $d = (4, 3, 2^6, 1)$.

Lemma 10. $\left\lfloor \frac{4(n-1)-\beta}{3} \right\rfloor = \max\left\{ 2\left\lfloor \frac{2(n-1-\beta)}{3} \right\rfloor + \beta, 2\left\lfloor \frac{2(n-1-(\beta+1))}{3} \right\rfloor + (\beta+1) \right\}$

Lemma 11. $2\left\lfloor \frac{2(n-1-\beta)}{3} \right\rfloor + \beta \geq 2\left\lfloor \frac{2(n-1-(\beta+2))}{3} \right\rfloor + (\beta+2)$.

Theorem 5. *A bi-cactus graph G with $n \geq 4$ and $\beta \geq 1$ has $m \leq \left\lfloor \frac{4(n-1)-\beta}{3} \right\rfloor$.*

Proof. Lemma 8 provides an upper bound for bi-cactus graphs. Also recall that $b \geq \beta$ by Lemma 6. Lemmas 10 and 11 imply that the bound is maximized either when $b = \beta$ or when $b = \beta + 1$. □

5   Realization of Bridge-Less Cactus and Bi-cactus Graphs

5.1   Bridge-Less Cactus Graph Realization

We give a characterization and a realization algorithm for bridge-less cacti. We first prove that a bridge-less cactus is a cactus with even degrees and vice versa.

Lemma 12. *A cactus G is bridge-less if and only if d_i is even, for every i.*

Proof. Suppose that G has no bridges. Consider a vertex v. Each cycle that contains v contributes exactly 2 to its degree. Hence, v's degree is even.

Suppose that d_i is even, for every i. Assume that G contains a bridge (x, y). Since $\deg(x)$ is even, it must be adjacent to another bridge (x, z). Consider the block-cutpoint graph $BC(G)$ of G. Recall that $BC(G)$ is a tree. A bridge node cannot be a leaf of $BC(G)$, since this means that there must be a vertex of degree 1 in G. Hence, all leaves of $BC(G)$ are cycle nodes. There must be a bridge node whose removal splits $BC(G)$ into two trees, one of which does not contains bridge nodes. Let this bridge be (x, y) in G. It follows that either x or y have an odd degree. A contradiction. □

Next we show that a degree sequence has a realization as a bridge-less cactus if and only if it satisfies the bound of Lemma 5 and it consists of even numbers.

Theorem 6. *Let d be a degree sequence of length $n \geq 3$. There is a bridge-less cactus realization of d if and only if $m \leq \lfloor 1.5(n-1) \rfloor$ and d_i is even, for every i.*

Proof. If there is a bridge-less cactus realization, then Lemma 12 and Lemma 5 imply that $m \leq \lfloor 1.5(n-1) \rfloor$ and d_i is even, for every i.

The converse is proved by induction on $m - \omega_2$. In the base case $d = (2^n)$, where $n \geq 3$, there is a realization of d consisting of one cycle that contains all vertices. For the inductive step, since $d \neq (2^n)$, it must be that $d_1 \geq 4$ since d_1 is even. Moreover, it must be that $n \geq 5$, since $m \geq 2 + 1(n-1) = n + 1 > \lfloor 1.5(n-1) \rfloor$, for $n \leq 4$. Also, since $\sum_i d_i < 3n$, there must be more than $n/2$ vertices of degree 2 in d. In particular, $d_n = d_{n-1} = 2$. Let d' be the sequence which is obtained by removing n and $n-1$ and subtracting 2 from d_1. Notice that $n' \geq 3$ because $n' = n - 2$. Also, since $2m = \sum_i d_i$, we have that

$$\sum_i d_i' = \sum_i d_i - 6 \leq 2 \lfloor 1.5(n-1) \rfloor - 6 = 2 \lfloor 1.5(n'-1) \rfloor.$$

In addition, $m' - \omega_2' \leq m - 3 - (\omega_2 - 2) = m - \omega_2 - 1$. By the induction hypothesis d' has a realization as a bridge-less cactus G'. We obtain a realization G for d by adding a triangle of the vertices 1, $n-1$, and n. □

A similar approach also works for triangulated cacti. This result already appeared in [30].

Theorem 7 ([30]). *Let d be a degree sequence of length $n \geq 3$. There is a triangulated cactus realization of d if and only if n is odd, $m = 1.5(n-1)$, and d_i is even, for every i.*

The proofs of Theorems 6 and 7 imply an algorithm that repeatedly forms a triangle composed of two vertices whose current degree is 2 and of one vertex whose current degree is at least 4 until only degree-2 vertices remain. Then, a cycle is created of all remaining vertices.

Corollary 3. *Let d be a degree sequence such that $n \geq 3$, $m \leq \lfloor 1.5(n-1) \rfloor$, and d_i is even, for every i. There is a linear time algorithm that computes a bridge-less cactus realization of d, where all cycles except maybe one are triangles.*

5.2 Bridge-Less Bi-cactus Graph Realization

We provide a characterization and a realization algorithm for bridge-less bi-cactus graphs. The approach is similar to the one for bridge-less cactus graphs, where the main difference is that we use cycles of length 4 and not triangles.

Theorem 8. *Let d be a degree sequence of length $n \geq 3$. There is a realization of d as a bridge-less bi-cactus if and only if $m \leq 2 \lfloor \frac{2(n-1)}{3} \rfloor$, m is even, and d_i is even, for every i.*

Proof. If there is a realization, then Lemma 12, Observation 7, and Lemma 9 imply that d_i is even, for every i, m is even, and $m \leq 2 \lfloor \frac{2(n-1)}{3} \rfloor$.

The converse is proved by induction on $m - \omega_2$. In the base case $d = (2^n)$. Since m is even, n is also even. There is a realization of d consisting of one cycle that contains all vertices.

For the inductive step, $d \neq (2^n)$, implies that $d_1 \geq 4$. It must be that $n \geq 7$, since otherwise $\sum_i d_i \geq 4 + 2(n-1) = 2(n+1) > 4 \left\lfloor \frac{2(n-1)}{3} \right\rfloor$. Also, there must be more than $2n/3$ vertices of degree 2 in d. In particular, $d_{n-2} = 2$. Let d' be the sequence obtained by removing $n-2$, $n-1$, and n and subtracting 2 from d_1. Notice that $n' = n - 3 \geq 4$ and $m' = m - 4$, which means that m is even. Also,

$$\sum_i d'_i = \sum_i d_i - 8 \leq 4 \left\lfloor \frac{2(n-1)}{3} \right\rfloor - 8 = 4 \left\lfloor \frac{2(n-1)}{3} - 2 \right\rfloor = 4 \left\lfloor \frac{2(n'-1)}{3} \right\rfloor.$$

In addition, $m' - \omega'_2 \leq m - 4 - (\omega_2 - 3) = m - \omega_2 - 1$. By the inductive hypothesis d' has a realization as a bridge-less bi-cactus G'. We obtain a realization G for d by adding the edges $(1, n-2)$, $(n-2, n-1)$, $(n-1, n)$, and $(1, n)$.   □

Corollary 4. *A degree sequence d where $n \geq 3$, $m \leq 2 \left\lfloor \frac{2(n-1)}{3} \right\rfloor$, m is even, and d_i is even for every i has a linear time algorithm that computes a bridge-less bi-cactus realization of d where all cycles except maybe one are of length 4.*

6   Realization of Core Cactus and Bi-cactus Graphs

6.1   Core Cactus Graph Realization

We provide a full characterization and a realization algorithm for core cactus graphs. As a first step, we observe that in core cacti we have that $\beta = \omega_1$.

Lemma 13. *If G is a core cactus, then $\omega_1 \geq \omega_{odd}$.*

Proof. By induction on the block point-cut graph $BC(G)$. The base case is a bridge-less cactus, in which $\omega_1 = \omega_{odd} = 0$. For the inductive step, we remove a bridge leaf from $BC(G)$. This amounts to removing a degree-1 vertex denoted j. This also lowers the degree of the vertex k on the other side of this bridge. Observe that $d_k \geq 2$. Let G' be the graph without j. By the inductive hypothesis we have that $\omega'_1 \geq \omega'_{odd}$. There are several options depending on d_k. If $d_k = 2$, then $\omega_1 = \omega'_1 \geq \omega'_{odd} = \omega_{odd}$. If $d_k \geq 4$ is even, then $\omega_1 = \omega'_1 + 1 \geq \omega'_{odd} + 1 = \omega_{odd} + 2$. If $d_k \geq 3$ is odd, then $\omega_1 = \omega'_1 + 1 \geq \omega'_{odd} + 1 = \omega_{odd}$.   □

We show that a degree sequence d has a realization as a core cactus if and only if it satisfies the upper bound of Theorem 4 and $\beta = \omega_1$.

Theorem 9. *Let d be a degree sequence such that $n \geq 3$ and $\sum_i d_i > 2n$. There is a core cactus realization of d if and only if $\omega_{odd} \leq \omega_1$ and $m \leq \left\lfloor \frac{3(n-1)-\omega_1}{2} \right\rfloor$.*

Proof. If there is a core cactus realization G of d, then $\omega_1 \geq \omega_{odd}$ by Lemma 13. Observe that in this case $\beta = \omega_1$ by definition. Hence, Theorem 4 implies that

$$m \leq \left\lfloor \frac{3(n-1)-\beta}{2} \right\rfloor = \left\lfloor \frac{3(n-1)-\omega_1}{2} \right\rfloor.$$

The converse is proved by induction on $\omega_1 + \omega_{odd}$. In the base case, $\omega_1 = \omega_{odd} = 0$, and thus d_i is even for every n. Since $m \leq \lfloor 1.5(n-1) \rfloor$, it follows by Theorem 6 that there exists a realization for d as a bridge-less cactus.

For the inductive step, there are two cases. If d contains an odd number $d_j > 1$, then since $\omega_{odd} \leq \omega_1$, it must be that $d_n = 1$. Let d' be the sequence obtained by subtracting 1 from d_j and removing n. Observe that $\omega'_{odd} = \omega_{odd} - 1 \leq \omega_1 - 1 = \omega'_1$. In addition. $\sum_i d'_i = \sum_i d_i - 2 > 2n - 2 = 2n'$, and

$$\sum_i d'_i = \sum_i d_i - 2 \leq 2 \left\lfloor \frac{3(n-1)-\omega_1}{2} \right\rfloor - 2 = 2 \left\lfloor \frac{3(n-1)-\omega_1-2}{2} \right\rfloor = 2 \left\lfloor \frac{3(n'-1)-\omega'_1}{2} \right\rfloor.$$

By the inductive hypothesis there is a realization G' of d' as a core cactus. We obtain a core cactus realization G of d by adding the edge (j, n).

Suppose that d does not contain an odd number $d_j > 1$, but $d_n = 1$. Then, it must be the case that $d_{n-1} = 1$ and $d_1 \geq 4$. Let d' be the sequence which is obtained by subtracting 2 from d_1 and removing $n-1$ and n. Observe that $0 = \omega'_{odd} \leq \omega'_1 = \omega_1 - 2$. Also, $\sum_i d'_i = \sum_i d_i - 4 > 2n - 4 = 2n'$, and

$$\sum_i d'_i = \sum_i d_i - 4 \leq 2 \left\lfloor \frac{3(n-1)-\omega_1}{2} \right\rfloor - 4 = 2 \left\lfloor \frac{3(n-1)-\omega_1-4}{2} \right\rfloor = 2 \left\lfloor \frac{3(n'-1)-\omega'_1}{2} \right\rfloor.$$

By the induction hypothesis there is a core cactus realization G' of d'. We obtain a core cactus realization G of d by adding the edges $(1, n-1)$ and $(1, n)$.   □

The algorithm which is implied by the proof of Theorem 9 initially connects 1-degree vertices to vertices with odd degree which is greater than 1. When $\omega_{odd} = 0$, it attaches two degree-1 vertices to a vertex with even degree which is larger than 2. When all degrees are even, it constructs a bridge-less graph.

Corollary 5. *A degree sequence d where $n \geq 3$, $n < m \leq \left\lfloor \frac{3(n-1)-\omega_1}{2} \right\rfloor$ and $\omega_{odd} \leq \omega_1$ has a linear time algorithm that computes a core cactus realization of d, where all cycles except maybe one are triangles, and all other edges are connected to cycle vertices.*

6.2   Core Bi-cactus Graph Realization

In this section we provide a characterization and a realization algorithm for core bi-cacti. The approach is similar to the one for core cactus graphs. One difference is that we use cycles of length 4 and not triangles. Another is that we sometimes need to use a correction as shown in Fig. 2.

We need the following technical lemma, whose proof is omitted.

Lemma 14. $\left\lfloor \frac{1}{2} \left\lfloor \frac{4(n-1)+1}{3} \right\rfloor \right\rfloor = \left\lfloor \frac{2(n-1)}{3} \right\rfloor.$

Theorem 10. *A degree sequence d where $n \geq 3$, $\sum_i d_i > 2n$ and $\omega_1 > 0$ has a core bi-cactus realization if and only if $\omega_1 \geq \omega_{odd}$ and $m \leq \left\lfloor \frac{4(n-1)-\omega_1}{3} \right\rfloor.$*

Proof. If there is a core bi-cactus realization of d, then by Lemma 13 we have that $\omega_1 \geq \omega_{odd}$. Moreover, Theorem 5 implies that $m \leq \left\lfloor \frac{4(n-1)-\beta}{3} \right\rfloor = \left\lfloor \frac{4(n-1)-\omega_1}{3} \right\rfloor$.

The converse is proved by induction on $\omega_1 + \omega_{odd}$. In the base case, d_i is even for every n, and m is even. Since $m \leq 2\lfloor 2(n-1)/3 \rfloor$, it follows by Theorem 8 that there is a realization for d as a bridge-less bi-cactus.

For the inductive step, there are two cases. If d contains an odd number $d_j > 1$, then since $\omega_{odd} \leq \omega_1$, it must be that $d_n = 1$. Let d' be the sequence obtained by subtracting 1 from d_j and removing n. Observe that $\omega'_{odd} = \omega_{odd} - 1 \leq \omega_1 - 1 = \omega'_1$, and $\sum_i d'_i = \sum_i d_i - 2 > 2n - 2 = 2n'$, In addition,

$$\sum\nolimits_i d'_i = \sum\nolimits_i d_i - 2 \leq 2 \left\lfloor \frac{4(n-1)-\omega_1}{3} \right\rfloor - 2 = 2 \left\lfloor \frac{4(n'-1)-\omega'_1}{3} \right\rfloor.$$

First, suppose that $\omega'_1 \geq 1$ or m' is even. By the inductive hypothesis there is a realization G' of d' as a core bi-cactus. We obtain a core cactus realization G of d by adding the edge (j, n). If $\omega'_1 = 0$ and m' is odd, we create a sequence d^* by removing $d'_{n'} = 2$. Observe that $n^* = n' - 1$, $m^* = m' - 1$, and $\omega'_1 = 0$. Hence, $\sum_i d^*_i = \sum_i d'_i - 2 > 2n' - 2 = 2n^*$, and

$$m^* = m' - 1 \leq \left\lfloor \frac{4(n'-1)}{3} \right\rfloor - 1 = \left\lfloor \frac{4(n^*-1)+1}{3} \right\rfloor.$$

Since m^* is even, we have that $m^* \leq 2 \left\lfloor \frac{1}{2} \left\lfloor \frac{4(n^*-1)+1}{3} \right\rfloor \right\rfloor$. By Lemma 14 it follows that $m^* \leq 2\lfloor 2(n^* - 1)/3 \rfloor$. By the inductive hypothesis there is a realization G^* of d^* as a core bi-cactus. We obtain a core bi-cactus realization G of d by adding the edges $(j, n-1)$ and $(n-1, n)$.

Suppose that d does not contain an odd number $d_j > 1$, but $d_n = 1$. Then, it must be that $d_{n-1} = 1$ and $d_1 \geq 4$. Let d' be the sequence obtained by subtracting 2 from d_1 and removing $n - 1$ and n. Observe that $0 = \omega'_{odd} \leq \omega'_1 = \omega_1 - 2$, and $\sum_i d'_i = \sum_i d_i - 4 > 2n - 4 = 2n'$. Also,

$$\sum\nolimits_i d'_i = \sum\nolimits_i d_i - 4 \leq 2 \left\lfloor \frac{4(n-1)-\omega_1}{3} \right\rfloor - 4 = 2 \left\lfloor \frac{4(n'-1)-\omega'_1}{3} \right\rfloor.$$

Suppose that $\omega'_1 \geq 1$ or m is even. By the inductive hypothesis there is a realization G' of d' as a core cactus. We obtain a core bi-cactus realization G of d by adding the edges $(1, n-1)$ and $(1, n)$.

If $\omega'_1 = 0$ and m is odd, continue as in the first case. We obtain a core bi-cactus realization G of d by adding the edges $(1, n-2)$, $(n-2, n-1)$, and $(1, n)$. □

Corollary 6. *Let d be a degree sequence such that $n \geq 3$, $\omega_1 \geq \max\{\omega_{odd}, 1\}$ and $n < m \leq \left\lfloor \frac{4(n-1)-\omega_1}{3} \right\rfloor$. Then, there is a linear time algorithm that computes a core bi-cactus realization of d, where all cycles except maybe one are of length 4. Also, all other edges, but maybe one, are connected to cycle vertices.*

7   Realization of Cactus and Bi-cactus Graphs

7.1   Cactus Graph Realization

Characterization and a realization algorithm for cactus graphs were given in [30].

Theorem 11 ([30]).   *Let d be a degree sequence such that $n \geq 3$ and $\sum_i d_i \geq 2n$. Then there is a cactus realization of d if and only if $m \leq \left\lfloor \frac{3(n-1)-\beta}{2} \right\rfloor$.*

The algorithm which is implied by our proof of Theorem 11 works as follows. If $\omega_1 \geq \omega_{odd}$ it constructs a core cactus. Otherwise, it connects 1-degree vertices to vertices with an odd degree which is greater than 1. When $\omega_1 = 0$, and as long as $\omega_{odd} > 0$, it adds a triangle consisting of two degree-2 vertices and a vertex j with the smallest odd degree. This is done until the degree of j becomes 1. If the volume becomes $2n$, then a unicyclic graph is constructed. Otherwise, a sequence consisting of even numbers is obtained, and a bridge-less cactus is created.

Corollary 7.   *Let d be a degree sequence such that $n \geq 3$, $n - 1 \leq m \leq \left\lfloor \frac{3(n-1)-\beta}{2} \right\rfloor$. There is a linear time algorithm that computes a cactus realization of d, where all cycles except maybe one are triangles.*

7.2   Bi-cactus Graph Realization

In this section we provide a full characterization and a realization algorithm for bicactus graphs. The approach is similar to the one for cactus graphs.

Theorem 12.   *Let d be a degree sequence such that $n \geq 3$, $\sum_i d_i > 2n$ and $\omega_{odd} + \omega_1 \geq 2$. There is a bi-cactus realization of d if and only if $m \leq \left\lfloor \frac{4(n-1)-\beta}{3} \right\rfloor$.*

Proof. If there is a bi-cactus realization of d, then $m \leq \left\lfloor \frac{4(n-1)-\beta}{3} \right\rfloor$ by Theorem 5.

For the other direction, suppose that $m \leq \left\lfloor \frac{4(n-1)-\beta}{3} \right\rfloor$. If $\omega_{odd} < \omega_1$, then Theorem 10 implies that there is a realization of d as a core bi-cactus.

Suppose that $\omega_{odd} \geq \omega_1$. That is, $\beta = (\omega_1 + \omega_{odd})/2$. We prove the claim by induction on $\omega_1 + \omega_{odd}$ and on ω_{odd}. In the base case, there are two options.

- If $\sum_i d_i = 2n$, then there exists a bi-unicyclic realization of d.
- If d_i is even for every n. Since $\sum_i d_i \leq 2 \left\lfloor \frac{2(n-1)}{3} \right\rfloor$, it follows by Theorem 8 that there exists a realization for d as a bridge-less bi-cactus.

For the inductive step, there are two cases. First, supposed that $\omega_1 > 0$. Since $\omega_{odd} \geq \omega_1$, the sequence d must contain an odd number $d_j \geq 3$. Let d' be the sequence which is obtained by subtracting 1 from d_j and removing n. Observe that $\omega'_{odd} = \omega_{odd} - 1 \geq \omega_1 - 1 = \omega'_1$, $\sum_i d'_i = \sum_i d_i - 2 > 2n - 2 = 2n'$, and that

$$\sum_i d'_i = \sum_i d_i - 2 \leq 2 \left\lfloor \frac{4(n-1) - \frac{1}{2}(\omega_1 + \omega_{odd})}{3} \right\rfloor - 2 = 2 \left\lfloor \frac{4(n'-1) - \frac{1}{2}(\omega'_1 + \omega'_{odd})}{3} \right\rfloor.$$

First, suppose that $\omega'_{odd} \geq 1$ or m' is even. By the induction hypothesis there is a bi-cactus realization G' of d'. We get a bi-cactus realization G of d by adding the edge (j, n).

Next, assume that $\omega'_{odd} = \omega'_1 = 0$ and m' is odd. In this case, we construct a sequence d^* as in the first case of Theorem 10. By the inductive hypothesis there is a realization G^* of d^* as a bi-cactus. We obtain a bi-cactus realization G of d by adding the edges $(j, n-1)$ and $(n-1, n)$.

The second case is when $\omega_1 = 0$. In this case, $\omega_{odd} \geq 2$. Let $d_j \geq 3$ be the smallest odd number in d. Since $\sum_i d_i \leq 2 \left\lfloor \frac{4(n-1)-\beta}{3} \right\rfloor$, it must be that $d_{n-2} = 2$. Let d' be the sequence which is obtained by removing $n-2$, $n-1$, and n and subtracting 2 from d_j. Observe that $\omega'_{odd} + \omega'_1 = \omega_{odd} + \omega_1$. In particular, if $d_j = 3$, then $\omega'_1 = 1$, and otherwise $\omega'_1 = 0$. Also, $\sum_i d'_i = \sum_i d_i - 8 > 2n - 8 = 2n' - 2$, namely, $\sum_i d'_i \geq 2n'$. In addition,

$$\sum_i d'_i = \sum_i d_i - 8 \leq 2 \left\lfloor \frac{4(n-1) - \frac{1}{2}(\omega_1 + \omega_{odd})}{3} \right\rfloor - 8 = 2 \left\lfloor \frac{4(n'-1) - \frac{1}{2}(\omega'_1 + \omega'_{odd})}{3} \right\rfloor.$$

By the induction hypothesis d' has a realization as a bi-cactus G'. We obtain a realization G for d by adding a cycle of the vertices j, $n-2$, $n-1$, and n.   □

The algorithm which is described in the proof of Theorem 12 works similarly to the one for cacti. Also, recall that the case where $\beta = 0$ (i.e., $\omega_1 = \omega_{odd} = 0$) is covered by Corollary 4.

Corollary 8. *Let d be a degree sequence such that $n \geq 3$, $\beta > 0$ and $n - 1 \leq m \leq \left\lfloor \frac{4(n-1)-\beta}{3} \right\rfloor$. There is a linear time algorithm that computes a bi-cactus realization of d, where all cycles except maybe one are of length 4.*

References

1. Adams, P., Nikolayevsky, Y.: Planar bipartite biregular degree sequences. Discr. Math. **342**, 433–440 (2019)
2. Agnarsson, G., Greenlaw, R.: Graph Theory: Modeling, Applications, and Algorithms. Prentice (2006)
3. Aigner, M., Triesch, E.: Realizability and uniqueness in graphs. Discr. Math. **136**, 3–20 (1994)
4. Arcak, M.: Diagonal stability on cactus graphs and application to network stability analysis. IEEE Trans. Autom. Control **56**(12), 2766–2777 (2011)
5. Bar-Noy, A., Böhnlein, T., Peleg, D., Ran, Y., Rawitz, D.: On key parameters affecting the realizability of degree sequences. In: 49th MFCS. LIPIcs, vol. 306, pp. 1:1–1:16 (2024)
6. Bar-Noy, A., Böhnlein, T., Peleg, D., Ran, Y., Rawitz, D.: Sparse graphic degree sequences have planar realizations. In: 49th MFCS. LIPIcs, vol. 306, pp. 18:1–18:17 (2024)
7. Bar-Noy, A., Böhnlein, T., Peleg, D., Ran, Y., Rawitz, D.: Approximate realizations for outerplanaric degree sequences. J. Comput. Syst. Sci. **148**, 103588 (2025)

8. Bar-Noy, A., Böhnlein, T., Peleg, D., Rawitz, D.: On realizing a single degree sequence by a bipartite graph. In: 18th SWAT. LIPIcs, vol. 227, pp. 1:1–1:17 (2022)
9. Bar-Noy, A., Böhnlein, T., Peleg, D., Rawitz, D.: On the role of the high-low partition in realizing a degree sequence by a bipartite graph. In: 47th MFCS. LIPIcs, vol. 241, pp. 14:1–14:15 (2022)
10. Bar-Noy, A., Böhnlein, T., Peleg, D., Rawitz, D.: Forcibly bipartite and acyclic (uni-)graphic sequences. Discr. Math. **346**(7), 113460 (2023)
11. Beineke, L.W., Schmeichel, E.F.: Degrees and cycles in graphs. Ann. N. Y. Acad. Sci. **319**, 64–70 (1979)
12. Bíyíkoglu, T.: Degree sequences of halin graphs, and forcibly cograph-graphic sequences. Ars Comb. **75** (2005)
13. Boesch, F.T., Harary, F.: Unicyclic realizability of a degree list. Networks **8**, 93–96 (1978)
14. Bose, P., et al.: A characterization of the degree sequences of 2-trees. J. Graph Theory **58**(3), 191–209 (2008)
15. Choudum, S.A.: Characterization of forcibly outerplanar graphic sequences. In: Combinatorics and Graph Theory, pp. 203–211 (1981)
16. Choudum, S.A.: A simple proof of the Erdös-Gallai theorem on graph sequences. Bull. Austral. Math. Soc. **33**(1), 67–70 (1991)
17. Dahl, G., Flatberg, T.: A remark concerning graphical sequences. Discr. Math. **304**(1–3), 62–64 (2005)
18. Erdős, P., Gallai, T.: Graphs with prescribed degrees of vertices [Hungarian]. Mat. Lapok (N.S.) **11**, 264–274 (1960)
19. Fanelli, S.: On a conjecture on maximal planar sequences. J. Graph Theory **4**(4), 371–375 (1980)
20. Fanelli, S.: An unresolved conjecture on nonmaximal planar graphical sequences. Discr. Math. **36**(1), 109–112 (1981)
21. Gale, D.: A theorem on flows in networks. Pacific J. Math. **7**, 1073–1082 (1957)
22. Gupta, G., Joshi, P., Tripathi, A.: Graphic sequences of trees and a problem of Frobenius. Czechoslovak Math. J. **57**, 49–52 (2007)
23. Hakimi, S.L.: On realizability of a set of integers as degrees of the vertices of a linear graph -I. SIAM J. Appl. Math. **10**(3), 496–506 (1962)
24. Havel, V.: A remark on the existence of finite graphs [in Czech]. Casopis Pest. Mat. **80**, 477–480 (1955)
25. Hawkins, A., Hill, A., Reeve, J., Tyrrell, J.: On certain polyhedra. Math. Gaz. **50**(372), 140–144 (1966)
26. Jao, K.F., West, D.B.: Vertex degrees in outerplanar graphs. J. Comb. Math. Comb. Comput. **82**, 229–239 (2012)
27. Lotker, Z., Majumdar, D., Narayanaswamy, N.S., Weber, I.: Sequences characterizing k-trees. In: Chen, D.Z., Lee, D.T. (eds.) COCOON 2006. LNCS, vol. 4112, pp. 216–225. Springer, Heidelberg (2006). https://doi.org/10.1007/11809678_24
28. Nishi, T., Chua, L.O.: Topological proof of the Nielsen-Willson theorem. IEEE Trans. Circuits Syst. **33**(4), 398–405 (1986)
29. Paten, B., et al.: Cactus graphs for genome comparisons. J. Comput. Biol. **18**(3), 469–481 (2011)
30. Ramachandra Rao, A.: Degree sequences of cacti. In: Combinatorics and Graph Theory. LNM, pp. 410–416 (1981)
31. Rao, S.B.: A survey of the theory of potentially p-graphic and forcibly p-graphic degree sequences. In: Combinatorics and Graph Theory. LNM, vol. 885, pp. 417–440 (1981)

32. Ryser, H.: Combinatorial properties of matrices of zeros and ones. Canad. J. Math. **9**, 371–377 (1957)

33. Schmeichel, E.F., Hakimi, S.L.: On planar graphical degree sequences. SIAM J. Appl. Math. **32**, 598–609 (1977)

34. Sysło, M.M.: Characterizations of outerplanar graphs. Discr. Math. **26**(1), 47–53 (1979)

35. Tripathi, A., Tyagi, H.: A simple criterion on degree sequences of graphs. Discr. Appl. Math. **156**(18), 3513–3517 (2008)

36. Tripathi, A., Venugopalan, S., West, D.B.: A short constructive proof of the Erdös-Gallai characterization of graphic lists. Discr. Math. **310**(4), 843–844 (2010)

37. Tripathi, A., Vijay, S.: A note on a theorem of Erdös & Gallai. Discr. Math. **265**(1–3), 417–420 (2003)

38. Wagner, D., Wolf, M.: The complexity of flow expansion and electrical flow expansion. In: Bureš, T., et al. (eds.) SOFSEM 2021. LNCS, vol. 12607, pp. 431–441. Springer, Cham (2021). https://doi.org/10.1007/978-3-030-67731-2_32

39. Wang, D., Kleitman, D.: On the existence of n-connected graphs with prescribed degrees ($n > 2$). Networks **3**, 225–239 (1973)

On Two Simple[st] Learning Tasks

Omrit Filtser[1]([✉]), Kien Huynh[3], Anastasia Lemetti[3], Joseph Mitchell[2],
Tatiana Polishchuk[3], and Valentin Polishchuk[3]

[1] The Open University of Israel, Ra'anana, Israel
omrit.filtser@gmail.com
[2] Stony Brook University, Stony Brook, USA
[3] Linköping University, Linköping, Sweden

Abstract. We consider two very basic problems – one in unsupervised and one in supervised learning. In the former, we are given a set of points and have to label half of the points red and half the points blue so as to maximize the red–blue separation, i.e., the length of a shortest bichromatic edge. In the latter, the data (points in the plane) are already labeled red and blue, and we seek a linear classifier (a separator of the two given point sets) that can be described using the smallest integers. We give algorithms for both problems. Our solutions are simple; the main contribution of the paper is highlighting the problems and their algorithmic solutions, which, to our knowledge, have not been presented previously, despite the problems being fundamental to the field. We also consider related problems.

Keywords: Computational geometry · Machine learning ·
Classification · Clustering · Exact algorithms

1 Introduction

Artificial intelligence (AI), convolutional neural networks (CNN) and machine learning (ML) are trendy topics both in [computer] science and in the society at large, heated by their outstanding performance in data processing. The two most basic tasks in data handling are clustering (splitting the data into well separated groups) and classification (discriminating between different data classes).

Clustering is an archetypal example of *unsupervised* learning: the input data comes without any labels, and the goal is to label the data by splitting it into the clusters. In contrast, classification is a classical *supervised* learning task: the data comes labeled (the name "supervised" comes from the vision that the labels are provided by the "supervisor"), and the goal is to build a classifier – a simple summary of the data (in AI parlance, provide data "generalization") so that the class of any new data point can be estimated by feeding the point into the classifier.

One way to formalize a simplest clustering problem is:

Problem 1. Given a set $S \subset \mathbb{R}^2$ of $2n$ points in the plane, split the points into n red points R and n blue points B so that the distance $\text{dist}(R, B)$ between R and B is maximized, where the distance between two sets $\text{dist}(R, B) = \min_{r \in R, b \in B} |rb|$ is the minimum distance between their members.

I. Finocchi and L. Georgiadis (Eds.): CIAC 2025, LNCS 15679, pp. 276–291, 2025.
https://doi.org/10.1007/978-3-031-92932-8_18

That is, the data is just a set of points in 2d – a simplest setup. Further, we want to separate the points into only two classes, which really makes the problem as simple as possible. This is in line with the general agenda of this paper: consider the simplest possible setup.[1] For instance, even though our algorithm for Problem 1 (Sect. 2) extends to more sophisticated scenarios (e.g., clustering into an arbitrary number r and $b = 2n - r$ red and blue points, partitioning into more than 2 classes, including hierarchical clustering), our main goal is to formulate and solve the most basic, simplest version. In particular, we are not competing with the advanced state-of-the-art clustering techniques, which work for multidimensional data in various metrics [25,50].

When it comes to classification, simplicity is at the heart of ML (with something as classical as Occam's Razor being cited as a grounding principle of ML [7,11,20, 24,45,59]; see also [62]): the very idea of "learning" lies in simplifying the data – the machine replaces the potentially complicated boundary between the classes by a "divider" or "discriminant" having low description complexity. The simplest setup here is finding a *linear* discriminant between *linearly separable* classes in 2d, i.e., a line that separates red points R from blue points B; moreover, to decrease the number of sought parameters, we consider linear separators without bias, i.e., lines of the form $y = kx$. A natural measure of the description complexity of such a separator is the maximum (absolute) value of the integers needed to represent $k = p/q$ as a rational number.[2] The above considerations crystallize into the following simplest 2-class linear separability problem:

Problem 2. Given red and blue points $R, B \subset \mathbb{R}^2$, find a line $y = \dfrac{p}{q}x$, with $p, q \in \mathbb{Z}$, that separates R and B and is such that $\max(|p|, |q|)$ is minimized.

The above problem and its solution can be extended to classifiers with bias and more generally to separating points in higher (but fixed) dimensions (Sect. 3.2). In addition to the theoretical interest (complexity of numbers representation arises as a subject of research dealing with computer word size [1,10,14], geometric rounding [32,35,39,51,52], stability of numerical algorithms and other areas; in particular, counting bits is central in distributed computing when studying communication complexity), computations with small numbers are faster and more energy-efficient [18], which is especially important for small chips (using large numbers may be even impossible on certain nanoscale hardware): in neuromorphic computing, *quantization* [42,44,57,63] is used to avoid high-precision numbers inside a deep NN. Because quantization errors propagate through the network, instead of uniformly quantizing the whole network (i.e., using the same precision everywhere), it makes sense to use a separate precision for each neuron, while ensuring that its output remains correct even after the quantized weights

---

[1] In its turn, this is in line with the folklore advice "If you can't solve a problem, there is a simpler problem you can't solve: find it" [58] which is attributed to John Conway quoting George Pólya in the foreword to Pólya's book "How to solve it" [53].

[2] This is in line with the margin note at p. 307 of the book "Concrete mathematics" [30], which refers to "a little-known Einsteinian assertion: 'God does not throw huge denominators at the universe'.".

are used. Note that the *input* numbers with which neuromorphic devices work (as well as the numbers output by the hardware), may be large (e.g., the input may come from a high-definition camera); it is the numbers used *on* the chip, encoded by various gates and fine-tuned properties of tiny areas of the chip, whose size should be kept small. Our setup reflects the above scenario: the input (the points in R and B) may use large numbers, and we seek to minimize the numbers used internally in the classifier (a single neuron).

Research on clustering and classification is vast and is impossible to summarize in this paper (we refer to dedicated surveys [6,25,40,50]); computational-geometry work (e.g., [2,8,31,34,46]) is surveyed in recent papers [27,28] and in the PhD thesis [56]. However, the specific problems that we consider here were not studied before. We also note that when considering geometric problems we work in realRAM model of computation – standard in computational geometry.

Roadmap. The next section addresses the clustering (Problem 1). Section 3 describes solutions to the classification (Problem 2); Sect. 3.2 discusses higher-dimensional extensions. Section 4 studies splitting into linearly separable clusters. We conclude with some open questions in Sect. 5.

2   Shortest Bichromatic Edge

The clustering can be done by following the algorithm of [33] for the *bottleneck graph partition* problem. Specifically, consider first the decision version of Problem 1:

> Is it possible to color the given $2n$ points S red and blue (n red, n blue) so that the minimum red–blue distance is larger than 1?

To solve the problem, build the *unit disk graph* G on S, which connects pairs of points of S that are within distance 1 from each other. Let $n_1, \ldots, n_k$ be the sizes of the connected components of G, where k is the number of the components. If the points S are colored so that there is no bichromatic edge of length at most 1, then all points in a connected component of the graph must have the same color. Hence, finding the balanced coloring amounts to partitioning the numbers $\{n_1, \ldots, n_k\}$ into two subsets so that the sum of the numbers in each subset is n. This is exactly the 2-PARTITION problem [26]. Since the total sum of the numbers $n_1 + \cdots + n_k = 2n$, the problem can be solved in $O(n \log^2 n)$ time [38, Theorem 3.1]. Finally, to find the largest possible balanced separation, we could solve the above decision problem for each of the $O(n^2)$ inter-point distances in S; to speed up the algorithm, compute the inter-point distances and do binary search on them – invoking the decision oracle only $O(\log n)$ times. The total running time is $O(n^2 + n \log^3 n) = O(n^2)$ (the bottleneck is computing the inter-point distances; selecting the kth largest distance for any $1 \leq k \leq \binom{n}{2}$ can be done, without sorting the $\binom{n}{2}$ distances, in time $O(n^{4/3} \text{polylog} \, n)$ – see [5, Section 8.4] and [13,29,36,55,61]):

Proposition 1. *Problem 1 can be solved in $O(n^2)$ time.*

2.1   Extensions

The above algorithm works for

- point sets $S \subset \mathbb{R}^d$ in arbitrary dimension d (there was nothing specific to $d = 2$) – the running time becomes $O(n^2 d)$,
- clustering into an arbitrary number r and $b = 2n - r$ of red and blue points (the algorithm in [38] works for any subset sum), and
- clustering into a constant number k of clusters of given sizes $c_1, \ldots, c_k$ (the running time of Proposition 1 no longer applies though, as k enters the exponent of the running time because instead of 2-PARTITION the problem is reduced to k-way partitioning [49]).

For non-constant k, it is NP-hard already to decide whether $3k$ given numbers $a_1, \ldots, a_{3k}$ can be split into k triples with equal sums – this is the 3-PARTITION problem, which is (strongly) NP-hard. We can reduce 3-PARTITION to splitting into k clusters by placing a_1 points of S at a common point $p_1 \in \mathbb{R}^2$, a_2 points at a point p_2, and so on. All points of S at a common location must get the same color. Choose well separated points $p_i, \ldots, p_{3k}$, so that the distance between any pair of them is larger than 1; then, no bichromatic edge is shorter than 1. The only question is whether the colors can be assigned in order to satisfy the cluster sizing constraints (equal sizes); this is the 3-PARTITION problem.

2.2   A Faster Algorithm

Consider disks of increasing radius ρ centered on points of S. Candidates for the optimum are among those ρ's for which the union of the disks merges two connected components (a candidate optimal distance is 2ρ, for a merging radius ρ). Such merging occurs along edges of the minimum spanning tree (MST) of S – this is the standard property of MST: each edge is a shortest edge connecting the subsets. We compute the MST and do the binary search on its edge lengths. For a given length ℓ, we remove from the MST all longer edges and identify the connected components (the trees in the forest); we then solve the corresponding 2-PARTITION problem. Computing the MST takes time $O(n \log n)$ in 2 dimensions, $O(n^{4/3})$ in 3 dimensions, and $O(n^{2 - \frac{2}{\lceil d/2 \rceil + 1}})$ in d dimensions [3,15] (the algorithms in [3,15] are randomized, but the need for randomness can be eliminated [4] using techniques from [47] – see e.g., footnote 1 in [23]); solving the partition takes $O(n \log^3 n)$ time. We thus have:

Theorem 1. *Problem 1 can be solved in time* $O(n \log^3 n)$ *in 2d,* $O(n^{4/3})$ *in 3d, and* $O(n^{2 - \frac{2}{\lceil d/2 \rceil + 1}})$ *in d dimensions.*

It is possible that our clustering problem cannot be solved faster than finding an MST. Indeed, it is known that constructing an MST is asymptotically equivalent to finding a bichromatic closest pair (BCP) [3,12,41]. On the one hand, after an MST is built, the shortest bichromatic edge is an edge of the tree. On the other hand, MST can be constructed by colorings (of subsets of S) based on "narrow cones" [3], covering [12] or packing [41], and using shortest bichromatic edges. Note that our problem is the coloring, with the added colors cardinality constraints. Even though this does not give a lower bound for our problem, it suggests that our clustering is related to BCP which is equivalent to computing MST. In Appendix A we show that BCP can be reduced to *weighted* balanced clustering.

3 Smallest Numbers

This section solves Problem 2: find a red–blue separating line $y = \frac{p}{q}x$ through the origin such that $p, q \in \mathbb{Z}$ and $\max(|p|, |q|)$ is minimized. Section 3.2 considers extensions to higher dimensions.

Start by computing the convex hulls of R and of B: if either of the convex hulls contains the origin (Fig. 1(a)), then the problem is infeasible. Otherwise, draw the tangents from the origin to the convex hulls: the tangents define 4 cones, and each pair of opposite cones forms an *hourglass*. If R and B are linearly separable, one of the hourglasses has all separating lines passing through it. If this hourglass contains the x-axis (Fig. 1(b)), then the x-axis is the solution (separating line $y = 0$). In what follows we will thus assume that the tangents have same-sign slopes, w.l.o.g. – positive (Fig. 1(c)). Let u, l be these slopes, with $u < l$. If $u \leq 1 \leq l$, then $y = x$ is the solution; thus we will assume w.l.o.g. that $0 < u < l < 1$ (the case $1 < u < l$ is symmetric). Problem 2 then reduces to finding a smallest-denominator rational number between u and l, or in geometric terms – finding the smallest-abscissa grid point $(q, p) \in \mathbb{Z}^2$ in the wedge W between the tangents.

Let $r^* \in R, b^* \in B$ be the extreme red and blue points that define u and l. If either of r^*, b^* is an integer grid point (say, $r^* = (a, b) \in \mathbb{Z}^2$), then the line $y = \frac{b}{a}x$ through O and r^* is a separator. That is, there exists a separator with denominator a, and hence, the search for a smallest-abscissa integer point in the wedge W can be confined to abscissas less than a: for every $i < a$ we can check whether there is an integer j between ui and li – if yes then $y = \frac{j}{i}x$ is a candidate separator with the denominator i; the solution is the candidate with the smallest i. Let N be the largest integer in the description of R and B; let $n = |R \cup B|$. Since the convex hulls can be computed in $O(n \log n)$ time, and since $a \leq N$, we have

Proposition 2. *If r^* or b^* is a grid point, Problem 2 can be solved in $O(n \log n + N)$ time.*

Suppose now that r^* and b^* are rational non-grid points (Fig. 1(d)):

$$r^* = \left(\frac{r_n^x}{r_d^x}, \frac{r_n^y}{r_d^y} \right) \quad , \quad b^* = \left(\frac{b_n^x}{b_d^x}, \frac{b_n^y}{r_d^y} \right),$$
$$r_n^x, r_d^x, r_n^y, r_d^y, b_n^x, b_d^x, b_n^y, b_d^y \in \mathbb{Z}^+$$

Then $\overline{r^*} = (r_n^x r_d^y, r_d^x r_n^y), \overline{b^*} = (b_n^x b_d^y, b_d^x b_n^y) \in \mathbb{Z}^2$ are grid points defining the same wedge W. Applying Proposition 2, to $\overline{r^*}, \overline{b^*} \in \mathbb{Z}^2$, we obtain

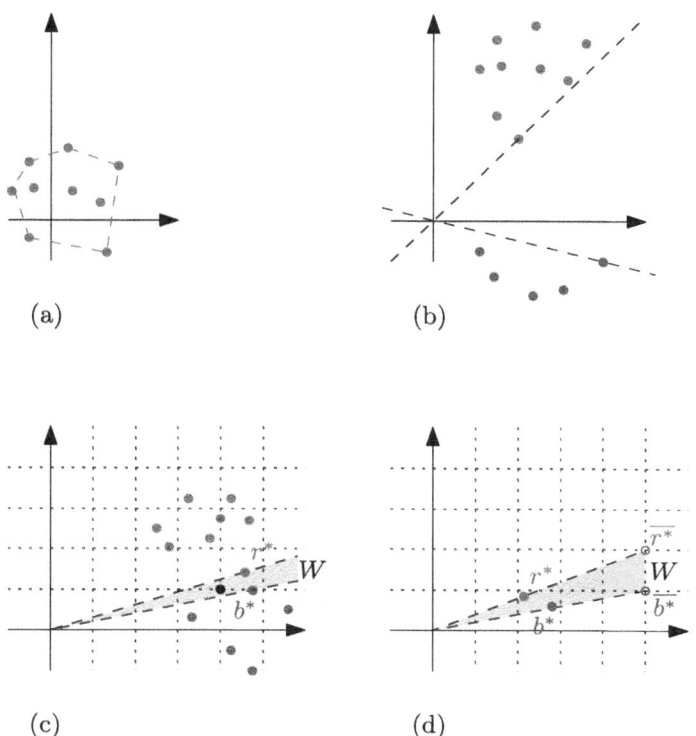

Fig. 1. (a) The convex hull of R (dashed) contains the origin: any line through the origin has red points on both of its sides. (b) $y = 0$ is the solution. (c) Grid lines are dotted; the solution is the line through the origin and the black point. (d) $r^*, b^* \in \mathbb{Q}^2, \overline{r^*}, \overline{b^*} \in \mathbb{Z}^2$. (Color figure online)

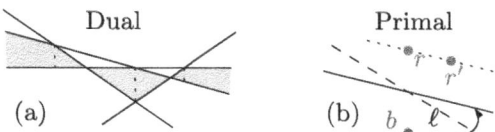

Fig. 2. (a) The median level of the arrangement is gray; its vertical decomposition D is dotted. (b) For max margin separation, ℓ (dashed) must pass through the midpoint of rb (so $\mathrm{dist}(\ell, r) = \mathrm{dist}(\ell, b)$), and is rotated as much as possible (solid) in the direction of the perpendicular bisector of rb while ℓ^* stays within the same cell of D, i.e., until ℓ becomes parallel to a line (dotted) through two points of S (if the bisector of rb separates R from B, then the bisector is the optimum – ℓ does not rotate past the bisector).

Proposition 3. *If $r^*, b^* \in \mathbb{Q}^2$ are rational points, Problem 2 can be solved in $O(n \log n + N^2)$ time, where N is the largest integer in the description of R and B.*

The algorithms in Propositions 2 and 3 run in pseudopolynomial time. A polynomial-time solution can be obtained by a "continued fraction" expansion of the number $m = (u + l)/2$. Specifically, the following definitions and properties of continued fractions may be found in any standard text, e.g., [54]. A *continued fraction* of m is

$$m \quad = \quad \cfrac{1}{a_1 + \cfrac{1}{a_2 + \cfrac{1}{a_3 + \cfrac{1}{a_4 + \cdots}}}}$$

where $a_1, \cdots \in \mathbb{Z}^+$. Rational numbers have finite continued fractions (every number has two expansions, which differ in the last term). A *convergent* of a continued fraction is the fraction truncated after a number of terms (e.g., the first two convergents of m are $1/a_1$ and $1/(a_1 + 1/a_2)$). The denominators of the convergents grow exponentially fast, which means that if m is a rational number with denominator N, then

- the expansion of m can be done in $O(\log N)$ time, and
- the number of integers between the denominators of two consecutive convergents is $O(\log N)$.

This implies that in $O(\log N)$ time we can

1. find the first convergent p/q that falls between u and l,
2. find the previous convergent $p^-/q^- \notin [u, l]$,
3. test, for every integer i between q^- and q, whether the wedge W contains a grid point with abscissa i.

The main property of continued fractions is that they give a best approximation with bounded denominators: a convergent is closer to m than any rational number whose denominator is at most that of the convergent. Thus, the above 3 steps solve Problem 2: the optimal denominator is the smallest found $i \in (q^-, q]$.

Proposition 4. *If $r^*, b^* \in \mathbb{Q}^2$ are rational points, Problem 2 can be solved in $O(n \log n + \log N)$ time, where N is the largest integer in the description of R and B.*

The algorithm above applies also to irrational r^*, b^*: irrational numbers have infinite continued fractions, but we can still do the expansion until a convergent p/q falls between u and l (Step 1), and perform Step 3. The running time is the time to find the optimal denominator:

Proposition 5. *Problem 2 can be solved in time $O(n \log n + \log q^*)$ where q^* is the optimum.*

Finally, in terms of n, the $O(n \log n)$ bottleneck of computing the convex hulls can be avoided by calculating the slopes of lines from O to points in R and in B: determining whether the origin is inside the convex hull of R can be done by a scan through the slopes (equivalently, it is a linear program in 1d—finding k such that $r_y \geq k r_x$ for all points $(r_x, r_y) \in R$—which can be solved in linear time [48]), and finding r^* and b^* is determining the extreme slopes. We thus have:

Theorem 2. *Problem 2 can be solved in output-sensitive time $O(n + \log q^*)$. If R and B are specified with integers smaller than N, then the problem can be solved in time $O(n + \log N)$ – linear in the input size.*

3.1   An IP Solution

The continued-fraction solution to Problem 2, described above, is very explicit and easily implementable; however, we do not see a way to extend it directly to higher dimensions (even though higher-dimensional generalizations of continued fractions, such as Stern–Brocot tree, do exist). A different way to solve Problem 2 is to reduce it to solving a constant number of integer programs (IPs) in constant dimensions. In a constant dimension d, IP can be solved in polynomial time (exponential in d, but in fact linear in the number of bits in the input) [43]. We now describe the IP-based solution to Problem 2.

We first give a "sloppy" solution that does not really work, but explains the idea; we then fix the little technicalities. The variables of the IP are $p, q, M \in \mathbb{Z}$ (M is the dummy variable, representing the objective function) and the constraints are:

1. For every red point $r = (x^r, y^r) \in R, y^r q - x^r p \geq 0$
2. For every blue point $b = (x^b, y^b) \in B, y^b q - x^b p \leq 0$
3. $M \geq p, M \geq q$

The first two constraints separate R from B; the last constraints imply that $M \geq \max(p, q)$. The objective is to minimize M, so in the optimal solution M will be equal to $\max(p, q)$.

There are a couple of issues with the above IP formulation. First, $p = q = M = 0$ is a feasible solution, which should not be the case. To deal with this, we test whether R and B can be separated by the x-axis ($p = 0, q = 1$) or by the y-axis ($p = 1, q = 0$) – this can be done by finding the extreme points of R and B in the corresponding direction. If such a separation is possible, we are done. Otherwise, we want to add the constraints that $p, q \neq 0$; however, such constraints are not linear. To linearize them, we add the constraints $p \geq 1, q \geq 1$ and scroll through all 4 assignments of the signs to p and q. For each assignment, we solve our IP after flipping the sign(s) of the corresponding coordinate(s) of the input points: e.g., if $p, q > 0$ we do not flip anything; if $p > 0, q < 0$, we flip the y-coordinates, etc. Our final solution is given by the best of the 4 IPs.

3.2   Higher Dimensions

In the d-dimensional generalization of Problem 2, we seek $d + 1$ integers $p_1, \ldots, p_d, q$ such that the hyperplane $p_1 x_1 + \cdots + p_d x_d = q$ separates R from B;

the goal is to minimize the maximum integer. The IP solution from the previous subsection extends to d dimensions for a constant d.

As in the 2d case, we scroll through all $2^d - 1$ proper subsets of $\{1, \ldots, d\}$ and for each subset S solve our problem assuming that $p_i = 0$ iff $i \in S$ – this simply reduces the number of the dimensions. In what follows we deal with the "full-dimensional" case ($S = \emptyset$): we add the constraints $p_1 \geq 1, \ldots, p_d \geq 1$. Also, as in Sect. 3.1, we scroll through all 2^d assignments of the signs to $p_1, \ldots, p_d$ and flip the signs of the input points along the dimension(s) i for which p_i is negative. The common constraints in all 2^d IPs is that for every red point $r = (x_1^r, \ldots, x_d^r) \in \mathbb{R}^d$ we add the constraint $p_1 x_1^r + \cdots + p_d x_d^r \geq q$; similar constraints but with $\leq$ are added for points in B. As in 2d, the constraints $M \geq p_1, \ldots, M \geq p_d, M \geq q$ ensure that minimizing M minimizes the maximum integer used in the separator. We thus obtain

Theorem 3. *In constant dimensions, Problem 2 can be solved in time linear in the size of the input.*

4   Linearly Separable Clustering

In this section we combine Problems 1 and 2, and study splitting a given set S of $2n$ points into 2 equal-size sets (red R and blue B) in order to maximize the separation between the clusters. We consider 2 measures of the separation:

Problem 3. The red–blue distance $\mathrm{dist}(R, B)$ – this is the same as in Problem 1, but with the constraint that the clusters are linearly separable.

Problem 4. The maximum width of the strip between R and B – this is maximum *margin* clustering, in the SVM (support vector machine) parlance [17]. We emphasize that just as elsewhere in the paper, our setup is much more basic than, for instance, in SVM clustering [9] where the data is mapped into a high-dimensional space, enclosed in a ball, etc.

We solve both problems by scrolling through all $O(n^{4/3})$ [19] ways of splitting S into linearly separable halves. A systematic way of going through all such partitions is to draw lines dual to points in S and build the *median level* of their arrangement A, i.e., points in the dual that have n lines above and n lines below (Fig. 2(a)). Such points correspond (via duality) to *halving lines* of S, i.e., lines having n points of S above (red) and n below (blue). In each cell of A, the red and blue subsets of S are the same for any halving line corresponding to the point in the cell.

To solve Problem 3 we march from cell to cell of the median level: in a new cell, two points of S swap their colors, so we can maintain the distance $\mathrm{dist}(R, B)$ in $O(n^\varepsilon)$ time per cell for any $\varepsilon > 0$ [22]. (Since red and blue are linearly separated, to find the initial distance between them we can compute their convex hulls and find the distance between them in $O(n \log n)$ time.)

For Problem 4 we additionally do the vertical decomposition D of the median level: the dual lines below/above any point ℓ^* in a cell σ of D are the same. In the primal plane any line ℓ, corresponding to a point $\ell^* \in \sigma$, hits the same points r, b of S when shifted or rotated while $\ell^* \in \sigma$. The best ℓ for $\ell^* \in \sigma$ can thus be determined in constant time: the optimal ℓ passes through the midpoint of rb (to equalize distances to r and b, and thereby make $\text{dist}(\ell, R) = \text{dist}(\ell, B)$) and is maximally rotated towards the bisector of rb (refer to Fig. 2(b)).

Since the median level can be built in $O(n^{4/3} \log n)$ time [21], we have:

Theorem 4. *Problems 3 and 4 can be solved in times* $O(n^{4/3+\varepsilon})$ *and* $O(n^{4/3} \log n)$ *respectively.*

The above algorithms generalize to higher dimensions, with the running times increasing to $O(n^{O(d)})$. Indeed, there are $O(n^d)$ halving hyperplanes (each defined by d points of S). For every hyperplane, it takes $O(n^2)$ time to find the red–blue distance, and to find the hyperplane maximizing the margin between R and B. For the latter, we use the support vector machinery from [17] that calculates the max-margin hyperplane between 2 given classes.

5   Conclusion

We formulated and solved simple basic problems in unsupervised and supervised learning. This opens up the door to a systematic study of exact algorithms for AI/NN/ML problems that are more advanced than ours, but are less advanced than the problems typically considered in the learning theory outlets (e.g., the Conference on Learning Theory) and books like [37], where the algorithms for such more complicated problems use probabilistic inference, deal with noise, etc. In particular, it may be of interest to identify and solve simplest problems in the *semi-supervised* learning [60,64].

Many open questions remain. One concrete open question is improving the efficiency of our algorithms or providing lower bounds. Also, we suspect that the classification problem is hard in high dimensions, but do not yet have a proof.

Finally, we mention another open problem related to classification of labeled non-linearly-separable data (we pose it as a decision problem):

Problem 5. Given red and blue points $R, B \subset \mathbb{R}^d$ and a number k, is there a subset $S \subset R \cup B$ of at most k points ($|S| \leq k$) such that each cell of the Voronoi diagram $VD(S)$ of S is monochromatic (i.e., any cell contains either only red and or only blue points)?

In machine learning terms, S is a small-size representative of the data: the color of every point in $R \cup B$ is the color of the closest site from S, and any new point $p \in \mathbb{R}^d$ may be classified as red or blue simply depending on which-color site is closest to p – this is the well known Nearest Neighbor (NN) classifier. (Eppstein [23] end Clarkson [16] gave efficient algorithms for removing points from $R \cup B$ while ensuring that *every* point in $\mathbb{R}^d$ keeps same-color NN that it had in S: the difference is that in Problem 5 we care only about points of $R \cup B$ being correctly classified by S, and also that [23] and [16] did not care about the number of "relevant" points that remain after the removal.) In Appendix B we show how to solve the problem for points on a line ($d = 1$); the complexity for dimensions $d \geq 2$ is open – we give an integer program (IP) for the problem (and present the output of the IP on several instances).

A   Bichromatic Closest Pair via Weighted Clustering

Suppose we are given two sets R, B of distinct points (no two coincide) and we seek to find dist(R, B). Let $b = |B|, r = |R|$, and assume w.l.o.g. that $b \leq r$. Let $b' = r + 1$; then, r and b' are relatively prime. Pick a point $s \in B$, and augment the set B by adding $b' - b$ ($b' - b \geq 1$) points to B, with all of them exactly at point s, so the set B now contains b' points. For each point $p \in B$, set its *weight* to be r, so the total weight of points in B is rb'. For each point $q \in R$, set its weight to be b', so the total weight of points in R is also rb'.

Let $S = R \cup B$, and consider the problem of splitting S into 2 equal-*weight* clusters in order to maximize the distance between the clusters – this is *weighted* balanced clustering, the weighted version of our Problem 1. The solution is forced to color B blue and R red: since r and b' are relatively prime, there is no other weight-balanced solution because it is not possible to replace $x < b'$ points of weight r from B by $y < r$ points of weight b' from R. Thus, if we know the distance between the clusters, we know the closest pair from R and B.

B   Representatives for NN Classification in 1D

The problem we solve is:

Problem 5-1: Problem 5 for points on a line. Given red and blue points $R, B \subset \mathbb{R}^1$ on the line and a number k, is there a subset $S \subset R \cup B$ of at most k points ($|S| \leq k$) such that each cell of the Voronoi diagram $VD(S)$ of S is monochromatic (i.e., any cell contains either only red and or only blue points)?

That is, the color of every point should coincide with the color of its cell in $VD(S)$.

The important notion for our solution to the problem is a *run* – a maximal sequence of points of the same color. Just to build up intuition, note that if the number of runs is less than $k/2$, then the answer is Yes (S may lie on the endpoints of the color runs); if the number of runs is larger than k, then the answer is No (because S must hit every run). So the problem is non-trivial only when the number of runs is between $k/2$ and k – we must decide which runs will have 1 point of S, and which 2.

The problem may be solved by searching for a shortest s-t path in the directed acyclic graph (DAG) G on $s \cup R \cup B \cup t$, where the *source* s and the *sink* t are auxiliary nodes. The arcs of the DAG, all pointing from left to right, are (Fig. 3):

– complete DAG on every color run (that is, for any two points u, v in the same color run, there is an arc $u \to v$ if u is left of v)

- between every bichromatic pair from neighboring runs, such that the bisector between the points falls between the runs
- from s to all points in the first run
- from all points in the last run to t

Any inter-run arc $r \rightarrow b$ of G connect feasible locations for points in S: if $r, b \in S$, then the r-b bisector correctly separates the colors between r and b. Thus the problem is equivalent to finding an s-t path through $k + 2$ nodes (including s and t): the internal nodes of the path are the points of S.

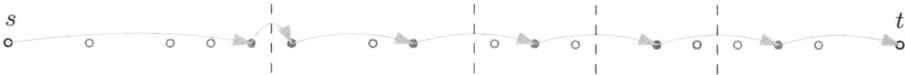

Fig. 3. A path in G through 6 internal vertices (filled circles) that form S; bisectors are dashed.

Since G has size $O(n^2)$ where $n = |R \cup B|$ is the number of points,

Theorem 5. *Problem 5-1 can be solved in $O(n^2)$ time.*

Our solutions applies also to a more general version with an arbitrary number of colors. One open problem is to decrease the running time in the above theorem; however, a more interesting problem is the complexity of the problem for points in dimension higher than 1. We conjecture that the problem is NP-hard for $d \geq 2$ and present an integer program (IP) for the problem (the optimization version, i.e., to minimize k); the IP works for any metric space (because it strips off the geometry immediately) and for any number of colors (we still describe the IP for 2 colors; the extension to arbitrary number of colors is immediate).

For every point $p \in R \cup B$, our IP has a binary variable x_p indicating weather p is picked into S ($x_p = 1$ iff $p \in S$); the objective is to minimize $\sum x_p$. For each point p, we sort all the points by their distance to p (assume all interpoint distances are distinct). For $r \in R, b \in B$, let $\ell_b^r = \{r' \in R : |rr'| < |rb|\}$ be the red points which are closer to r than b. If b is in S, then one of the points in ℓ_b^r should also be in S (Fig. 4). This can be written as a constraint

$$\sum_{r' \in \ell_b^r} x_{r'} \geq x_b$$

On the contrary, if the constraints like above are satisfied for all bichromacic pairs (r, b) (and symmetric constraints for all bichromatic pairs (b, r)), then the points with $x_p = 1$ form a valid set S (because for any point p, the closest point from S has the same color as p). We also add the constraint that some points should be picked into S: $\sum x_p \geq 1$ (otherwise, $x_p = 0 \, \forall p$ is the optimal solution). Altogether, the constraints are necessary and sufficient to model our problem as an IP.

Figure 5 illustrates optimal solutions found by our implementation of the IP. Figure 6 lists the MATLAB code of the implementation.

1   2   4   3   5   7   6   8   9

Fig. 4. Points sorted by distance to red point $1 \in R$ (an imaginary example; we do not exhibit the gemeotric instance). If the blue point 6 is in S, then also one of the red points in $\ell_5^1 = \{1, 4, 3, 7\}$ should be in S: we add the constraint $x_1 + x_4 + x_3 + x_7 \geq x_6$. (Color figure online)

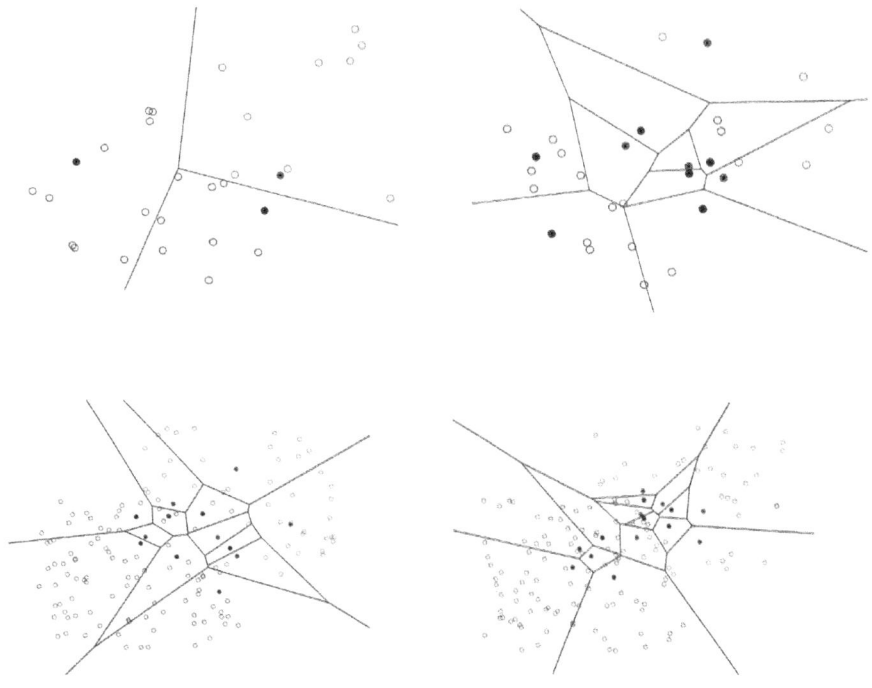

Fig. 5. S are filled circles, $VD(S)$ is back.

```
1     close all;clear%min # of Voronoi centers to classify red/blue
2     n=50;R=rand(n,2)+[.25 .25];B=rand(2*n,2)-[.25 .25];nr=size(R,1);nb=size(B,1);%don't have to be equal-size
3     RB=pdist2(R,B);RR=pdist2(R,R);BB=pdist2(B,B);f=ones(1,nr+nb);
4     A=[-ones(1,nr) zeros(1,nb); zeros(1,nr) -ones(1,nb)];rhs=[-1;-1];%at least 1red, 1blue
5     for r=1:nr,for b=1:nb, closerreds=find(RR(r,:)<RB(r,b));%for any red, there shoukd be a red clsoer than chosen blue
6     if ~isempty(closerreds),row=zeros(1,nr+nb);row(closerreds)=-1;row(nr+b)=1;
7     A=[A;row]; rhs=[rhs;0];end;end;end
8     for b=1:nb,for r=1:nr, closerblues=find(BB(b,:)<RB(r,b));%for any blue, there shoukd be a blue clsoer than chosen red
9     if ~isempty(closerblues),row=zeros(1,nr+nb);row(nr+closerblues)=-1;row(r)=1;
10    A=[A;row]; rhs=[rhs;0];end;end;end
11    x = intlinprog(f,[1:nr+nb],A,rhs,[],[],zeros(nr+nb,1),ones(nr+nb,1));
12    figure;hold on;plot(R(:,1),R(:,2),'or');plot(B(:,1),B(:,2),'ob');
13    chosenRind=find(x(1:nr));chosenBind=find(x(nr+1:end));
14    chosenR=R(chosenRind,:);chosenB=B(chosenBind,:);chosen=[chosenR;chosenB];
15    plot(chosenR(:,1),chosenR(:,2),'or','MarkerFaceColor','red');plot(chosenB(:,1),chosenB(:,2),'ob','MarkerFaceColor','b');
16    if size(chosen,1)>2,voronoi(chosen(:,1),chosen(:,2),'k');end;axis off;
17    sum(x)
```

Fig. 6. IP code listing. Line 2 generates the data: R and B. Lines 3–4 prepare the objective function and the constraints matrix for the IP; lines 5–10 fills the constraints matrix. Line 11 solves the IP, and the other lines generate the graphical output.

References

1. Afshani, P., Arge, L., Larsen, K.G.: Higher-dimensional orthogonal range reporting and rectangle stabbing in the pointer machine model. In: Symposium on Computational Geometry, pp. 323–332 (2012)
2. Agarwal, P.K., Aronov, B., Koltun, V.: Efficient algorithms for bichromatic separability. ACM Trans. Algorithms (TALG) 2(2), 209–227 (2006)
3. Agarwal, P.K., Edelsbrunner, H., Schwarzkopf, O., Welzl, E.: Euclidean minimum spanning trees and bichromatic closest pairs. Discret. Comput. Geom. 6(5), 407–422 (1991). Preliminary version: SoCG 1990
4. Agarwal, P.K., Matoušek, J.: Dynamic half-space range reporting and its applications. Algorithmica 13(4), 325–345 (1995)
5. Agarwal, P.K., Sharir, M.: Efficient algorithms for geometric optimization. ACM Comput. Surv. (CSUR) 30(4), 412–458 (1998)
6. Aggarwal, C.C.: Data Classification: Algorithms and Applications. Chapman and Hall/CRC (2014)
7. Antal, B.B., Chesebro, A.G., Strey, H.H., Mujica-Parodi, L.R., Weistuch, C.: Achieving occam's razor: deep learning for optimal model reduction (2023)
8. Aronov, B., Garijo, D., Núñez-Rodríguez, Y., Rappaport, D., Seara, C., Urrutia, J.: Minimizing the error of linear separators on linearly inseparable data. Discret. Appl. Math. 160(10–11), 1441–1452 (2012)
9. Ben-Hur, A., Horn, D., Siegelmann, H.T., Vapnik, V.: Support vector clustering. J. Mach. Learn. Res. (2001)
10. Bille, P., Gørtz, I.L., Stordalen, T.: Predecessor on the ultra-wide word ram. Algorithmica (2024)
11. Bonawitz, E.B., Chang, I.Y., Clark, C., Lombrozo, T.: Ockham's razor as inductive bias in preschooler's causal explanations. In: 2008 7th IEEE International Conference on Development and Learning, pp. 7–12. IEEE (2008)
12. Callahan, P.B., Kosaraju, S.R.: Faster algorithms for some geometric graph problems in higher dimensions. In: SODA, vol. 93, pp. 291–300 (1993)
13. Chan, T.M.: On enumerating and selecting distances. In: Proceedings of the Fourteenth Annual Symposium on Computational Geometry, pp. 279–286 (1998)
14. Chan, T.M., Larsen, K.G., Pătraşcu, M.: Orthogonal range searching on the ram, revisited. In: Symposium on Computational Geometry, pp. 1–10 (2011)
15. Chan, T.M., Zheng, D.W.: Hopcroft's problem, log-star shaving, 2D fractional cascading, and decision trees. ACM Trans. Algorithms (2022)
16. Clarkson, K.L.: More output-sensitive geometric algorithms. In: Proceedings 35th Annual Symposium on Foundations of Computer Science, pp. 695–702. IEEE (1994)
17. Cortes, C., Vapnik, V.: Support-vector networks. Mach. Learn. 20(3), 273–297 (1995)
18. Demaine, E.D., Lynch, J., Mirano, G.J., Tyagi, N.: Energy-efficient algorithms. In: ITCS 2016, pp. 321–332 (2016)
19. Dey, T.K.: Improved bounds for planar k-sets and related problems. DCG 19, 373–382 (1998)
20. Domingos, P.: The role of occam's razor in knowledge discovery. Data Min. Knowl. Disc. 3, 409–425 (1999)
21. Edelsbrunner, H., Welzl, E.: Constructing belts in two-dimensional arrangements with applications. SIAM J. Comput. 15(1), 271–284 (1986)

22. Eppstein, D.: Dynamic euclidean minimum spanning trees and extrema of binary functions In: DCG (1995)

23. Eppstein, D.: Finding relevant points for nearest-neighbor classification. In: Symposium on Simplicity in Algorithms (SOSA), pp. 68–78. SIAM (2022)

24. Esmeir, S., Markovitch, S.: Occam's razor just got sharper. In: IJCAI, pp. 768–773. Citeseer (2007)

25. Ezugwu, A.E., et al.: A comprehensive survey of clustering algorithms: State-of-the-art machine learning applications, taxonomy, challenges, and future research prospects. Eng. Appl. Artif. Intell. (2022)

26. Garey, M.R., Johnson, D.S.: Computers and intractability. Feeman, San Francisco (1979)

27. Glazenburg, E., Staals, F., van Kreveld, M.: Robust classification of dynamic bichromatic point sets in R2. arXiv preprint arXiv:2406.19161 (2024)

28. Glazenburg, E., van der Horst, T., Peters, T., Speckmann, B., Staals, F.: Robust bichromatic classification using two lines. arXiv preprint arXiv:2401.02897 (2024)

29. Goodrich, M.T.: Geometric partitioning made easier, even in parallel. In: Proceedings of the Ninth Annual Symposium on Computational Geometry, pp. 73–82 (1993)

30. Graham, R.L., Knuth, D.E., Patashnik, O.: Concrete Mathematics. Addison-Wesley, Reading (1989)

31. Har-Peled, S., Koltun, V.: Separability with outliers. In: International Symposium on Algorithms and Computation, pp. 28–39. Springer, Cham (2005)

32. Hershberger, J.: Stable snap rounding. CGTA (2013)

33. Hochbaum, D.S., Pathria, A.: The bottleneck graph partition problem. Networks 28(4), 221–225 (1996)

34. Houle, M.F.: Algorithms for weak and wide separation of sets. Discret. Appl. Math. 45(2), 139–159 (1993)

35. Kahan, S., Snoeyink, J.: On the bit complexity of minimum link paths: superquadratic algorithms for problems solvable in linear time. In: Symposium on Computational Geometry, pp. 151–158 (1996)

36. Katz, M.J., Sharir, M.: An expander-based approach to geometric optimization. SIAM J. Comput. 26(5), 1384–1408 (1997)

37. Kearns, M.J.: The Computational Complexity of Machine Learning. MIT Press, Cambridge (1990)

38. Koiliaris, K., Xu, C.: Faster pseudopolynomial time algorithms for subset sum. TALG 15(3), 1–20 (2019)

39. Kostitsyna, I., Löffler, M., Polishchuk, V., Staals, F.: The complexity of minlink path problems. In: SGG 2016, pp. 1–16 (2016)

40. Krishnaiah, V., Narsimha, G., Chandra, N.S.: Survey of classification techniques in data mining. Int. J. Comput. Sci. Eng. 2(9), 65–74 (2014)

41. Krznaric, D., Levcopoulos, C., Nilsson, B.J.: Minimum spanning trees in d dimensions. Nordic J. Comput. 6(4), 446–461 (1999)

42. Kwon, D., et al.: Adaptive weight quantization method for nonlinear synaptic devices. IEEE Trans. Electron Devices 66(1), 395–401 (2018)

43. Lenstra, H.W., Jr.: Integer programming with a fixed number of variables. Math OR 8(4), 538–548 (1983)

44. Liu, F., Liu, C.: Towards accurate and high-speed spiking neuromorphic systems with data quantization-aware deep networks. In: Design Automation Conference, pp. 1–6 (2018)

45. Mari, C., Mari, E.: Occam's razor, machine learning and stochastic modeling of complex systems: the case of the Italian energy market. Qual. Quant. **58**(2), 1093–1111 (2024)
46. Matheny, M., Phillips, J.M.: Approximate maximum halfspace discrepancy. In: 32nd International Symposium on Algorithms and Computation (ISAAC 2021). Schloss-Dagstuhl-Leibniz Zentrum für Informatik (2021)
47. Matoušek, J.: Approximations and optimal geometric divide-and-conquer. In: Proceedings of the Twenty-Third Annual ACM Symposium on Theory of Computing, pp. 505–511 (1991)
48. Megiddo, N.: Linear programming in linear time when the dimension is fixed. JACM **31**(1), 114–127 (1984)
49. Mertens, S.: The easiest hard problem: Number partitioning. In: Percus, A., Istrate, G., Moore, C. (eds.) Computational Complexity and Stat Physics. Oxford University Press (2006)
50. Oyewole, G.J., Thopil, G.A.: Data clustering: application and trends. AI Rev. **56**(7), 6439–6475 (2023)
51. Packer, E.: Iterated snap rounding with bounded drift. In: Symposium on Computational Geometry, pp. 367–376 (2006)
52. Packer, E.: Controlled perturbation of sets of line segments in R2 with smart processing order. CGTA (2011)
53. Polya, G.: How to Solve It. Princeton Press (2004)
54. Rockett, A.M., et al.: Continued Fractions. World Scientific (1992)
55. Salowe, J.S.: L-infinity interdistance selection by parametric search. Inf. Process. Lett. **30**(1), 9–14 (1989)
56. Seara, C.: On geometric separability. Ph.D. thesis, Universitat Politècnica de Catalunya, Barcelona (2002)
57. Song, C., Liu, B., Wen, W., Li, H., Chen, Y.: A quantization-aware regularized learning method in multilevel memristor-based neuromorphic computing system. In: NVMSA, pp. 1–6 (2017)
58. https://math.stackexchange.com/questions/2086285/did-p%C3%B3lya-say-can-or-cannot
59. Sun, K., Nielsen, F.: A geometric modeling of occam's razor in deep learning. arXiv preprint arXiv:1905.11027 (2019)
60. Van Engelen, J.E., Hoos, H.H.: A survey on semi-supervised learning. Mach. Learn. (2020)
61. Wang, H.: Unit-disk range searching and applications. In: Czumaj, A., Xin, Q. (eds.) 18th Scandinavian Symposium and Workshops on Algorithm Theory (SWAT 2022). Leibniz International Proceedings in Informatics (LIPIcs), vol. 227, pp. 32:1–32:17, Dagstuhl, Germany (2022). Schloss Dagstuhl – Leibniz-Zentrum für Informatik
62. Webb, G.I.: Further experimental evidence against the utility of occam's razor. J. Artif. Intell. Res. **4**, 397–417 (1996)
63. Yang, Q., Li, H., Wu, Q.: A quantized training method to enhance accuracy of reram-based neuromorphic systems. In: ISCAS, pp. 1–5. IEEE (2018)
64. Zhu, X., Goldberg, A.B.: Introduction to Semi-supervised Learning. Springer, Cham (2022)

Dynamic Filter and Retrieval with One Access to Modifiable Memory

Ioana O. Bercea[1], Guy Even[2(✉)], Tomer Even[3],
and Gabriel Marques Domingues[2]

[1] KTH Royal Institute of Technology, Stockholm, Sweden
bercea@kth.se
[2] School of Electrical Engineering, Tel-Aviv University, Tel Aviv, Israel
guy@eng.tau.ac.il, gm@mail.tau.ac.il
[3] Department of Computer Science, Technion, Haifa, Israel
tomer.even@campus.technion.ac.il

Abstract. We present two constant-time dynamic data-structures that support insertions, deletions, and queries with one-sided errors: a space-efficient dynamic (key-only) filter and a compact dynamic data-structure that combines retrieval and filtering (called a *key-value filter*). A one-sided error occurs when a query for a key not in the dataset is issued and the outcome is wrong, i.e., a "yes" in a filter or a non-null in the key-value filter. The response to a query with a key in the dataset always returns the correct answer, i.e., a "yes" in a filter and the correct value in a key-value filter. The probability of the one-sided error in our data-structures is $\Omega(1/\mathsf{poly}(\log n))$, where n is the maximum cardinality of the dataset, and the probability space is over the random bits of the data-structure (i.e., random choice of hash function). The computational framework is the Word RAM model.

We differentiate between accesses to non-modifiable memory (i.e., read-only memory that stores the program instructions, hash function seed or tables, etc.) and accesses to modifiable memory (i.e., read-write memory that stores the representation of the dataset). We are not aware of previous works that make this distinction in the context of data-structures.

Our dynamic filter design requires only a single access to the modifiable memory per operation in the worst-case. We also present a dynamic key-value filter for values of $O(\log \log n)$ bits that requires $1 + o(1)$ accesses to the modifiable memory per operation in expectation. Previous dynamic filter designs require, in the worst case, at least two accesses to modifiable memory for queries with keys not in the dataset. Previous dynamic retrieval data-structure designs always require two dictionary accesses for queries with keys not in the dataset even for single bit values.

We prove bounds on the number of balls that overflow in a dynamic balls-into-bins random process for a range of bin capacities that extends the Iceberg Lemma of [Bender et al., JACM 2023]. The correctness of the key-value filter is based on the previously unstudied natural case of unit-capacity bins with more bins than balls.

I. Finocchi and L. Georgiadis (Eds.): CIAC 2025, LNCS 15679, pp. 292–309, 2025.
https://doi.org/10.1007/978-3-031-92932-8_19

Finally, we observe that the splitting technique for achieving succinct representation of hash functions is not necessary for our data-structures.

Keywords: Balls into Bins · Approximate membership queries · Bloom Filter · Bloomier Filter · Retrieval data-structure

1   Introduction

Filters and retrieval data-structures are two important probabilistic data-structures that exhibit interesting trade-offs between accuracy and space [2, 4,7,9,10,13–15,29–31,33]. We begin by presenting a brief description of their functionality.

A *dynamic filter* is a data-structure that supports insertions and deletions of keys and answers membership queries with a one-sided error.[1] If the queried key is in the dataset, then the output must be "yes"; if the queried key is not in the dataset (i.e., a no-query), then the output can be "yes" with probability at most ε, where ε is a tunable parameter. The event of an output "yes" to a queried key that is not in the dataset is called a *false-positive error*. The probability space is induced by the random bits of the data-structure and neither the dataset nor the queried key should influence the upper bound on the probability of a false-positive error.

A *dynamic retrieval* data-structure supports insertions and deletions of key-value pairs as well as value queries.[2] If the queried key is in the dataset, then the output must equal the value of the key. If the queried key is not in the dataset (i.e., a null-query), then there is no guarantee on the output (of course, a null output is preferable in such a case).

We assume that upon creation of a filter or a retrieval data-structure: (1) the upper bound n is specified on the number of keys that may be inserted and not deleted at every point of time, (2) an upper bound ε on the false-positive probability is specified, and (3) the random hash function is chosen.

The desiderata in data-structure design consist of achieving constant time per operation, reducing space as much as possible, and reducing the number of memory words accessed per operation.

Memory Access Model. We propose to differentiate between accesses to non-modifiable memory (i.e., read-only memory that stores the program instructions, hash function seed or tables, etc.) and accesses to modifiable memory (i.e., read-write memory that stores the state of the data-structure). The Word RAM model

---

[1] In the dynamic setting, one may not insert an existing key nor delete a non-existing key (see [1]).

[2] One may not insert a key-value pair with an existing key nor delete a non-existing key.

is applied to both types of memories with respect to the number of bits per word and the operations over words that can be performed in constant time.

One motivation for minimizing accesses to modifiable memory is related to the distributed deployment of multiple instances of the data-structure. Non-modifiable memory can be stored locally by each instance, whereas modifiable memory requires coherency mechanisms to maintain consistency.

Another motivation is driven by special-purpose hardware implementations of data-structures (as opposed to data-structures programmed on von Neumann architectures). Such implementations may offer high throughput in applications such as: network processors, database acceleration, and firewalls. See, for example, the implementation of a dynamic filter in [20,21]. High throughput is achieved because the operations are executed within a few clock cycles by special-purpose circuits rather than by a CPU that executes a program stored in memory. In particular, lookup tables and hash functions representations are stored in cheap and fast read-only memory (ROM) and are evaluated by special-purpose circuits. On the other hand, the state of the data-structure (e.g., representation of the dataset in a filter) is stored in slow read-write memory (RAM) that has limited throughput. It is therefore desirable to reduce the number of accesses to the RAM.

This paper explores techniques for supporting operations in filters and retrieval data-structures that require only accessing a single word from the modifiable memory. We consider a dynamic setting that allows for insertions, deletions, and queries.

1.1   Previous Work

Bloom [9] introduced the first filter, which is an incremental filter that supports both queries and insertions, but not deletions. The Bloom filter is compact but not space efficient[3] and requires $O(\log 1/\varepsilon)$ accesses to modifiable memory. Let n denote an upper bound on the cardinality of the dataset. A lower bound of $n \log_2(1/\varepsilon)$ bits on the space of static filters was proved in [10]. The lower bound for the space was increased to $C_\varepsilon n \log_2(1/\varepsilon)$ bits in [30] in the incremental setting (i.e., queries and insertions, but no deletions), where $C_\varepsilon > 1$ for constant ε. In [29], a lower bound of $n \log(1/\varepsilon) + \Omega(n)$ bits is presented for the space of a dynamic filter for all values of $1 > \varepsilon > 0$. In addition, an incremental filter with $n \log(1/\varepsilon) + o(n)$ bits for the space is presented for $\varepsilon = o(1)$. For $\log(1/\varepsilon) = \omega(\log \log n)$, a succinct dynamic filter with constant time per operation is obtained by combining results in [13,14].[4] For $\log(1/\varepsilon) = \omega(\varepsilon \log n)$ (e.g., $\log(1/\varepsilon) \geq \log \log n$), a succinct dynamic filter with constant time per operations

---

[3] Let opt denote the optimal space of a data-structure for a given task. A data-structure is *succinct* if it requires $(1 + o(1)) \cdot$ opt space. A data-structure is *compact* if it requires $O(\text{opt})$ space. A data-structure is *space-efficient* if it requires $(1 + o(1))\text{opt} + O(n)$ space.

[4] The construction stores fingerprints in a dynamic retrieval data-structure that employs dynamic perfect hashing.

follows from the extendable filter in [33] (i.e., an upper bound on the cardinality of the dataset is not set upfront). A space-efficient incremental filter with constant time operations is presented in [2].[5] In [4], a constant time space-efficient dynamic filter that is adaptive is presented (i.e., false-positive errors are not repeated). A dynamic space-efficient filter with constant time per operation is presented in [7]. These dynamic filters are organized in two levels that are stored in modifiable memory, where the second level stores overflows from the first level. In the worst case, a "no" response to a query requires accessing both levels. An incremental filter with $1 + o(1)$ accesses per operation, in expectation, to modifiable memory is presented in [22]. A constant-time space-efficient hashtable (and hence an incremental filter) with $1 + o(1)$ accesses per operation, in expectation, to modifiable memory appears in [3].[6][7]

Data-structures for retrieval were initiated in [11] with the name Bloomier filter. Chazelle et al. presented a static data-structure and a lower bound for the dynamic case. In [31], a stronger lower bound and a compact dynamic retrieval data-structure are presented. The dynamic retrieval data-structure in [31] requires $O(\ell + \log \log(u/n))$ bits per key-value pair, where ℓ denotes the number of bits per value and u denotes the cardinality of the universe from which keys are taken. By employing two constant-time dynamic dictionaries (c.f. [2]) (one storing distinct fingerprints and the other storing keys whose fingerprints collide with other fingerprints), the retrieval data-structure requires constant time per operation in the worst case. In particular, it always requires accessing both dictionaries for null-queries (even if $\ell = 1$). Further results on static retrieval appear in [14,15].

This paper explores the theory of data-structures and the results are asymptotic. Many practical designs with interesting ideas have been developed. We list a very partial list of such results: static filters [18,26], incremental filters [28,34], dynamic filters [5,17,23,32]. A practical retrieval data-structure appears in [12] in which the response to a query may return multiple values, only one of which is correct.

1.2   Contributions

The following theorem summarizes the properties of our single memory access dynamic filter. We use n to denote an upper bound on the number of inserted keys that have not been deleted. Deletion and query operations do not fail and the probability of failure per insertion operation is $o(1/\mathsf{poly}(n))$. Hence, the dynamic filter does not fail with probability $1 - 1/\mathsf{poly}(n)$ over a sequence of operations that contains $\mathsf{poly}(n)$ insertions. The space of the data-structure refers to the sum of the space in the modifiable and non-modifiable memories. The

---

[5] The construction stores fingerprints in an incremental dictionary.

[6] The description of the hashtable in [2,3] does not include duplicate keys, a feature that helps in the design of a dynamic filter, see [32].

[7] Note that more than one access in expectation implies at least two accesses in the worst case.

adversary is oblivious, and the probability space is induced by the random bits used by the dynamic filter (the sequence of operations does not influence the probability space).

Theorem 1 (dynamic filter with a single memory word access). *For every n and for $\varepsilon = \Omega(1/poly\log n)$,[8] there exists a probabilistic dynamic filter $F(n, \varepsilon)$ for datasets of cardinality at most n over a polynomial universe with a false-probability error that is bounded by $\varepsilon + 1/poly\log n$. The probability of failure per insert operation is $o(1/poly(n))$ if the adversary is oblivious. The space required to store $F(n, \varepsilon)$ is $(1 + o(1)) \cdot n \log_2(1/\varepsilon) + O(n)$ bits. Every operation is executed in constant time in the worst case in the Word RAM Model, and every operation requires accessing only a single word in the modifiable memory.*

The following theorem summarizes the properties of our dynamic key-value filter (in short, KV-filter). A KV-filter is a data-structure that combines a retrieval data-structure and a filter. Namely, if a queried key is in the dataset, then its value is returned (as in a retrieval data-structure). On the other hand, if a queried key is not in the dataset, then a null value is returned with probability at least $1 - \varepsilon$ (i.e., the false-positive probability is bounded by ε).[9] Note that a KV-filter that is constructed from a retrieval data-structure and a filter may require accessing both, thus incurring at least two accesses to modifiable memory.[10]

Theorem 2 (dynamic KV-filter with almost single memory word access). *For every n and for $\varepsilon = 1/poly\log n$, there exists a probabilistic dynamic KV-filter KV-Filter(n, ℓ, ε) for datasets of cardinality at most n, where keys are $O(\log n)$ bits long and values are $\ell = O(\log \log n)$ bits long. For every insert operation, the probability of a failure is $o(1/poly(n))$ if the adversary is oblivious. The space required to store KV-Filter(n, ℓ, ε) is $O(n \log \log n)$ bits. Every operation is executed in constant time in the worst case in the Word RAM Model and requires accessing $1 + 1/(poly\log n)$ words in expectation in the modifiable memory. The response to every value-query for a key not in the dataset is null with probability at least $1 - \frac{1}{poly\log n}$.*

The main tool used for proving the correctness of the presented data-structures is an analysis of overflows in a dynamic balls-into-bins random process. The Iceberg Lemma in [3] proves bounds that hold with respect to bins of up to poly-logarithmic capacity and an over-provisioning factor $1 + \delta$, where $\delta < 1$. In Theorem 4, we present bounds that hold for a wider range of parameters.

---

[8] That is, $\log(1/\varepsilon) = O(\log \log n)$.

[9] We emphasize that employing a dictionary to store key-value pairs would require $\Omega(\log(u/n))$ bits per, while the theorem requires only $O(\log \log n)$ bits per key-value pair.

[10] If the filter is accessed first on queries and the queried key is in the dataset, then the retrieval has to queried to find the value. If the retrieval is accessed first on a query, then the filter has to be also accessed to bound the probability of a non-null response for queries with keys not in the dataset.

In particular, our bound holds for the natural case of $n\mathsf{poly}(\log n)$ unit-capacity bins.[11]

We conclude with an observation that mitigates a standard technique for employing succinct and efficient hash functions. Theorem 9 shows that, for our dynamic filter, we can circumvent the method of "splitting" data-structures to multiple small instances [2,3,6,16]. A similar observation holds for our dynamic KV-filter.

1.3   Techniques

The organization of dynamic filters used in previous years uses two levels, where the (smaller) second level stores overflows from the first level [4,7,8,13]. As soon as a bin in the first level is no longer full (due to deletions), one may relocate elements from the second level back to the first level [2,4,7]. Alternatively, one may leave elements in the second level even if their bin in the first level is no longer full [3,8]. The organization of a filter using two levels stored in modifiable memory leads to at least two accesses if both levels need to be accessed.

In this paper, elements remain at the level in which they were initially inserted. The analysis of the dynamic balls-into-bins random process in Theorem 4 is based on partitioning the overflow into two parts: (1) The incremental part which is due to excess balls in bins with respect to the "current" dataset. (2) The dynamic part which is due to overflow (upon insertion) of balls of the "current" dataset due to "historical" balls. These "historical" balls caused the bin to be full at the time of the insertion of a "current" ball and are deleted after the overflowed ball is inserted but before the current time. The analysis of the incremental part appears in [2,3,8,22]. The analysis of the dynamic part is new and relies on identifying negatively associated random variables (that help bound the dynamic overflow) and applying a Chernoff bound on them.

Our dynamic filter $F(n, \varepsilon)$ deviates from previous two-level designs by discarding the second level. Instead, each bin in the first level counts the number of elements that "leaked". The correctness proof is based on analyzing the overflows in a dynamic balls-into-bins random process [3].

The design of the dynamic key-value filter deviates from previous retrieval data-structure design [13,31] in the order in which dictionaries are accessed. The correctness proof relies on bounding the overflows of a dynamic balls-into-bins random process in which bins have unit capacity and there are more bins than balls. Indeed, the dynamic key-value filter maps keys to a range of $n\mathsf{poly}(\log n)$ fingerprints (each fingerprint behaves as a unit-capacity bin). Colliding fingerprints are sent to a global key-value dictionary. The maximum occupancy of the global key-value dictionary is bounded by applying Theorem 4 to the special case of unit-capacity bins.

Splitting data-structures is a technique that allows for succinct hash functions [2,3,6,16]. The designs of both the dynamic filter and the dynamic KV-filter

---

[11] Namely, a sparse case with more bins than balls.

are based on arrays of bin dictionaries. Hence, there is no need to split, and it suffices to conceptually consider splitting only within the proof itself.

Organization. Notation, definitions, and the computational model are presented in Sect. 2. The bound on the number of overflows in a balls-into-bins random process is presented in Sect. 3. The dynamic filter is presented in Sect. 4. The key-value filter is presented in Sect. 5. Succinct hashing without the splitting technique is presented in Sect. 6. Due to space limitations, some proofs and algorithms are deferred to the full version.

2   Preliminaries

For $k \in \mathbb{R}$, let $[k]$ denote the set $\{0, \ldots, \lceil k \rceil - 1\}$. Let $\mathcal{U} \triangleq [u]$ denote the *universe* from which *keys* are chosen. We assume that $u = \mathsf{poly}(n)$.[12] Logarithms $\log()$ are base 2 and $\ln()$ is the natural logarithm.

Dynamic Sets. We consider two types of datasets: *key-only datasets* $\mathcal{D} \subseteq \mathcal{U}$ and *key-value datasets* $\widehat{\mathcal{D}} \subseteq \mathcal{U} \times \{0, 1\}^\ell$. Let (x, v) denote a pair in a key-value dataset. We refer to the first coordinate x as the *key* and to the second coordinate as the *value*. We require that the keys in a key-value dataset are distinct.

A dynamic (key-only) dataset is defined by the operations $\mathsf{insert}(x)$ and $\mathsf{delete}(x)$, where $x \in \mathcal{U}$, as follows.

Definition 1 (Dynamic Dataset). *A sequence* $\sigma \triangleq \{\mathsf{op}_t\}_{t=1}^T$ *of insert and delete operations defines a* dynamic dataset $\mathcal{D} = \{\mathcal{D}(t)\}_t$ *as follows:*

$$\mathcal{D}(t) \triangleq \begin{cases} \emptyset & \text{if } t = 0 \\ \mathcal{D}(t-1) \cup \{x_t\} & \text{if } \mathsf{op}_t = \mathsf{insert}(x_t) \\ \mathcal{D}(t-1) \setminus \{x_t\} & \text{if } \mathsf{op}_t = \mathsf{delete}(x_t) \end{cases} \tag{1}$$

We say that the cardinality of a dynamic set is bounded by n *if* $n \geq \max_{t \leq T} |\mathcal{D}(t)|$.

Similarly, a dynamic dataset of key-value pairs is defined by a sequence of operations $\mathsf{insert}(x, v)$ and $\mathsf{delete}(x)$. In the context of key-value datasets, let $\mathcal{D}(t)$ denote the set of keys that appear in a key-value dataset $\widehat{\mathcal{D}}(t)$.

For a key-value dataset $\widehat{\mathcal{D}}$ (in which the keys are distinct), we define a function $\mathsf{val} : \mathcal{U} \to \{0, 1\}^\ell \cup \{\mathsf{null}\}$ as follows. If $x \in \mathcal{D}(t)$, then let $v \in \{0, 1\}^\ell$ such that $(x, v) \in \widehat{\mathcal{D}}(t)$ and define $\mathsf{val}(x) \triangleq v$; otherwise, $\mathsf{val}(x) \triangleq \mathsf{null}$. [13]

---

[12] Indeed, this standard assumption can be satisfied by hashing the keys to a polynomial universe using a universal family of hash functions. The hashing introduces a collision with probability $1/\mathsf{poly}(n)$, and hence we can assume that with probability $1 - 1/\mathsf{poly}(n)$, no collision is introduced during an input sequence that contains a polynomial number of insertions.

[13] Values of keys may change over time. Consider, a subsequence of operations over a key x: $\mathsf{insert}_{t_1}(x, v_1)$, $\mathsf{delete}_{t_2}(x)$, $\mathsf{insert}_{t_3}(x, v_2)$. For $t < t_1$, $\mathsf{val}_t(x) = \mathsf{null}$, for $t \in [t_1, t_2)$, $\mathsf{val}_t(x) = v_1$, for $t \in [t_2, t_3)$, $\mathsf{val}_t(x) = \mathsf{null}$, and for $t \geq t_3$, $\mathsf{val}_t(x) = v_2$.

Definition 2 (Dynamic Filter). *A dynamic filter $F(n, \varepsilon)$ is a probabilistic data-structure that supports insertions, deletions, and approximate membership-queries over a dynamic key-only dataset. A query(x) operation at time t returns a bit out that satisfies the following guarantees:*[14]

$$\Pr\left[out = 1 \mid x \in \mathcal{D}(t)\right] = 1, \qquad \Pr\left[out = 1 \mid x \notin \mathcal{D}(t)\right] \leq \varepsilon.$$

Definition 3 (Dynamic KV-Filter). *A dynamic KV-Filter(n, ℓ, ε) is a probabilistic data-structure that supports insertions, deletions, and approximate value-queries over a dynamic key-value dataset of cardinality at most n. A val-query(x) operation at time t returns val$'(x)$ that satisfies the following guarantees:*[14]

$$\Pr\left[val'(x) = val(x) \mid val(x) \neq null\right] = 1, \quad \Pr\left[val'(x) \neq null \mid val(x) = null\right] \leq \varepsilon.$$

Dynamic KV-Filters have been previously called *dynamic Bloomier filters* [11, 31]. Note that the term *dynamic Retrieval data-structure* [14,15] refers to a dynamic KV-Filter$(n, \ell, 1)$ (i.e., no guarantee for val-query(x) when $x \notin \mathcal{D}(t)$).

The distinction between a KV-Filter$(n, \ell, 1)$ and a KV-Filter(n, ℓ, ε) makes sense when a single memory word access per operation is sought. Indeed, a KV-Filter$(n, \ell, 1)$ can be turned into a KV-Filter(n, ℓ, ε) by adding a dynamic filter $F(n, \varepsilon)$ to approximately check whether the queried key is in the dataset.

Assumption 3 (Restrictions on the input) *The adversary is oblivious (i.e., the input sequence is independent of the random bits of the data-structure). Moreover, the input sequence must satisfy the following conditions: (1) No duplicate insertions: if $op_t = insert(x_t)$, then $x_t \notin \mathcal{D}(t-1)$. (2) Only deletions of current dataset elements are allowed: if $op_t = delete(x_t)$, then $x_t \in \mathcal{D}(t-1)$. (3) At most n elements: for every t, $|\mathcal{D}(t)| \leq n$.*

An analogous assumption on the input is used in the key-value setting.

Word RAM Model. Following Fredman and Willard [25], we consider a computational model in which computer instructions manipulate *words* (see also [27]). Let w denote the number of bits per word. We assume that $w = \Theta(\log u)$, where u is the size of the universe so that keys "fit" in a word. The following operations can be completed in constant time: read or write a word in memory, addition, subtraction, multiplication, division, shifting, and bitwise operations (AND, OR, XOR).

3   Bound on Number of Overflows

In this section, we present a random process that captures the dynamics of a two-level data-structure that employs random hashing. The analysis of the occupancy

---

[14] We emphasize that the probability space is over the random bits of the data-structure and does not depend on the dataset nor on x.

of the second level in this random process is summarized in Theorem 4. This random process was introduced in [8] in the context of designing a dynamic filter.[15] The Iceberg Lemma in [3] bounds the number of elements stored in the backyard for the typical setting in which bin capacities are $O(\text{poly} \log n)$. Concrete bounds for a specific hardware design appear in [20].

Balls-into-Bins Random Process. Consider the following balls-into-bins experiment. Balls are elements in the universe $[u]$. There are m bins indexed by $[m]$. Let $B \triangleq \frac{n}{m}$. Each bin is capable of storing at most $(1 + \delta)B$ balls. The (oblivious) adversary issues a sequence σ of insert and delete operations that satisfy Assumption 3.

The algorithm picks a fully random function $\text{bin} : [u] \rightarrow [m]$ (i.e., the random variables $\{\text{bin}(x)\}_{x \in [u]}$ are independent and distributed uniformly). The algorithm maintains the following sets: m bins and a backyard that are initially empty. The algorithm processes the operations as follows:

1. $\text{insert}(x)$. If $\text{bin}(x)$ is full (i.e., contains $(1 + \delta)B$ balls), then store x in the backyard. If $\text{bin}(x)$ is not full, then store x in $\text{bin}(x)$.
2. $\text{delete}(x)$. Delete *any* ball y with $\text{bin}(y) = \text{bin}(x)$. The ball y can be stored in $\text{bin}(x)$ or in the backyard.

Note that balls stay where they are initially stored until deleted. A ball is not moved from the backyard to its bin even if the bin is no longer full.

Statement of Theorem 4. Consider the balls-into-bins random process with the input sequence σ. Let $N(t, \sigma)$ denote the random variable that equals the number of balls stored in the backyard at time t. The theorem proves an upper bound on $N(t, \sigma)$, namely, an upper bound on the occupancy of the backyard at time t. We compare Theorem 4 with the Iceberg Lemma [3] in Sect. 3.2.

Parameters. The following parameters are used in Theorem 4. Here, $\delta \geq 0$ is a free parameter. Let $\tau \triangleq (1 + \delta)B \geq 1$, $\gamma \triangleq \min((1 + \delta)^{-1}, e^{-\min(\delta^2, \delta)B/3})$, and $\theta \leq 1/(2e\gamma)$. The following theorem bounds the probability that many balls are stored in the backyard. See Table 1 for applications of the theorem to different bin capacities. In preparing Table 1, we expressed the relations with θ as the free parameter. Namely, $\delta^* \geq \max(2e\theta - 1, 3\ln(2e\theta)/B)$ and $\delta = \max(\sqrt{\delta^*}, \delta^*)$.

The full proof of the following theorem is deferred to the full version.

Theorem 4. *For every input sequence σ that satisfies Assumption 3, and for every time t, the following holds:*

$$\Pr[N(t, \sigma) \geq 2\tau \cdot n/\theta] \leq 2^{-n/\theta} + 2^{-\tau^2 n/\theta^2}. \tag{2}$$

---

[15] The proof in [8] holds in a restricted setting in which deleted elements are not re-inserted.

Proof (sketch). For the analysis, we fix an input sequence σ and a time t^*. Consider the dataset at time t^*, namely, $\mathcal{D}(t^*) = \{x_0, \ldots, x_{n'-1}\}$, for $n' \leq n$.

Let $n_b(t) \triangleq |\{x \in \mathcal{D}(t) \mid \mathsf{bin}(x) = b\}|$. Let t_i denote the time in which x_i was inserted to $\mathcal{D}(t^*)$. We consider two separate phenomena:

1. Incremental overflows: Define $W_b \triangleq \max(n_b(t^*) - \tau, 0)$, i.e., the excess occupancy of bin b at time t^*. $\mathbb{E}[W_b] \leq \gamma B$ and $\sum_{b \in [m]} W_b$ is a 1-Lipschitz function of $\{\mathsf{bin}(x)\}_{x \in [u]}$. By a McDiarmid inequality, $\Pr\left[\sum_{b \in [m]} W_b \geq \tau \cdot \frac{n}{\theta}\right] \leq 2^{-\tau^2 n/\theta^2}$.

2. Dynamic overflows: Define $L_b \triangleq \bigvee_{i \in [n']}[(\mathsf{bin}(x_i) = b) \wedge (n_b(t_i - 1) \geq \tau)]$, i.e., L_b indicates whether there exists an $x_i \in \mathcal{D}_b(t^*)$ such that more than τ balls are mapped to $\mathsf{bin}(x_i)$ when x_i is inserted. $\{L_b\}_{b \in [m]}$ are negatively associated random variables with $\mathbb{E}[L_b] \leq \gamma B$. By a Chernoff bound $\Pr\left[\sum_{b \in [m]} L_b \geq \frac{n}{\theta}\right] \leq 2^{-n/\theta}$.

The contribution of the balls mapped to bin b to $N(t^*, \sigma)$ is bounded by $\tau \cdot L_b + W_b$. Hence, $N(t^*, \sigma) \leq \sum_{b \in [m]}(\tau \cdot L_b + W_b)$. The proof follows by a union bound.

3.1   Application of Theorem 4

In this section, we apply Theorem 4 with different values of B (i.e., average bin occupancies). Equation (2) has the form:

$$\Pr[N(t, \sigma) \geq \Lambda] \leq 2^{-\mathcal{E}}, \tag{3}$$

where $\Lambda = 2\tau \cdot n/\theta$ and $\mathcal{E} = \Theta(\frac{n}{\theta} \cdot \min(1, \tau^2/\theta))$. Table 1 lists values of Λ and $\mathcal{E}$ implied by Theorem 4 for various values of B and δ.[16]

3.2   Comparison with the Iceberg Lemma

Below we compare the Iceberg Lemma in [3] with Theorem 4.

1. The Iceberg Lemma proves comparable results to Theorem 4 for the cases $B \in \left\{2, \log n/\log \log n, \log^d n\right\}$ with $\delta < 1$ in rows 3–5 of Table 1.

2. The proof of the Iceberg Lemma requires that $\delta < 1$. In Table 1, rows 1–3 deal with cases in which $\delta > 1$. The first row deals with the extreme setting of unit bin capacities (i.e., $(1 + \delta)B = 1$) with n/α bins. We use this setting for the proof of the KV-Filter (namely, bounding the number of overflows of FPDict).

3. The Iceberg Lemma does not apply for $B = \omega(\mathsf{poly} \log n)$, e.g., the last two rows of Table 1. For such large bin capacities, [3] suggest to use Chernoff bounds to prove that $N(t, \sigma) = 0$ with high probability (i.e., with probability $1 - \gamma n$). For the last two rows in Table 1, [3] achieve a polynomial error bound, while Theorem 4 achieves a super-polynomial bound on the error (i.e., $\mathcal{E} = \omega(\mathsf{poly} \log(n))$).

---

[16]   Note that the total bin capacity $(1 + \delta) B \cdot m$ in first 2 rows is $\omega(n)$, and the total bin capacity in third row ($B = 2$, $\delta = O(1)$) is $\Theta(n)$. The total bin capacity in remaining rows is $(1 + o(1)) \cdot n$.

302   I. O. Bercea et al.

Table 1. Application of Theorem 4 for various bin capacities. The columns for "B" and "δ" are concrete (i.e., no extra constants). The remaining columns are asymptotic big-Θ. The parameters q, c, d are constants that satisfy $0 < q < 1$ and $c > d > 0$. Note that in the first row, bin capacities are unit (i.e. $(1+\delta)B = 1$) and the number of bins is $m = n/\alpha$, where the parameter α may depend on n (e.g., $\alpha = 1/\mathsf{poly}(\log\log(n))$).

B	δ	θ	Λ	$\mathcal{E}$
$\alpha \leq \frac{1}{2}$	$\frac{1}{\alpha} - 1$	$\frac{1}{\alpha}$	$2\alpha n$	$\alpha^2 n$
2	$3c\ln(\log n)/2$	$(\log n)^c$	$\dfrac{n \log\log n}{(\log n)^c}$	$\dfrac{n(\log\log n)^2}{(\log n)^{2c}}$
2	$3c\ln(2)/2$	2^c	$c \cdot n/2^c$	$c^2 \cdot n/2^{2c}$
$\dfrac{\log n}{\log\log n}$	$\sqrt{3(c+1)\ln(\log n)/B}$	$(\log n)^{c+1}$	$\dfrac{n}{(\log n)^c \log\log n}$	$\dfrac{n}{(\log n)^{2c}(\log\log n)^2}$
$(\log n)^d$	$\sqrt{3(c+d)\ln(\log n)/B}$	$(\log n)^{c+d}$	$\dfrac{n}{(\log n)^c}$	$\dfrac{n}{(\log n)^{2c}}$
n^{1-q}	$\sqrt{3(1-q/2)\ln(n)/B}$	$n^{1-\frac{q}{2}}$	$n^{1-\frac{q}{2}}$	$n^{q/2}$
$\dfrac{n}{(\log n)^{c+d}}$	$\sqrt{3\ln(n/\log^d n)/B}$	$\dfrac{n}{(\log n)^d}$	$\dfrac{n}{(\log n)^c}$	$(\log n)^d$

4 A Dynamic Filter that Accesses a Single Modifiable Memory Word per Operation

In this section, we prove Theorem 1. The design of the dynamic filter $F(n, \varepsilon)$ uses the following parameters.

Since $\varepsilon = \Omega(1/\mathsf{poly}\log n)$, it follows that $\log(1/\varepsilon) = O(\log\log n)$ (we may assume that $\varepsilon \leq 1/2$). Let $w = \Theta(\log n)$ denote the RAM word length in bits.

Parametrization. Define $B \triangleq \dfrac{w}{8\log(1/\varepsilon)}$, $m \triangleq n/B$, and $\delta \triangleq \sqrt{3(c+1)(\ln\log n)/B}$ (for a constant $c > 0$).[17]

Architecture. The dynamic filter $F(n, \varepsilon)$ employs a pair of hash functions $\mathsf{bin} : [u] \to [m]$ and $\mathsf{fp} : [u] \to [B/\varepsilon]$. For a key $x \in [u]$, we refer to $\mathsf{bin}(x)$ as the *bin assignment* of x and to $\mathsf{fp}(x)$ as the *fingerprint* of x. We assume that these functions are fully random (see [6] for succinct hash functions).

The dynamic filter consists of an array of m bin-dictionaries $\{\mathsf{Dict}_b\}_{b\in[m]}$. Each Dict_b can store a multi-set of up to $(1+\delta)B$ fingerprints using Fano-Elias encoding [7,19,24]. In addition, each Dict_b contains an $O(\log\log n)$-bit counter that counts the number of overflows in Dict_b. We refer to the counter used by Dict_b as $\mathsf{counter}_b$.

---

[17] Note that $B = \Omega(\log n/\log\log n)$ and $B = O(\log n)$.

Execution of Operations.

1. Initially, for every $b \in [m]$, Dict_b stores an empty set and $\mathsf{counter}_b \leftarrow 0$.
2. $\mathsf{insert}(x)$. If $\mathsf{bin}(x)$ is not full, then store $\mathsf{fp}(x)$ in $\mathsf{Dict}_{\mathsf{bin}(x)}$. Else, increment $\mathsf{counter}_{\mathsf{bin}(x)}$.
3. $\mathsf{delete}(x)$. If $\mathsf{fp}(x)$ is stored in $\mathsf{Dict}_{\mathsf{bin}(x)}$, then delete it. Else, decrement $\mathsf{counter}_{\mathsf{bin}(x)}$.
4. $\mathsf{query}(x)$. If $\mathsf{counter}_{\mathsf{bin}(x)} > 0$, return $\mathsf{out} \leftarrow 1$. Else, query $\mathsf{Dict}_{\mathsf{bin}(x)}$ for $\mathsf{fp}(x)$, if found return $\mathsf{out} \leftarrow 1$, else return $\mathsf{out} \leftarrow 0$.

4.1   Correctness Proof for Dynamic Filter

Let $n_b(t)$ denote the number of balls in bin b at time t. Let $\mathsf{counter}_b(t)$ denote the value of $\mathsf{counter}_b$ at time t. Let $s_b(t)$ denote the number of fingerprints stored in bin b at time t.

Invariant 5 *At every time t, $\mathsf{counter}_b(t) = \max\{0, n_b(t) - s_b(t)\}$.*

The only cause of a failure is if $\mathsf{counter}_{\mathsf{bin}(x)}$ becomes too big as a result of $\mathsf{insert}(x)$.

Lemma 1. *If $\log(B \log n) = O(\log \log n)$ bits are used per counter, then every insertion operation fails with probability $o(1/\mathsf{poly}(n))$.*

Proof. Note that $\mathsf{counter}_b(t) \leq n_b(t)$ and $\mathbb{E}[n_b(t)] \leq B$. Hence, by a Chernoff bound on $n_b(t)$, $\Pr[\mathsf{counter}_b(t) \geq B \log n] \leq n^{-B}$, and the lemma follows.

In the following lemma, we assume that $\delta = \sqrt{3(c+1)(\ln \log n)/B} \leq 1$.

Lemma 2. *For every time t and every bin b, $\Pr[\mathsf{counter}_b(t) > 0] \leq O(\log^{-c} n)$.*

Proof (sketch). Note that if $\mathsf{counter}_b(t) > 0$ then there exists an x_i in $\mathcal{D}(t)$ such that there were $(1 + \delta) \cdot B$ balls in bin b when x_i was inserted, which increased the counter. This event is captured by the 0-1 random variable L_b in the proof of Theorem 4. Namely, $\Pr[\mathsf{counter}_b(t) > 0] = \mathbb{E}[L_b]$. In the full version we prove that $\mathbb{E}[L_b] \leq B \cdot \exp(-\delta^2 B/3) \leq O(\log^{-c} n)$, and the lemma follows.

If $x \in \mathcal{D}(t)$, then either $\mathsf{fp}(x)$ is stored in $\mathsf{Dict}_{\mathsf{bin}(x)}$ or $\mathsf{counter}_{\mathsf{bin}(x)} > 0$. In both cases $\mathsf{out} = 1$, as required. The probability of a false-positive event is bounded in the following lemma.

Lemma 3. *For a $\mathsf{query}(x)$ operation at time t, $\Pr[\mathsf{out} = 1 \mid x \notin \mathcal{D}(t)] \leq \varepsilon + \frac{1}{\mathsf{poly} \log n}$.*

Proof. A false-positive event may occur due to one of two events. 1. A collision: $\mathsf{bin}(x) = \mathsf{bin}(y)$ and $\mathsf{fp}(x) = \mathsf{fp}(y)$ for $y \in \mathcal{D}(t)$; or 2. a bin overflow: $\mathsf{counter}_{\mathsf{bin}(x)} > 0$. The probability of a collision is bounded by ε. By Lemma 2, the probability that $\mathsf{counter}_{\mathsf{bin}(x)} > 0$ is $O(\log^{-c} n)$, and the lemma follows.

Lemma 4. *Every bin dictionary fits in a word.*

Proof. Every Dict_b stores at most $(1 + \delta) B$ fingerprints and an $\log(B \log n)$-bit counter. Fano-Elias encoding requires $\log_2(1/\varepsilon) + 2$ bits per fingerprint. Hence the space per bin-dictionary is $(1 + \delta) B \cdot (\log(1/\varepsilon) + 2) + \log(B \log n) \leq w$ bits.

In the following lemma, we assume that the evaluation of the hash function does not require accessing the non-modifiable memory.

Lemma 5. *Every operation in the dynamic filter requires accessing a single memory word in the modifiable memory and is completed in constant time in the worst case.*

Proof. Every operation accesses a single bin dictionary, and hence a single word. By applying rank and select computations over a word, operations over a Fano-Elias encoding that fits in a word can be executed in constant time. The same holds for reading, incrementing, and decrementing the counter. We conclude that every operation can be executed in constant time in the worst case while accessing a single memory word.

Lemma 6. *The space of $F(n, \varepsilon)$ is $(1 + o(1)) \cdot n \log_2(1/\varepsilon) + O(n)$.*

Proof. Every bin dictionary requires $(1 + \delta) B \cdot (\log(1/\varepsilon) + 2) + O(\log \log n)$ bits. There are m bins. The lemma follows because $B = O(\log n)$ and $B = \Omega(\log n / \log \log n)$, thus $\delta = O(\sqrt{(\log \log n)/B}) = o(1)$.

This completes the proof of Theorem 1.

5   Almost Single Modifiable Memory Access Dynamic Key-Value Filter (Retrieval)

In this section we describe the dynamic key-value filter $\mathsf{KV\text{-}Filter}(n, \ell, \varepsilon)$. The proof of Theorem 2 appears in the full version of the paper.

Parametrization Set $\varepsilon = 1/\log^d n$ (for a constant $d \geq 5$) and $\ell = O(\log \log n)$. Define $B \triangleq \Theta(w/\log(1/\varepsilon))$, $m \triangleq n/B$, and $\delta \triangleq \sqrt{3(c+1)(\ln \log n)/B}$ (for a constant $c \geq 3$).[18]

---

[18] The constant in the definition of B should be set so that $(1 + \delta) B \cdot (\log(1/\varepsilon) + 3 + \log(B \log n) + \ell) + \log(B \log n) \leq w$ (i.e., every bin-dictionary fits in a single word).

Notation. Let bin : $[u] \rightarrow [m]$ and fp : $[u] \rightarrow [B/\varepsilon]$ denote fully random hash functions. Let $\mathsf{FP}(x) \triangleq (\mathsf{bin}(x), \mathsf{fp}(x))$. Let $\mathcal{D}(t)$ denote the set of keys that appear in the key-value dataset $\widehat{\mathcal{D}}(t)$.

Definition 4 (FP-family). *An FP-family for a fingerprint $\varphi \in [m] \times [B/\varepsilon]$ and time t is the set $\{x \in \mathcal{D}(t) \mid \mathsf{FP}(x) = \varphi\}$. We denote the FP-family for φ at time t by $\Phi(\varphi, t)$.*

5.1   KV-Filter(n, ℓ, ε) Design

The KV-filter consists of three parts (see Fig. 1): (1) An array of m bin-dictionaries $\{\mathsf{Dict}_b\}_{b \in [m]}$; (2) A global fingerprint-value dictionary FPDict; and (3) A global key-value dictionary GDict.

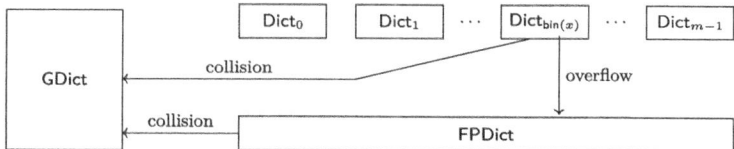

Fig. 1. A block diagram for KV-Filter. The overflow arrow refers to insertion of a record to a full bin-dictionary. A collision arrow refers to a collision of an inserted fingerprint with an existing one.

Each bin dictionary Dict_b can store a set of up to $(1 + \delta) B$ "fp-records" using a variation of Fano-Elias encoding in which values are attached to the fingerprints. Each fp-record is a 4-tuple $(\mathsf{fp}(x), \mathsf{val}(x), \mathsf{cc}(x), \mathsf{valid}(x))$, where (1) $\mathsf{cc}(x)$ is a collision counter that counts the size of the FP-family $\Phi(\mathsf{FP}(x), t)$, and (2) $\mathsf{valid}(x) \in \{0, 1\}$ indicates whether the record belongs to an element that is in $\mathcal{D}(t)$. Note that the $\mathsf{fp}(x)$ field serves as a unique key within each bin dictionary. In addition, each bin dictionary Dict_b contains a $O(\log \log n)$-bit counter denoted $\mathsf{counter}_b$.

The global fingerprint-value dictionary FPDict can store a set of $O(n / \log^3 n)$ "FP-records". Each FP-record is a 4-tuple $(\mathsf{FP}(x), \mathsf{val}(x), \mathsf{cc}(x), \mathsf{valid}(x))$. Note that the $\mathsf{FP}(x) = (\mathsf{bin}(x), \mathsf{fp}(x))$ field serves as a unique key in FPDict.

The global key-value dictionary GDict can store $O(\varepsilon n)$ key-value pairs of the form $(x, \mathsf{val}(x))$.[19] Both FPDict and GDict can be designed using any compact constant-time dynamic dictionary (e.g., the succinct dictionaries of [2,3]).

---

[19] Theorem 4 (i.e. the first line in Table 1) states that with probability at least $1 - 2^{-\Omega(\varepsilon^2 n)}$, at most $2\varepsilon n$ key-value pairs are stored in GDict.

Invariants. The following two invariants state that every FP-family has exactly one representative in the bin dictionaries or FPDict and that the counter $cc(x)$ in such a record equals the size of the FP-family $\Phi(FP(x), t)$.

Invariant 6 *At every time t, for every $x \in \mathcal{D}(t)$, exactly one (possibly invalid) record of a key $y \in \Phi(FP(x), t)$ appears in $\mathsf{Dict}_{bin(x)} \cup \mathsf{FPDict}$.*

Invariant 7 *At every time t, for every record in $\mathsf{Dict}_{bin(x)} \cup \mathsf{FPDict}$, it holds that $cc(x) = |\Phi(FP(x), t)|$.*

The following invariant is analogous to Invariant 5.

Invariant 8 *At every time t, $\mathsf{counter}_b$ equals the number of keys in $\mathcal{D}(t)$ that were mapped to the b'th bin yet stored in FPDict.*

5.2   Execution of Operations

A detailed listing of the operations appears on the full version.

$\mathsf{insert}(x, \mathsf{val})$. If the FP-family $\Phi(FP(x), t-1)$ is not empty, then the key-value is stored in GDict. If $\Phi(FP(x), t-1)$ is empty, then the corresponding record is stored in $\mathsf{Dict}_{bin(x)}$ if there is a vacant slot; otherwise, it is stored in FPDict. If $\mathsf{Dict}_{bin(x)}$ or FPDict contain an invalid record for a fingerprint from the FP-family of $FP(x)$, then x overtakes the record by validating it and increasing the collision counter.

$\mathsf{delete}(x)$. A deletion of a key x looks for $\mathsf{fp}(x)$ in $\mathsf{Dict}_{bin(x)}$. Suppose that a record r with key $\mathsf{fp}(x)$ is found in $\mathsf{Dict}_{bin(x)}$. The difficulty is that record r might belong to another key $y \in \Phi(FP(x), t)$. This ambiguity is resolved by searching for x in GDict if the collision counter in record r is greater than 1. If x is found in GDict, simply delete it. However, if x is not found in GDict, then the record in $\mathsf{Dict}_{bin(x)}$ is indeed x's record, but we cannot delete from $\mathsf{Dict}_{bin(x)}$ if $cc(x) > 1$. In this case, we mark the record as invalid ($\mathsf{valid}(x) \leftarrow 0$) and it acts as a "placeholder" for $\Phi(FP(x), t)$. If $\mathsf{fp}(x)$ is not found in $\mathsf{Dict}_{bin(x)}$ and the bin overflowed (i.e., $\mathsf{counter}_{bin(x)} > 0$), then we need to search for x's record also in FPDict. The handling of this case is analogous to the case of $\mathsf{Dict}_{bin(x)}$.

$\mathsf{val\text{-}query}(x)$. A val-query for a key x is similar to deletion (because deletion requires searching). The difference is that a deleted key must be in the dataset, but a queried key might not (in which case, a null is returned).

6   Succinct Hash Functions Without Splitting

In this section, we discuss how to use succinct hash functions for the dynamic filter $F(n, \varepsilon)$ and the dynamic KV-filter $\mathsf{KV\text{-}Filter}(n, \ell, \varepsilon)$.

Splitting. The standard technique for designing dynamic filters and dictionaries with succinct hash functions involves "splitting" the data-structure to many parts [2,3,8]. We outline this splitting technique for the case of a dynamic filter, and then show that it can be circumvented.

Consider the task of designing a filter $F(n, \varepsilon)$. Splitting takes place by instantiating M dynamic filters $\{F_i(n', \varepsilon)\}_{i \in [M]}$. A family $\mathcal{H}$ of hash functions $h = (h_1, h_2) : [u] \to [M] \times [n'/\varepsilon]$ is used for an operation over $x \in [u]$, as follows. The first coordinate $h_1(x)$ maps x to $F_{h_1(x)}(n', \varepsilon)$. The second coordinate $h_2(x)$ determines $\mathsf{bin}(x)$ and $\mathsf{fp}(x)$ within $F_{h_1(x)}(n', \varepsilon)$. Previous works used the family of hash functions of Dietzfelbinger and Wölfel [16] for splitting.

Circumventing Splitting. In the following claim we argue that splitting can be circumvented using the property of *local uniformity* of tornado tabulation hashing.[20] Let $\mathcal{H}$ denote a given family of tornado tabulation hash function. We have that:

Theorem 9. *Let $N \triangleq n \cdot (1 + \sqrt{M/n} \cdot \log n)$ for $M = n^\rho$, where $\rho \in (0, 1)$. The dynamic filter $F(N, \varepsilon)$ with a hash function chosen randomly from $\mathcal{H}$ can support datasets of cardinality at most n with a false positive probability of $\varepsilon + 1/\mathsf{poly} \log(n)$.*

Proof (sketch). Let h be a tornado tabulation hash function chosen uniformly from $\mathcal{H}$. We think of the m bins of $F(N, \varepsilon)$ as being partitioned into M super-bins, where each super-bin consists of a block of m/M bins. We emphasize that this partitioning is done merely for the purposes of the analysis, and does not affect the design.

For correctly chosen parameters, we meet the conditions for local uniformity from [6]. Within each super-bin, the family $\mathcal{H}$ is fully-random with probability $1 - \mathsf{poly}(n)$. Hence, Theorem 4 can be applied to every super-bin separately. The bound of probability of failure increases additively by $1/\mathsf{poly}(n)$.

References

1. Agarwala, A., Even, G.: A space lower bound for approximate membership with duplicate insertions or deletions of nonelements (2024). https://arxiv.org/abs/2412.19249
2. Arbitman, Y., Naor, M., Segev, G.: Backyard cuckoo hashing: constant worst-case operations with a succinct representation. In: FOCS 2010, pp. 787–796 (2010). see also arXiv:0912.5424v3
3. Bender, M.A., Conway, A., Farach-Colton, M., Kuszmaul, W., Tagliavini, G.: Iceberg hashing: optimizing many hash-table criteria at once. JACM **70**(6), 1–51 (2023)

---

[20] It is also possible to employ hashing from [16].

4. Bender, M.A., Farach-Colton, M., Goswami, M., Johnson, R., McCauley, S., Singh, S.: Bloom filters, adaptivity, and the dictionary problem. In: FOCS 2018, pp. 182–193 (2018). https://doi.org/10.1109/FOCS.2018.00026

5. Bender, M.A., et al.: Don't thrash: how to cache your hash on flash. PVLDB **5**(11), 1627–1637 (2012). https://doi.org/10.14778/2350229.2350275

6. Bercea, I.O., Beretta, L., Klausen, J., Houen, J.B.T., Thorup, M.: Locally uniform hashing. In: FOCS 2023. pp. 1440–1470 (2023)

7. Bercea, I.O., Even, G.: Fully-dynamic space-efficient dictionaries and filters with constant number of memory accesses (2019). http://arxiv.org/abs/1911.05060

8. Bercea, I.O., Even, G.: A dynamic space-efficient filter with constant time operations. In: SWAT 2020. LIPIcs, vol. 162, pp. 1–17 (2020). https://doi.org/10.4230/LIPIcs.SWAT.2020.11

9. Bloom, B.H.: Space/time trade-offs in hash coding with allowable errors. Commun. ACM **13**(7), 422–426 (1970). https://doi.org/10.1145/362686.362692

10. Carter, L., Floyd, R., Gill, J., Markowsky, G., Wegman, M.: Exact and approximate membership testers. In: STOC 1978. pp. 59–65 (1978)

11. Chazelle, B., Kilian, J., Rubinfeld, R., Tal, A.: The Bloomier filter: an efficient data structure for static support lookup tables. In: SODA 2004, pp. 30–39 (2004)

12. Conway, A., Farach-Colton, M., Johnson, R.: SplinterDB and Maplets: improving the tradeoffs in key-value store compaction policy. Proc. ACM Manage. Data **1**(1), 1–27 (2023)

13. Demaine, E.D., Meyer auf der Heide, F., Pagh, R., Pătraşcu, M.: De dictionariis dynamicis pauco spatio utentibus. In: LATIN 2006, pp. 349–361 (2006)

14. Dietzfelbinger, M., Pagh, R.: Succinct data structures for retrieval and approximate membership. In: ICALP 2008, pp. 385–396 (2008)

15. Dietzfelbinger, M., Walzer, S.: Constant-time retrieval with $o(\log m)$ extra bits. In: STACS 2019, pp. 1–16 (2019). https://doi.org/10.4230/LIPIcs.STACS.2019.24

16. Dietzfelbinger, M., Woelfel, P.: Almost random graphs with simple hash functions. In: STOC 2003, pp. 629–638 (2003)

17. Dillinger, P.C., Manolios, P.: Fast, all-purpose state storage. In: Model Checking Software: 16th International SPIN Workshop 2009, Proceedings 16, pp. 12–31 (2009)

18. Dillinger, P.C., Walzer, S.: Ribbon filter: practically smaller than Bloom and Xor. CoRR **abs/2103.02515** (2021).https://arxiv.org/abs/2103.02515

19. Elias, P.: Efficient storage and retrieval by content and address of static files. JACM **21**(2), 246–260 (1974)

20. Even, G., Marques Domingues, G.: A micro-architecture that supports the Fano–Elias encoding and a hardware accelerator for approximate membership queries. Microprocessors Microsyst. **105** (2024). https://doi.org/10.1016/j.micpro.2023.104992

21. Even, G., Marques Domingues, G., Toutian, P.: Brief announcement: a parallel architecture for dynamic approximate membership. In: SPAA 2023, pp. 291–294 (2023)

22. Even, T., Even, G., Morrison, A.: Prefix filter: practically and theoretically better than Bloom. Proc. VLDB Endowment **15**(7), 1311–1323 (2022)

23. Fan, B., Andersen, D.G., Kaminsky, M., Mitzenmacher, M.: Cuckoo filter: practically better than bloom. In: CoNEXT, pp. 75–88. ACM (2014)

24. Fano, R.M.: On the number of bits required to implement an associative memory. memorandum 61. Computer Structures Group, Project MAC, MIT, Cambridge, Mass (1971)

25. Fredman, M.L., Willard, D.E.: Blasting through the information theoretic barrier with fusion trees. In: STOC 1990, pp. 1–7 (1990)
26. Graf, T.M., Lemire, D.: Xor filters: Faster and smaller than bloom and cuckoo filters. CoRR **abs/1912.08258** (2019). http://arxiv.org/abs/1912.08258
27. Hagerup, T.: Sorting and searching on the word RAM. In: STACS 1998, pp. 366–398 (1998)
28. Kirsch, A., Mitzenmacher, M.: Less hashing, same performance: building a better Bloom filter. In: ESA 2006, pp. 456–467 (2006)
29. Kuszmaul, W., Walzer, S.: Space lower bounds for dynamic filters and value-dynamic retrieval. In: STOC 2024, pp. 1153–1164 (2024). https://doi.org/10.1145/3618260.3649649
30. Lovett, S., Porat, E.: A lower bound for dynamic approximate membership data structures. In: FOCS 2010, pp. 797–804 (2010)
31. Mortensen, C.W., Pagh, R., Pătraşcu, M.: On dynamic range reporting in one dimension. In: STOC 2005, pp. 104–111 (2005)
32. Pagh, A., Pagh, R., Rao, S.S.: An optimal Bloom filter replacement. In: SODA 2005, pp. 823–829 (2005)
33. Pagh, R., Segev, G., Wieder, U.: How to approximate a set without knowing its size in advance. In: FOCS 2013. pp. 80–89 (2013). see also arXiv:1304.1188v2
34. Putze, F., Sanders, P., Singler, J.: Cache-, hash-, and space-efficient Bloom filters. J. Exp. Algorithmics **14**, 4–4 (2010)

Longest Path Transversals in Claw-Free and P_5-Free Graphs

Paloma T. Lima$^{(\boxtimes)}$ and Amir Nikabadi$^{(\boxtimes)}$

IT University of Copenhagen, Copenhagen, Denmark
{palt,amir}@itu.dk

Abstract. For a connected graph G, the *longest path transversal number* of G, denoted by $lpt(G)$, is the minimum cardinality of a set of vertices that intersects all longest paths in G. It is an open problem whether any graph admits a longest path transversal of constant size. This question remains open even when restricted to claw-free graphs and P_5-free graphs. In this work, we investigate these two graph classes. We show that, given a connected graph G, $lpt(G) = 1$ if G is a (P_5, H)-free graph, when H is a triangle, a paw, or a diamond. We also provide a complete characterization of the graphs H on at most five vertices for which for any $(claw, H)$-free graph G it holds that $lpt(G) = 1$. Moreover, in each of these cases, we present a polynomial-time algorithm which finds a vertex in G that belongs to all its longest paths.

Keywords: Longest path transversal · claw-free graphs · P_5-free graphs

1 Introduction

In 1966, Gallai [10] proposed the following question: *do all longest paths in a connected graph share at least one vertex?* A few years later, the question was answered in the negative by a graph due to Walther [20]. This graph has the property that, for each of its vertices, there exists a longest path in the graph that does not contain that vertex. A minimal example was later provided independently by Voss and Walther [19] and Zamfirescu [21], and is depicted in Fig. 1. Gallai's question and its negative answer sparked the interest in the problem of determining the size of a smallest vertex set that intersects all longest paths of a graph. Such a set is called a *longest path transversal* of a graph G, and its size is denoted by $lpt(G)$. A sublinear upper bound on $lpt(G)$, for any graph G, was provided by Long Jr, Milans and Munaro [15]. However, it remains an open question whether this upper bound could be brought down to a constant. In particular, it is still unknown whether there exists a graph G for which $lpt(G) > 4$.

On the other hand, graphs belonging to multiple graph classes have been shown to admit longest path transversals of size one. This is the case, for

I. Finocchi and L. Georgiadis (Eds.): CIAC 2025, LNCS 15679, pp. 310–325, 2025.
https://doi.org/10.1007/978-3-031-92932-8_20

instance, of circular-arc graphs [3,14], dually chordal graphs [13], bipartite permutation graphs [6], graphs of treewidth at most 2 [8], P_4-sparse graphs [4] and graphs of matching number at most 3 [7]. Graph classes defined by a finite family of forbidden induced subgraphs have also been investigated. In particular, all $2P_2$-free graphs [12] and $(claw, P_5)$-free graphs [4] were shown to admit longest path transversals of size one. More recently, Long Jr., Milans, and Munaro [16] initiated a systematic study of the problem on H-free graphs. They characterized the graphs H of size at most four for which *any* H-free graph admits a longest path transversal of size one. In particular, they showed that there exists a claw-free graph G for which $lpt(G) > 1$, while Gao and Shan [11] identified multiple subclasses of claw-free graphs that admit longest path transversals of size one, including $(claw, P_6)$-free graphs. It remains a challenging and interesting open problem to determine if there exists a constant c such that any H-free graph admits a longest path transversal of size c, even when H is a claw or a path on at least five vertices.

In this work, motivated by these recent results and open questions, we focus on subclasses of claw-free graphs and P_5-free graphs that are defined by two forbidden induced subgraphs. For claw-free graphs, we prove a theorem that completes the characterization of the graphs H on at most five vertices for which any $(claw, H)$-free graph admits a longest path transversal of size one. We do so by proving the theorem below, which together with the results of Gao and Shan [11] and Long Jr., Milans, and Munaro [16], yields the classification.

Theorem 1. *Let H be a graph in $\{P_3 + 2P_1, K_3 + 2P_1, 2P_2 + P_1, P_2 + 3P_1\}$. Let $\mathcal{G}$ be the class of $(claw, H)$-free graphs. Then for any $G \in \mathcal{G}$, $lpt(G) = 1$, and a longest path transversal of G of size one can be computed in polynomial time.*

We then shift our attention to P_5-free graphs, and prove the following.

Theorem 2. *Let H be one of the following graphs: triangle, paw, or diamond. Let $\mathcal{G}$ be the class of (P_5, H)-free graphs. Then for any $G \in \mathcal{G}$, $lpt(G) = 1$, and a longest path transversal of G of size one can be computed in polynomial time.*

Theorem 2, together with the results from Long Jr., Milans, and Munaro [16] leaves open only two cases ($H = K_4$ and $H = C_4$) towards showing that for any graph H of size at most four, any (P_5, H)-free graph admits a longest path transversal of size one.

This paper is organized as follows. In Sect. 2 we define the notation to be used in the paper and state previous results that are relevant to our proofs. In Sect. 3, we consider families of claw free graphs and prove Theorem 1. In Sect. 4, we consider P_5-free graphs and show Theorem 2. The proofs of the statements marked with ♠ have been omitted due to space constraints.

2   Preliminaries

Graphs in this paper are simple and connected. Let $G = (V(G), E(G))$ be a graph. A *clique* in G is a set of pairwise adjacent vertices. A *stable set* or *independent set* in G is a set of pairwise non-adjacent vertices. A *linear forest* is a forest whose components are paths. We let $P := p_1 p_2 \ldots p_k$ denote a path in G. We call the vertices p_1 and p_k the endpoints of P and say that P is a path from p_1 to p_k. For a vertex $x \in V(G)$, we say x is *between* p_i *and* p_j, if x is in the path $p_i \ldots p_j$. We also say x *has no neighbor from* p_i *to* p_j, if there is no vertex y between p_i and p_j such that xy is an edge.

A set $S \subseteq V(G)$ is a *longest path transversal* if every longest path of G has at least one vertex in S. We denote the size of a minimum longest path transversal of G by $lpt(G)$. We say a vertex in a graph is a *Gallai vertex* if it belongs to all of its longest paths. This name is due to the fact that this vertex attests a positive answer to the initital question proposed by Gallai. Note that G has a Gallai vertex if and only if $lpt(G) = 1$.

A path (cycle) in a graph is *Hamiltonian*, if it contains all vertices of the graph. A graph is k-*connected* if it has at least $k + 1$ vertices and no vertex separator of size at most $k - 1$. A *diamond* is a graph obtained from K_4 by removing an edge. A *triangle* is a K_3. If $X \subseteq V(G)$ we denote the subgraph induced on X by $G[X]$. For disjoint $X, Y \subseteq G$, we say that X is *complete* to Y if every vertex in X is adjacent to every vertex in Y, and X is *anticomplete* to Y if there are no edges between X and Y. For given graphs G and H, we say that G is H-*free* if G does not contain H as an induced subgraph. We say G is (H_1, H_2)-free if G do not contain H_1 and H_2 as an induced subgraphs. We let P_n, C_n, and K_n denote the chordless path, chordless cycle, and the complete graph on n vertices. For integer $t \geq 1$, we denote by tP_n the graph obtained from the disjoint union of t copies of the n-vertex path, and for graphs G_1, G_2, we write $G_1 + G_2$ to denote the disjoint union of G_1 and G_2. A subset $D \subseteq V(G)$ is a *dominating set* if each vertex of G either belongs to D or is adjacent to some vertex of D.

Observation 1 ([4]). *Let $D \subseteq V(G)$ be a dominating set in a graph G. Then, $lpt(G) \leq |D|$.*

Observation 2. *Let G be a connected graph and $P := p_1 \ldots p_k$ be a longest path of G. If $p_1 p_k \in E(G)$, then P is a Hamiltonian path.*

Proof. Suppose P is not a Hamiltonian path. Since G is connected, there exists $x \notin V(P)$ that has a neighbor in $V(P)$. Let p_i be such a neighbor. Then $x p_i \ldots p_k p_1 \ldots p_{i-1}$ is a path longer than P in G which is a contradiction.

□

The following results (see also [12] for the case of $2P_2$-free graphs) will partly play a role in our results:

Theorem 3 ([16]). *Let H be a graph and let $\mathcal{F}$ be the class of H-free graphs. If $lpt(G) = 1$ for every $G \in \mathcal{F}$, then H is a linear forest on at most 9 vertices.*

Theorem 4 ([16]). *Let H be a graph on at most four vertices and let $\mathcal{F}$ be the class of H-free graphs. Then $lpt(G) = 1$ for any $G \in \mathcal{F}$ if and only if H is a linear forest.*

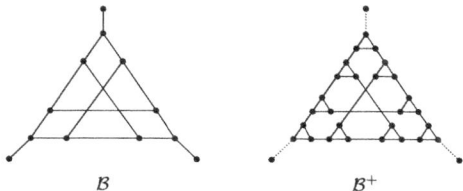

Fig. 1. Left: The graph $\mathcal{B}$ due to Voss and Walther [19] and Zamfirescu [21] with no Gallai vertex. Right: A claw-free graph $\mathcal{B}^+$ obtained from $\mathcal{B}$ (dotted lines show paths of length at least one).

3 (claw, H)-Free Graphs

We begin the section with two observations. The graph $\mathcal{B}$ depicted in Fig. 1 is the minimal example due to Voss and Walther [19] and Zamfirescu [21] answering Gallai's question in the negative. In particular, the graph $\mathcal{B}$ is constructed from the Petersen graph (which has no Hamiltonian cycle) by blowing up an arbitrary vertex into a set of three vertices, each of degree 1.

Long Jr., Milans and Munaro [16] showed that one can construct a claw-free graph $\mathcal{B}^+$ from the graph $\mathcal{B}$ (by replacing each vertex of degree three in $\mathcal{B}$ with a triangle) such that $\mathcal{B}^+$ has no Gallai vertex. This implies that:

Observation 3. *There exists a claw-free graph G with $lpt(G) \geq 2$.*

Second, Gao and Shan [11] proved that the family of (H_1, H_2)-free graphs such that H_1 and H_2 are connected and every 2-connected (H_1, H_2)-free graph has a Hamiltonian cycle, admits a longest path transversal of size one. Let us state this precisely. For integers $i, j, k \geq 0$, we denote by $N_{i,j,k}$ the graph obtained from a triangle by appending disjoint paths of length i, j, k at each vertex of the triangle. A *paw* is a $N_{1,0,0}$, and a *bull* is a $N_{1,1,0}$.

Theorem 5 ([11]). *Let H be a graph in $\{P_4, P_5, P_6, bull, N_{2,1,0}, N_{i,0,0}\}$, with $0 \leq i \leq 3$. If G is an H-free graph, then $lpt(G) = 1$.*

Observe that the classes considered in Theorem 1 are not contained in any of the classes covered by Theorem 5. The rest of the section is then devoted to proving Theorem 1. We will need the following observation for our proofs in this section:

Observation 4. *Let G be a claw-free graph, $P = p_1 p_2 \ldots p_k$ be a longest path of G, and let $u \in V(G)$ be such that $u \notin V(P)$. Then the following holds.*

- *If $up_i \in E(G)$, then $p_{i-1} p_{i+1} \in E(G)$;*
- *Moreover, if u has a neighbor $v \notin V(P)$, then $\{u, v, p_{i-1}, p_{i+1}\}$ induces a $2P_2$ in G.*

Proof. For the first part of Observation 4; see that $up_{i-1} \notin E(G)$, otherwise $p_1 \ldots p_{i-1}up_i \ldots p_k$ is a path longer than P. The same holds for the edge up_{i+1}. Hence we conclude $p_{i-1}p_{i+1} \in E(G)$, otherwise $\{p_{i-1}, p_i, p_{i+1}, u\}$ induces a claw in G, a contradiction. For the second part; suppose u has a neighbor $v \notin V(P)$. If $p_{i-1}v \in E(G)$, then $p_1 \ldots p_{i-1}vup_i \ldots p_k$ is a path longer than P, again, a contradiction. The same holds if $p_{i+1}v \in E(G)$.

□

Theorem 6. *Let G be a $(claw, P_3 + 2P_1)$-free graph. Then $lpt(G) = 1$.*

Proof. For a connected $(claw, P_3 + 2P_1)$-free graph G, observe that if G does not have an induced subgraph isomorphic to $P_3 + P_1$, by Theorem 4, G has a Gallai vertex. Hence, let $D \subset V(G)$ be a set inducing a $P_3 + P_1$. Note that D must be a dominating set, otherwise G would have an induced $P_3 + 2P_1$. Let $D = \{x, y, z, w\}$, where xyz is a P_3 and w is an isolated vertex in D. Let $P := p_1p_2 \ldots p_k$ be a longest path in G. We claim the following:

Claim 1 (♠). $V(P) \cap \{x, y, z\} \neq \emptyset$.

We proceed by showing that y is a Gallai vertex. Assume for the sake of contradiction that $y \notin V(P)$, then by Claim 1 either of x or z must be in $V(P)$. By symmetry we may assume $x \in V(P)$. Let a and b be the vertices appearing before and after x in P, respectively. Since $y \notin V(P)$, by Observation 4, $ab \in E(G)$. Clearly, p_1y cannot be an edge, otherwise we can extend P to y. Moreover, $p_1x \notin E(G)$, otherwise $yxp_1 \ldots ab \ldots p_k$ is a longer path in G. In a similar way, we conclude $p_1z \notin E(G)$. Indeed, if that was the case, z would be in P, and the same arguments would yield a longer path. The same holds for p_k: it cannot be adjacent to any vertex in $\{x, y, z\}$. Consider the set $\{x, y, z, p_1, p_k\}$. Since G is $(P_3 + 2P_1)$-free, we have $p_1p_k \in E(G)$. Then $yxb \ldots p_kp_1 \ldots a$ is a path longer than P in G, a contradiction. Hence, y is common to all longest paths in G. This finishes the proof of Theorem 6.

□

Observation 5. *Let G be a $(claw, P_3 + 2P_1)$-free graph. There exists a polynomial-time algorithm which finds a Gallai vertex in G.*

Proof. Consider the following simple algorithm: in polynomial time we can check whether G contains an induced copy of $P_3 + P_1$. If the answer is yes, that is, if G contains $P_3 + P_1$ as an induced subgraph, then we return the middle vertex of the P_3, say v. It follows from the proof of Theorem 6 that v is a Gallai vertex. If the answer is no, that is, if G is $(claw, P_3 + P_1)$-free, then by the result of [16] (see Theorem 6 in [16]), every vertex of degree at least $\Delta(G) - 1$ is a Gallai vertex, and such vertex can be found in polynomial time.

□

A similar proof as above works for the case in which G is a $(claw, K_3 + 2P_1)$-free graph, in which case we would have that $\{x, y, z\}$ induces a triangle instead. Hence, we have the following:

Theorem 7. *Let G be a $(claw, K_3 - 2P_1)$-free graph. Then $lpt(G) = 1$.*

To apply the argument of Observation 5 for Theorem 7, note that the class of (claw, $N_{3,0,0}$)-free graphs contains the class of (claw, $K_3 + P_1$)-free graphs. Every graph G in the former class has $lpt(G) = 1$, by [11]. We leave it to the reader to check that the proof of [11] for (claw, $N_{3,0,0}$)-free graphs is constructive.

Next, we prove the following:

Theorem 8. *Let G be a (claw, $2P_2 + P_1$)-free graph. Then $lpt(G) = 1$.*

Proof. Let G be a connected (claw, $2P_2 + P_1$)-free graph. If G does not have an induced $2P_2$, then it follows by Theorem 4 that G has a Gallai vertex. Hence, let $\{x, y, z, w\} \subset V(G)$ be a set inducing a $2P_2$, where $xy \in E(G)$ and $zw \in E(G)$. We may assume $d(x) \geq d(y)$. We will now show that x is a Gallai vertex in G. Let $P = p_1 \ldots p_k$ be a longest path in G and suppose for a contradiction that $x \notin V(P)$.

Claim 2 (♠). $V(P) \cap \{x, y\} \neq \emptyset$.

Observe that, since $x \notin V(P)$, by Claim 2, we have $y \in V(P)$. Moreover, if a and b are the vertices before and after y in P, by Observation 4, we have $ab \in E(G)$.

Claim 3 (♠). $\{x, y, p_1, p_2\}$ *induces a* $2P_2$ *in* G.

Now consider the set $\{x, y, p_1, p_2, p_k\}$. By Claim 3, $\{x, y, p_1, p_2\}$ induces a $2P_2$. If $xp_k \in E(G)$, P could be extended to x. If $yp_k \in E(G)$, then $xyp_k \ldots ba \ldots p_1$ is a path longer than P. By Observation 2, $p_k p_1 \notin E(G)$. Then we must have $p_k p_2 \in E(G)$, otherwise $\{x, y, p_1, p_2, p_k\}$ induces a $2P_2 + P_1$ in G. Now we use again an argument similar to the one in the proof of Claim 3, where we had the edge yp_2. This time, instead we will use the fact that $p_k p_2 \in E(G)$. Since $d(x) \geq d(y)$, and $\{a, b\} \subset N(y) \setminus N(x)$, x must have a neighbor x' that is not a neighbor of y. If $x' \notin V(P)$, then $x'xyb \ldots p_k p_2 \ldots a$ is a path longer than P. Hence, $x' \in V(P)$. We may assume x' lies between p_1 and a in P (if not, a similar argument would work with the edge $p_{k-1} p_k$). Let c and d be the vertices before and after x' in P, respectively. Consider the set $\{x, y, c, d, p_1\}$. By Observation 4, $cd \in E(G)$. Note that $\{x, y, c, d\}$ induces a $2P_2$ in G. Similarly, as before, $p_1 x, p_1 y \notin E(G)$, otherwise longer paths could be obtained. If $p_1 c \in E(G)$, then $p_3 \ldots cp_1 p_2 p_k \ldots x'x$ is a path longer than P. So $p_1 c \notin E(G)$. Then we must have $p_1 d \in E(G)$, otherwise $\{x, y, c, d, p_1\}$ induces a $2P_2 + P_1$. To conclude the proof, consider the set $\{x, y, c, d, p_k\}$. Similarly to p_1, we cannot have edges between p_k and x or y. Let c' be the vertex before c in P and d' be the vertex after d. If $p_k c \in E(G)$, then $xx'cp_k \ldots dp_1 \ldots c'$ is a path longer than P in G. If $p_k d \in E(G)$, then $d' \ldots p_k dp_1 \ldots x'x$ is a path longer than P. Therefore we conclude $\{x, y, c, d, p_k\}$ induces a $2P_2 + P_1$ in G, a contradiction. This implies that x must belong to any longest path P of G, hence finishes the proof of Theorem 8. $\qquad\square$

Observation 6. *Let G be a (claw, $2P_2 + P_1$)-free graph. There exists a polynomial-time algorithm which finds a Gallai vertex in G.*

The proof of Observation 6 is is quite similar to the proof of Observation 5. We remind the reader that in a connected $2P_2$-free graph, every vertex of maximum degree is common to all longest paths (see Theorem 1 in [12]).

The remainder of this section is devoted to settling Theorem 1 for (claw, $P_2 + 3P_1$)-free graphs. This will complete our picture for the longest path transversal of (claw, H)-free graphs with $|H| \leq 5$. Although the big picture behind the proof is similar to the proof of Theorem 6 to Theorem 8, we are compelled to take extra steps to mold the details. Recall that a *cut vertex* $x \in V(G)$ of G is a vertex such that $G - x$ is disconnected.

Lemma 1 (♠). *Let G be a (claw, $P_2 + 3P_1$)-free graph with three cut vertices $\{x, y, z\} \subseteq V(G)$ inducing a triangle in G. Then x, y and z are Gallai vertices.*

Theorem 9. *Let G be a (claw, $P_2 + 3P_1$)-free graph. Then $lpt(G) = 1$.*

Proof. Let G be a connected (claw, $P_2 + 3P_1$)-free graph. By Lemma 1, we may assume that G has no three cut vertices inducing a triangle. If G is a ($P_2 + 2P_1$)-free graph, then by Theorem 4, G has a Gallai vertex. Hence, we may assume G has a set $D \subset V(G)$ inducing a $P_2 + 2P_1$. Let $D = \{x, y, z, w\}$, with $xy \in E(G)$. Note that since G is ($P_2 + 3P_1$)-free, D must be a dominating set in G. We may assume, without loss of generality, that $d_G(x) \geq d_G(y)$. We wish to show that x is a Gallai vertex in G. As before, let $P := p_1 \ldots p_k$ be a longest path of G and suppose for the sake of contradiction that $x \notin V(P)$. We begin with a sequence of claims, starting with Claim 4.

Claim 4 (♠). $|V(P) \cap V(D)| \geq 2$

Claim 5. $V(P) \cap \{x, y\} \neq \emptyset$.

Proof. Suppose not. Let $y' \in V(G)$ be a vertex in $V(P)$ that is closest to $\{x, y\}$ and let Q be a shortest path between $\{x, y\}$ and y' (as G is connected). We may assume that y and y' are the endpoints of this path. Let $\hat{y}$ be the vertex adjacent to y' in Q. Note that it may be that $\hat{y} = y$. Let c and d be the vertices before and after y' in P. By Claim 4, we have $z \in V(P)$ and $w \in V(P)$. Let a and b be the neighbors of z in P. Observe that neither $p_1 x \in E(G)$ nor $p_1 y \in E(G)$, as otherwise P could be extended to x or y. Analogously, $p_k x \notin E(G)$ and $p_k y \notin E(G)$. Since D is a dominating set, p_1 and p_k must be dominated by $\{z, w\}$. Therefore we have two cases:

Case 1. There exists a vertex $q \in \{z, w\}$ such that both p_1 and p_k are dominated by q.

Suppose, without loss of generality, that $q = z$ and z is between y' and p_k in P. Since p_1 and p_k dominated by z, we may assume that $ab \notin E(G)$, otherwise the path $xy \ldots y' \ldots p_1 z p_k \ldots ba \ldots d$ is a longer path. We will show that $\{x, y, a, b, c\}$ induces a $P_2 + 3P_1$. First, if $xa \in E(G)$, then $yxa \ldots p_1 z b \ldots p_k$ is a longer path (symmetrically, $bx \notin E(G)$) and if $ya \in E(G)$, then $xya \ldots p_1 z b \ldots p_k$ is a

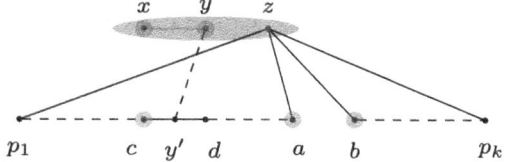

Fig. 2. Proof of Claim 5, Case 1: p_1, p_k dominated by z. Dashed lines show paths of arbitrary length.

longer path (symmetrically, $by \notin E(G)$). Second, $ac, bc \notin E(G)$, since otherwise $xy \ldots y' \ldots ac \ldots p_1 zb \ldots p_k$, and $xy \ldots y' \ldots azp_k \ldots bc \ldots p_1$ are longer paths. Third, $cx \notin E(G)$, since otherwise $p_1 \ldots cxy \ldots y' \ldots p_k$ is a longer path. We may also assume that $cy \notin E(G)$, as $y \notin V(P)$. Hence, the fact that $ab \notin E(G)$ implies that $\{x, y, a, b, c\}$ form a $P_2 + 3P_1$ (red vertices in Fig. 2). This yields a contradiction and shows that Claim 5 holds for Case 1.

Case 2. p_1 and p_k are dominated by distinct vertices among $\{z, w\}$.

Suppose, without loss of generality, that p_1 is dominated by z and p_k is dominated by w. We have two possibilities:

(a) y' is between p_1 and z (symmetrically, y' is between p_k and w),
(b) y' is between z and w.

To prove Case 2(a) we show that (see Fig. 3):

(1) $p_1 a \in E(G)$.

Proof. Recall that $a, b \in V(G)$ are the neighbors of z in P, and that $\hat{y}$ is the vertex of Q adjacent to y' (and it can be that $\hat{y} = y$). We show that $\{\hat{y}, y', p_1, p_k, a\}$ induces a $P_2 + 3P_1$, unless $p_1 a \in E(G)$. It is easy to see that, $cd \in E(G)$ by Observation 4, and $\hat{y}p_1, \hat{y}p_k \notin E(G)$ as $\hat{y} \notin V(P)$. First, we may assume that $p_1 y' \notin E(G)$ otherwise, $xy \ldots y' p_1 \ldots cd \ldots p_k$ is a longer path. Similarly, $p_k y' \notin E(G)$ otherwise, $xy \ldots y' p_k \ldots dc \ldots p_1$ is a longer path. Second, neither of $\hat{y}a$ and $y'a$ is an edge in $E(G)$ since otherwise in the former case, $xy \ldots \hat{y}a \ldots p_1 zb \ldots p_k$ is a longer path and in the latter case, $xy \ldots y'a \ldots dc \ldots p_1 zb \ldots p_k$ is a longer path. From Observation 2 we have $p_1 p_k \notin E(G)$. Moreover, $p_k a \notin E(G)$ otherwise, $xy \ldots y' \ldots ap_k \ldots bzp_1 \ldots c$ is a longer path. This shows that $\{\hat{y}, y', p_1, p_k, a\}$ induces a $P_2 + 3P_1$ which is a contradiction, hence $p_1 a \in E(G)$. This proves (1).

□

Now we claim that $\{x, y, z, c, p_k\}$ induces a $P_2 + 3P_1$ in G. Observe that $xc \notin E(G)$ since otherwise $p_1 \ldots cxy \ldots y' \ldots p_k$ is a longer path and $yc \notin E(G)$ otherwise, we can extend P to y. Moreover, since $ap_1 \in E(G)$ by (1), neither of cz and cp_k is an edge in $E(G)$ since otherwise in the former case, $xy \ldots y' \ldots ap_1 \ldots cz \ldots p_k$ is a longer path, and in the latter case, $xy \ldots y' \ldots ap_1 \ldots cp_k \ldots bz$ is a longer path. Moreover, by our assumption, neither of xp_k and

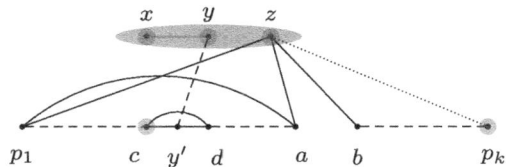

Fig. 3. Proof of Claim 5, Case 2(a): y' is between p_1 and z. Dashed lines show paths of arbitrary length, dotted lines show non-edges.

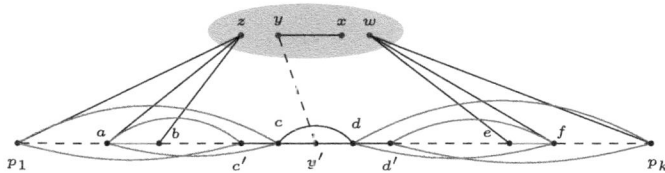

Fig. 4. Proof of Claim 5, Case 2(b): y' is between z and w, outcomes of (2). Dashed lines show paths of arbitrary length.

yp_k is an edge in $E(G)$. Note that $zp_k \notin E(G)$ by the assumption of Case 2. Then, it follows that $\{x, y, z, c, p_k\}$ induces a $P_2 + 3P_1$ (red vertices in Fig. 3) which is a contradiction. Therefore, Claim 5 holds for Case 2(a). We leave it to the reader to check that the case y' is between p_k and w can be shown symmetrically.

We proceed with Case 2(b), that is, when y' is between z and w. We remind the reader that by Lemma 1, we assume G has no set of three vertices that are cut vertices and induce a triangle. Our goal is to show that, in this case, such vertices exist, leading to a contradiction. Note that, $w \in V(P)$ and $cd \in E(G)$ by Observation 4. Let $e, f \in V(G)$ be the neighbors of w in P. Let c' be the vertex before c, and d' be the vertex after d in P (see Fig. 4).

(2) *The edges* $ac, df, ac', p_1c', p_kd, p_kd', p_1c, d'f, ab,$ *and* ef *exist in* G. (♠)

For the rest of the proof of Case 2(b), we wish to show that $\{c, y', d\}$ are three cut vertices that induce a triangle in G (see Fig. 4). Let $w' \in V(G)$ be any vertex from p_1 to y', and $z' \in V(G)$ be any vertex from y' to p_k. We say ℓ is a *crossing edge* in G if either $\ell = zz'$ or $\ell = ww'$.

We claim the following:

(3) *G has no crossing edges.* (♠)

(4) x *and* y *have no neighbor in* P *other than* y'. (♠)

(5) y' *has no neighbor from* p_1 *to* c', *and from* p_k *to* d'. (♠)

(6) *There is no edge* $e = \alpha\beta$ *with* α *from* p_1 *to* c', *and* β *from* d' *to* p_k. (♠)

It is now easy to conclude from (3)–(6) that the vertices in $K_3 = cy'd$ are cut vertices in G. This violates the assumption that G has no three cut vertices inducing a triangle, hence completing the proof of Claim 5.   □

We are ready to finish the proof of Theorem 9. Recall that our desire is to show that x is a Gallai vertex. Since $x \notin V(P)$ (by the assumption), Claim 5 implies that $y \in V(P)$, say $y = p_i$ for $1 < i < k$. As $d_G(x) \geq d_G(y)$, x must have a private neighbor in G, say x'. Recall that, by Observation 4 we have $p_{i-1}p_{i+1} \in E(G)$ as $x \notin V(P)$. We have two possibilities:

- If $x' \notin V(P)$ then we show that $\{p_1, p_2, x', p_k, y\}$ induces a $P_2 + 3P_1$. First, neither of p_1x' and p_2x' is an edge; otherwise, in the former case $yxx'p_1 \ldots p_{i-1}$ $p_{i+1} \ldots p_k$ is a longer path, and in the latter case $yxx'p_2 \ldots p_{i-1}p_{i+1} \ldots p_k$ is a longer path, symmetrically, $p_kx' \notin E(G)$. Observe that since x' is a private neighbor of x, $yx' \notin E(G)$. Second, if $p_2p_k \in E(G)$ then the path $x'xyp_{i-1} \ldots p_2p_k \ldots p_{i+1}$ is a longer path. Third, neither of p_1y and p_2y is an edge, since otherwise, in the former case $x'xyp_1 \ldots p_{i-1}p_{i+1} \ldots p_k$ is a longer path, and in the latter case the path $x'xyp_2 \ldots p_{i-1}p_{i+1} \ldots p_k$ is a longer path, symmetrically, $p_ky \notin E(G)$. This shows that $\{p_1, p_2, x', p_k, y\}$ induces a $P_2 + 3P_1$ in G which is impossible.
- By the first bullet point we may suppose that $x' \in V(P)$. Let x' be a vertex between p_{i+1} and p_k, and observe that the case for when x' is between p_i and p_1 is symmetric. We denote by x^-, x^+ the vertices before and after x' in P. We show that $\{x, y, p_1, p_k, x^-\}$ induces a $P_2 + 3P_1$ in G. Since G is claw-free we have $x^-x^+ \in E(G)$ otherwise, $\{x^-, x', x^+, x\}$ induces a claw. Moreover, it follows from Observation 4 that $\{x, y, x^-, x^+\}$ induces a $2P_2$ in G. Note that, $\{x, y, p_1\}$ and $\{x, y, p_k\}$ induces a $P_2 + P_1$ in G since otherwise we could extend P to x ($p_{i-1}p_{i+1} \in E(G)$ by Observation 4). If $p_1x^- \in E(G)$ then $p_{i+1} \ldots x^-p_1 \ldots p_{i-1}yxx' \ldots p_k$ is a longer path, and if $p_kx^- \in E(G)$ then $p_{i+2} \ldots x^-p_k \ldots x'xyp_{i+1}p_{i-1} \ldots p_1$ is a longer path. This shows that $\{x, y, p_1, p_k, x^-\}$ induces a $P_2 + 3P_1$ which is a contradiction.

The above two bullet points conclude that x must be a common vertex to all longest paths in G. This finishes the proof of Theorem 9.   □

Observation 7. *Let G be a (claw, $P_2 + 3P_1$)-free graph. There exists a polynomial-time algorithm which finds a Gallai vertex in G.*

The proof of Observation 7 is quite similar to the proof of Observation 5. We remind the reader that in a connected $(P_2 + 2P_1)$-free graph, every vertex of maximum degree is a Gallai vertex (see Proposition 7 in [16]).

Let us now put everything together; Theorems 6, 7, 8, along with Theorem 9 yields Theorem 1. Now Theorem 1, combined with Theorem 5 and Theorem 4 implies a complete classification, for all graphs H of size at most five, of (claw, H)-free graphs that admit a longest path transversal of size one. More precisely (Fig. 5):

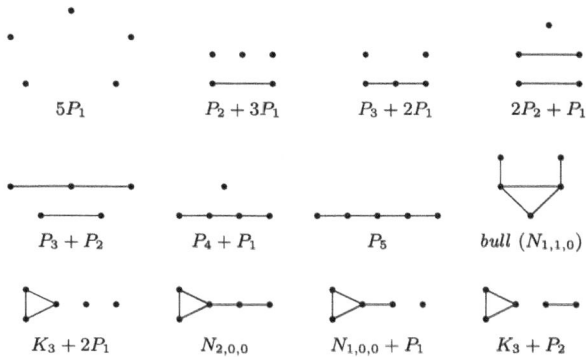

Fig. 5. All five-vertex induced subgraphs of $\mathcal{B}^+$.

Corollary 1 (♠). *Let H be a graph on at most five vertices. Let $\mathcal{G}$ be a class of $(claw, H)$-free graphs. Then for every graph $G \in \mathcal{G}$, $lpt(G) = 1$ if and only if one of the following hold:*

- H *is a linear forest;*
- H *is one of the graphs bull, $K_3 + 2P_1$, $N_{2,0,0}$, $N_{1,0,0} + P_1$ or $K_3 + P_2$.*

4 (P_5, H)-Free Graphs

We begin the section with proving Theorem 2 for $H = K_3$. The reader may observe that the class of $(P_5, \text{diamond})$-free graphs contains the class of $(P_5, \text{triangle})$-free graphs. However, we include a proof for $(P_5, \text{triangle})$-free case as it is a vital piece in the proof of (P_5, paw)-free case.

Lemma 2 ([2]). *A graph G is (P_5, C_5)-free if and only if every induced subgraph of G contains a dominating clique.*

A 5-*ring* is a graph whose vertices can be partitioned into five non-empty stable sets $S_1, \ldots, S_5$ such that for each i (in modulo 5) S_i is complete to $S_{i-1} \cup S_{i+1}$ and anticomplete to $S_{i-2} \cup S_{i-2}$. Sumner [18] characterized the structure of $(P_5, \text{triangle})$-free graphs.

Lemma 3 ([18]). *Let G be a $(P_5, \text{triangle})$-free graph. Then each component of G is either a 5-ring or bipartite.*

Note that a 5-ring graph is $2P_2$-free. We are now ready to prove the following:

Theorem 10. *Let G be a $(P_5, \text{triangle})$-free graph. Then $lpt(G) = 1$.*

Proof. Let G be a connected $(P_5, \text{triangle})$-free graph. If G contains an induced C_5, then it follows from Lemma 3 that G is a 5-ring, hence $2P_2$-free, and Theorem 4 implies that G has a Gallai vertex. Therefore, we may assume that G is a $(P_5, C_5, \text{triangle})$-free graph. Then by Lemma 2, G contains a dominating clique, which has size at most two, since G is triangle-free. If G has a dominating vertex, by Observation 1, G has a Gallai vertex. Let $D = \{x, y\}$ with $xy \in E(G)$ be a dominating edge in G. We may suppose, without loss of generality, that $|N_G(y)| \leq |N_G(x)|$. We wish to show that x is a Gallai vertex in G. Let $P := p_1 p_2 \ldots p_k$ be a longest path, and assume for the sake of contradiction that x does not belong to P. Since D is a dominating set, p_1 must be adjacent to at least one vertex among $\{x, y\}$. Since P is a longest path and $x \notin V(P)$, $p_1 x \notin E(G)$ and so $p_1 y \in E(G)$. Similarly $p_k y \in E(G)$. We have $y \in V(P)$, otherwise we could extend P to y. Let $y = p_i$ for some $1 < i < k$. We claim that $|N_P(y)| \geq |N_P(x)|$. To see this, observe that for each $1 < j < k$, either of $p_j x$ or $p_j y$ is an edge. Since G is triangle-free if $p_j x \in E(G)$ then neither $p_{j-1} x \in E(G)$ nor $p_{j+1} x \in E(G)$. This combined with the fact that $y \in V(P)$, and $p_1 y, p_k y \in E(G)$ proves our claim. Now consider the set $\{p_{i-2}, p_{i-1}, p_i, p_{i+1}, p_{i+2}\}$. Observe that $p_{i-1} p_{i+1} \notin E(G)$; otherwise, $\{p_{i-1}, p_i, p_{i+1}\}$ induces a triangle. Since G is triangle-free, neither of $p_{i-2} p_i$ and $p_{i+2} p_i$ is an edge. Suppose $p_{i-1} p_{i+2} \in E(G)$. Since $|N_P(y)| \geq |N_P(x)|$ and $|N_G(y)| \leq |N_G(x)|$, there is a vertex $u \in N_G(x)$ such that $u \notin V(P)$. But then $uxyp_1 \ldots p_{i-1} p_{i+2} \ldots p_k$ is a longer path, as it excludes p_{i+1} and includes u, x in P. Analogously, if $p_{i+1} p_{i-2} \in E(G)$ then $uxyp_1 \ldots p_{i-2} p_{i+1} \ldots p_k$ is a longer path, as it excludes p_{i-1} and includes u, x in P. Therefore, neither of $p_{i-1} p_{i+2}$ and $p_{i+1} p_{i-2}$ is an edge. Moreover, observe that since G is C_5-free we have $p_{i-2} p_{i+2} \notin E(G)$. This implies that $\{p_{i-2}, p_{i-1}, p_i, p_{i+1}, p_{i+2}\}$ induces a P_5 in G which is a contradiction, hence x is common to all longest paths in G, as desired. $\qquad\square$

Observation 8 (♠). *Let G be a $(P_5, \text{triangle})$-free graph. There exists a polynomial-time algorithm which finds a Gallai vertex in G.*

Next, we give a short argument for the case $H = \text{paw}$ of Theorem 2. Recall that a paw is a $N_{1,0,0}$, the following result describes the structure of paw-free graphs.

Lemma 4 ([17]). *A graph G is paw-free if and only if each component of G is triangle-free or complete multipartite.*

If G is a (P_5, paw)-free graph, then by Lemma 4, G is either $(P_5, \text{triangle})$-free or a complete multipartite. In the former case, as we have shown in Theorem 10, G has a Gallai vertex. In the latter case, the fact that multipartite graphs are $2P_2$-free, combined with Theorem 4 implies that G has a Gallai vertex. The following is immediate:

Corollary 2. *Let G be a (P_5, paw)-free graph. Then $lpt(G) = 1$.*

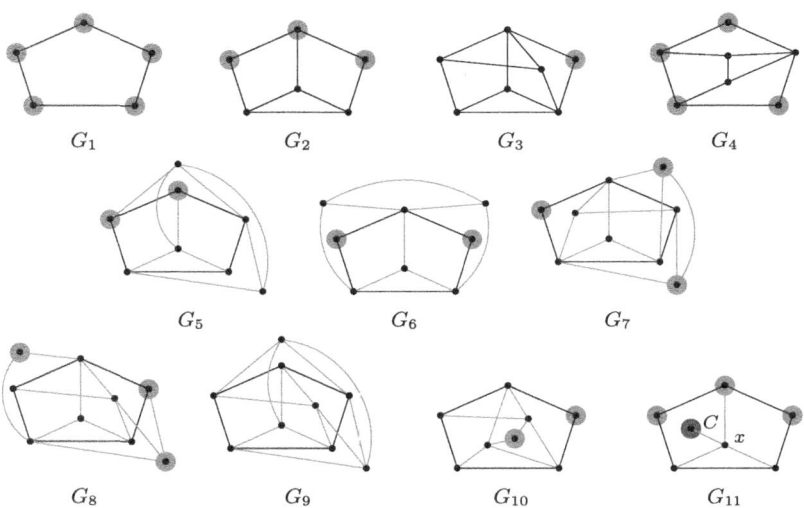

Fig. 6. The graphs used in Lemma 7.

We now plunge into the $(P_5, \text{diamond})$-free case. We start with further definitions. Let G be a graph. Let $A, B \subseteq V(G)$ be two vertex disjoint subsets. We denote by $[A, B]$ the set of edges of G with one vertex in A and the other in B. We also say $[A, B]$ is complete if every vertex in A is adjacent to every vertex in B. We use $[A]$ to denote the subgraph induced by vertices of A.

Let $\mathcal{G}_1, \mathcal{G}_2, \mathcal{G}_3$ be the families of graphs. For a graph G, let

1. $G \in \mathcal{G}_1$ if G contains a dominating clique K such that:
   - every component of $G \setminus K$ is complete,
   - for each component D in $G \setminus K$, there is a unique vertex $x \in K$ such that $[V(D) \cup \{x\}]$ is complete,
   - the above be the only edges in G.
2. $G \in \mathcal{G}_2$ if $V(G)$ can be partitioned into two sets X and Y with the following specifications:
   - $[X], [Y]$ are complete,
   - $[X, Y]$ is a non-empty set of independent edges.
3. $G \in \mathcal{G}_3$ if G contains a $2P_2$-free bipartite graph H with an edge $e = xy$ dominates G such that $[V(H) \setminus e, V(G) \setminus V(H)] = \emptyset$, and for all components D_i of $G \setminus V(H)$ the following holds:
   - for each i, D_i is complete,
   - for each i, either $[V(D_i), \{x\}]$ or $[V(D_i), \{y\}]$ is complete,
   - for at most one i, $[V(D_i), \{x, y\}]$ is complete,
   - the above be the only edges in G.

We need the following decomposition for $(P_5, C_5, \text{diamond})$-free graphs:

Lemma 5 ([9]). *A graph G is $(P_5, C_5, diamond)$-free if and only if $G \in \mathcal{G}_1 \cup \mathcal{G}_2 \cup \mathcal{G}_3$.*

We mention two remarks on the families $\mathcal{G}_1, \mathcal{G}_2$. A graph is *star-like* if it is an intersection graph of substars of a star [5]. The following result is due to Cerioli and Lima [4]:

Lemma 6 ([4]). *Let $\mathcal{H}$ be a connected graph in which $V(\mathcal{H})$ can be partitioned into $k + 1$ sets $K, V_1, \ldots, V_k$, for some $k \in \mathbb{N}$, such that the following holds*

- *K is a clique;*
- *For all $x \in V_i$ and $y \in V_j$, $i \neq j$, it holds that $xy \notin E(\mathcal{H})$;*
- *The vertices of V_i can be ordered $v_{i1}, v_{i2}, \ldots, v_{i|V_i|}$ in such a way that for all $x \in K$, if $xv_{ij} \in E(\mathcal{H})$, then $xv_{ik} \in E(\mathcal{H})$ for all $k < j$;*
- *For all $x \notin K$, there exists $y \in K$ such that $xy \in E(\mathcal{H})$.*

Then $\mathcal{H}$ has a Gallai vertex.

The reader may observe that the family $\mathcal{G}_1$ is contained in the class of star-like graphs. Also it is not hard to see that the family $\mathcal{G}_2$ is Hamiltonian.

We need one more lemma to define our next result precisely. An *expansion* of a graph H is any graph G such that the vertices of G can be partitioned into $|V(H)|$ nonempty sets S_x, for $x \in V(H)$, such that $G[S_x \cup S_y]$ is complete if $xy \in E(H)$, and $G[S_x \cup S_y] = \emptyset$ if $xy \notin E(H)$. An expansion of a graph is a *stable set expansion* if each S_x is a stable set, and is a *P_3-free expansion* if each S_x induces a P_3-free graph. The following characterization for $(P_5, diamond)$-free graphs is due to Arbib and Mosca [1] (see also Theorem 2 in [9]):

Lemma 7 ([1]). *G is a $(P_5, diamond)$-free graph that contains an induced C_5 if and only if G is obtained from one of the graphs among $G_1, \ldots, G_{11}$ by stable set expansion of each blue vertex, and P_3-free expansion of the red vertex.*

We are now ready to prove the following result:

Theorem 11. *Let G be a $(P_5, diamond)$-free graph. Then $lpt(G) = 1$.*

Proof. Let G be a $(P_5, diamond)$-free graph. We first show that the claim of Theorem 11 holds for when G has no induced C_5. Let G be a $(P_5, C_5, diamond)$-free graph, then it follows from Lemma 5 that $G \in \mathcal{G}_1 \cup \mathcal{G}_2 \cup \mathcal{G}_3$. Since for every graph $G \in \mathcal{G}_1$ we have $lpt(G) = 1$, and $\mathcal{G}_2$ is Hamiltonian, the claim holds if $G \in \mathcal{G}_1 \cup \mathcal{G}_2$. Therefore, we may assume that $G \in \mathcal{G}_3$. We let $\mathcal{D}_x$ denote those components of $G \setminus H$ that are complete to only x, analogously, $\mathcal{D}_y$ denote those components of $G \setminus H$ that are complete to only y.

Claim 6 (♠). *For every graph $G \in \mathcal{G}_3$, $lpt(G) = 1$.*

So far we have shown that the class of $(P_5, C_5, diamond)$-free graphs admit $lpt = 1$. It remains to discuss the case when G contains an induced C_5. Then by Lemma 7, either G is obtained from $G_1, \ldots, G_{10}$ by stable set expansion of each blue vertex, or G is obtained from G_{11} by stable set expansion of each blue vertex and P_3-free expansion of the red vertex. For the former case, it is easy to check that G_i is $2P_2$-free for $1 \leq i \leq 10$ (see Fig. 6). Thus, it follows by Theorem 4 that G has a Gallai vertex if it is obtained from $G_1, \ldots, G_{10}$ by a stable set expansion of each blue vertex. For the latter case, we claim that:

Claim 7 (♠). *Let $\mathfrak{g}$ be a class of graphs obtained from G_{11} by a stable set expansion of each blue vertex and P_3-free expansion of the red vertex. Then for every graph $G \in \mathfrak{g}$, $lpt(G) = 1$.*

Now, Claim 6 combined with Claim 7 finishes the proof of Theorem 11.   □

Observation 9 (♠). *Let G be a $(P_5, diamond)$-free graph. There exists a polynomial-time algorithm which finds a Gallai vertex in G.*

Finally, we restate and prove Theorem 2:

Theorem 12. *Let H be one of the graphs triangle, paw, or diamond. Let $\mathcal{G}$ be a class of graphs restricted to (P_5, H)-free graphs. Then for every graph $G \in \mathcal{G}$, $lpt(G) = 1$, and there exists a polynomial-time algorithm which finds Gallai vertices on graphs in $\mathcal{G}$.*

Proof. The first part of Theorem 12 follows directly from the proofs of Theorem 10, Corollary 2, and Theorem 11. The second part follows from Observation 8 and Observation 9. Note that the algorithm for (P_5, paw)-free graphs is the same as $(P_5, diamond)$-free case. This completes the proof of Theorem 12.   □

5   Open Problems

One of the main open problem related to our work is whether P_5-free graphs admit a longest path transversal of size one. This question was also posed in [4, 12, 16]. We conjecture the answer to be affirmative. Towards proving it, it would be interesting to consider (P_5, C_4)-free graphs and (P_5, K_4)-free graphs, which are the last cases of (P_5, H)-free graphs with $|H| \leq 4$ that remain open. Note that in the latter case, it is easy to give a constant upper bound on the size of a longest path transversal. Indeed, a result of Bacsó and Tuza [2] states that every P_5-free graph contains a dominating clique or a dominating P_3. By Observation 4, the size of a dominating set upper bounds the size of a longest path transversal. Since in (P_5, K_4)-free graphs, a dominating clique has size at most 3, the bound follows. Beyond H-free graphs, it is still an open question whether there exists a constant c such that *any* graph admits a longest path transversal of size c. An upper bound that is sublinear on the number of vertices is currently known [15], but there are no examples of graph families that match the upper bound. In fact, it is not known whether there exists a graph G for which $lpt(G) \geq 4$.

Acknowledgments. We acknowledge the support of the Independent Research Fund Denmark grant agreement number 2098-00012B.

References

1. Arbib, C., Mosca, R.: On $(P_5, \text{diamond})$-free graphs. Discret. Math. **250**(1–3), 1–22 (2002)
2. Bacsó, G., Tuza, Z.: Dominating cliques in P_5-free graphs. Period. Math. Hung. **21**(4), 303–308 (1990)
3. Balister, P.N., Győri, E., Lehel, J., Schelp, R.H.: Longest paths in circular arc graphs. Combin. Probab. Comput. **13**(3), 311–317 (2004)
4. Cerioli, M.R., Lima, P.T.: Intersection of longest paths in graph classes. Discret. Appl. Math. **281**, 96–105 (2020)
5. Cerioli, M.R., Szwarcfiter, J.L.: Characterizing intersection graphs af substars of a star. Ars Combin. **79**, 21–31 (2006)
6. Cerioli, M.R., Fernandes, C.G., Gómez, R., Gutiérrez, J., Lima, P.T.: Transversals of longest paths. Discret. Math. **343**(3), 111717 (2020). https://doi.org/10.1016/j.disc.2019.111717
7. Chen, F.: Nonempty intersection of longest paths in a graph with small matching number. Czechoslov. Math. J. **65**, 545–553 (2015)
8. Chen, G., et al.: Nonempty intersection of longest paths in series-parallel graphs. Discret. Math. **340**(3), 287–304 (2017)
9. Choudum, S.A., Karthick, T.: First-fit coloring of $\{P_5, K_4 - e\}$-free graphs. Discret. Appl. Math. **158**(6), 620–626 (2010)
10. Erdős, P., Katona, G. (eds.): Theory of Graphs. Proceedings of the Colloquium held at Tihany, Hungary, September 1966, Academic Press, New York (1968). problem 4 (T. Gallai), p. 362
11. Gao, Y., Shan, S.: Nonempty intersection of longest paths in graphs without forbidden pairs. Discret. Appl. Math. **304**, 76–83 (2021)
12. Golan, G., Shan, S.: Non-empty intersection of longest paths in $2K_2$-free graphs. Electr. J. Comb. **25**, P2.37 (2018)
13. Jobson, A.S., Kézdy, A.E., Lehel, J., White, S.C.: Detour trees. Discret. Appl. Math. **206**, 73–80 (2016)
14. Joos, F.: A note on longest paths in circular arc graphs. Discussiones Mathematicae Graph Theory **35**(3), 419–426 (2015)
15. Long, J.A., Milans, K.G., Munaro, A.: Sublinear longest path transversals. SIAM J. Discret. Math. **35**(3), 1673–1677 (2021). https://doi.org/10.1137/20M1362577
16. Long Jr, J.A., Milans, K.G., Munaro, A.: Non-empty intersection of longest paths in H-free graphs. Electr. J. Comb., p. P1.32 (2023)
17. Olariu, S.: Paw-fee graphs. Inf. Process. Lett. **28**(1), 53–54 (1988)
18. Sumner, D.P.: Subtrees of a graph and chromatic number. In: Chartrand, G. (ed.) The Theory and Applications of Graphs. John Wiley & Sons, New York **557**, 576 (1981)
19. Voss, H.J., Walther, H.: Über Kreise in Graphen. VEB Deutscher Verlag der Wissenschaften (1974)
20. Walther, H.: Über die nichtexistenz eines knotenpunktes, durch den alle längsten wege eines graphen gehen. J. Comb. Theory **6**(1), 1–6 (1969)
21. Zamfirescu, T.: On longest paths and circuits in graphs. Math. Scand. **38**(2), 211–239 (1976)

Realizing Graphs with Cut Constraints

Vítor Gomes Chagas[1] , Samuel Plaça de Paula[1] ,
Greis Yvet Oropeza Quesquén[1] , Lucas de Oliveira Silva[1(✉)] ,
and Uéverton dos Santos Souza[2,3]

[1] Instituto de Computação, Universidade Estadual de Campinas, Campinas, Brazil
{vitor.chagas,greis.quesquen,lucas.oliveira.silva}@ic.unicamp.br,
s233554@dac.unicamp.br
[2] IMPA, Instituto de Matemática Pura e Aplicada, Rio de Janeiro, Brazil
ueverton.souza@impatech.org.br
[3] Instituto de Computação, Universidade Federal Fluminense, Niterói, Brazil

Abstract. Given a finite non-decreasing sequence $\mathbf{d} = (d_1, \ldots, d_n)$ of
natural numbers, the GRAPH REALIZATION problem asks whether $\mathbf{d}$ is
a graphic sequence, i.e., there exists a labeled simple graph such that
$(d_1, \ldots, d_n)$ is the degree sequence of this graph. Such a problem can be
solved in polynomial time due to the Erdős and Gallai characterization
of graphic sequences. Since vertex degree is the size of a trivial edge cut,
we consider a natural generalization of GRAPH REALIZATION, where we
are given a finite sequence $\mathbf{d} = (d_1, \ldots, d_n)$ of natural numbers (representing the trivial edge cut sizes) and a list of nontrivial cut constraints
$\mathcal{L}$ composed of pairs (S_j, ℓ_j) where $S_j \subset \{v_1, \ldots, v_n\}$, and ℓ_j is a natural
number. In such a problem, we are asked whether there is a simple graph
with vertex set $V = \{v_1, \ldots, v_n\}$ such that v_i has degree d_i and $\partial(S_j)$
is an edge cut of size ℓ_j, for each $(S_j, \ell_j) \in \mathcal{L}$. We show that such a
problem is polynomial-time solvable whenever each S_j has size at most
three. Conversely, assuming $P \neq NP$, we prove that it cannot be solved
in polynomial time when $\mathcal{L}$ contains pairs with sets of size four, and our
hardness result holds even assuming that each d_i of $\mathbf{d}$ equals 1.

Keywords: Graph realization · Degree sequence · Graph factor

1 Introduction

Graph realization is a fundamental combinatorial problem in the field of Graph
Theory, and its studies have fostered interest in and understanding of the discrete
structure of graphs. Nowadays, graph realization is a topic commonly covered
in introductory Graph Theory courses, providing those new to the world of
graphs with many insights into their combinatorial properties. The Handshaking
Lemma, for example, is usually one of the first statements that beginners come
across when they begin studying graphs. Although simple to understand, it is
the gateway to a world of more intriguing graph-related questions. By observing
that the sum of the degrees of all vertices is equal to twice the number of edges

in the graph, it follows that not every sequence of n natural numbers can be the degree sequence of some graph with n vertices, and it becomes natural to ask when a sequence of n natural numbers admits a graph with n vertices whose degree sequence corresponds to the given sequence; the topic related to such a question is called *graph realization*.

Given a sequence d of n natural numbers that satisfy the Handshaking Lemma (i.e., the sum of its values is even), it is a simple exercise to verify that it is possible to construct a multigraph (parallel edges and loops are allowed) whose degree sequence corresponds to d. However, this question becomes more intriguing when the goal is to realize a simple graph where parallel edges and loops are not allowed. A non-decreasing sequence $d = (d_1, \ldots, d_n)$ of natural numbers is said to be *graphic* if it is realizable by a simple graph, that is, if there exists a labeled simple graph G with n vertices such that d is its degree sequence. Formally, the classical GRAPH REALIZATION problem is stated as follows:

GRAPH REALIZATION

Input: A non-decreasing sequence $d = (d_1, \ldots, d_n)$ of natural numbers.

Question: Is d a graphic sequence?

In 1960, Erdős and Gallai provided necessary and sufficient conditions for a sequence of non-negative integers to be graphic, proving the following theorem.

Theorem 1 (Erdős and Gallai [15]). *A non-decreasing sequence $d = (d_1, \ldots, d_n)$ of natural numbers is graphic if and only if*

1. $\sum_{i=1}^{n} d_i$ *is even, and*

2. $\sum_{i=1}^{k} d_i \le k(k-1) + \sum_{i=k+1}^{n} \min\{d_i, k\}$, *for every* $1 \le k \le n$.

It is not difficult to see that the conditions presented by Erdős and Gallai are necessary. However, the sufficiency proof provided by Erdős and Gallai is quite elaborated. Several alternative proofs for this sufficiency condition have been shown since then until recently, such as Harary [23] in 1969, Berge [9] in 1973, Choudum [13] in 1986, Aigner and Triesch [2] in 1994, Tripathi and Tyagi [34] in 2008, and Tripathi, Venugopalan, and West [35] in 2010. From Theorem 1, it follows that GRAPH REALIZATION can solved in polynomial time. Additionally, Tripathi, Venugopalan, and West [35] presented a simple constructive proof of Theorem 1 that allows us to obtain the graph to be realized in $\mathcal{O}(n \cdot \sum_{i=1}^{n} d_i)$ time. Furthermore, algorithms like the one provided by Havel and Hakimi [21,24], which iteratively reduce the degree sequence while maintaining its realizability, also give a constructive approach to finding such graphs if one exists. Havel and Hakimi's algorithms run in $\mathcal{O}(\sum_{i=1}^{n} d_i)$ time, which is optimal.

Variants of the GRAPH REALIZATION problem requiring that the realizing graph belongs to a particular graph class have also been studied in the literature. Examples of already studied classes included trees [20], Split graphs [11,22],

Chordal, interval, and perfect graphs [12]. Besides that, sequence pairs representing the degree sequences of a bipartition in a realizing bipartite graph were also
studied in [10]. Surprisingly, the question regarding GRAPH REALIZATION for
the class of bipartite graphs appears to remain open for over 40 years [6,33].
In addition, the problem of determining whether a given sequence defines a
unique realizing simple graph was studied in [2,26,30], and Bar-Noy, Peleg, and
Rawitz [8] introduced the vertex-weighted variant of GRAPH REALIZATION where
we are given a sequence $\mathbf{d} = (d_1, d_2, \ldots, d_n)$ representing a "weighted degree"
sequence, and a vector $\mathbf{w} = (w_1, w_2, \ldots, w_n)$ representing vertex weights, and
asked whether there is a graph with vertex set $V = \{v_1, v_2, \ldots, v_n\}$ such that
for each v_i the sum of the weights of its neighbors is equal to d_i.

In today's interconnected world, many fields face the challenge of structuring
systems with specific connectivity requirements. For instance, in social network
analysis, building a network where individuals (vertices) have a fixed number
of connections (degree) is crucial for analyzing influence, community structures,
and information diffusion. Similarly, urban planners confront similar issues when
designing road networks, where intersections must be connected with a specific
number of roads to optimize traffic flow. These examples highlight the significance of addressing connectivity challenges in various fields where specific connectivity patterns must be achieved. One of the most studied problems in this
context is the realization problem that deals with degree sequences. According to
Bar-Noy, Böhnlein, Peleg, and Rawitz [7], GRAPH REALIZATION and its variants
have interesting applications in network design, randomized algorithms, analysis
of social networks, and chemical networks.

In this paper, in the same flavor as Bar-Noy, Peleg, and Rawitz [8], we introduce another natural variant of the GRAPH REALIZATION, which we propose
calling GRAPH REALIZATION WITH CUT CONSTRAINTS. First, we start with
some definitions. For a vertex set $V = \{v_1, \ldots, v_n\}$, a cut list is defined as a list
of pairs $\mathcal{L} = \{(S_1, \ell_1), \ldots, (S_m, \ell_m)\}$, where each pair $(S_j, \ell_j) \in \mathcal{L}$ consists of a
nonempty, proper subset $S_j \subset V$ and a natural number ℓ_j. Given a cut list $\mathcal{L}$ for
a set V and a graph G with vertex set V, we say that G realizes $\mathcal{L}$ if, for every
pair $(S_j, \ell_j) \in \mathcal{L}$, the edge cut $\partial(S_j)$ has size ℓ_j. For a cut list $\mathcal{L}$, we denote by
$w(\mathcal{L}) = \max_j |S_j|$ the largest size among the subsets S_j in $\mathcal{L}$. Now, we define our
problem:

GRAPH REALIZATION WITH CUT CONSTRAINTS (GR-C)

Input: A cut list $\mathcal{L}$ for a set of vertices $V = \{v_1, \ldots, v_n\}$, and a non-
decreasing sequence $\mathbf{d} = (d_1, \ldots, d_n)$ of natural numbers.

Question: Does there exist a simple graph with vertex set V such that,
for every i, v_i has degree d_i and G realizes $\mathcal{L}$?

Recall that the degree of a vertex v_i of a graph G is the size of the trivial
edge cut $\partial(\{v_i\})$ in G. Therefore, one can see GRAPH REALIZATION as given a
cut list $\mathbf{d}$ of all trivial edge cut sizes $(\{v_i\}, d_i)$, decide whether there is a graph G
realizing $\mathbf{d}$. In GR-C, we assume that $\mathcal{L}$ is a list of some nontrivial edge cut sizes

for the realizing graph, i.e., each S_j has a size of at least two and at most $n-2$. If $\mathcal{L} = \emptyset$, then the problem becomes the original GRAPH REALIZATION problem. So, through this work, we always consider $\mathcal{L} \neq \emptyset$ and $w(\mathcal{L}) \geq 2$.

Although our problem, as we have defined, has never been explored before, it was motivated by an active research topic with several recent results [5,27] that aims to learn an unknown graph G or properties of G via cut-queries. In this context, given a graph $G = (V, E)$ with a known vertex set but an unknown edge set, the objective is to reconstruct G or compute some property of G with a minimal number of queries. A cut-query receives $S \subseteq V$ as input and returns the size of the edge cut $\partial(S)$. One of the main driving interests in this model is its connection with submodular function minimization [29]. Furthermore, these active learning questions have applications in fields like computational biology [19] and relate to data summarization, where queries reveal "relevant information" about the graph. More generally, this type of question can be viewed as a means of determining a property of an unknown object via indirect queries about it [1,14].

Within this framework, our problem can be viewed as a validity check to test whether the cut queries are consistent and if there is some graph that satisfies them. Also, it can be viewed as a variant, where we cannot choose the queries, but rather, we are given cut constraints and want to find one satisfying candidate graph. Concerning cut-queries, knowing $\partial(\{u\}), \partial(\{v\})$, and $\partial(\{u,v\})$, we find out whether or not there is an edge between u and v in G. Thus, with at most $\binom{n}{2} + n$ queries, one can always obtain the edge set of G. Similarly, regarding GR-C, if $\mathcal{L}$ contains all possible sets of size two, then the problem is trivial. Therefore, in this paper, we are mainly interested in the case where $\mathcal{L}$ has polynomial size with respect to n and does not contain all sets of size two.

Our Contribution. In this work, we study the GR-C problem and provide a comprehensive characterization of its computational complexity, focusing on the size of the cut sets involved. We show that it is polynomial-time solvable for instances where $w(\mathcal{L}) \leq 3$. Specifically, when $w(\mathcal{L}) = 2$, the problem reduces to the classic f-factor problem, which can be solved in polynomial time. Additionally, we show that cuts of size three, surprisingly, do not increase complexity. Instances involving such cuts can be transformed into equivalent ones where size-three constraints are replaced by cut constraints involving only sets of size at most two, all while preserving the same realizability. This shows that even with cut constraints using sets of size three, the problem remains polynomial-time solvable. On the other hand, we also prove that when cut sets of size four or larger are allowed, the problem cannot be solved in polynomial time unless P = NP. This provides a complete dichotomy regarding the computational complexity of the problem and the size of the cut sets. In addition, we also prove the NP-completeness for $w(\mathcal{L}) = 6$ when the cut constraints restrict the possibility graph, formally defined in Sect. 2, to be subcubic and bipartite. In contrast, when the cut constraints restrict the possibility graph to be a tree, the problem is solvable in polynomial time.

Related Work. Several other generalizations and related problems exist in the study of degree sequences and graph realizability. Aigner and Triesch [2] explored the realizability and uniqueness of graphs based on two types of invariants (degree sequences and induced subgraph sizes), focusing on both directed and undirected graphs and their computational complexity. Similarly, Erdős and Miklós [17] discussed the complexity of degree sequence problems, focusing on the second-order degree sequence problem, which is shown to be strongly NP-complete. Erdős et al. [16] presented a skeleton graph structure for a more general restricted degree sequence problem, studying two cases with specific edge restrictions and examining the connectivity of the realization space. Iványi [25] explored conditions and algorithms for determining if a sequence is the degree sequence of an (a, b, n)-graph, which is a (directed or undirected) graph whose vertices degrees are in the $[a, b]$ range.

Another field in Graph Theory that is closely related to the GRAPH REAL-IZATION problem is the study of graph factors and factorizations. A *factor* of a graph G is simply a spanning subgraph of G. There have been several studies on graph factors under different constraints, such as conditions on their degrees or restrictions on the classes that they must belong to. Here, we are particularly interested in graph factors described by their degrees, which we call degree factors. In this context, given an integer k, a *k-factor* of a graph G is a k-regular spanning subgraph of G. This generalizes many problems, for instance a 1-factor is the same as a perfect matching, and studies in this area date back to the 19th century when Petersen [31] gave one of the first sufficient conditions for a 1-factor.

The concept of k-factors has been generalized to consider other values of degrees rather than a fixed number. Given two functions $g, f \colon V \to \mathbb{N}$ such that $g \leq f$, a spanning subgraph H of the graph $G = (V, E)$ is a (g, f)-*factor* if for every vertex v, it holds that $g(v) \leq d_H(v) \leq f(v)$. If $g = f$, then it is simply called an f-*factor*. The problem of determining if a graph admits a (g, f)-factor is known to be solvable in polynomial time [4]. As will be shown later, the GR-C problem generalizes the f-factor problem. Classical results in this field include Tutte's theorem on f-factors [36] and Lovasz's characterization of (g, f)-factors [28]. For a detailed treatment of this topic, we refer the reader to the surveys of Akiyama and Kano [3] and Plummer [32].

Organization of the Text. The remainder of this paper is organized as follows. In Sect. 2, we introduce key definitions and notations related to the GR-C problem, including some conditions for the realizability of a GR-C instance. In Sect. 3, we investigate the GR-C problem with cut sizes restricted to three, while Sect. 4 focuses on instances with cuts of size at least four. Finally, in Sect. 5, we summarize our results and discuss potential extensions of this work. Due to space constraints, some proofs have been omitted.

2   Preliminaries

Let S be a set of vertices of a graph $G = (V, E)$. We denote the total degree of vertices in S by $d(S) = \sum_{u \in S} d_u$, where d_u represents the degree of vertex u. A simple observation is that the size of the edge cut $\partial(S)$ is determined by the degree of the vertices in S and the edges between vertices of S. If there are k edges between vertices of S, then $|\partial(S)| = d(S) - 2k$. Since the number of edges in S may vary from 0 to $\binom{|S|}{2}$, a necessary condition for the realizability of an GR-C instance is as follows.

Remark 1. A GR-C instance $(\mathsf{d}, \mathcal{L})$ is realizable only if, for each cut $(S, \ell) \in \mathcal{L}$, we have $\ell \in \{d(S) - 2k : 0 \le k \le \binom{|S|}{2}\}$.

Since this condition is easily verifiable, we assume henceforth that it holds for any GR-C instance. In particular, for cuts of size two, this observation implies that only two feasible values are possible, determining whether an edge must exist between the corresponding vertices, as detailed below.

Remark 2. Given an instance $I = (\mathsf{d}, \mathcal{L})$ of GR-C, in any realization G of I, if $(\{u, v\}, d_u + d_v - 2) \in \mathcal{L}$, then $uv \in E(G)$, and if $(\{u, v\}, d_u + d_v) \in \mathcal{L}$, then $uv \notin E(G)$.

Based on this, we say that an edge uv is *fixed* if $(\{u, v\}, d_u + d_v - 2) \in \mathcal{L}$ and is *forbidden* if $(\{u, v\}, d_u + d_v) \in \mathcal{L}$. We apply similar terminology when constructing an instance of GR-C. Given an instance $(\mathsf{d}, \mathcal{L})$ of GR-C, to *fix* or *forbid* an edge uv means adding the cut $(\{u, v\}, d_u + d_v - 2)$ or $(\{u, v\}, d_u + d_v)$ to $\mathcal{L}$, respectively.

Remark 2 implies that the GR-C problem, when limited to cuts of size two, is equivalent to the GR problem with added constraints: a subset of edges is fixed, and another disjoint one is forbidden. Moreover, we can simplify the problem by focusing only on forbidden edges by reducing the degree of vertices incident to fixed edges and then marking those edges as forbidden. Formally, given an instance $(\mathsf{d}, \mathcal{L})$ and a cut $(\{u, v\}, d_u + d_v - 2) \in \mathcal{L}$, in which case the edge uv is fixed, we can produce an equivalent instance $(\mathsf{d}', \mathcal{L}')$ as follows. For all $i \notin \{u, v\}$ set $d_i' = d_i$. Reduce $d_u' = d_u - 1$ and $d_v' = d_v - 1$; and $\mathcal{L}'$ is obtained from $\mathcal{L}$ by replacing $(\{u, v\}, d_u + d_v - 2)$ with $(\{u, v\}, d_u + d_v)$.

The resulting instance $(\mathsf{d}', \mathcal{L}')$ has a realization if and only if $(\mathsf{d}, \mathcal{L})$ has a realization. If $G = (V, E)$ is a realization of $(\mathsf{d}, \mathcal{L})$, then, as discussed above, we must have $uv \in E$, and $G - uv$ is a realization of $\mathcal{L}'$. Conversely, if $G' = (V, E')$ is a realization of $(\mathsf{d}', \mathcal{L}')$, then necessarily $uv \notin E'$ due to the cut $(\{u, v\}, d_u + d_v)$, and $G' + uv$ is a realization of $(\mathsf{d}, \mathcal{L})$.

Thus, cut restrictions involving sets of size two can be simply reinterpreted as forbidding edges. Let F be the set of all forbidden edges that cannot appear in any realization of instance $(\mathsf{d}, \mathcal{L})$. Then $\mathcal{G} = K_n - F$ is what we call the *possibility graph*, which must be a supergraph of any valid realization of $(\mathsf{d}, \mathcal{L})$.

3   Small Cuts

In this section, we show that the GR-C problem can be solved in polynomial time for instances $(d, \mathcal{L})$ where $w(\mathcal{L}) \leq 3$. Reinterpreting the size-two cuts of $\mathcal{L}$ as forbidden edges allows us to transform the GR-C problem into an equivalent formulation of the classic f-factor problem whenever $w(\mathcal{L}) = 2$. This leads us to the following conclusion.

Lemma 1. *Any instance $I = (d, \mathcal{L})$ of the GR-C problem can be solved in polynomial time if $w(\mathcal{L}) = 2$.*

Proof. Given the instance I, we apply the aforementioned method to fixed edges to produce an equivalent instance $I' = (d', \mathcal{L}')$ containing only forbidden edges. The problem then reduces to finding a subgraph of the possibility graph $\mathcal{G}$ of I' that realizes the degree list d'. By interpreting d' as a function $f \colon V \to \mathbb{N}$, the problem becomes finding an f-factor of $\mathcal{G}$, which is solvable in cubic time using, for example, the algorithm of Anstee [4].   □

Interestingly, the GR-C problem remains solvable in polynomial time even when cuts of size three are present. This is because cuts of size three actually have no more restraining power on realizability than cuts of size two, in the sense that we can construct an equivalent instance containing only cuts of size at most two that maintain the same realizability as the original instance.

Theorem 2. *Any instance $I = (d, \mathcal{L})$ of the GR-C problem can be solved in polynomial time if $w(\mathcal{L}) = 3$.*

Proof. We will show that it is possible to construct, in polynomial time, an instance $(d', \mathcal{L}')$ such that $w(\mathcal{L}') = 2$ and $(d', \mathcal{L}')$ is realizable if and only if $(d, \mathcal{L})$ is realizable. This will complete our proof by applying Lemma 1 to $(d', \mathcal{L}')$. To achieve this, consider a cut $(S, \ell) \in \mathcal{L}$ where $S = \{u, v, w\}$. From Remark 1, we know there are exactly four possible values for ℓ: $d(S)$, $d(S) - 2$, $d(S) - 4$, and $d(S) - 6$. In each case, we show that (S, ℓ) can be replaced by cuts of size two, along with, possibly, some additional vertices. Recall that forbidding or fixing an edge uv is a constraint that we can express through a cut constraint $(\{u, v\}, d_u + d_v)$ or $(\{u, v\}, d_u + d_v - 2)$, respectively.

Case 1: $\ell = d(S)$. In this case, all edges incident to S must be included in the edge cut $\partial(S)$. So, this cut effectively forbids the edges uv, uw, and vw, as shows Fig. 1a.

Case 2: $\ell = d(S) - 6$. This case is similar to Case 1, but we require here all three edges between vertices in S to be present. Therefore, we fix the edges uv, uw, and vw, as in Fig. 1b.

Case 3: $\ell = d(S) - 2$. This cut enforces that exactly one edge within S must be included in any realization. Equivalently, this constraint requires selecting two vertices from S to decrease their degrees by 1 each.

To eliminate this cut from $\mathcal{L}$ (see Fig. 1c), we proceed as follows. We create a new vertex x, set $d_x = 2$, and forbid all edges between x and vertices outside S.

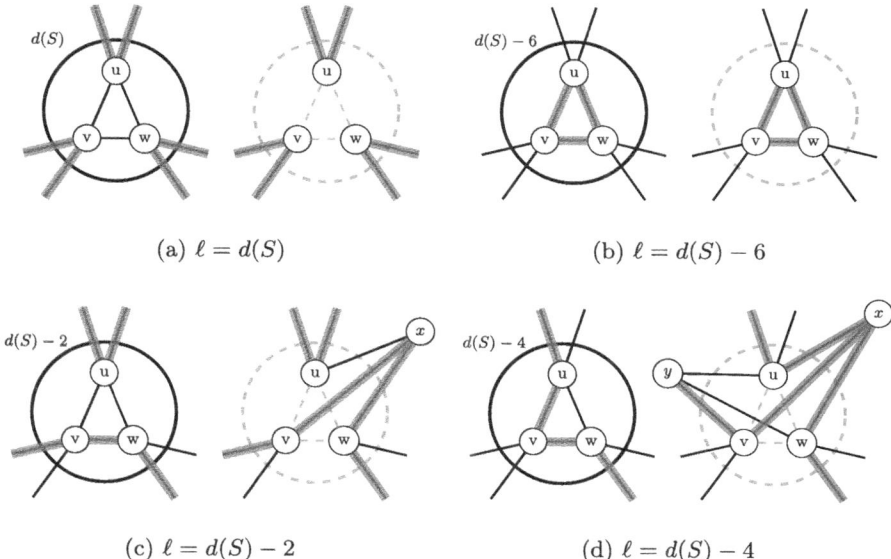

(a) $\ell = d(S)$

(b) $\ell = d(S) - 6$

(c) $\ell = d(S) - 2$

(d) $\ell = d(S) - 4$

Fig. 1. Illustration of all cases for a cut (S, ℓ) with $S = \{u, v, w\}$, assuming $d_u = d_v = d_w = 2$ (so $d(S) = 6$). Solid edges represent possible edges, dashed edges are forbidden, and blue-highlighted edges belong to a realization. In each case, the left image shows a realization satisfying (S, ℓ), while the right image shows the equivalent realization of the modified instance without the cut.

Additionally, we forbid the edges between the vertices within S. In this setting, the two vertices in S adjacent to x will simulate the selection of an edge in a realization of the original instance. Assume, without loss of generality, that a realization G of the original instance exists with $vw \in E(G)$. Then, in the modified instance, a realization G' exists in which x is adjacent to both v and w and vw is not present. The converse also holds, ensuring that this modification to $\mathcal{L}$ preserves the realizability of the instance.

Case 4: $\ell = d(S) - 4$. In this case, exactly two edges within S must be included in any realization. Following the same rationale as in the previous case, this amounts to the degrees of two vertices in S being reduced by 1, while the degree of the remaining vertex is reduced by 2. Note that since we only have three vertices and, therefore, three possible edges, the choice of the two edges can be defined by selecting which vertex of S will have its degree decreased by 2.

This can be equivalently accomplished by proceeding as follows (see Fig. 1d). We create two new vertices, x and y, and set $d_x = 3$ and $d_y = 1$. We fix all three edges from x to S, forbid all edges between y and vertices outside S, and forbid the edges within S. Note that the fixed edges from x to S reduce the degree of each vertex in S by 1, while the vertex in S that connects to y will have its degree reduced by an additional 1, simulating the required decrease of 2.

Therefore, without loss of generality, there is a realization G of the original instance such that $uv, vw \in E(G)$ if and only if there is a realization G' of the modified instance with $xu, xv, xw, yv \in E(G')$ and $uv, vw \notin E(G')$.

We apply these modification rules to each cut (S, ℓ) of size three in $\mathcal{L}$, resulting in a new instance $I' = (\mathsf{d}', \mathcal{L}')$ with $w(\mathcal{L}') = 2$ and the same realizability as $(\mathsf{d}, \mathcal{L})$. In Cases 1 and 2, each cut (S, ℓ) is replaced by three smaller cuts, while in Cases 3 and 4, $\mathcal{O}(n)$ additional cuts are required. Nevertheless, the total size of $\mathcal{L}'$ and the number of vertices are only increased polynomially. Therefore, by applying Lemma 1 to I', we solve our original instance in polynomial time.   □

4   Large Cuts

Now we discuss the GR-C with $w(\mathcal{L}) \geq 4$. Interestingly, we get a dichotomy and can no longer solve GR-C within polynomial time unless P = NP. Our hardness result holds even if d is a sequence of ones. Additionally, we explore restrictions over the possibility graph $\mathcal{G}$ and show that GR-C is NP-complete for $w(\mathcal{L}) = 6$ even if $\mathcal{G}$ is bipartite and subcubic. In contrast, if $\mathcal{G}$ is a tree, we argue how the problem can be efficiently solved.

4.1   Cuts of Size Four

Regarding the size of cuts, one might initially think that the approach of Theorem 2, which reduces an instance with $w(\mathcal{L}) = 3$ to one with $w(\mathcal{L}) = 2$, could be extended to larger cuts. However, this extension is not feasible. For cuts (S, ℓ) of size three, the number of edges within S uniquely determines how much the degrees of each vertex are affected. In contrast, when $|S| = 4$, this property already does not hold. Consider, for instance, a cut (S, ℓ) where $S = \{u, v, w, x\}$ and $\ell = d(S) - 4$. In any realization, there must be exactly two edges within S. If, in a realization G, these edges are disjoint (e.g., $uv, wx \in E(G)$), then each vertex in S has its degree decreased by 1. On the other hand, the edges might not be disjoint (e.g., $uv, uw \in E(G)$). In this case, one vertex has its degree decreased by 2, two vertices have a reduction of 1, and one vertex remains unchanged. Since it is impossible to determine beforehand which of these configurations applies, the reduction strategy used in Theorem 2 cannot be generalized.

In fact, we show that when $w(\mathcal{L}) = 4$, the GR-C problem cannot be solved in polynomial time unless P = NP. We use a restricted variant of the 1-in-3-SAT problem [18] in our proof. In our case, we consider propositional formulas in conjunctive normal form where every variable appears exactly three times, two times as a positive literal (i.e., not negated) and one time as a negative literal (i.e., negated). We ask if it is possible to find a satisfying assignment such that each clause has exactly one literal that evaluates to true while the rest are false. Additionally, we require that each clause has two or three literals (it is trivial to handle clauses with only one literal, so we assume they are preprocessed away). We call this variant 1-in-3-SAT$_{(2,1)}$, and although pretty restricted, this problem remains NP-complete.

1-IN-3-SAT$_{(2,1)}$

Input: A set of variables X and a formula ϕ in conjunctive normal form over X such that:

- each variable of X occurs twice as a positive literal and once as a negative literal;
- each clause of ϕ has two or three literals.

Question: Is there a truth assignment of X such that exactly one literal in every clause of ϕ is true?

Lemma 2. *1-in-3-SAT$_{(2,1)}$ is NP-complete.*

Proof. Notice that the problem is in NP. To prove that it is NP-hard, we show a reduction from Positive 1-in-3-SAT, the monotone version of 1-in-3-SAT in which all literals are positive. This problem is known as NP-complete [18].

Let (X, ϕ) be an input of the Positive 1-in-3-SAT problem. We will show how to add clauses and variables to (X, ϕ) to obtain an equivalent instance (X', ϕ') of 1-in-3-SAT$_{(2,1)}$. We start with $X' = X$ and $\phi' = \phi$. Let $x \in X$. First, consider that x only occurs once in ϕ (we know it appears as a positive literal). Then we add to ϕ' the redundant clause $(x + \overline{x})$. Now x occurs in ϕ' twice as a positive literal and once as a negative one, and ϕ' is equivalent to ϕ.

Now suppose x has two appearances in ϕ. Then we create a new variable a; we add the clause $(x + \overline{a})$ to set a logical equivalence between x and a (since exactly one of x and $\overline{a}$ must be true in a satisfying truth assignment, $x \equiv a$ in any feasible assignment). This allows us to replace the second occurrence of x with its equivalent variable a. The resulting ϕ' is equivalent to ϕ, and by also adding the redundant clause $(a + \overline{x})$ to ϕ, both x and the new variable a occur twice as a positive literal and once as a negative one.

Finally, we generalize this last idea. Lets say that x occurs t times, $t \geq 2$. We create $t - 1$ variables, $a_1, a_2, \ldots, a_{t-1}$, which will be all equivalent to x. To this end, we add t clauses to ϕ': $(x + \overline{a}_1), (a_1 + \overline{a}_2), \ldots, (a_{t-2} + \overline{a}_{t-1}), (a_{t-1} + \overline{x})$. Now x and the new variables $a_1, \ldots, a_{t-1}$ are all equivalent, i.e., they must have the same truth value in any assignment that satisfies exactly one literal of every clause. In ϕ', we maintain the first appearance of x, but the second one is replaced by a_1, the third is replaced by a_2, and so forth. The resulting ϕ' is still equivalent to the original ϕ, and if we do this for every variable in X, we obtain an instance (X', ϕ') that is an instance of 1-in-3-SAT$_{(2,1)}$. Furthermore, we remark that in ϕ', every clause that comes from ϕ has three literals, and every clause that we created for the reduction has two literals; thus, ϕ' only has clauses of sizes two and three. □

To argue the hardness of GR-C with $w(\mathcal{L}) = 4$, it will be useful to know beforehand how many variables in a 1-in-3-SAT$_{(2,1)}$ instance must be set to true. To this end, we consider a more restricted problem, which remains hard.

k-TRUE 1-IN-3-SAT$_{(2,1)}$

Input: A tuple (X, ϕ, k), where (X, ϕ) is an instance of 1-in-3-SAT$_{(2,1)}$ and k is a nonnegative integer.

Question: Is there a feasible solution to (X, ϕ) in which exactly k variables are assigned to true?

Lemma 3. *k-True 1-in-3-SAT$_{(2,1)}$ cannot be solved in polynomial time unless* $P = NP$.

Proof. Follows directly from Lemma 2. Suppose that $\mathcal{A}$ is an algorithm that decides k-True 1-in-3-SAT$_{(2,1)}$ in polynomial time. Given an instance (X, ϕ) of 1-in-3-SAT$_{(2,1)}$ with n variables, we run $\mathcal{A}$ on inputs $(X, \phi, 0), \ldots, (X, \phi, n)$. If $\mathcal{A}$ accepts any of these, we determine that the answer to ϕ is YES; otherwise, it is NO. Therefore, we can solve 1-in-3-SAT$_{(2,1)}$ in polynomial time, implying that $P = NP$.   □

Equipped with the k-True 1-in-3-SAT$_{(2,1)}$ problem and by knowing that it cannot be solved in polynomial time unless $P = NP$, we can proceed to show the hardness of the GR-C problem when $w(\mathcal{L}) = 4$.

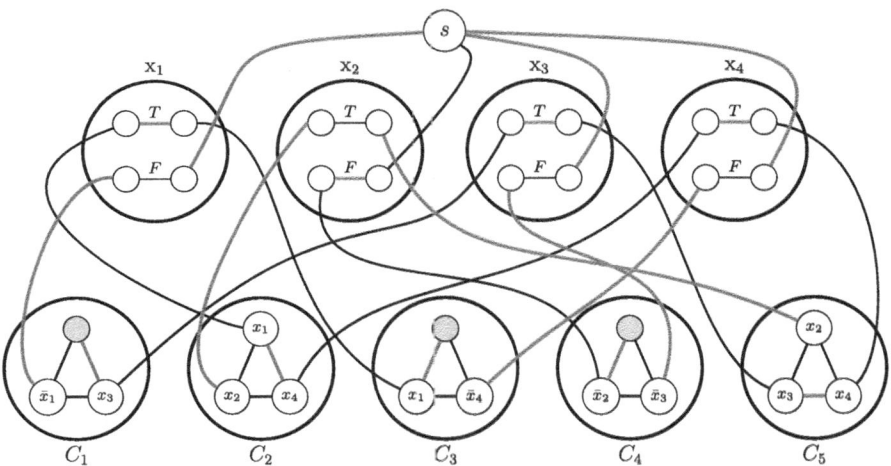

Fig. 2. Illustration of the possibility graph $\mathcal{G}$ built from an instance (X, ϕ, k) of k-True 1-in-3-SAT$_{(2,1)}$ with $X = \{x_1, x_2, x_3, x_4\}$, $\phi = (\bar{x}_1 + x_3)(x_1 + x_2 + x_4)(x_1 + \bar{x}_4)(\bar{x}_2 + \bar{x}_3)$ $(x_2 + x_3 + x_4)$ and $k = 1$. Gray vertices represent artificial vertices created for clauses with only two literals. The highlighted edges show an example of a feasible realization for such an instance.

Theorem 3. *The GR-C problem cannot be solved in polynomial time unless* $P = NP$ *even when* $w(\mathcal{L}) = 4$ *and all degrees in the degree sequence* d *are 1.*

Proof. We present a reduction from the k-True 1-in-3-SAT$_{(2,1)}$ to the GR-C problem with the desired properties. To this end, let (X, ϕ, k) be an instance of the k-True 1-in-3-SAT$_{(2,1)}$. We now describe the building of the instance $I = (d, \mathcal{L})$ of GR-C. Refer to Fig. 2 for an illustrative example. We start with d and $\mathcal{L}$ empty, and we let V be the corresponding set of vertices and $\mathcal{G}$ be its possibility graph. First, for each variable x_i in X, we build a *variable gadget* as follows. Let X_i be a set of four vertices, namely $X_i = \{x^i_{T_1}, x^i_{T_2}, x^i_{F_1}, x^i_{F_2}\}$. We add X_i to V and set $d_u = 1$ for each $u \in X_i$, we add the edges $x^i_{T_1} x^i_{T_2}$ and $x^i_{F_1} x^i_{F_2}$ to $\mathcal{G}$, and we add the cut $(X_i, 2)$ to $\mathcal{L}$. Moreover, for each clause C_j of ϕ, we define its *clause gadget*. We create a new vertex for each literal that occurs in C_j. If C_j has only two literals, we create another artificial one. Let Y_j be this set of three vertices. We add Y_j to V and set $d_v = 1$ for each $v \in Y_j$, we add all the edges between vertices of Y_j to $\mathcal{G}$, and we add the cut $(Y_j, 1)$ to $\mathcal{L}$.

To conclude the definition of the vertices V, we create a vertex s and set $d_s = n - k$. The set V of our instance is thus composed of s and the vertices of each variable and clause gadget. To finish $\mathcal{G}$'s construction, we join the vertex and clause gadgets as follows. For each variable x_i, let C_{i_1} and C_{i_2} be the two clauses where x_i appears as a positive literal, and C_{i_3} the clause in which it appears as a negative literal. We connect $x^i_{T_1}$, $x^i_{T_2}$ and $x^i_{F_1}$ to the vertex that corresponds to its literal in C_{i_1}, C_{i_2} and C_{i_3}, respectively, while $x^i_{F_2}$ is connected to s. The final cut list $\mathcal{L}$ is given by the aforementioned cuts in the vertex and clause gadgets, plus the ones defining $\mathcal{G}$. Now we show that there is a realization for such $(d, \mathcal{L})$ if and only if (X, ϕ) is satisfiable using exactly k variables as true.

Let $\hat{x}$ be a feasible solution to (X, ϕ) using exactly k variables as true. Let G be a spanning subgraph of $\mathcal{G}$, initially with no edges. For each $\hat{x}_i$ from $\hat{x}$, if $\hat{x}_i = T$, we add to G the edge $x^i_{F_1} x^i_{F_2}$ along with the edges from $x^i_{T_1}$ and $x^i_{T_2}$ to their corresponding positive literals in the clause gadgets. Similarly, if $\hat{x}_i = F$, then we add to G the edge $x^i_{T_1} x^i_{T_2}$ along with the edges from $x^i_{F_1}$ to its corresponding negative literal in the clause gadget and from $x^i_{F_2}$ to s. At last, for each clause C_j in ϕ, let $\hat{x}_{j_1}$ and $\hat{x}_{j_2}$ be the corresponding vertices of the two literals in C_j that were evaluated as false in $\hat{x}$ in case C_j has three literals, or the literal evaluated to false and the artificial vertex added in the clause gadget of C_j, otherwise. We then add the edge $\hat{x}_{j_1} \hat{x}_{j_2}$ to G for each clause C_j. In both cases, in G we have that $\partial(X_i) = 2$. Since $\hat{x}_i$ is a feasible solution, in each clause C_j, exactly one variable is evaluated to true, which implies that $\partial(Y_j) = 1$ in G. Furthermore, the degree of all vertices except for s is 1, and since there are exactly k variables in $\hat{x}$ assigned to true, there are $n - k$ edges in G from vertices $x^i_{F_2}$ to s, thus respecting d_s. Therefore, G is a realization of $(d, \mathcal{L})$.

Conversely, let G be a realization of $(d, \mathcal{L})$. Since $(X_i, 2) \in \mathcal{L}$ for each variable x_i, $d_u = 1$ for each $u \in X_i$, and $E(\mathcal{G}[X_i]) = \{x^i_{T_1} x^i_{T_2}, x^i_{F_1} x^i_{F_2}\}$. It holds that either $x^i_{T_1}, x^i_{T_2}$ or $x^i_{F_1}, x^i_{F_2}$ have neighbors outside X_i in G. Therefore, we define an assignment $\hat{x}$ to (X, ϕ) as follows: $x_i = T$ if $x^i_{T_1}, x^i_{T_2}$ have neighbors outside X_i in G, otherwise $x_i = F$. As $(Y_j, 1) \in \mathcal{L}$ for each clause C_j, it follows that each clause of ϕ has exactly one literal assigned to true in $\hat{x}$. Given that $d(s) = n - k$, the vertex s has $n - k$ neighbors in G. By construction, each neighbor of s in G is a $x_{F_2}^i$ vertex for some i. Thus, $\hat{x}$ has exactly $n - k$ negative literals evaluating true, and therefore $\hat{x}$ is a feasible solution to (X, ϕ) in which exactly k variables are assigned to true.

To have all degrees in the degree sequence d equal 1 is enough to modify the construction, replacing s by $n - k$ copies each with desired degree equals one in d. Thus, from this reduction, we conclude that if the GR-C problem is solvable in polynomial time, then we can also solve the k-True 1-in-3-SAT$_{(2,1)}$ problem in polynomial time, which would imply that P = NP due to Lemma 3.   □

4.2   Restricted Possibility Graph

We now move our attention to particular instances $(d, \mathcal{L})$ of GR-C in which $w(\mathcal{L})$ is not bounded, but the possibility graph $\mathcal{G}$ belongs to a restricted graph class. If $\mathcal{G}$ is a tree, we can solve the GR-C in polynomial time. For instance, we can construct a candidate realizing graph G by processing $\mathcal{G}$'s leaves iteratively, ensuring at each step that the degree constraints are met. If a violation occurs or the final vertex has a nonzero degree, we return NO; otherwise, we can verify whether G realizes $\mathcal{L}$ in polynomial time.

Proposition 1. *Given an instance $(d, \mathcal{L})$ of GR-C with a tree possibility graph $\mathcal{G}$, we can decide if there is a solution in polynomial time.*

As it turns out, if we relax the restrictions on $\mathcal{G}$ and allow a bipartite graph, we get a NP-complete problem. To show this, we will make use of the 3-Dimensional Matching problem, which is defined next.

3-DIMENSIONAL MATCHING – 3DM

Input: 3 disjoint sets X, Y, and Z with $|X| = |Y| = |Z| = n$, and a set of triples $T \subseteq X \times Y \times Z$.

Question: Is there a subset $M \subseteq T$ such that $|M| = n$ and no two triples of M intersect?

This problem remains NP-complete if no element occurs in more than three triples [18]. We refer to this particular case as 3DM-3.

Theorem 4. *The GR-C problem is NP-complete when the possibility graph $\mathcal{G}$ is subcubic and bipartite, even when $w(\mathcal{L}) = 6$ and d is a sequence of ones.*

Proof. The GR-C is clearly in NP. We show a reduction from 3DM-3 to prove its hardness. Consider an instance (X, Y, Z, T) of 3DM-3 where $|X| = |Y| = |Z| = n$. Without loss of generality, assume that every element in X, Y, and Z appears in at least one triple in T (see Fig. 3 for an illustration of the reduction).

To construct the vertex set V for our GR-C instance, we proceed as follows: for each element $x_i \in X$, we create a corresponding vertex x_i in V. For each element $y_j \in Y$, we construct a group of vertices V_j, determined by the number of triples in T containing y_j. If y_j appears in l triples, where $1 \leq l \leq 3$, we create $2l$ vertices labeled $y_{1,a}^j, \ldots, y_{l,a}^j$ and $y_{1,b}^j, \ldots, y_{l,b}^j$, and group these into two sets, $Y_a^j = \{y_{1,a}^j, \ldots, y_{l,a}^j\}$ and $Y_b^j = \{y_{1,b}^j, \ldots, y_{l,b}^j\}$. Define $Y_a = \bigcup_j Y_a^j$ and $Y_b = \bigcup_j Y_b^j$. Lastly, for each element $z_k \in Z$, we create a vertex z_k in V. In total, this construction yields $2(|T| + n)$ vertices, where $V = X \cup Y_a \cup Y_b \cup Z$.

We define the degree sequence d such that each vertex in V has a degree exactly one. This degree constraint ensures that each vertex is matched with only one other vertex, guaranteeing that any feasible solution forms a matching. Next, we construct the cut list $\mathcal{L}$. For each group V_j, we add the pair $(V_j, 2)$ to $\mathcal{L}$. This is the largest cut with a size of at most six, enforcing exactly two edges connecting vertices in V_j to vertices outside of V_j. We will later argue that they specifically connect to a vertex of X and a vertex of Z.

We then add cuts of size two, as per Remark 2, to $\mathcal{L}$ to prohibit all edges except those allowed by the following rules. For each $y_j \in Y$, let $(x_{i_1}, y_j, z_{k_1}), \ldots,$ (x_{i_l}, y_j, z_{k_l}) denote the l triples of T in which y_j appears. We only allow edges from x_{i_u} to $y_{u,a}^j$, from $y_{u,a}^j$ to $y_{u,b}^j$, and from $y_{u,b}^j$ to z_{k_u}, for each $1 \leq u \leq l$. The construction encodes the selection of a triple (x_{i_v}, y_j, z_{k_v}) by including the edges $x_{i_v} y_{v,a}^j$ and $y_{v,b}^j z_{k_v}$ in the realization, while the remaining vertices in V_j forms a matching. Observe that this instance's possibility graph $\mathcal{G}$ is bipartite and subcubic. Each vertex of $Y_a \cup Y_b$ has degree two, while the vertices in $X \cup Z$ have a degree at most three, as no element occurs in more than three triples.

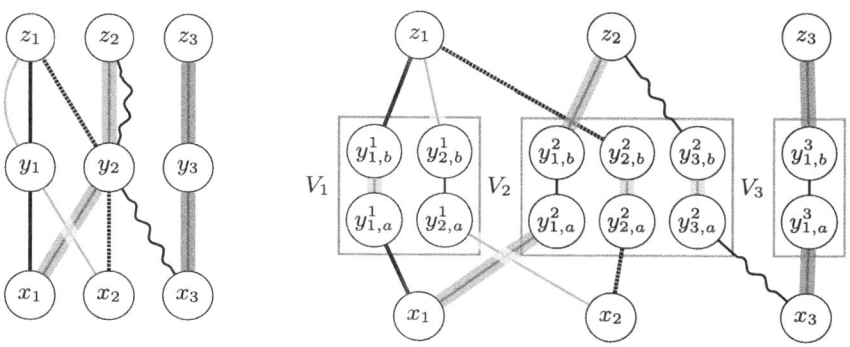

(a) A 3DM-3 input instance   (b) The corresponding GR-C instance

Fig. 3. A 3DM-3 instance example where $T = \{(x_1, y_1, z_1), (x_1, y_2, z_2), (x_2, y_1, z_1), (x_2, y_2, z_1), (x_3, y_2, z_2), (x_3, y_3, z_3)\}$. Distinct edge types are assigned to each triple, with a solution highlighted. The right image depicts the possibility graph of the reduced GR-C instance, with a feasible realization highlighted.

If a feasible matching $M \subseteq T$ exists in the 3DM-3 instance, we can map it directly to the edges of a valid realization G for the constructed GR-C instance. For each $(x_{i_u}, y_j, z_{k_u}) \in M$, using the uth occurrence of y_j, we add the edges

$x_{i_u} y_{u,a}^j$ and $y_{u,b}^j z_{k_u}$ to G. For each $(x_{i_v}, y_j, z_{k_v}) \in T \backslash M$, we add the edge $y_{v,a}^j y_{v,b}^j$. Since M is a solution, each vertex in $X \cup Z$ has degree one, satisfying the degree constraints. Additionally, within each group V_j, exactly two vertices—$y_{u,a}^j$ and $y_{u,b}^j$ from a triple in M—connect to vertices in X and Z, respectively. All other vertices within V_j correspond to triples not included in M, forming a matching within V_j. Therefore, G fulfills both the degree sequence $\mathbf{d}$ by assigning degree one to every vertex and the cut list $\mathcal{L}$, meeting all the required constraints for a valid realization.

Conversely, if a graph G exists that realizes both the degree sequence $\mathbf{d}$ and the cut list $\mathcal{L}$, we can construct a feasible matching $M \subseteq T$ for the 3DM-3 instance. Since $\mathbf{d}$ specifies a degree of one for each vertex, the edges of G form a matching. Additionally, exactly two vertices within each group V_j are matched to vertices of $X \cup Z$, meaning the remaining vertices within each V_j form an internal matching. These two externally matched vertices must correspond to the same triple in T; otherwise, the remaining vertices in V_j could not be paired and meet the type $(V_j, 2)$ cut constraint. Let M consist of the triples in T for which the associated $y_{u,a}^j$ and $y_{u,b}^j$ vertices in V_j are connected to vertices in X and Z, respectively. Thus, by construction, vertex x_i connects to $y_{u,a}^j$ and $y_{u,b}^j$ to z_k if and only if the triple (x_i, y_j, z_k) of T belongs to M. So M contains exactly one triple per V_j, covering each element of Y exactly once, thus $|M| = n$. Since G realizes $\mathcal{L}$, no edges exist between vertices in X and Z. Hence, given that G is a matching, each vertex in X connects to exactly one vertex in Y_a, and each vertex in Z connects to exactly one vertex in Y_b. Consequently, M constitutes a valid matching for the 3DM-3 instance. □

5   Final Remarks

We introduced the GRAPH REALIZATION WITH CUT CONSTRAINTS problem in this work. This problem is interesting because it combines different graph theory concepts, including degree sequence, cut constraints, f-factors, and graph realization. We provide a detailed characterization of its computational complexity based on the size of the cuts. Our results show that the problem can be solved in polynomial time when the cuts are small enough (size at most three). However, the complexity significantly increases when the cuts are larger, and we proved that it becomes NP-hard. An interesting direction for future work is identifying other graph classes where the possibility graph $\mathcal{G}$ of a GR-C instance ensures polynomial-time solvability. For example, the idea of Proposition 1 might extend to cactus or, more generally, to graphs with bounded degeneracy or treewidth. The case of a planar possibility graph also deserves further investigation. We also ask about the complexity of 1-in-3 SAT$_{(2,2)}$, the variant of 1-in-3 SAT where each variable occurs exactly four times, twice positive and twice negative.

Acknowledgments.. This work was started during the 6th edition of WoPOCA, which took place in Campinas, São Paulo, Brazil. We thank the organizers and the agencies CNPq (process number 404315/2023-2) and FAEPEX (process number

2422/23). We thank Esther Arkin, Soumya Banerjee, Rezaul Alam Chowdhury, Mayank Goswami, Dominik Kempa, Joseph Mitchell, Valentin Polishchuk, and Steven Skiena for some discussions prior the event, which helped motivate this work. This research has received funding from Rio de Janeiro Research Support Foundation (FAPERJ) under grant agreement E-26/201.344/2021, the National Council for Scientific and Technological Development (CNPq) under grant agreements 309832/2020-9 and 163645/2021-3, the São Paulo Research Foundation (FAPESP) under grant agreement 2022/13435-4, and the Brazilian Federal Agency for Support and Evaluation (CAPES) with process numbers 88887.646008/2021-00 and 88887.647870/2021-00.

Disclosure of Interests. The authors have no competing interests to declare that are relevant to the content of this article.

References

1. Aigner, M.: Combinatorial Search. John Wiley & Sons Inc., Hoboken (1988)
2. Aigner, M., Triesch, E.: Realizability and uniqueness in graphs. Disc. Math. **136**(1–3), 3–20 (1994)
3. Akiyama, J., Kano, M.: Factors and factorizations of graphs–a survey. J. Graph Theory **9**(1), 1–42 (1985)
4. Anstee, R.: An algorithmic proof of tutte's f-factor theorem. J. Algor. **6**(1), 112–131 (1985)
5. Apers, S., Efron, Y., Gawrychowski, P., Lee, T., Mukhopadhyay, S., Nanongkai, D.: Cut query algorithms with star contraction. In: FOCS 2022, pp. 507–518 (2022)
6. Bar-Noy, A., Böhnlein, T., Peleg, D., Rawitz, D.: On realizing a single degree sequence by a bipartite graph. In: SWAT 2022 (2022)
7. Bar-Noy, A., Böhnlein, T., Peleg, D., Rawitz, D.: On vertex-weighted realizations of acyclic and general graphs. Theor. Comput. Sci. **922**, 81–95 (2022)
8. Bar-Noy, A., Peleg, D., Rawitz, D.: Vertex-weighted realizations of graphs. Theor. Comput. Sci. **807**, 56–72 (2020)
9. Berge, C., Minieka, E.: Graphs and Hypergraphs. North-Holland Publishing Company, Amsterdam (1973)
10. Burstein, D., Rubin, J.: Sufficient conditions for graphicality of bidegree sequences. SIAM J. Disc. Math. **31**(1), 50–62 (2017)
11. Chat, B.A., Pirzada, S., Iványi, A.: Recognition of split-graphic sequences. Acta Universitatis Sapientiae, Informatica **6**(2), 252–286 (2014)
12. Chernyak, A.A., Chernyak, Z.A., Tyshkevich, R.I.: On forcibly hereditary p-graphical sequences. Disc. Math. **64**(2–3), 111–128 (1987)
13. Choudum, S.A.: A simple proof of the erdos-gallai theorem on graph sequences. Bull. Aust. Math. Soc. **33**(1), 67–70 (1986)
14. Du, D., Hwang, F.: Combinatorial Group Testing and Its Applications. Applied Mathematics, World Scientific, Singapore (2000)
15. Erdős, P., Gallai, T.: Gráfok előírt fokszámú pontokkal. Mat. Lapok (N.S.) **11**, 264–274 (1960)
16. Erdős, P.L., Hartke, S.G., Van Iersel, L., Miklós, I.: Graph realizations constrained by skeleton graphs. Electron. J. Comb. **24**(2) (2017)
17. Erdős, P.L., Miklós, I.: Not all simple looking degree sequence problems are easy. J. Comb. **9**(3), 553–566 (2018)

18. Garey, M.R., Johnson, D.S.: Computers and Intractability: A Guide to the Theory of NP-Completeness. W.H. Freeman, New York (1979)
19. Grebinski, V., Kucherov, G.: Reconstructing a hamiltonian cycle by querying the graph: application to dna physical mapping. Disc. Appl. Math. **88**(1), 147–165 (1998)
20. Gupta, G., Joshi, P., Tripathi, A.: Graphic sequences of trees and a problem of frobenius. Czechoslov. Math. J. **57**, 49–52 (2007)
21. Hakimi, S.L.: On realizability of a set of integers as degrees of the vertices of a linear graph. I. J. Soc. Ind. Appl. Math. **10**, 496–506 (1962)
22. Hammer, P.L., Simeone, B.: The splittance of a graph. Combinatorica **1**, 275–284 (1981)
23. Harary, F.: Graph theory. In: Clustering Algorithm, p. 1975 (1969)
24. Havel, V.: Poznámka o existenci konečných grafů. Časopis pro pěstování matematiky **080**(4), 477–480 (1955)
25. Iványi, A.: Degree sequences of multigraphs. In: Annales University of Science Budapest., Security and Computation, vol. 37, pp. 195–214. ELTE (2012)
26. Koren, M.: Sequences with a unique realization by simple graphs. J. Comb. Theory Ser. B **21**(3), 235–244 (1976)
27. Liao, H., Chakrabarty, D.: Learning spanning forests optimally in weighted undirected graphs with cut queries. In: Vernade, C., Hsu, D. (eds.) Proceedings of the 35th International Conference on Algorithmic Learning Theory. Proceedings of Machine Learning Research, vol. 237, pp. 785–807. PMLR (2024)
28. Lovász, L.: Subgraphs with prescribed valencies. J. Comb. Theory **8**(4), 391–416 (1970)
29. McCormick, S.T.: Submodular function minimization. In: Aardal, K., Nemhauser, G., Weismantel, R. (eds.) Discrete Optimization, Handbooks in Operations Research and Management Science, vol. 12, pp. 321–391. Elsevier (2005)
30. Pak, I., Vilenchik, D.: Constructing uniquely realizable graphs. Disc. Comput. Geom. **50**(4), 1051–1071 (2013)
31. Petersen, J.: Die Theorie der regulären graphs. Acta Math. **15**, 193–220 (1900)
32. Plummer, M.D.: Graph factors and factorization: 1985–2003: a survey. Disc. Math. **307**(7), 791–821 (2007)
33. Rao, S.B.: A survey of the theory of potentially P-graphic and forcibly P-graphic degree sequences. In: Rao, S.B. (ed.) Combinatorics and Graph Theory. LNM, vol. 885, pp. 417–440. Springer, Heidelberg (1981). https://doi.org/10.1007/BFb0092288
34. Tripathi, A., Tyagi, H.: A simple criterion on degree sequences of graphs. Disc. Appl. Math. **156**(18), 3513–3517 (2008)
35. Tripathi, A., Venugopalan, S., West, D.B.: A short constructive proof of the erdős-gallai characterization of graphic lists. Disc. Math. **310**(4), 843–844 (2010)
36. Tutte, W.T.: The factors of graphs. Can. J. Math. **4**, 314–328 (1952)

On The Computational Complexity of Games with Uncertainty

Bruce M. Kapron[(✉)] [iD] and Koosha Samieefar[(✉)] [iD]

Department of Computer Science, University of Victoria, Victoria, Canada
bmkapron@uvic.ca, koosha021@gmail.com

Abstract. We investigate the computational complexity of several equilibrium problems involving uncertain data in distribution-free models. Primarily, we focus on L/F-equilibrium in multi-leader-follower games and resilient Nash equilibrium-concepts that are traditionally known for their lack of general equilibrium solutions. In the presence of uncertain data, we adopt an approach rooted in robust optimization that considers the robust version of these equilibrium notions. We also present a more general formulation that encompasses existing remedial approaches that address the problem of the lack of solutions for multi-leader-follower games. Building on recent advancements in the computational complexity of generalized quasi-variational inequalities, we prove these equilibrium problems are PPAD-complete under reasonable assumptions.

Keywords: Distribution-Free Models · Multi-Leader-Follower Games · Computational Complexity · Equilibrium Problems · Remedial Solutions

1 Introduction

Game theory is the mathematical study of strategic decision-making in competitive situations, where an agent's choices depend on others. It encompasses various formats, including normal-form games, extensive-form games, and sequential games, each providing insights into interactions. These formats have emerged due to their applications across fields, reflecting the need to model complex real-world situations involving strategic behavior. The different formats of games in game theory include normal-form (matrix form) games, extensive-form games, sequential games, repeated games, and stochastic games. Further extensions of these game formats may include games in which players have different roles, such as *Stackelberg games*. Stackelberg games are sequential games in which the leader makes the first move, followed by the followers who observe and respond to the leader's decision. This setup reflects hierarchical relationships and strategic advantages, as the leader's actions directly influence the followers' choices. These games find applications in various fields, including business strategy and military tactics [4,55,56,61]. *Multi-leader-follower* games are a generalization of

© The Author(s), under exclusive license to Springer Nature Switzerland AG 2025
I. Finocchi and L. Georgiadis (Eds.): CIAC 2025, LNCS 15679, pp. 343–360, 2025.
https://doi.org/10.1007/978-3-031-92932-8_22

Stackelberg games that include multiple leaders. The concept of multi-leader-follower games has a variety of applications that arise from situations where multiple oligopoly firms operate in the market [12,32,33,50].

Games with uncertain data, introduce unpredictability as players often lack complete knowledge about their opponents' strategies or the game's state. This uncertainty requires strategic decision-making under ambiguity, arising from sources such as hidden information, randomness, or incomplete knowledge of the game. Assuming the Bayesian hypothesis, Harsanyi [28–30] investigated games featuring incomplete information, where players lack full knowledge of key game parameters. By assuming that all players share common knowledge of these probability distributions, the game can ultimately be re-conceptualized as one with complete information, referred to as the Bayes equivalent of the original game. Furthermore, stochastic optimization techniques are shown to be suitable for addressing Stackelberg games involving uncertain data, enabling probabilistic decision-making. Detailed insights into stochastic Stackelberg and multi-leader-follower games and their applications in numerous fields are available in [9,13,16,42,45,54]. In addition to probability distribution models, recent years have seen a rise in the adoption of distribution-free models based on worst-case scenarios [2,13,34,35,38,41]. In these models, each player's decision-making process is guided by the principles of robust optimization. Robust optimization operates under the assumption that uncertain data lie within an uncertainty set, seeking solutions that account for the worst-case outcomes in terms of objective function values and/or constraints. For example, Hu et al. [34] use the robust optimization technique for multi-leader-follower games under uncertainty assumptions. A recent paper also applies theoretical developments to explore the impact of uncertainty on robust game solutions [14].

Nash equilibrium (NE) [46,47] is a central concept in game theory, requiring that each player's strategy is optimal given the strategies of the other players. Nash equilibrium does not always ensure the best outcome for all players which led to variants that guarantee properties such as high social welfare [24]. Also, generalized Nash equilibrium (GNE) is another notion that extends the concept of Nash equilibrium that has emerged to address the applications in which a player's strategy set may depend on the strategies chosen by players [15,18]. The solution concept of *L/F-(Nash) equilibrium* is an extension of the Nash equilibrium and Stackelberg equilibrium specifically defined for multi-leader-follower games. Finding equilibrium in multi-leader-follower games is a well-defined mathematical problem but is computationally intractable. Specifically, these problems resemble equilibrium problems for a more complex form of GNE, requiring each leader to solve a non-convex mathematical program with equilibrium constraints [6,50,59]. To address the issue, [33,50] have proposed "remedial models" that aim to obtain meaningful equilibrium solutions by making reasonable assumptions.

Robust Nash equilibrium is a solution concept that addresses uncertainty in game parameters, ensuring each player's strategy remains optimal despite uncertainty about others' strategies or the game's conditions. This concept guarantees the stability of strategies in the face of unpredictability or incomplete information. Hayashi et al. explored the problem of robust Nash equilibria in bi-matrix games and presented some existence results under specific assumptions on the uncertainty sets [31]. Aghassi and Bertsimas [2] studied robust Nash equilibria in n-player games with bounded polyhedral uncertainty sets, where each player solves a linear programming problem, and also proposed a method for finding such equilibria. In contrast, more recent work by [48] investigates a more general NE with uncertain data, where each player solves an optimization problem with a nonlinear objective function. Moreover, under mild assumptions on the uncertainty sets, the authors presented results regarding the existence, and uniqueness of robust Nash equilibria in this setting. Hu and Fukushima [34] expanded their earlier work [33] by applying robust optimization techniques to multi-leader-follower games under uncertainty. They introduced the concept of *robust L/F-equilibrium* and established its existence and uniqueness for a specific class of these games with a particular structure.

Resilient Nash equilibrium, introduced in [1], is a fundamental notion in applications such as secret sharing and multi-party computation. It ensures stability not only against unilateral deviations but also against deviations by coalitions. Informally, an equilibrium of this type not only ensures that each player has no incentive to unilaterally deviate from their strategy given the strategies of others, but it also remains stable against deviations by any coalition of players. We can further restrict our analysis to coalitions of size at most t. This restriction is practical in scenarios such as networks, where large coalitions are difficult to coordinate, leading to the concept of t-*resilient Nash equilibrium* [1].

Our Contribution

The class PPAD plays a significant role in studying equilibria and related problems in game theory and economics. This paper aims to develop a comprehensive computational framework for various approximate equilibrium problems with uncertainty in distribution-free models and to establish their PPAD-completeness under reasonable assumptions. We first focus on the computational complexity of the robust Nash equilibrium notion. Next, considering the proposed remedial solutions for L/F-equilibrium in multi-leader-follower games in [33, 50], we propose a more general definition that encompasses both remedial models dealing with uncertain data. Then, we carefully establish the PPAD-completeness of an appropriate computational version. Building on the significant efforts in game theory to address deviations by coalitions of players-dating back to Aumann's foundational work [3] and [8], we extend the PPAD-completeness to t-resilient Nash equilibrium under uncertainty with reasonable assumptions. Specifically, we introduce the notion of a *robust and resilient* equilibrium that combines robustness to uncertainty with resilience to coalition deviations. To our knowledge, this concept has not been explored, despite its potential applications.

The main technical challenge in proving PPAD-completeness here is establishing inclusion in PPAD as PPAD-hardness can be implied by the hardness of simpler equilibrium problems such as Nash equilibrium. Our approach builds upon newly formulated computational definitions for variational inequalities which offer a broad framework capable of accommodating the intricacies of games with substantial variations in constraints and formats [39]. Note that our results for multi-leader followers are limited to remedial models while there are a variety of hardness results that even hold for Stackelberg games (e.g. [45]). Our proposed definition also extends the results of [50] to the case where uncertainty is involved by embracing an approach similar to that of [34].

Related Work

The complexity of problems where Kakutani's fixed-point theorem ensures solutions, especially in games such as those of Debreu and Rosen [15,18,53], has only recently been explored [7,22,39,40,51]. A key challenge is representing convex sets explicitly and succinctly. The computational approach in [51], based on computational convex geometry [25], was used to determine the complexity of approximating equilibria in Rosen-style games[1]. Unlike prior works [2,7,43], we consider the most general formulations. Simple methods, such as convex hulls or polytopes defined by linear inequalities [52], often fail to capture crucial applications of Kakutani's theorem, including its role in these games [51].

The technical approach of [39] builds on the relationship between the existence of a solution of variational inequalities and Kakutani's fixed-point theorem (see [10]) establishing PPAD-completeness of variational inequalities under simple convexity assumptions using the machinery provided in [51]. The primary drawback of this approach lies in the technical difficulties that arise from addressing general convex sets. An alternative computational approach to variational inequalities exists [7], but it does not cover the generalized version needed for games with uncertainty, such as robust Nash equilibrium. Notably, it is restricted to polyhedral constraints. There are multiple motivating reasons for using variational inequalities in game theory problems. For example, they naturally centralize the role of dynamics in understanding the behavior of interacting agents [60]. Furthermore, applying well-known approaches to solving variational inequality problems is useful for problems in game theory [50].

2   Preliminaries

2.1   Basic Game Theory

Definition 1. *Suppose that we have* k *players, where each player* i, $1 \leq i \leq k$ *has a set* S_i *of possible strategies. Each also player has a* loss *function* $\theta_i : S_1 \times \cdots \times S_k \to \mathbb{R}$. *A game can specified by the* $\Theta = (\theta_1, \ldots, \theta_k)$ *and* $\mathcal{S} = (S_1, \ldots, S_k)$. *Player* i's *loss when each player* j, $1 \leq j \leq k$ *plays* s_j *is shown* $\theta_i(s_1, \ldots, s_k)$.

---

[1] The exact problem was studied in [21].

Definition 2. *A ϵ-approximate Nash equilibrium is a strategy profile[2] $s = (s_1^*, \ldots, s_k^*)$ such that for $1 \leq i \leq k$ and*

$$\theta_i(s^*) - \epsilon \leq \theta_i(s_i, s_{-i}^*), \qquad\qquad \forall s_i \in S_i \qquad\qquad (1)$$

where s_{-i}^* includes all strategies except i.

Remark 1. Throughout the paper, we assume that each S_i is a bounded subset of $\mathbb{R}^{n_i}$ and each player i controls $dim(S_i) = n_i$ variables of the strategy profile. Furthermore, $\sum_{i=1}^{k} n_i = n$.

Definition 3. *A correspondence from a set X to a set Y is a function $\mathcal{R} : X \to 2^Y$. Sometimes we denote such a mapping by $\mathcal{R} : X \rightrightarrows Y$.*

A natural way to limit the notion of a player's best response is to suppose that player i, when considering responses to s_{-i}^*, is constrained to those belong to some subset of S_i, determined by s_{-i}^*. This constraint might be given as a correspondence $\mathcal{R}_i : S_{-i} \rightrightarrows S_i$, where S_{-i} denotes $S_1 \times \cdots \times S_{i-1} \times S_{i+1} \times \cdots \times S_k$.

Definition 4 ([15,18])**.** *A generalized Nash (also called constrained/social) equilibrium is a strategy profile $s = (s_1^*, \ldots, s_n^*)$ which satisfies $s_i^* \in \mathcal{R}_i(s_{-i}^*)$ and $(**)$[3] .*

$$\forall i \in [n], \forall s_i \in \mathcal{R}_i(s_{-i}^*), \quad \theta_i(s_i^*, s_{-i}^*) \leq \theta_i(s_i, s_{-i}^*) \qquad\qquad (**)$$

Remark 2. We may assume that player i's constraints are represented by the non-empty set $\mathcal{R}_i(\mathbf{x}_{-i}) \equiv \{x_i \in S_i \mid g_i(x_i, \mathbf{x}_{-i}) \leq 0, h_i(x_i) = 0\}$ where $x_i \in S_i$ and $g_i : \Pi_{j=1}^{k} S_j \to \mathbb{R}^{m_i}$ and $h_i : S_i \to \mathbb{R}^{l_i}$ are continuously differentiable convex functions. Note that $g_i(\cdot, \mathbf{x}_{-i})$ is convex assuming $\mathbf{x}_{-i}$ is fixed. m_i and l_i are some given constants (for more information, see [18,50]).

The notion of resilient Nash equilibrium is more restricted compared to well-known definitions such as *strong Nash* [3] and *coalition-proof Nash* [8]. Consider a k-player game and let $\mathcal{J}$ be the set of proper subsets of $\{1, \ldots, k\}$. $J \in \mathcal{J}$ denotes a *coalition* and $S_J \equiv \prod_{j \in J} S_j$ is *the possible strategy set* for the coalition J. The approximate version can be defined similarly.

Definition 5 ([1])**.** *Given a nonempty set $J \subseteq [k]$, $s_J^* \in S_J$ is a group best response for J to $s_{-J}^* \in S_{-J}$ if, for all $s_J' \in S_J$ and all $j \in J$, we have:*

$$\theta_j(s^*) \leq \theta_j(s_J', s_{-J}^*).$$

A strategy profile $s^* \in S$ is a t-resilient Nash equilibrium if, $\forall J \subseteq [k]$ with $|J| \leq t$, s_J^* is a group best response for J to s_{-J}^*[4].

---

[2] A *strategy profile* is a vector that includes all players' strategies.
[3] Rosen provides an alternative formulation under weaker conditions [53].
[4] We call a strategy *strongly resilient* if it is t-resilient for all $t \leq k - 1$.

2.2   Variational Inequalities

Definition 6. *Given a correspondence* $\mathcal{F} : \mathbb{R}^m \rightrightarrows \mathbb{R}^m$ *and a correspondence* $\mathcal{R} :$ $\mathbb{R}^m \rightrightarrows \mathbb{R}^m$, *an approximate solution to generalized quasi-variational inequality* *(GQVI)* $(\mathcal{R}, \mathcal{F})$ *consists of two vectors* $x^* \in \mathcal{R}(x^*)$ *and* $w^* \in \mathcal{F}(x^*)$ *such that:*

$$(y - x^*)^T w^* + \epsilon \geq 0, \quad \forall y \in \mathcal{R}(x^*)$$

GQVI extends the notion of *quasi-variational inequality* (QVI), specifically, when $\mathcal{F}$ is a function, GQVI becomes quasi-variational inequality (QVI). When $\mathcal{R}(x)$ is independent of x, $\mathcal{R}(x) = \mathcal{R}$, the QVI will be *variational inequality* (VI).

2.3   Technical Definitions for Representation

Strong Separation Oracles. For addressing the limitations of Kakutani's fixed-point definition, the paper [51], assumes an implicit representation of convex sets using a polynomially-sized circuit that computes (weak/strong) separation oracles. While this formulation has been proven valuable in many game-theoretic applications, it introduces technical challenges [51]. We follow the assumption that the sets are bounded by a box that is a subset of $\mathbb{R}^m$ and denote this by $\mathbb{R}^{m*}$. A strong separation oracle (SO) is defined as follows:

A STRONG SEPARATION ORACLE (VIA A CIRCUIT $C_{\mathcal{R}(x)}$)

Input: A vector $z \in \mathbb{Q}^m \cap \mathbb{R}^{m*}$ as an input:
Output: $(a, b) \in \mathbb{Q}^m \times \mathbb{Q}$ such that $b \in [0, 1] \cap \mathbb{Q}$ denotes the membership of z in $\mathcal{R}(x)$:

- If $z \in \mathcal{R}(x)$ then $b > \frac{1}{2}$ and the vector $a \in \mathbb{Q}^m$ will be $\perp$. In other words, a is meaningful only when $b \leq \frac{1}{2}$
- $b \leq \frac{1}{2}$ and vector a , with $\|a\|_\infty = 1$ which defines a separating hyperplane $\mathcal{H}(a, z) := \{y \in \mathbb{R}^{m*} : \langle a, y - z \rangle = 0\}$ between the vector z and the set $\mathcal{R}(x)$ such that $\langle a, y - z \rangle \leq 0$ for every $y \in \mathcal{R}(x)$.

Linear Arithmetic Circuits. Linear arithmetic circuits can approximate any well-behaved polynomially computable function, including polynomials, and exhibit properties such as Lipschitzness [19]. A linear arithmetic circuit is differentiable if it represents an affine function. However, for loss functions expressed by such circuits, we can compute a member of the subgradient vector [5].

Definition 7. *A* linear arithmetic circuit C *is a circuit represented as a directed acyclic graph with nodes labeled either as input nodes or as output nodes or as gate nodes with one of the possible gates* $\{+, -, \min, \max, \times\zeta\}$, *where the* $\times\zeta$ *gate refers to the multiplication by a constant. Also, rational constant gates are allowed. We use* $\mathrm{size}(C)$ *to refer to the number of nodes of* C.

2.4   Multi-leader-Follower Games

In a multi-leader-follower game, leaders engage in non-cooperative Nash games at the upper level, considering followers' responses, while followers participate in a game at the lower level, treating leaders' strategies as parameters. The L/F-(Nash) equilibrium is defined as a set of strategies where no player-leader or follower-can reduce their loss by unilaterally changing their strategy. Stackelberg games can be seen as a specific instance of mathematical programs with equilibrium constraints (MPEC). In this case, the followers' problems are replaced by a constraint given by their optimality conditions. In a broader context, an MPEC is an optimization problem that encompasses two sets of variables, namely decision variables and response variables [20,44,49]. The mathematical framework frequently employed to depict the multi-leader-follower game is known as the equilibrium problem with equilibrium constraints (EPEC). An EPEC [17,23,36,37,57,58] is essentially an equilibrium problem that includes multiple parametric MPECs. Achieving equilibria in an EPEC involves solving all the embedded MPECs simultaneously. For simplicity, we examine games with two leaders, denoted by I and II, and k followers, denoted as $i \in [k]$. The strategy domains of leaders I and II are denoted by X^{I} and X^{II} respectively. The leaders' loss functions are denoted by $\phi_{\mathrm{I}}(x_{\mathrm{I}}, x_{\mathrm{II}}, y)$ and $\phi_{\mathrm{II}}(x_{\mathrm{I}}, x_{\mathrm{II}}, y)$. This notation implies that the loss of each leader is influenced by not only their own strategies but also the strategies of the opposing leader, as well as the strategies of the followers, represented by y.

Throughout the paper, we assume that external variables of the optimization problems are denoted by $\star$ and $-$ for the leaders' and followers' strategies. For each follower $i = 1, \ldots, k$, let $\theta_i(x_{\mathrm{I}}, x_{\mathrm{II}}, y)$ and $\mathcal{R}_i(x_{\mathrm{I}}, x_{\mathrm{II}}, y_{-i})$ denote i's loss function and available constrained strategy set, respectively. This constrained strategy set of the followers depends on the pair of strategies $(x_{\mathrm{I}}, x_{\mathrm{II}}) \in X^{\mathrm{I}} \times X^{\mathrm{II}}$ of the leaders. For each pair $(x_{\mathrm{I}}^*, x_{\mathrm{II}}^*)$, the followers' problem is modeled by a generalized Nash equilibrium problem parameterized by the leaders' strategies. Let $Y_{sol}(x_{\mathrm{I}}^*, x_{\mathrm{II}}^*) \subseteq Y(x_{\mathrm{I}}^*, x_{\mathrm{II}}^*)$ denote the set of such solutions from the plausible set of strategies Y (not necessarily a singleton). Each $\bar{y} \in Y_{sol}(x_{\mathrm{I}}^*, x_{\mathrm{II}}^*) \subseteq \Pi_{i=1}^k S_i$ is a tuple $(\bar{y}_i)_{i=1}^k$ where for each follower i, $\bar{y}_i$ is a solution of problem (2). The tuple $(x_{\mathrm{I}}^*, x_{\mathrm{II}}^*, \bar{y}_{-i})$ is external to the minimization program (2), and y_i is the primary variable that needs to be computed.

$$\begin{aligned}
&\text{Min } \theta_i \left(x_{\mathrm{I}}^*, x_{\mathrm{II}}^*, \bar{y}_{-i}, y_i \right) \\
&\text{s.t } y_i \in \mathcal{R}_i \left(x_{\mathrm{I}}^*, x_{\mathrm{II}}^*, \bar{y}_{-i} \right),
\end{aligned} \tag{2}$$

We are now ready to define the concept of equilibrium in multi-leader-follower games. A pair $(x_{\mathrm{I}}^*, x_{\mathrm{II}}^*) \in X^{\mathrm{I}} \times X^{\mathrm{II}}$ is called a *L/F-equilibrium*, if there exists $(y_{\mathrm{I}}^*, y_{\mathrm{II}}^*)$ such that $(x_{\mathrm{I}}^*, y_{\mathrm{I}}^*)$ is an optimal solution of leader I's problem, which tries to find a pair $(x_{\mathrm{I}}, y_{\mathrm{I}})$ to the following[5]:

$$\mathrm{Min}\, \phi_{\mathrm{I}}\,(x_{\mathrm{I}}, x_{\mathrm{II}}^*, y_{\mathrm{I}})$$
$$\text{s.t } x_{\mathrm{I}} \in X^{\mathrm{I}} \tag{3}$$
$$\text{and } y_{\mathrm{I}} \in Y_{sol}\,(x_{\mathrm{I}}, x_{\mathrm{II}}^*)$$

and $(x_{\mathrm{II}}^*, y_{\mathrm{II}}^*)$ is an optimal solution of leader II's problem, which tries to find a pair $(x_{\mathrm{II}}, y_{\mathrm{II}})$ to the following problem:

$$\mathrm{Min}\, \phi_{\mathrm{II}}\,(x_{\mathrm{I}}^*, x_{\mathrm{II}}, y_{\mathrm{II}})$$
$$\text{s.t } x_{\mathrm{II}} \in X^{\mathrm{II}} \tag{4}$$
$$\text{and }\quad y_{\mathrm{II}} \in Y_{sol}\,(x_{\mathrm{I}}^*, x_{\mathrm{II}})$$

In this definition, the followers' equilibrium strategies, y_{I}^* and y_{II}^*, belong to the same set $Y_{\mathrm{sol}}(x_{\mathrm{I}}^*, x_{\mathrm{II}}^*)$. However, they are not required to be the same (i.e., $y_{\mathrm{I}}^* = y_{\mathrm{II}}^*$). This flexibility arises because these strategies are based on anticipations by different leaders of how the followers collectively respond to the pair of strategies $(x_{\mathrm{I}}^*, x_{\mathrm{II}}^*)$. One could introduce another variant of this problem by enforcing $y_{\mathrm{I}}^* = y_{\mathrm{II}}^*$. However, even if $Y_{\mathrm{sol}}(x_{\mathrm{I}}, x_{\mathrm{II}})$ has a unique response for all $(x_{\mathrm{I}}, x_{\mathrm{II}})$, an L/F-equilibrium may not exist [50].

2.5   Robust Nash Equilibrium

Robust Nash Equilibrium for General Games. Robust Nash equilibrium, introduced by [2,31], addresses uncertainty in general games. While this concept was initially developed for simpler settings, [48] extended it to k-player non-cooperative games with nonlinear loss functions. To address the uncertainty that arises in this setting, a specific uncertainty parameter of the form $u_i \in U^i \subset \mathbb{R}^{l_i}$ is introduced for each player i, and it becomes part of the player's loss function, allowing a representation of the uncertainty. Consequently, the loss function can now be expressed as $\theta_i : S_i \times S_{-i} \times U^i \to \mathbb{R}$. Although player i does not know the exact value of parameter u_i yet, the player knows that it must belong to a given nonempty set $U^i \subseteq \mathbb{R}^{l_i}$. Finally, player i tries to find x_i solving the following optimization problem with parameter $u_i \in U^i$ and x_{-i}^* as an external parameter:

$$\mathrm{Min}\, \theta_i\,(x_i, x_{-i}^*, u_i)$$
$$\text{s.t } x_i \in S_i \tag{5}$$

By the robust optimization paradigm, we assume that each player i tries to minimize their worst-case loss function. Under this assumption, each player i

---

[5] Similar to [50], we assume a leader's strategy set is independent of other leaders.

considers the worst-case loss function $\tilde{f}_i : S_i \times S_{-i} \rightarrow \mathbb{R}$, defined as $\tilde{f}_i (x_i, x_{-i}) :=$ $\sup \{ \theta_i (x_i, x_{-i}, u_i) \mid u_i \in U^i \}$, and tries to solve the following problem:

$$\text{Min } \tilde{f}_i \left(x_i, x^*_{-i} \right)$$
$$\text{s.t } x_i \in S_i \qquad (6)$$

This indeed is a Nash equilibrium problem with complete information. We now define the resulting notion of equilibrium with uncertainty. The definition can be extended to the approximate version and the generalized (Nash) equilibrium.

Definition 8. *A strategy profile x^* is called a* robust Nash equilibrium *of the game of the optimization program (5) if x^* is a (Nash) solution of (6).*

Robust Equilibrium for Multi-leader-Follower Games. Next, we consider extending the solution concept of robust Nash equilibrium to multi-leader-follower games. We consider only two types of uncertainty caused by observation errors of the leaders towards other leaders and also the followers similar to [34]. In a multi-leader-follower with two leaders I and II, leader I tries to find x_{I} and y_{I} solving the following optimization problem:

$$\text{Min } \phi_{\text{I}} (x_{\text{I}}, x^*_{\text{II}}, y_{\text{I}}, u_{\text{I}})$$
$$\text{s.t } x_{\text{I}} \in X^{\text{I}} \qquad (7)$$
$$\text{and } y_{\text{I}} \in Y_{sol} (x_{\text{I}}, x^*_{\text{II}})$$

where the loss function $\phi_{\text{I}} : X^{\text{I}} \times X^{\text{II}} \times \Pi^k_{i=1} S_i \times U^i \rightarrow \mathbb{R}$ has an uncertainty parameter $u_{\text{I}} \in \mathbb{R}^{l_{\text{I}}}$. As described above, although leader I does not know the exact value of parameter u_i, the leader knows that it must belong to a given nonempty set U^{I}. We can similarly write the optimization problem for leader II:

$$\text{Min } \phi_{\text{II}} (x^*_{\text{I}}, x_{\text{II}}, y_{\text{II}}, u_{\text{II}})$$
$$\text{s.t } x_{\text{II}} \in X^{\text{II}} \qquad (8)$$
$$\text{and } y_{\text{II}} \in Y_{sol} (x^*_{\text{I}}, x_{\text{II}})$$

The optimization program (2) only examines the followers' problem solely from the perspective of the followers while the concept of equilibrium in multi-leader-follower games reflects a hierarchical structure. From the perspective of the leaders, they cannot exactly anticipate the response of the followers due to uncertainty. In conclusion, leader I estimates that follower i will try to find $y_{(i,\text{I})}$ which solves the following optimization problem:

$$\text{Min } \theta_i \left(x^*_{\text{I}}, x^*_{\text{II}}, \bar{y}_{(\text{I},-i)}, y_{(\text{I},i)}, e_{\text{I}} \right)$$
$$\text{s.t } y_{(\text{I},i)} \in \mathcal{R}_i \left(x^*_{\text{I}}, x^*_{\text{II}}, \bar{y}_{(\text{I},-i)} \right) \qquad (9)$$

where each element $\bar{y}_{\text{I}} \in Y_{sol}(x^*_{\text{I}}, x^*_{\text{II}}) \subseteq Y(x^*_{\text{I}}, x^*_{\text{II}}) \subseteq \Pi^k_{i=1} S_i$ is a tuple $(\bar{y}_{(\text{I},i)})^k_{i=1}$ such that for each follower i, $\bar{y}_{\text{I}}$ is a solution of (9). The tuple $(x^*_{\text{I}}, x^*_{\text{II}}, \bar{y}_{(\text{I},-i)})$ is external to (9), and $y_{(\text{I},i)}$ is the variable that must be computed. Here $\bar{y}_{(\text{I},-i)}$

is the estimated strategy of leader I of all followers expect i. Similarly, leader II estimates that follower i will try to find $y_{(i,\mathrm{II})}$ which solves:

$$
\begin{aligned}
&\operatorname{Min} \theta_i \left(x_\mathrm{I}^*, x_\mathrm{II}^*, \bar{y}_{(\mathrm{II},-i)}, y_{(\mathrm{II},i)}, e_\mathrm{II} \right) \\
&\text{s.t } y_{(\mathrm{II},i)} \in \mathcal{R}_i \left(x_\mathrm{I}^*, x_\mathrm{II}^*, \bar{y}_{(\mathrm{II},-i)} \right),
\end{aligned}
\tag{10}
$$

We considered the fact that each leader's anticipation of the followers can be different. Also, the leaders do not have any information about the uncertainty parameters e_I and e_II but they know that they must belong to the sets $E^\mathrm{I} \subset \mathbb{R}^{k_\mathrm{I}}$ and $E^\mathrm{II} \subset \mathbb{R}^{k_\mathrm{II}}$ respectively[6]. To define robust L/F-Nash equilibrium, we reformulate the problem for leader I with uncertainty parameters e_I and u_I[7]:

$$
\begin{aligned}
&\operatorname{Min} \phi_\mathrm{I} \left(x_\mathrm{I}, x_\mathrm{II}^*, y_\mathrm{I}(x_\mathrm{I}, x_\mathrm{II}^*, e_\mathrm{I}), u_\mathrm{I} \right) \\
&\text{s.t } x_\mathrm{I} \in X^\mathrm{I}
\end{aligned}
\tag{11}
$$

For leader II, we have:

$$
\begin{aligned}
&\operatorname{Min} \phi_\mathrm{II} \left(x_\mathrm{I}^*, x_\mathrm{II}, y_\mathrm{II}(x_\mathrm{I}^*, x_\mathrm{II}, e_\mathrm{II}), u_\mathrm{II} \right) \\
&\text{s.t } x_\mathrm{II} \in X^\mathrm{II}
\end{aligned}
\tag{12}
$$

where $y_\mathrm{I}(x_\mathrm{I}, x_\mathrm{II}^*, e_\mathrm{I})$ and $y_\mathrm{II}(x_\mathrm{I}^*, x_\mathrm{II}, e_\mathrm{II})$ represent one of solution of Eq. (9) and (10) respectively and belong to Y_{sol}. Utilizing the robust optimization, we define a new game $\tilde{\mathcal{G}}$ with worst-case loss functions $\tilde{\psi}_\mathrm{I} : X^\mathrm{I} \times X^\mathrm{II} \to \mathbb{R}$ and $\tilde{\psi}_\mathrm{II} : X^\mathrm{I} \times X^\mathrm{II} \to \mathbb{R}$ as follows:

$$
\begin{aligned}
\tilde{\psi}_\mathrm{I} \left(x_\mathrm{I}, x_\mathrm{II}^* \right) &:= \sup \left\{ \phi_\mathrm{I} \left(x_\mathrm{I}, x_\mathrm{II}^*, y_\mathrm{I}(x_\mathrm{I}, x_\mathrm{II}^*, e_\mathrm{I}), u_\mathrm{I} \right) \mid u_\mathrm{I} \in U^\mathrm{I}, e_\mathrm{I} \in E^\mathrm{I} \right\} \\
\tilde{\psi}_\mathrm{II} \left(x_\mathrm{I}^*, x_\mathrm{II} \right) &:= \sup \left\{ \phi_\mathrm{II} \left(x_\mathrm{I}^*, x_\mathrm{II}, y_\mathrm{II}(x_\mathrm{I}^*, x_\mathrm{II}, e_\mathrm{II}), u_\mathrm{II} \right) \mid u_\mathrm{II} \in U^\mathrm{II}, e_\mathrm{II} \in E^\mathrm{II} \right\}
\end{aligned}
\tag{13}
$$

Definition 9. *Assume that we have a multi-leader-follower game $\mathcal{G}$ with k followers with the format (11) and (12) with loss functions ϕ_I and ϕ_II. A strategy profile $(x_\mathrm{I}^*, x_\mathrm{II}^*)$ is called a robust L/F-(Nash) equilibrium of $\mathcal{G}$ if $(x_\mathrm{I}^*, x_\mathrm{II}^*)$ is a L/F-equilibrium of $\tilde{\mathcal{G}}$ with loss functions $\tilde{\psi}_\mathrm{I}$ and $\tilde{\psi}_\mathrm{II}$ defined in (13).*

3 Main Results

3.1 Complexity of Robust Remedial L/F Equilibrium

Recall that finding an L/F-equilibrium in a multi-leader-follower game is challenging due to the complexity of the leaders' optimization problems and the

---

[6] Recall that $\mathcal{R}_i \left(x_\mathrm{I}^*, x_\mathrm{II}^*, \bar{y}_{(\mathrm{I},-i)} \right)$ is a nonempty, closed and convex correspondence. In [34,48], they assume that $\mathcal{R}$ is a set and is independent from $y_{(\mathrm{I},-i)}$ and $y_{(\mathrm{II},-i)}$.

[7] This reformulation is well-known in the literature, see [34,35]. Informally, remedial solutions try to characterize Y_{sol} in a computationally friendly manner.

potential non-convexity of the solution set Y_{sol}, which can prevent the existence of a solution [50]. The algebraic relaxation/restriction approach (*remedial solutions*) introduced in [34,50] addresses this issue. We introduce a general formulation inspired by both approaches that encompasses both remedial models. These approaches may consider *Karush-Kuhn-Tucker conditions* (KKT) and introduce various relaxations/restrictions. More information on the approaches and how they are considered in our approach is available in the full version.

Our General Remedial Approach. First, we provide a mathematical definition for a general remedial approach to multi-leader-follower games with uncertainty. Assuming two leaders and k followers, we first consider the following optimization programs from the perspective of leaders.

$$
\begin{aligned}
&\text{Min } \phi_{\text{I}}\left(x_{\text{I}}, x_{\text{II}}^*, y, u_{\text{I}}\right) := \omega_{\text{I}}\left(x_{\text{I}}, x_{\text{II}}^*, u_{\text{I}}\right) + \pi_{\text{I}}\left(x_{\text{I}}, y\right) \\
&\text{s.t } g_{\text{I}}\left(x_{\text{I}}, x_{\text{II}}^*\right) \le 0, \quad h_{\text{I}}\left(x_{\text{I}}\right) = 0. \\
&\text{Min } \phi_{\text{II}}\left(x_{\text{I}}^*, x_{\text{II}}, y, u_{\text{II}}\right) := \omega_{\text{II}}\left(x_{\text{I}}^*, x_{\text{II}}, u_{\text{II}}\right) + \pi_{\text{II}}\left(x_{\text{II}}, y\right) \\
&\text{s.t } g_{\text{II}}\left(x_{\text{I}}^*, x_{\text{II}}\right) \le 0, \quad h_{\text{II}}\left(x_{\text{II}}\right) = 0.
\end{aligned}
\tag{14}
$$

where $u_{\text{I}} \in U^{\text{I}}$ and $u_{\text{II}} \in U^{\text{II}}$ are uncertainty parameters of the leaders. The uncertainty sets are convex and compact. Function $\omega_{\text{I}} : \mathbb{R}^{n_{\text{I}}+n_{\text{II}}} \to \mathbb{R}$ is convex for fixed x_{II} and u_{I} (similar assumption holds for ω_{II}). Also, for each i $\in \{\text{I}, \text{II}\}$, we have the assumption that $g_{\text{i}} : X^{\text{I}} \times X^{\text{II}} \to \mathbb{R}^{m_{\text{i}}}$ is a convex function for fixed $x_{-\text{i}}$ and $h_{\text{i}} : X^{\text{i}} \to \mathbb{R}^{l_{\text{i}}}$ is an affine function.

Instead of $y_i \in \mathcal{R}_i(\bar{y}_{-i})$ as in traditional L/F-equilibrium, we take a similar approach combining the ideas of [33,34,50] restricting or relaxing followers' feasible solutions. We assume that a valid solution $\bar{y}_i$ for follower i belongs to a set Z_i that we call the *restricted or relaxed feasible set of follower i*: Assuming $Z = \Pi_{i=1}^k Z_i$, we can either consider $Z \subseteq Y_{sol}$ or $Y_{sol} \subseteq Z$ similar to the approach of [50]. Finally, from the perspective of the followers, each follower $i \in [k]$ will try to find $y_i \in Z_i$ given the strategies of the leaders and other followers $(\bar{y}_{-i})$:

$$
\begin{aligned}
&\text{Min } \theta_i(x_{\text{I}}^*, x_{\text{II}}^*, y) := \gamma_i(y) - \pi_{\text{I}}\left(x_{\text{I}}^*, y\right) - \pi_{\text{II}}\left(x_{\text{II}}^*, y\right) \\
&\text{s.t } g_i'(x_{\text{I}}^*, x_{\text{II}}^*, y) \le 0, \quad h_i'\left(y_i\right) = 0
\end{aligned}
$$

where $y = (y_i, \bar{y}_{-i})$. For follower i, we consider γ_i, g_i' and h_i' to have an explicit representation[8]. As mentioned, each leader's anticipation of the strategies of their followers could be different. Also, the uncertainty parameter of the followers needs to be taken into account. In conclusion, leaders I and II anticipate that each follower will solve the following optimization problems respectively:

---

[8] For more information, see the full version, which also includes a simpler case without uncertainty.

$$\text{Min } \theta_{(\mathrm{I},i)}(x_{\mathrm{I}}^*, x_{\mathrm{II}}^*, y_{\mathrm{I}}, e_{\mathrm{I}}) := \gamma_i(y_{\mathrm{I}}, e_{\mathrm{I}}) - \pi_{\mathrm{I}}(x_{\mathrm{I}}^*, y_{\mathrm{I}}) - \pi_{\mathrm{II}}(x_{\mathrm{II}}^*, y_{\mathrm{I}})$$
$$\text{s.t } y_{(\mathrm{I},i)} \in Z_i$$
$$\text{Min } \theta_{(\mathrm{II},i)}(x_{\mathrm{I}}^*, x_{\mathrm{II}}^*, y_{\mathrm{II}}, e_{\mathrm{II}}) := \gamma_i(y_{\mathrm{II}}, e_{\mathrm{II}}) - \pi_{\mathrm{I}}(x_{\mathrm{I}}^*, y_{\mathrm{II}}) - \pi_{\mathrm{II}}(x_{\mathrm{II}}^*, y_{\mathrm{II}}) \tag{15}$$
$$\text{s.t } y_{(\mathrm{II},i)} \in Z_i$$

where $y_{\mathrm{I}} = (y_{(\mathrm{I},i)}, \bar{y}_{(\mathrm{I},-i)})$ and $y_{\mathrm{II}} = (y_{(\mathrm{II},i)}, \bar{y}_{(\mathrm{II},-i)})$. g_i' and h_i' for $i \in [k]$ have similar assumptions such as convexity. Each remedial approach has a specific feasible set and relaxation method, aiming to simplify the followers' problem into a feasible solution set that is convex. In conclusion, a remedial version (such as approaches in [34,50]) can be formulated in the format of a generalized (Nash) equilibrium problem efficiently (in polynomial time).

$$\text{Min } F_{\mathrm{I}}(x_{\mathrm{I}}, x_{\mathrm{II}}^*, y_{\mathrm{I}}, u_{\mathrm{I}}, e_{\mathrm{I}})$$
$$\text{s.t } (x_{\mathrm{I}}, y_{\mathrm{I}}) \in G_{\mathrm{I}}(x_{\mathrm{II}}^*, e_{\mathrm{I}})$$
$$\text{Min } F_{\mathrm{II}}(x_{\mathrm{I}}^*, x_{\mathrm{II}}, y_{\mathrm{I}}, u_{\mathrm{II}}, e_{\mathrm{II}}) \tag{16}$$
$$\text{s.t } (x_{\mathrm{II}}, y_{\mathrm{II}}) \in G_{\mathrm{II}}(x_{\mathrm{I}}^*, e_{\mathrm{II}})$$

By using the robust optimization approach, one can construct the robust counterpart defining $\tilde{F}_{\mathrm{I}} : X^{\mathrm{I}} \times X^{\mathrm{II}} \rightrightarrows \mathbb{R}$ and $\tilde{F}_{\mathrm{II}} : X^{\mathrm{I}} \times X^{\mathrm{II}} \rightrightarrows \mathbb{R}$:

$$\text{Min } \tilde{F}_{\mathrm{I}}(x_{\mathrm{I}}, x_{\mathrm{II}}^*) := Max \left\{ F_{\mathrm{I}}(x_{\mathrm{I}}, x_{\mathrm{II}}^*, y_{\mathrm{I}}, u_{\mathrm{I}}, e_{\mathrm{I}}) \mid u_{\mathrm{I}} \in U^{\mathrm{I}}, e_{\mathrm{I}} \in E^{\mathrm{I}} \right\}$$
$$\text{s.t } (x_{\mathrm{I}}, y_{\mathrm{I}}) \in G_{\mathrm{I}}(x_{\mathrm{II}}^*, e_{\mathrm{I}})$$
$$\text{Min } \tilde{F}_{\mathrm{II}}(x_{\mathrm{I}}^*, x_{\mathrm{II}}) := Max \left\{ F_{\mathrm{II}}(x_{\mathrm{I}}^*, x_{\mathrm{II}}, y_{\mathrm{II}}, u_{\mathrm{II}}, e_{\mathrm{II}}) \mid u_{\mathrm{II}} \in U^{\mathrm{II}}, e_{\mathrm{II}} \in E^{\mathrm{II}} \right\} \tag{17}$$
$$\text{s.t } (x_{\mathrm{II}}, y_{\mathrm{II}}) \in G_{\mathrm{II}}(x_{\mathrm{I}}^*, e_{\mathrm{II}})$$

We approximate Sup with Max. The following proposition is a simple extension of [34] transforming the above-mentioned problem into GQVI[9].

Proposition 1. *In a multi-leader-follower game, the problem of finding a remedial L/F-equilibrium of the optimization programs in (17) with convex and compact uncertainty sets can be transformed into a GQVI problem.*

Computational Formulation. We now provide a computational definition for remedial solutions to robust L/F-equilibrium in multi-leader-follower games and establish the PPAD-completeness. Given the original game $\mathcal{G}$, the remedial game $\tilde{\mathcal{G}}$ can be constructed in polynomial time by pre-processing.

---

[9] They reduced their specific remedial equilibrium problem to GVI (see [10]), not GQVI while we need this extension to possibly include the remedial approach of [50]. The proof is available in the full version.

TRANSFORMED REMEDIAL L/F ROBUST EQUILIBRIUM

Input: Transformed remedial game $\tilde{\mathcal{G}}$ as input with all the following:

- Two linear arithmetic circuits representing the leaders' convex loss functions $(\tilde{F}_{\mathrm{I}}, \tilde{F}_{\mathrm{II}})$,
- Two linear arithmetic circuits representing strong separation oracles for G_{I} and G_{II} that represents the *remedial constrained domain* of the strategies of the leaders I, II, and followers $i \in [k]$ that are two non-empty, convex-valued, and compact correspondences,
- An accuracy parameter ϵ.

Output: One of the following cases:

- (Violation of non-emptiness): A certificate indicating at least one of the following cases is empty:
    • $G_{\mathrm{I}}(x, e)$ for some $x \in X^{\mathrm{II}}$ or $e \in E^{\mathrm{I}}$
    • $G_{\mathrm{II}}(x, e)$ for some $x \in X^{\mathrm{I}}$ or $e \in E^{\mathrm{II}}$
- (Violation of convexity): A certificate showing the loss functions are not convex.
- (Approximate minimization): Vectors $(x_{\mathrm{I}}^{*}, y_{\mathrm{I}}^{*}) \in G_{\mathrm{I}}(x_{\mathrm{II}}^{*}, e_{\mathrm{I}})$ and $(x_{\mathrm{II}}^{*}, y_{\mathrm{II}}^{*}) \in G_{\mathrm{II}}(x_{\mathrm{I}}^{*}, e_{\mathrm{II}})$ having the following relationship:
    • $\tilde{F}_{\mathrm{I}}(x_{\mathrm{I}}^{*}, x_{\mathrm{II}}^{*}) \leq \epsilon + \underset{x_{\mathrm{I}}}{\mathrm{Min}}\ \tilde{F}_{\mathrm{I}}(x_{\mathrm{I}}, x_{\mathrm{II}}^{*})$
    • $\tilde{F}_{\mathrm{II}}(x_{\mathrm{I}}^{*}, x_{\mathrm{II}}^{*}) \leq \epsilon + \underset{x_{\mathrm{II}}}{\mathrm{Min}}\ \tilde{F}_{\mathrm{II}}(x_{\mathrm{I}}^{*}, x_{\mathrm{II}})$

For simplicity and clarity of inclusion in PPAD, we focused on the robust counterpart problem. For generality, we need to show that $\tilde{F}_{\mathrm{I}}$ and $\tilde{F}_{\mathrm{II}}$ are computable by linear arithmetic circuits in polynomial time. For finite uncertainty sets, the problem is straightforward. The discussion of infinite convex uncertainty sets is not straightforward and thoroughly investigated the full version.

Remark 3. The case (Violation of convexity) is meaningful as an output whenever the form of input is explicitly given otherwise, convexity holds as a promise. (Violation of non-emptiness) is an exception can be given due to the definition of the computational version of GQVI that has a direct relationship with Kakutani's problem and the ellipsoid algorithm (see [39]). For this exception, the type of exception (first or second case) can be distinguished by a simple modification of the ellipsoid algorithm (see [39]).

Theorem 1. *Finding an approximate solution for the robust remedial L/F-equilibrium in a multi-leader-follower game with the above-mentioned assumptions for strong separation oracles is PPAD-complete.*

Proof (sketch). Inclusion in PPAD can be implied by careful adaptation of Proposition 1 as this problem can be transformed into a GQVI problem. The transformation can be done in polynomial time due to to construction of the GQVI instance in Proposition 1. The PPAD-hardness of this problem can be implied by the hardness of finding a mixed Nash equilibrium (see [11]) by which we can construct a game in which two leaders have loss functions that represent the expected loss of their mixed strategies. The followers will have only one strategy and no restrictions and all the uncertainty sets are singleton.

3.2   Complexity of Robust and Resilient Nash Equilibrium

We now proceed to the computational complexity of the robust version of resilient Nash equilibrium. Assume that we have game $\mathcal{G}$ with k players and given uncertainty sets $U = (U_1, \ldots, U_k)$. We assume that either the number of players in the game we discuss is constant or the parameter t is constant[10]. Similarly, we can transform the game $\mathcal{G}$ consider the robust counterpart $\tilde{\mathcal{G}}$ and consider the following problem (see Eq. (6)):

TOTAL ROBUST AND t-RESILIENT NASH

Input: We receive the robust counter part $\tilde{\mathcal{G}}$ as input all the following:

- k linear arithmetic circuits representing k convex loss functions,
- k linear arithmetic circuits representing strategy sets of S_i for $i \in [k]$,
- An accuracy parameter ϵ.

Output: One of the following cases:

- (Violation of k-multi-convexity): A subset $J \in \mathcal{J}$ of at most size k, one index j, vectors $\mathbf{x}_J, \mathbf{y}_J \in S_J$ and $\mathbf{x}_{-J} \in S_{-J}$ s.t for some $\alpha \in (0,1)$:

$$\tilde{\theta}_j \left(\alpha \mathbf{x}_J + (1-\alpha) \mathbf{y}_J, \mathbf{x}_{-J} \right) > \alpha \cdot \tilde{\theta}_j \left(\mathbf{x}_J, \mathbf{x}_{-J} \right) + (1-\alpha) \cdot \tilde{\theta}_j \left(\mathbf{y}_J, \mathbf{x}_{-J} \right)$$

- One vector s^* representing the strategy profile of all players that satisfies:

$$\left(\forall J \in \mathcal{J} \text{ s.t } |J| \le t \right), \forall s'_J \in S_J, \forall j \in J : \tilde{\theta}_j \left(\mathbf{s}^* \right) - \epsilon \le \tilde{\theta}_j \left(\mathbf{s}'_J, \mathbf{s}^*_{-J} \right)$$

Theorem 2 *The problem of finding a total robust and t-resilient Nash equilibrium is PPAD-complete.*

Proof (sketch). We adopt a more generalized version of GQVI called MGQVI in [39]. Briefly, for inclusion in PPAD, the paper [39], generalized the techniques of [51] by considering different versions of robust Berge's maximum theorem and the sub-gradient ellipsoid central cut method to obtain a version appropriate for t-resilient Nash equilibrium without uncertainty handling multi-concave vector-valued functions. This stems from the fact that for a t-resilient Nash equilibrium, we have more complex conditions (immunity from deviation from

---

[10] Otherwise, the problem is unlikely to be in PPAD. For more information, see [39].

coordinated actions of any possible coalition) and assumptions such as multi-concavity. Furthermore, the transformation from the t-resilient Nash equilibrium problem without uncertainty to MGQVI is more complex. This transformation can be done by extending the famous transformation introduced by [27]. We can similarly adopt this approach by considering the robust version of the resilient Nash with uncertainty. For PPAD-hardness, we consider the problem of finding a mixed approximate Nash equilibrium (1-resilient Nash equilibrium) in which the expected loss is convex with singleton uncertainty sets.

Remark 4. We can similarly extend this definition for multi-leader-follower games and introduce the concept of robust and resilient L/F-equilibrium.

4   Conclusion and Future Work

We have examined the computational aspects of various general equilibrium problems in the presence of uncertainty, where players can speculate on their payoffs based on the knowledge that their opponents' uncertainty parameters belong to some set. We established the PPAD-completeness of several equilibrium problems with uncertainty by carefully constructing their robust counterparts. Obtaining these results required leveraging the general computational approach of [39] for generalized variational inequalities and general convex optimization tools of [25]. Below, we highlight several potential research directions.

Some recent research in machine learning incorporates ideas from game theory where the goal is to develop algorithms that are robust in the long term in the presence of rational agents in an unpredictable environment. Our findings and insights could be of use in deriving computational complexity results for various learning problems noted in the introduction. An interesting extension of this work could be investigating games with smart contracts and their applications to permissionless blockchains such as Ethereum [26]. This is because contracts are a generalization of the concept of multi-leader-follower games. Another obvious direction could be investigating the exact version of the problems we studied (see [21]). Inspecting the computational complexity of other game formats and different uncertainty models also could be of interest.

Acknowledgments. We would like to acknowledge the support of NSERC discovery grant RGPIN-2021-02481.

References

1. Abraham, I., Dolev, D., Gonen, R., Halpern, J.: Distributed computing meets game theory. In: PODC 2006, ACM Press, New York (2006)
2. Aghassi, M., Bertsimas, D.: Robust game theory. Math. Program. **107**(1–2), 231–273 (2006)
3. Aumann, R.J.: 16. Acceptable Points in General Cooperative n-Person Games, pp. 287–324. Princeton University Press, Princeton (1959)

4. Baar, T., Olsder, G.: Dynamic Noncooperative Game Theory. Academic Press, New York (1982)
5. Barton, P.I., Khan, K.A., Stechlinski, P., Watson, H.: Computationally relevant generalized derivatives: theory, evaluation and applications. Optim. Meth. Softw. **33**(4–6), 1030–1072 (2018)
6. Ben-Ayed, O.: Bilevel linear programming. Comput. Oper. Res. **20**(5), 485–501 (1993)
7. Bernasconi, M., Castiglioni, M., Celli, A., Farina, G.: On the role of constraints in the complexity of min-max optimization (2024)
8. Bernheim, B., Peleg, B., Whinston, M.D.: Coalition-proof Nash Equilibria i. Concepts J. Econ. Theor. **42**(1), 1–12 (1987)
9. Birge, J., Louveaux, F.: Introduction to Stochastic Programming. Springer Series in Operations Research and Financial Engineering, 2nd edn. Springer, New York (2011)
10. Chan, D., Pang, J.S.: The generalized quasi-variational inequality problem. Math. Oper. Res. **7**(2), 211–222 (1982)
11. Chen, X., Deng, X., Teng, S.H.: Settling the complexity of computing two-player Nash equilibria. J. ACM **56**(3), 1–57 (2009)
12. Chen, Y., Hobbs, B.F., Leyffer, S., Munson, T.S.: Leader-follower equilibria for electric power and no x allowances markets. CMS **3**(4), 307–330 (2006)
13. Conejo, A., Carrion, M., Morales, J.: Decision Making Under Uncertainty in Electricity Markets. Springer, New York (2010)
14. Crespi, G.P., Radi, D., Rocca, M.: Insights on the theory of robust games. Comput. Econ. (2023)
15. Debreu, G.: A social equilibrium existence theorem. Proc. Natl. Acad. Sci. **38**(10), 886–893 (1952)
16. DeMiguel, V., Xu, H.F.: A stochastic multiple-leader stackelberg model: analysis, computation, and application. Oper. Res. **57**, 1220–1235 (2009)
17. Ehrenmann, A.: Equilibrium problems with equilibrium constraints and their applications in electricity markets. Ph.D. thesis, Cambridge University (2005)
18. Facchinei, F., Kanzow, C.: Generalized Nash equilibrium problems. 4OR **5**(3), 173–210 (2007)
19. Fearnley, J., Goldberg, P.W., Hollender, A., Savani, R.: The complexity of gradient descent: CLS = PPAD ∩ PLS. In: STOC. ACM (2021)
20. Ferris, M., Dirkse, S., Meeraus, A.: Mathematical programs with equilibrium constraints: automatic reformulation and solution via constrained optimization. Technical report (2002)
21. Filos-Ratsikas, A., Hansen, K.A., Høgh, K., Hollender, A.: FIXP-Membership via convex optimization: games, cakes, and markets. SIAM J. Comput., FOCS21-30–FOCS21-84 (2023)
22. Filos-Ratsikas, A., Hansen, K.A., Høgh, K., Hollender, A.: PPAD-membership for Problems with exact rational solutions: a general approach via convex optimization. In: STOC 2024. ACM (2024)
23. Gabriel, S.A., Conejo, A.J., Fuller, J.D., Hobbs, B.F., Ruiz, C.: Equilibrium problems with equilibrium constraints. In: Complementarity Modeling in Energy Markets, pp. 263–321. Springer, New York (2012)
24. Gilboa, I., Zemel, E.: Nash and correlated equilibria: some complexity considerations. Games Econ. Behav. **1**(1), 80–93 (1989)
25. Grötschel, M., Lovász, L., Schrijver, A.: Geometric Algorithms and Combinatorial Optimization, vol. 2. Springer, Berlin (2012)

26. Hall-Andersen, M., Schwartzbach, N.I.: Game theory on the blockchain: a model for games with smart contracts. In: Caragiannis, I., Hansen, K.A. (eds.) SAGT 2021. LNCS, vol. 12885, pp. 156–170. Springer, Cham (2021). https://doi.org/10.1007/978-3-030-85947-3_11

27. Harker, P.T.: Generalized Nash games and quasi-variational inequalities. Eur. J. Oper. Res. **54**(1), 81–94 (1991)

28. Harsanyi, C.: Games with incomplete information played by Bayesian players, part I. the basic model. Manag. Sci. **14**, 159–182 (1967)

29. Harsanyi, J.C.: Games with incomplete information played by Bayesian players, part II. Bayesian equilibrium points. Manag. Sci. **14**, 320–340 (1968)

30. Harsanyi, J.C.: Games with incomplete information played by Bayesian players, part III. the basic probability distribution of the game. Manag. Sci. **14**, 486–502 (1968)

31. Hayashi, S., Yamashita, N., Fukushimay, M.: Robust Nash equilibria and second-order cone complementarity problems. J. Nonlinear Convex Anal. **6** (2005)

32. Hobbs, B., Metzler, C., Pang, J.S.: Strategic gaming analysis for electric power systems: an MPEC approach. IEEE Trans. Power Syst. **15**(2), 638–645 (2000)

33. Hu, M., Fukushima, M.: Variational inequality formulation of a class of multi-leader-follower games. J. Optim. Theory Appl. **151**(3), 455–473 (2011)

34. Hu, M., Fukushima, M.: Existence, uniqueness, and computation of robust Nash equilibria in a class of multi-leader-follower games. SIAM J. Optim. **23**(2), 894–916 (2013)

35. Hu, M., Fukushima, M.: Multi-leader-follower games: models, methods and applications. J. Oper. Res. Soc. Japan **58**(1), 1–23 (2015)

36. Hu, X.: Mathematical programs with complementarity constraints and game theory models in electricity markets. Ph.D. thesis, university of Melbourne (2003)

37. Hu, X., Ralph, D.: Using EPECs to model bilevel games in restructured electricity markets with locational prices. Oper. Res. **55**(5), 809–827 (2007)

38. Jiang, H., Netessine, S., Savin, S.: Robust newsvendor competition under asymmetric information. Oper. Res. **59**(1), 254–261 (2011)

39. Kapron, B.M., Samieefar, K.: The computational complexity of variational inequalities and applications in game theory (2024)

40. Kapron, B.M., Samieefar, K.: On the computational complexity of quasi-variational inequalities and multi-leader-follower games. In: Proceedings of the 23rd International Conference on Autonomous Agents and Multiagent Systems, AAMAS 2024, pp. 2324–2326. International Foundation for Autonomous Agents and Multiagent Systems, Richland (2024)

41. Karde, E., Ordóñez, F., Hall, R.W.: Discounted robust stochastic games and an application to queueing control. Oper. Res. **59**(2), 365–382 (2011)

42. Lin, G.H., Fukushima, M.: Stochastic equilibrium problems and stochastic mathematical programs with equilibrium constraints: a survey. Pac. J. Optim. **6**, 455–482 (2010)

43. Lingyuan Liu, Y.C., Dang, C.: A differentiable path-following method to compute Nash equilibria in robust normal-form games. Optimization, 1–40 (2023)

44. Luo, Z.Q., Pang, J.S., Ralph, D.: Mathematical Programs with Equilibrium Constraints. Cambridge University Press, Cambridge (1996)

45. Marchesi, A.: Leadership games: multiple followers, multiple leaders, and perfection. In: Geraci, A. (ed.) Special Topics in Information Technology. SAST, pp. 107–118. Springer, Cham (2021). https://doi.org/10.1007/978-3-030-62476-7_10

46. Nash, J.: Non-cooperative games. Ann. Math. **54**(2), 286–295 (1951)

47. Nash, J.F.: Equilibrium points in n-person games. Proc. Natl. Acad. Sci. **36**(1), 48–49 (1950)
48. Nishimura, R., Hayashi, S., Fukushima, M.: Robust Nash equilibria in N-person non-cooperative games: uniqueness and reformulations. Pac. J. Optim. **5**, 237–259 (2009)
49. Outrata, J., Kočvara, M., Zowe, J.: Nonsmooth Approach to Optimization Problems with Equilibrium Constraints. Springer, Cham (1998)
50. Pang, J.S., Fukushima, M.: Quasi-variational inequalities, generalized Nash equilibria, and multi-leader-follower games. CMS **2**(1), 21–56 (2005)
51. Papadimitriou, C., Vlatakis-Gkaragkounis, E.V., Zampetakis, M.: The computational complexity of multi-player concave games and kakutani fixed points. In: EC 2023. ACM, New York (2023)
52. Papadimitriou, C.H.: On the complexity of the parity argument and other inefficient proofs of existence. J. Comput. Syst. Sci. **48**(3), 498–532 (1994)
53. Rosen, J.B.: Existence and uniqueness of equilibrium points for concave N-person games. Econometrica **33**(3), 520–534 (1965)
54. Shanbhag, U.V., Infanger, G., Glynn, P.W.: A complementarity framework for forward contracting under uncertainty. Oper. Res. **59**, 810–834 (2011)
55. Sherali, H.D.: A multiple leader stackelberg model and analysis. Oper. Res. **32**(2), 390–404 (1984)
56. von Stackelberg, H.: The Theory of Market Economy. Oxford University Press, Oxford (1952)
57. Su, C.L.: Equilibrium Problems With Equilibrium Constraints: Stationarities, Algorithms, And Applications. Ph.D. thesis, Stanford University (2005)
58. Su, C.L.: Analysis on the forward market equilibrium model. Oper. Res. Lett. **35**(1), 74–82 (2007)
59. Vicente, L.N., Calamai, P.H.: Bilevel and multilevel programming: a bibliography review. J. Global Optim. **5**(3), 291–306 (1994)
60. Wadia, N.S., Dandi, Y., Jordan, M.I.: A gentle introduction to gradient-based optimization and variational inequalities for machine learning (2024)
61. Wilczyski, A., Jakbik, A., Kolodziej, J.: Stackelberg security games: models, applications and computational aspects. J. Telecommun. Inf. Technol. **3**, 70–79 (2016)

Author Index

I. Finocchi and L. Georgiadis (Eds.): CIAC 2025, LNCS 15679, pp. 361–362, 2025.
https://doi.org/10.1007/978-3-031-92932-8

The manufacturer's authorised representative in the EU is Springer
Nature Customer Service Centre GmbH, Europaplatz 3, 69115 Heidelberg,
Germany. If you have any concerns regarding our products, please
contact ProductSafety@springernature.com

Printed and bound by CPI Group (UK) Ltd, Croydon, CR0 4YY

28/04/2026

02098518-0008